The Proceedings of the
Fifth IFIP-TC6 International Conference on

Mobile and Wireless Communications Networks

editors

Khaldoun Al Agha
LRI, University of Paris XI, France

Cambyse Guy Omidyar
A Star, Institute for Infocomm Research, Singapore

Singapore 27–29 October 2003

The Proceedings of the
Fifth IFIP-TC6 International Conference on

Mobile and Wireless Communications Networks

Institute for Infocomm Research
A*STAR

IFIP TC6

World Scientific

NEW JERSEY · LONDON · SINGAPORE · SHANGHAI · HONG KONG · TAIPEI · BANGALORE

Published by

World Scientific Publishing Co. Pte. Ltd.
5 Toh Tuck Link, Singapore 596224
USA office: Suite 202, 1060 Main Street, River Edge, NJ 07661
UK office: 57 Shelton Street, Covent Garden, London WC2H 9HE

British Library Cataloguing-in-Publication Data
A catalogue record for this book is available from the British Library.

MOBILE AND WIRELESS COMMUNICATIONS NETWORKS
Proceedings of the Fifth IFIP-TC6 International Conference

ISBN 981-238-686-6

Printed by FuIsland Offset Printing (S) Pte Ltd, Singapore

Message from Conference General Chair

Welcome to the beautiful garden city, Singapore, a wonderful place to be in, while attending the Fifth IFIP-TC6 International Conference on Mobile and Wireless Communications Networks MWCN 2003. The conference has received a large number of submissions and after careful review and selection, the conference was organized around a three-track and was complemented with keynote and invited papers. The conference has been arranged with one day for tutorials and two days for presentation. This conference attracted a large number of submissions due to its Technical Committee efforts. The Fifth IFIP-TC6 International Conference on Mobile and Wireless Communications Networks is a gathering among scientists, industry and developers to report on their finding discuss their experiences and lesson learned from one another. The MWCN 2003 conference is a forum for tutorials, discussions and presentations of the new developments in Mobile and Wireless Communications research and industry. Mobile and Wireless communications networks have created a major breakthrough in new telecom services in local and wide area networks.

MWCN 2003 conference was organized this year by the Institute for Infocomm Research and arranged in co-operation with IFIP-TC6 WG6.8. We hope we have done our best to bring you an outstanding MWCN 2003 program and that you will enjoy it.

Cambyse Guy Omidyar
Conference and Steering Committee General Chair

Message from the Technical Program Committee Chair

On behalf of the Technical Program Committee, I welcome you to the 5th Mobile Wireless Communications Networks (MWCN 2003) in Singapore.

After the huge success of 2002 edition, MWCN continue to receive and produce a high quality of research papers in the Mobile and Wireless communications areas. This year, we received 146 papers and accepted 60 papers for oral presentation and 14 posters. Each paper was distributed to three reviewers from the 59 experts of the program committee.

The conference covers all area concerning mobility and wireless communications. Accepted papers deal with cellular networks (2G, 3G and 4G), wireless networks (IEEE802.11, Bluetooth, and sensor networks), security, quality of service and applications.

Four editions of MWCN were organized. The first edition, MWCN 1997, took place at Sorbonne University, Paris, France May 12–15, 1997. The second edition, MWCN 2000, took place at the Cite des Sciences, La Villette, Paris, France May 16–17, 2000. The third edition, MWCN 2001, was held in Recife, Brazil in August 2001. The MWCN 2002 was held in Stockholm, Sweden September 9–11 2002. At MWCN 2003 in Singapore, we will announce the venue of next year MWCN 2004.

I want to thank the authors for their submissions. Also, I address my gratitude to the reviewers for their multiple reviews during the selection process. I give a special thank to Boon Sain Yeo, Azzedine Boukerche and Khaled Elsayed to co-chair the technical program committee and to their strong participation to the paper selection. I have to thank Guillaume Vivier and Khaled Boussetta from the program committee for their useful help when constructing the technical program.

See you in Singapore.

Khaldoun Al Agha
LRI, University of Paris XI, France

MWCN 2003 Executive Committee

Conference General Chair
Dr. Guy Omidyar
Institute for Infocomm Research, Singapore
Email: gomidyar@i2r.a-star.edu.sg

Conference Technical Program Chair
Khaldoun Al-Agha, LRI, University of Paris XI, France
Email: alagha@lri.fr

Conference Technical Program Co-chairs and Area Coordinators
Asia-Pacific: Boon Sain Yeo
Institute for Infocomm Research, Singapore
Email: boonyeo@ieee.org

North and South Americas: Azzedine Boukerche
University of North Texas, USA
Email: boukerche@cs.unt.edu

Middle-East and Far East: Khaled Elsayed
Cairo University, Egypt
Email: khaled@ieee.org

Tutorial, Workshop, Panel Chair
Dr. Guy Omidyar
Institute for Infocomm Research, Singapore
Email: gomidyar@i2r.a-star.edu.sg

Finance Chair
S Kunaselvam
Institute for Infocomm Research, Singapore
Email: kuna@i2r.a-star.edu.sg

Web Chair
Gek Hiong Tan
Institute for Infocomm Research, Singapore
Email: tangh@i2r.a-star.edu.sg

Steering Committee
Guy Omidyar, I2R, Singapore (Chair)
Mario Gerla, UCLA, USA
Guy Pujolle, Université Paris 6, France
Boon Sain Yeo, I2R, Singapore
Djamel Sadok, Universidade Federal de Pernambuco, Brazil
Khaldoun Al-Agha, LRI, University of Paris XI, France

Technical Scientific Program Committee

Albert Zomaya, University of Western Australia
Adam Wolisz, Technische Universitat Berlin, Germany
Adolf Finger, Technical University of Dresden, Germany
Afonso Ferreira, CNRS INRIA, France
André-Luc Beylot, IRIT - France
Andrew T. Campbell, Columbia University, USA
Antonio A.F. Loureiro, Federal University of Minas Gerais, Brazil
Ayman Naguib, Qualcomm, USA
Azzedine Boukerche, University of North Texas, US
Boon Sain Yeo, Institute for Infocomm Research, Singapore
Branimir Vojcic, George ashington University, USA
Christopher Rose, Associate Director, Rutgers WINLAB, USA
Colin Chandler, Qualcomm Europe, France
Cormac Sreenan, University College Cork, Ireland
Djamel Sadok, Universidade Federal de Pernambuco, Brazil
Elena Pagani, Universita' degli Studi di Milano, Italy
Elisabeth Royer, University of California, Santa Barbara
Fabrizio Sestini, European Commission, Belgium
Farouk Kamoun, ENSI, Tunisia
Francois Chin, Institute for Infocomm Research, Singapore
Gerald Q. Maguire, Jr., KTH, Wireless Center, Sweden
Geraldo Robson Mateus, Federal University of Minas Gerais, Brazil
Guillaume Vivier, Motorola Labs, France
Guy Omidyar, Institute for Infocomm Research, Singapore
Guy Pujolle, Université Paris 6, France
Hamid Aghvami, Kings College of London, UK
Hossam Afifi, INT, France
Ivan Vukovic, Motorola, USA
Jamshid Khun Jush, Ericsson, Germany
Jan SLAVIK, TESTCOM, Czech Republic
Jon W. Mark, University of Waterloo, Canada
Kassem Saleh, American Uni of Sharjah, UAE
Kaveh Pahlavan, Worcester Polytechnic Institute, USA
Kelvino S. Sousa, University of Toronto, Canada
Khaldoun Al Agha, LRI, University of Paris XI, France
Khaled Ben Letaief, Hong Kong University, HK
Khaled Boussetta, University of Paris VI, France
Khaled Elsayed, Cairo University, Egypt
Lajos Hanzo, University of Southampton, UK
Lila Boukhatem, LRI, University of Paris XI, France
Luigi Fratta, Politecnico di Milano, Italy

Mahmoud Naghshineh, IBM Watson Research Center, USA
Marie-Jose Montpetit, Nokia, USA
Mario Gerla, UCLA, USA
Markus Endler, IME/USP, Brazil
Martin Haardt, Ilmenau University of Technology, Germany
Mirela Notare, Barddal University, Brasil
Olga Casals, Universitat Politecnica de Catalunya, Spain
Parviz Kermani, IBM Watson Research Center, USA
Pascal Lorenz, Universite de Haute Alsace, France
Ramjee Prasad, Aalborg University, Denmark
Sajal K. Das, University Texas at Arlington, USA
Serge Fdida, Laboratoire d'informatique de Paris 6, France
Sergio Palazzo, University of Catania, Italy
Sumei Sun, Institute for Infocomm Research, Singapore
Tjeng T. Tjhung, Institute for Infocomm Research, Singapore
Theo Kanter, Ericsson Research, Sweden
Witold Krzymien, University of Alberta, Canada
Xavier Lagrange, ENST Bretagne, France
Zygmunt J. Haas, Cornell University, USA

Contents

Message from Conference General Chair v

Message from the Technical Program Committee Chair vii

MWCN 2003 Executive Committee ix

Technical Scientific Program Committee xi

SESSION 1A: QOS IN AD HOC NETWORKS

Design of A Hybrid Routing Architecture 3
Guillaume Chelius and Eric Fleury (France)

A Framework to Relieve Wireless Hot-Spot Congestion by Means of Ad Hoc Connections 7
Jiancong Chen, Jingyi He and S.-H. Gary Chan (Hong Kong)

Mobile Ad Hoc Networks and QoS Demanding Applications 11
Leila Boukhalfa, Pascale Minet, Laurent George and Serge Midonnet (France)

QoS for Ad Hoc Networking Based on Multiple Metric: Bandwidth and Delay 15
Hakim Badis, Anelise Munaretto, Khaldoun Al Agha and Guy Pujolle (France)

SESSION 1B: BEYOND 3G AND 4G

Service Provision and Reconfiguration Management in 4G Mobile Networks 21
Vangelis Gazis, Nikos Houssos, Nancy Alonistioti and Lazaros Merakos (Greece)

GMSME: An Architecture for Heterogeneous Collaboration with Mobile Devices 25
Sangyoon Oh, Sunghoon Ko and Geoffrey C. Fox (USA)

3G and WLAN Interworking QoS Solution 29
Eero Wallenius, Timo Hämäläinen, Timo Nihtilä, Jyrki Joutsensalo and Hannu Koivisto (Finland)

Reconfigurable Equipment Management: An Agent Based Framework 34
Karim El-Khazen, Guillaume Vivier and Nicolas Lhuillier (France)

SESSION 2A: WIRELESS NETWORKS — IEEE 802.11

Improving the Performance of IEEE 802.11 in Wireless Multi-Hop Networks 41
Gang Lu and Cauligi S. Raghavendra (USA)

An Enhanced SuperPoll Protocol for IEEE 802.11 PCF with Rate Differentiation 45
Juki Wirawan Tantra, Teck Meng Lim, Chuan Heng Foh and Bu Sung Lee (Singapore)

Fast Handover Support in Wireless LAN Based Networks 49
Norbert Jordan, Reinhard Fleck and Christian Ploninger (Austria)

Study of IEEE 802.11 Fairness and Its Interaction with Routing Mechanism 53
Zhifei Li, Sukumar Nandi and Anil K. Gupta (Singapore)

SESSION 2B: SECURITY

Mutual Authentication and Group Key Agreement for Low-Power Mobile Devices 59
Emmanuel Bresson, Olivier Chevassut, Abdelilah Essiari and David Pointcheval (USA)

Light TLS: Extended Handshake for Wireless Communications 63
Mohamad Badra and Ahmed Serhrouchni (France)

A Novel Secure Handoff Approach for Wireless LANs 67
S. Nanduri and S. Zeadally (USA)

Modelling a PKI-Based Network System using Multi-Agents Systems Engineering 71
Gustavo A. Santana Torrellas, Leonid Sheremetov and Luis A. Villa Vargas (Mexico)

SESSION 3A: QUALITY OF SERVICE

Predictive FH Control For FH-OFDMA Systems 77
Bongseog Jang Changhoi Koo, D. S. Park and Hyeonwoo Lee (South Korea)

A Class-Based Admission Control Algorithm for Shared Wireless Channels Supporting QoS Services 81
Patrick Hosein (USA)

A Framework for End-to-End QoS Provisioning in Mobile Internet Environment 86
Salem Lepaja, Arben Lila, Natalia Kryvinska and Hoang M. Nguyen (Singapore)

Adaptive Quality of Service for A Mobile Ad Hoc Network 90
Antonios Dimakis, Linhai He, John Musacchio, Hoi-Sheung Wilson So, Teresa Tung and Jean Walrand (USA)

SESSION 3B: ENERGY CONSUMPTION

Energy Balanced Communication in Wireless Sensor Networks 97
Amol Bakshi and Viktor K. Prasanna (USA)

Adaptive Path Energy Conserving Routing in MANETs 101
Hasnaa Moustafa and Houda Labiod (France)

A Rate-Adaptive Power Control Scheme for Wireless Microsensor Networks 105
Caimu Tang and Cauligi S. Raghavendra (USA)

Energy Awareness and Ad Hoc Routing 109
Erol Gelenbe and Ricardo Lent (USA)

SESSION 4A: TOPOLOGY IN AD HOC NETWORKS

Introduction of Node Promiscuity Listen Mode and Optimizations in LAR Protocol for Wireless Ad Hoc Networks 115
Floriano De Rango, Gianluca Aloi and Salvatore Marano (Italy)

An Information Theoretic Approach to Cluster Head Selection in Mobile Ad Hoc Networks 120
Philips Koshy and S. V. Raghavan (India)

A Localized Topology Control Algorithm for Power-Efficient Routing on Mobile Ad Hoc Networks 124
Guo Xiaoguang and Hsu Wen Jing (Singapore)

Group Registration in Wireless Cellular Systems with Short-Range Radio Networks 129
Hyunku Jeong, Hyeon-Ju Yoon, Cheolgi Kim, Joongsoo Ma and Joonwon Lee (Republic of Korea)

Optimizing Mobile Network Datagram Routing through Programmable Proxying 134
Peter Tabery, Christian Bachmeir and Xiaohe Li (Germany)

SESSION 4B: ROUTING

A Distributed Algorithm for Multiple-Metric Link State QoS Routing Problem 141
Hakim Badis and Khaldoun Al Agha (France)

ZONEPOW: A Novel Approach to Power Control in Ad Hoc Networks 145
Shachindra Sharma and R. S Sreenivas (India)

A Load Aware, Location Aided, Ad Hoc On-Demand Routing Algorithm 150
Tahmid A. Quazi and Stephen McDonald (South Africa)

Gateway-Based Multicast Protocol for Mobile Ad Hoc Networks 155
Jiawei Xie, Sukumar Nandi and Anil K. Gupta (Singapore)

DIP3S: A Distributed Routing Algorithm for Power-Conserving Broadcasting in Wireless Ad Hoc Networks 159
Vikas P. Verma, Amit A. Chandak and Hung Q. Ngo (USA)

SESSION 5A: SMALL AREA NETWORKS — PICONET

A Novel Traffic and Energy Aware Bridge Management Policy for Bluetooth Scatternets 165
Ranganath Duggirala, Roy L. Ashok and Dharma P. Agrawal (USA)

Implementing a Self-Organizing Wireless Sensor Network: Experiences and Challenges 169
Joo Ghee Lim, Kim Loon Chee, Hee Boon Leow, Yew Kuan Chong, P. K. Sivaprasad and S. V. Rao (Singapore)

A Greedy Algorithm to Configure and Reconfigure Bluetooth Scatternets 173
Wanderley Ravagnani Junior, Edmundo R. M. Madeira and Jacques Wainer (Brazil)

Potential Interference of Ultra-Wide Band Transmitters on HipeLAN/2 177
N. P. Simeonidis, K. Haider and Hamed S. Al-Raweshidy (UK)

SESSION 5B: APPLICATIONS

Media Synchronization Quality of Packet Sceduling Algorithms for Wireless and Mobile Networks 183
Azzedine Boukerche and Harold Owens (USA)

Proactive Energy-Aware Video Streaming to Mobile Handheld Devices 187
Shivajit Mohapatra and Nalini Venkatasubramanian (USA)

Developing PDA for Low-Bitrate Low-Delay Video Delivery 191
K.-W. Roger Cheuk, S.-H. Gary Chan, K.-W. Alice Mong, C.-M. Martin Lee and S.-S. Shirley Sy (Hong Kong)

Feature Extraction in Wireless Personal and Local Area Networks 195
Rene Mayrhofer, Harald Radi and Alois Ferscha (Austria)

SESSION 6A: MOBILITY

Policy Based Mobile IP Handoff Decision (POLIMAND) using Generic Link Layer Information 201
Stefan Aust, Daniel Proetel, Nikolaus A. Fikouras, Carmelita Görg and *Cornel Pampu (Germany)*

Minimising Handover Overhead in Host Based Routing IP Wireless Access Networks 205
Mudar A. O. Barry and Hamed S. Al-Raweshidy (UK)

Achieving Seamless IP Handover and Dynamic Neighborhood Discovery in IEEE 802.11 Networks 209
Paul Tan, QunYing Xie, Hoang M. Nguyen, Jaya Shankar and Winston Seah (Singapore)

New Mechanism for Mobile IPv6 Fast Handover in Wireless LAN 213
Pyung Soo Kim, Young Keun Kim and M. E. Lee (Korea)

A Mobility-Based Adjustment of Binding Update Lifetime for Signaling Cost Reduction in Hierarchical Mobile IPv6 217
Myung-Kyu Yi and Chong-Sun Hwang (Republic of Korea)

SESSION 6B: ARCHITECTURE

Service Management Architecture for Providing AAA and Differentiated Service in Mobile IPv6 223
Shoubin Dong, Luliang Jiang, Liqun Tang, Jian Ma and Ling Zhang (China)

Persistent Connectivity and Reliable Message Delivery in Component-Based Mobile Applications 227
Radu Litiu and Atul Prakash (USA)

Service and Personal Mobility in Next Generation Networks 231
Natalia Kryvinska, Salem Lepaja and Hoang M. Nguyen (Austria)

Composite Simulations for B3G Service and Network Management Platform 235
Guillaume Vivier, Karim El-Khazen, Panagiotis Demestichas and Pierre Kuonen (France)

Multicasting in UMTS: Effect of Mobility on Tree Maintainance 239
Nilanjan Banerjee and Sajal K. Das (USA)

SESSION 7A: PHYSICAL LAYER

A Hardware Testbed for the Implementation and Evaluation of MIMO Algorithms 245
Patrick Murphy, Feifei Lou and J. Patrick Frantz (USA)

A Comparison of MC-CDMA and SC-CDMA with SCC Schemes in Rayleigh Fading Channels 249
T. Detert (Germany)

M-ary PPM UWB Radio in Fading Channels 253
Mohammed Abdel-Hafez, Fatih Alagoz and Matti Hamalainen (United Arab Emirates)

Comparison of Several Multiple Antenna Multiuser Detectors in Wireless CDMA Systems 258
S. M. Razavizadeh, V. T. Vakili and P. Azmi (Iran)

SESSION 7B: SECURITY II

Heterogeneous Network Establishment Assisted by Cellular Operators 265
Marc Danzeisen, Torsten Braun, Daniel Rodellar and Simon Winiker (Switzerland)

An Efficient Authentication and Key-Exchange Protocol for Low Power Clients 270
Amol Dabholkar and Yow Kin Choong (Singapore)

A Secure Location Service in Mobile Ad Hoc Networks 274
Zhou Zhi, Yow Kin Choong and Sukumar Nandi (Singapore)

Certification and Authentication Services for Securing MANET Routing Protocols 278
Ricardo Staciarini Puttini, Ludivic Me and Rafael Timóteo de Sousa (Brazil)

POSTER SESSION P1

Performance Analysis of Admission Control Algorithms for CDMA Networks 285
Kuo-Chung Chu, Frank Yeong-Sung Lin and Shin-Fu Wang (Taiwan)

A Performance Enhancing Proxy for Mobile Satellite Internet 289
Giovanni Ciccarese, Luigi Patrono, Sebastiano Elia, Giuseppe Tomasicchio and Mario De Blasi (Italy)

Mobile Firewall 293
Guangrui Fu, Daichi Funato, Jonathan Wood and Toshiro Kawahara (USA)

Macrodiversity Evaluation of WCDMA with Radio over Fibre Access Networks 297
Nazem Khashjori and H. S. Al-Raweshidy (UK)

The Performance Evaluation of Handoff Methods using H.323 301
Youngsin Lee (Korea)

A Context Information Service for Mobile Applications 305
Feng Wang and Paddy Nixon (UK)

POSTER SESSION P2

A Simple and Efficient MAC for Implementation in a Wireless Sensor Networks 311
Kim Loon Chee, Joo Ghee Lim, S. V. Rao and P. K. Sivaprasad (Singapore)

OLSR Routing for Miltary Ad Hoc Networks 315
Thierry Plesse, Jerome Lecomte, Cedric Adjih, Marc Badel, Philippe Jacquet, Anis Laouiti, Pascale Minet, Paul Muhlethaler and Adokoe Plakoo (France)

Bluetooth Network Performance in Nakagami-m Fading Channels 319
João Henrique Kleinschmidt, Marcelo E. Pellenz and Edgard Jamhour (Brazil)

QoS Support in Wireless LANs using Dynamic Adaptation of Channel Access Parameters 323
Liqiang Zhang and Sherali Zeadally (USA)

A Reliable Region Based Hierarchical Mobile Multicast Protocol for IPv6 Network 327
Sun Limin, Liao Yong and Wu Zhimei (China)

Impact of Security Architectures on Wireless Network Performance 331
Ray Hunt, John Vargo and Jenne Wong (New Zealand)

Asymmetric UMTS for Spectrum Efficient Asymmetric Services Delivery 335
L. Vignali, F. Malavasi, D. Grandblaise, D. Lacroix-Penther and J.-P. Javaudin (Italy)

Architecture Design of Extensible Multi-Media Micro-Browser Software 339
Toshihiko Yamakami (Japan)

SESSION 1A:
QOS IN AD HOC NETWORKS

DESIGN OF A HYBRID ROUTING ARCHITECTURE

G. CHELIUS
É. FLEURY
Ares Inria - Citi - INSA de Lyon
21 avenue Jean Capelle
F69621 Villeurbanne Cedex - France. Tel: +33 472 436 421. Fax: +33 472 436 227.

This paper deals with the design of hybrid architectures, *i.e.*, merging two different mobility concepts: ad hoc and cellular. It presents the architecture and briefly reviews some solutions proposed in order to extend the last hop of a wireless LAN with an ad hoc network. The goals are to provide ad hoc connectivity beyond the range of fixed and cellular infrastructures, to integrate ad hoc networks into the hierarchical Internet and to support the migration of mobile nodes from the Internet into and out of ad hoc networks. The paper tackles the issues that arise from the merging of the different mobility protocols and proposes several solutions.

1 Introduction

Wireless communication is rapidly becoming a critical aspect of computer networks and offers open solutions for providing mobility as well as essential network services. With wireless communication technologies, the Internet becomes very pervasive and starts being more and more bound to cellular networks. Research on mobile wireless networking has roughly been concentrating on two distinct areas. The first theme is seeking the way to extend the edge of the traditional network infrastructures with a last wireless hop. Wireless connectivity is provided by base stations and mobile users are able to switch between base stations while moving around. The second theme corresponds to the spontaneous self organization of networks of devices. Such networks are called ad hoc networks.

Research efforts put into integrating these two fields of wireless networking – wireless access to a fixed network and ad hoc networking – are increasing. The goal of this architecture is to provide connectivity to a global network infrastructure for an isolated ad hoc node. The aim of the paper is to highlight the problems encountered while merging the two concepts and to propose several solutions. In order to increase the flexibility of the infrastructure networks, we consider fixed wireless links in the access network.

2 Multipolar topology

Proposals [9,12,2] for routing in an hybrid network usually divide the network into one or two logical zones. In a flat architecture, the ad hoc routing protocol is spread over the whole ad hoc network. Ad hoc mobility and micro-mobility are both handled by the ad hoc routing protocol. In a bipolar architecture, two different protocols are used. On one side, an ad hoc routing protocol manages mobility between ad hoc nodes, *e.g.* *DSR* [7], *AODV* [11], and on the other side, a micro-mobility protocol manages the infrastructure related mobility of ad hoc nodes, *e.g.* *Cellular IP* [3]. Examples are given in [12,2].

Flat and bipolar architectures have a common main drawback. They do not scale well. Indeed, the ad hoc routing protocol is spread over the whole ad hoc network, inducing much control traffic and huge delays in the ad hoc mobility management. To reduce this traffic, we propose a multipolar architecture based on Ananas [5,4]. Each base station defines its zone of influence and ad hoc routing is performed independently in each zone. A zone of influence is composed by the set of ad hoc nodes that have chosen a common base station as gateway. Inter-zone routing is performed using the infrastructure network. Zone interpenetration allows border nodes to directly access an adjacent zone. Pictures 1(a), 1(b) and 1(c) depict the three different architectures and the associated routing processes. Figure 2 compares the number of control packet broadcasts for a flat and a multipolar architecture. In this simulation, the network is composed of 9 infrastructure nodes and a variable number of ad hoc nodes in a $1000 * 500m$ rectangle. The radio range is $150m$ and the maximum mobile speed is $10m/s$. Ad hoc routing is performed using OLSR.

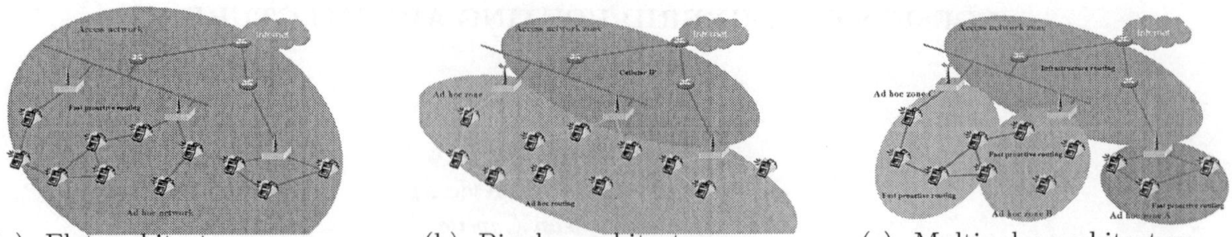

(a): Flat architecture. (b): Bipolar architecture. (c): Multipolar architecture.

Figure 1. Several hybrid architectures.

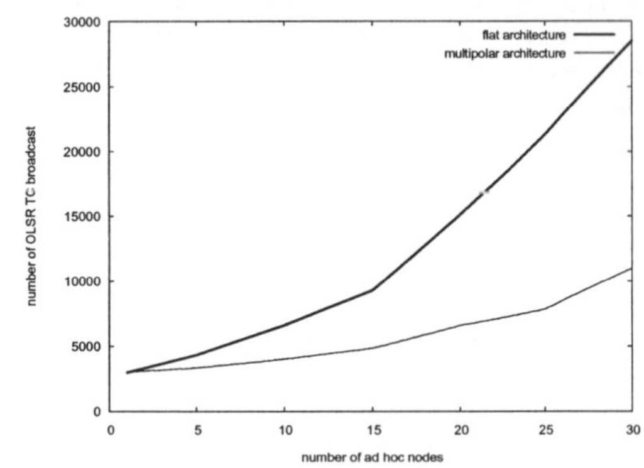

Figure 2. Number of TC packet broadcasts depending on the hybrid architecture.

3 Fast ad hoc routing

Once the hybrid network has been divided into one infrastructure and several ad hoc zones, ad hoc routing can be performed into each ad hoc zone. If most of the current works [12,2] use reactive protocols (respectively AODV and DSR), we prefer using a proactive one, *e.g.* OLSR [8] or TBRPF [10]. Our choice is justified by the fact that reactive protocols present, in a micro-mobility environment, several drawbacks that have been clearly identified in previous works [9,6]. First, it induces a large latency at external communication startup. Second, reactive protocols use route breakage for mobility detection. But route reconstruction is a long process, too long to enable a fast handoff support.

To enable fast handoffs and a fast mobility support, the ad hoc routing protocol must be adapted. The Fast-OLSR [1] extension of OLSR is an example. However, it is not possible to speed up the ad hoc routing protocol in all ad hoc nodes without overloading the network with control traffic. We propose a particular scheme where the fast mobility support degrades according to an approximative distance to the gateway. The frequency at which a node emits routing messages depends on its distance to the gateway. Knowledge of this distance is based upon route computation and Hello packets. Distance to the gateway are advertised in Hello packets. Each time a node recomputes its routes, it considers its computed distance to the gateway to determine its routing speed. Each time a node receives a Hello packet with an advertised distance d smaller than its own, it considers $d+1$ to determine its routing speed. Table 1 gives the OLSR constants we used for our simulations. We compared the control traffic load and the data delivery achieved by a classical OLSR with short control message periods (values given by the first line in table 1) and our adaptive OLSR. The network is a $700 * 700$ square, the radio range is $150m$. The network is composed of 1 base station and a variable number of ad hoc nodes which all transmit 1024 bytes to the gateway every $0.1s$. Results are presented in figures 3(a) and 3(b).

4 Load balancing handoff policy

Ad hoc nodes define ad hoc zones by choosing which ad hoc routing protocol instance they want to participate to. Each instance is related to a base station which is used for inter-zone and external routing. A node is on the frontier of a zone if it has one neighbor in another zone. A border node may decide to handoff: to change its ad hoc gateway and thus its ad hoc zone. The handoff decision is performed according to a handoff policy. In a multi hop wireless envi-

Table 1. OLSR constants for an adaptive fast mobility support.

Gateway distance	Hello period	Hello timeout	TC period	TC timeout	Min TC period
1	0.1s	0.3s	0.25s	0.55s	0.1s
2	0.2s	0.6s	0.5s	1.1s	0.2s
3	0.4s	1.2s	1s	2.2s	0.4s
4	0.6s	1.8s	1.5s	3.3s	0.6s
5+	0.8s	2.4s	2s	4.4s	0.8s

ronment, classical handoff policies, *e.g. Lazy Cell Switching (LCS), Eager Cell Switching (ECS)*, are inappropriate. Other policies have to be defined such as the *MIPMANet Cell Switching (MMCS, see* [9]*)* or the *Shortest Cell Switching (SCS)*.

In addition to the gateway distance, more complex parameters are considered for the definition of a handoff policy. An interesting one is load balancing. A node may decide to change its ad hoc zone to balance the number of ad hoc nodes between two adjacent zones. Knowledge of the zone load is given by the ad hoc gateway inside the routing process. Border nodes know the load of all adjacent zones and decide wether they handoff or not in a distributed fashion. This rule defines a new handoff policy, the *Load Balanced Cell Switching (LBCS)*.

5 Infrastructure routing

Infrastructure routing is crucial regarding the network performance. In the case of CIP, the routing tree is far from optimal in term of distance as it is reactively built. To optimize routing in the infrastructure, other routing schemes can be used. As the access network is composed of wireless links, it may be problematic to use a classical routing protocol. We prefer to use a proactive MANet protocol that ensures route optimality. Since the access network is almost not subject to mobility, we slow down the protocol so that the control traffic it induces is negligible in regard to the data traffic. We also set timeouts so that routing is collision proof.

6 Mobility notification

Given a routing scheme in each ad hoc zone and another one in the infrastructure zone, the last task consists in notifying the ad hoc nodes location to the infrastructure nodes. This notification is performed by the ad hoc gateways. They periodically advertise the set of ad hoc nodes present in their zone. However, each gateway notification does not systematically advertise all ad hoc nodes. If a node does not change its zone, the delay between two consecutive advertisements of this node increases. When a node enters a new zone, its advertisement period is reinitialized to a small base value, similar to a slow start mechanism. This differential advertisement limits the volume of control traffic while preventing misrouting due to control packet loss. The notification period sequence we considered for simulations is $0.1s, 0.1s, 0.2s, 0.2s, 0.4s, 0.4s, 0.8s, 0.8s, 1.2s....$ Simulations were performed using the parameters described in section 2 and they show that the ratio of node location notifications between a constant ($0.1s$) notification period and our differential one is approximatively 11.

The notification process is based on CIP. Notifications are dispatched to a subset of the infrastructure nodes, all nodes on a shortest path from the ad hoc gateway to the infrastructure gateway. They are unicasted. Unicast offers reliability in the transmission of control messages and prevents from route loss and misrouting. Of course, as for CIP, routing between infrastructure nodes is far from optimal as routes follow a routing tree rooted at the infrastructure gateway.

7 Conclusion

This paper provides an overview and solutions for several main issues raised by the integration of ad hoc networks and access networks. It is more focused on the architecture and fast mobility support aspects.

By merging and integrating ad hoc networks with wireless access to the Internet, we must be able to achieve a more flexible network with multiple ac-

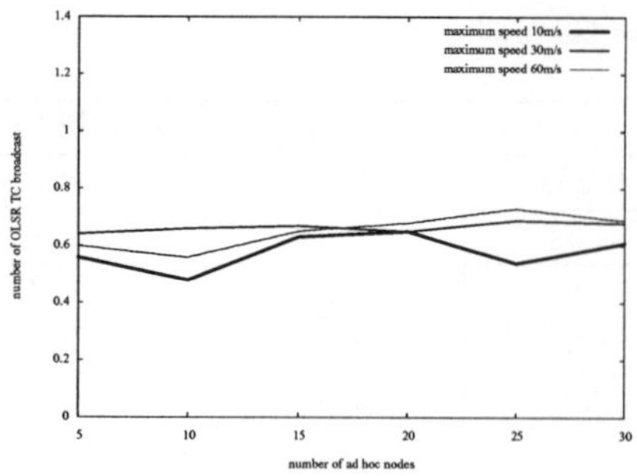

(a): Ratio between the control traffic load induced by our adaptive mobility support and by OLSR.

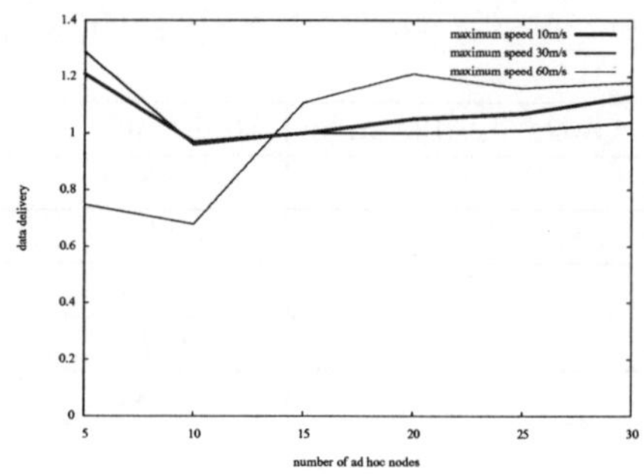

(b): Ratio between the volume of data delivered by our adaptive mobility support and by OLSR.

Figure 3. Simulation results.

cess options providing users the capability to connect and communicate easily and seamlessly. This ultra-connectivity deals with either future or current research activities. It appears that new MAC protocols must be designed. Typically, designing a reliable MAC broadcast is a necessary feature to offer better performances within the mobility notification mechanism. It would allow optimization of the infrastructure routes while keeping the routing protocol reliable. Cooperation between routing protocols and MAC protocols must be set up. Security issues also remain a major concern. The merging of cellular technologies and ad hoc networks may provide elegant solutions. Indeed, the usually completely infrastructureless ad hoc network can, in this configuration, count on a fixed infrastructure that may play an important/determinant role for security.

References

1. M. Benzaid, P. Minet, and K. Al Agha. Integrating fast mobility in the OLSR routing protocol. In *Proceedings of MWCN 2002*, Stockholm, Sweden, 2002. IEEE Communications Society.
2. R. Bhan, A. Croswell, K. Dedhia, W. Lin, M. Manteo, S. Merchant, A. Pajjuri, and J. Thomas. Adding ad hoc Network Capabilities to Cellular IP. www.columbia.edu.
3. A. Campbell, J. Gomez, S. Kim, C.-Y. Wan, Z. Turanyi, and A. Valko. Comparison of IP Micromobility Protocols. *IEEE Wireless Communications*, February 2002.
4. G. Chelius and Fleury E. IPv6 Addressing Architecture Support for mobile ad hoc networks. Internet Draft, September 2002.
5. G. Chelius and Fleury. Ananas: A Local Area Ad hoc Network Architectural Scheme. In *Proceedings of MWCN 2002*, Stockholm, Sweden, September 2002. IEEE Communications Society.
6. M. Ergen and A. Puri. MEWLANA-Mobile IP Enriched Wireless Local Area Network Architecture. In *IEEE VTC*, 2002.
7. Y. Hu, J. Jetcheva, D. Johnson, and D. Maltz. The Dynamic Source Routing Protocol for Mobile Ad Hoc Networks (DSR). Internet Draft, April 2003.
8. P. Jacquet, P. Muhlethaler, A. Qayyum, A. Laouiti, L. Viennot, and T. Clausen. Optimized Link State Routing Protocol. Internet Draft, July 2003.
9. U. Jönsson, F. Alriksson, T. Larson, P. Johansson, and G. Maguire. MIPMANET - Mobile IP for Mobile Ad hoc Networks. In *IEEE/ACM Workshop on mobile and ad hoc networking and computing*, 2000.
10. R. Ogier, F. Templin, B. Bellur, and M. Lewis. Topology Broadcast Based on Reverse-Path Forwarding (TBRPF). Internet Draft, July 2003.
11. C. Perkins, E. Royer, and S. Das. Ad Hoc On-Demand Distance Vector (AODV) Routing. RFC 3561, July 2003.
12. V. Typpö. Mobility within Wireless Ad Hoc Networks: Towards Hybrid Wireless Multihop Networks, 2001.

A FRAMEWORK TO RELIEVE WIRELESS HOT-SPOT CONGESTION BY MEANS OF AD HOC CONNECTIONS

Jiancong Chen[†] Jingyi He[‡] S.-H. Gary Chan[†]

[†]Department of Computer Science

[‡]Department of Electrical and Electronic Engineering

Hong Kong University of Science and Technology

Clear Water Bay, Kowloon, Hong Kong

Email: kentchen@cs.ust.hk eehjy@ust.hk gchan@cs.ust.hk

Abstract— **In recent years, wireless public networks have been widely deployed to provide high-speed Internet access to mobile users. Due to the dynamic and uneven distribution of user requests, some access points (the so-called "hot-spots") may be congested by many users. In order to alleviate such congestion, we introduce an ad hoc relaying wireless LAN framework in this paper. A user can be served by a neighboring access point through an ad hoc connection in such a network. We propose a framework to establish such ad hoc connections. Using simulations, we show that the framework is efficient to reduce the service failure rate as well as to improve channel utilization in hot-spot networks.**

I. INTRODUCTION

Wireless LANs (WLANs) are being widely deployed in offices, homes, as well as public areas like airports and malls, to provide high-speed Internet access to mobile users. A WLAN is set up by attaching to the wired network a base station (BS) (also called an access point, or AP), which relays the traffic between the wireless users and the wired network. In public areas covered by a number of APs, the user load is in general dynamic and unevenly distributed among the APs. It is possible that some APs (so-called "hot-spots") are congested by many users and cannot admit more requests, while their neighboring ones are serving few users and the bandwidth is under-utilized. In this paper, we propose and study serving a mobile user blocked in a hot-spot by a neighboring AP through ad hoc connections, thus alleviating the congestion at the hot-spot as well as improving the bandwidth utilization of the whole network.

We illustrate in Fig. 1 the basic idea of the proposed scheme. In the example, three APs are considered, in which AP 1 is a hot-spot. The mobile host (MH) 1 in the coverage of AP 1 is blocked because there is insufficient bandwidth to satisfy its request requirement. As AP 3 is lightly-loaded and has enough free bandwidth, MH 1 is directed to set up an ad hoc connection to MH 3 which is in the coverage of AP 3. MH 3 maintains the ad hoc connection to MH 1 as well as its connection to AP 3. In addition to communicating with AP 3 for its own traffic, MH 3 also operates as a relay agent for the communication between MH 1 and AP 3. In this way, MH 1 is served by (i.e., associated to) AP 3 in a two-hop manner, instead of being blocked. The major issue of the scheme is to find the appropriate relay agent to establish such an ad hoc connection for the blocked MH.

This work was supported, in part, by the Competitive Earmarked Research Grant of Research Grant Council in Hong Kong (HKUST6014/01E and HKUST6199/02E).

†Correspondence author: Jiancong Chen

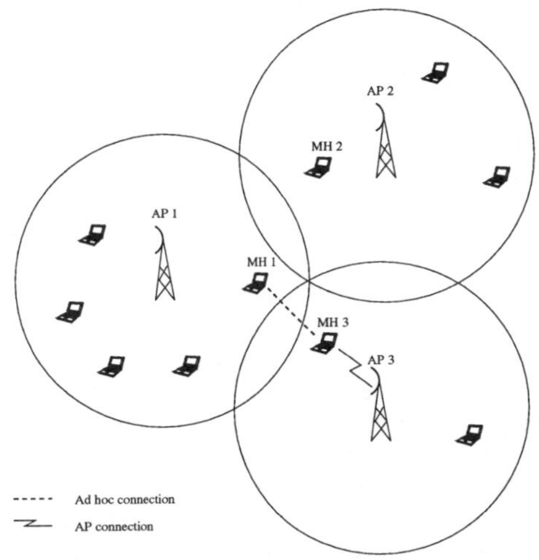

Fig. 1. An example of relieving hot-spot congestion through ad hoc connection.

In this paper, we propose and study a scheme called "AP-Assisted Passive Agent Selection" (APAS) for agent discovery and ad hoc connection establishment. With APAS, the neighboring APs cooperate to find out some possible agents for the blocked MH while the MH just waits for the connections from these agents and selects a proper one. The proposed scheme is based on the assumption that a mobile host (MH) can switch between infrastructure mode (i.e., communicating with the AP) and ad hoc mode (i.e., direct communication with other MHs), which has been shown able to significantly improve the performance of the WLAN [1].

We implement the framework of ad hoc relaying network in NS2 and use simulations to evaluate the performance. The performance metrics of interest are the service failure rate defined as the percentage of requests that cannot get their services and the bandwidth utilization in each AP. Our results show that with ad hoc relaying connections between neighboring APs, the service failure rate is reduced significantly and the system utilization is increased accordingly.

We briefly discuss previous work as follows. The congestion problem at WLAN hot spots has been studied in [2]. The authors proposed two approaches to relieve the congestion, namely, explicit channel switching and network-directed roaming. However, both of them have their limitations. Explicit channel switching only works for MHs in the overlap area of multiple APs, while network-directed roaming requires MHs to move around to get the access, which is neither convenient for the users nor reliable as network state might have changed after the roaming. Our proposed scheme of using ad hoc connections between MHs to balance the load between APs does not suffer from these limitations. A similar idea of using ad hoc connections to do load balancing between cells has been proposed for cellular networks [3]. As the cellular system uses different frequency band than the mobile ad hoc network, the performance improvement by introducing ad hoc connections to deal with congestion is obvious. Our study differs primarily in that we are considering a unified wireless system in one frequency band, in which the bandwidth is shared by all the communications, and the performance improvement by introducing ad hoc connections is not as obvious and deserves a separate study.

The rest of this paper is organized as follows. We present in Sect. II a scheme for agent discovery and ad hoc connection establishment. Simulation results illustrating the performance of the scheme is presented in Sect. III. We conclude in Sect. IV.

II. Ad hoc relaying wireless LANs

In the proposed ad hoc relaying wireless LANs, when a mobile host sends a service request to the access point, the AP has admission control to check whether it can provide the service or not. If not, instead of blocking the request, the AP tries to find a neighboring AP that can provide the service and divert the MH to it by an ad hoc connection. Several issues need to be addressed in this process, which we discuss in the following.

First, the original AP (hot-spot) should know whether there is a neighboring AP with sufficient resources for the request that it cannot admit. This is achieved by maintaining and exchanging state information among APs. Since the number of neighboring APs is small and all APs are connected to the wired network, the communication overhead is limited and does not affect the performance of the wireless network.

Second, if there is such an AP that can provide the service, the MH should be able to connect to it. Since the cover ranges of the APs are limited and may not be overlapping (depending on the deployment and layout of the public wireless networks), the MH may not be in the range of the neighboring AP, i.e., a direct (one-hop) connection is not possible. Therefore, an agent is needed to relay the traffic in between. Hence finding such an agent to establish an ad hoc connection is the major issue in ad hoc relaying WLANs.

Moreover, after the connection is established between the MH and the neighboring AP through an agent, since the connection between the MH and the agent works in ad hoc mode while the agent connects to its AP in infrastructure mode, a good packet forwarding method is important in order to achieve high performance. As discussed in our previous paper, the difference between infrastructure mode and ad hoc mode in the IEEE 802.11 standard is several parameters in the frames, and the behavior of the transmissions [1]. The agent is hence able to send and receive packets in both modes. Upon receiving a packet from the AP to the MH, it forwards the packet in ad hoc mode. Likewise, the packets from the MH are forwarded in infrastructure mode to the AP.

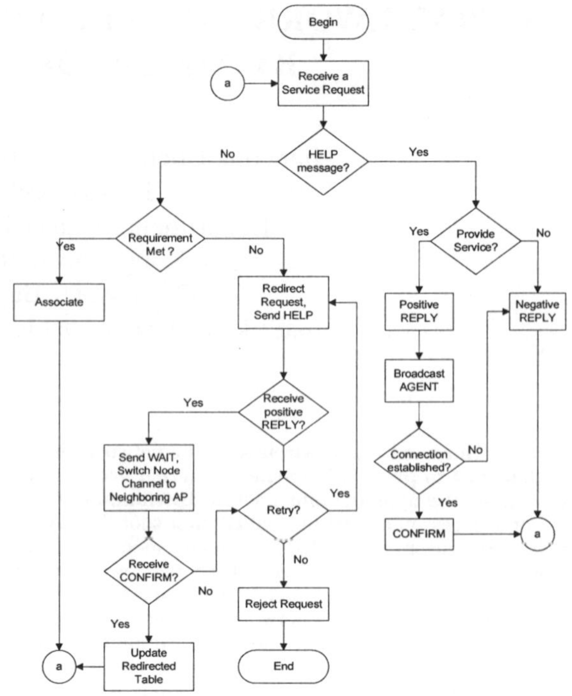

Fig. 2. The flowchart of ad hoc connection establishment algorithm at APs.

A. *AP-Assisted Passive Agent Selection (APAS)*

In this paper, we mainly focus on the agent discovery and connection establishment schemes, illustrated in Figs. 2 and 3, including the operation procedures at both the APs and the MHs. With APAS, the hot-spot AP forwards the (bandwidth) request of the blocked MH to its neighboring APs, which check if they have enough bandwidth to satisfy the request through a two-hop connection. Those having enough bandwidth indicate the MHs in their coverage to advertise the availability. The blocked MH listens to the advertisements for a while and selects the best one (which offers the largest bandwidth or is nearest in space) from the advertising agents within the communication range as the relay agent, and set up the ad hoc connection.

As mentioned before, when an AP receives a service request from a MH, it first checks whether it has sufficient resources (termed residual bandwidth in this paper) to meet the requirement. If it is able to provide to service to the user, the user is then associated to it. If so, it sends "HELP" messages to its neighboring APs, including the information of the service requirement. The APs cooperate to find out which APs adjacent to the hot-spot AP can provide the service instead.

Upon receiving a "HELP" message, an neighboring AP responds with a positive "REPLY" message if it has enough resource to serve the request by a two-hop connection, or a negative one if otherwise. Then it broadcasts an "AGENT" message to the MHs associated to it. After a MH acting as an agent establishes an ad hoc connection to the redirected user, the AP sends a "CONFIRM" message back to the hot-spot AP.

After the hot-spot AP receives a "REPLY," it sends to the MH a "WAIT" message containing the channel information of the neighboring AP. The MH then switches to the channel and waits for the connection messages from the agents. Once the ad hoc connection is established, the hot-spot AP will receive the

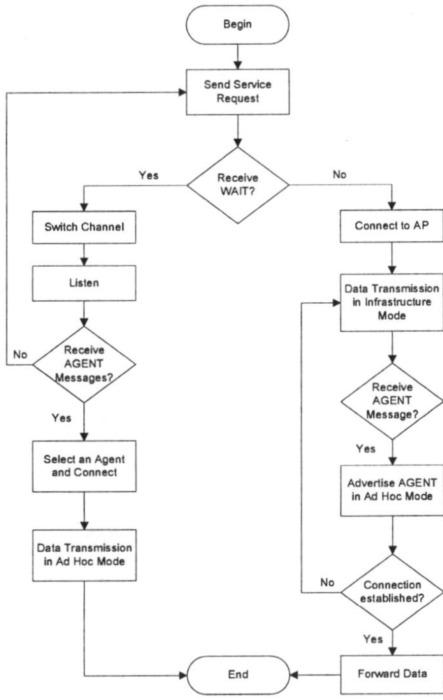

Fig. 3. The flowchart of ad hoc connection establishment algorithm at hosts.

"CONFIRM" message from the neighboring AP. Otherwise, it repeats the above procedures to find another AP for the MH.

At a mobile host, if it receives a "WAIT" message after it sends a service request to an AP, it switches to the channel given by the message and listens in ad hoc mode for some time. After it receives some "AGENT" messages, it selects the agent with the strongest signal power to connect to the neighboring AP. On the other hand, when an agent receives an "AGENT" message from its AP, it advertises the message in ad hoc mode so that the redirected MH can receive it.

As an example, refer to Fig. 1 again, the detailed steps of the connection establishment procedures are listed as follows.

1) A user (MH 1) sends a service request to its AP (AP 1);
2) AP 1 finds out that the requirement can not be met;
3) AP 1 sends "HELP" messages containing the request information to its neighboring APs;
4) Once an AP (AP 3) receive a "HELP" message, it checks whether it can provide service to the request by a two-hop connection and responds with a "REPLY";
5) After AP 1 receives a positive "REPLY" message, it sends a "WAIT" message to MH 1;
6) MH 1 switches to the channel given by the "WAIT" message and listens;
7) After sending a positive "REPLY" to AP 1, AP 3 sends a "AGENT" message to the agents within the coverage;
8) MH 3 advertises the "AGENT" message;
9) Upon receiving the "AGENT" message, MH 1 selects an agent (MH 3). The connection is hence established;
10) After the connection is established, AP 3 sends a "CONFIRM" message to AP 1.

III. Illustrative Numerical Results

In this section, we present the performance evaluation using NS2 (ns-2.26) with the CMU wireless extension. We use 11 Mbps channel bitrate to simulate IEEE 802.11b WLAN environment. Packet size is 512 bytes unless otherwise specified. For the radio propagation model, a two-ray path loss model is used. We do not consider fading and mobility in our simulations. We assume the transmission range of all nodes including the access point is 250 m. All simulation results are the average of 10 runs and each simulation runs for 10 minutes of simulation time. We use the pure WLAN performance as the baseline to compare with.

For network topology, we generate a network with three APs, the locations of which are as shown in Fig. 1 (The distances among are $250\sqrt{2}\ m$.), and a number of users of random location within the transmission range of each AP. Such users arrive at each AP according to the Poisson distribution with different mean inter-arrival time τ_i seconds, which is smaller in the hot-spot AP such that the users arrive more frequently. The holding time t of the users is uniformly distributed with a mean of H. We use UDP traffic from the users to a AP with bandwidth requirement B Kbps. Two different traffic conditions are generated to simulate a lightly congested network and a heavy loaded one. In this paper, we compare the performance of WLAN with and without ad hoc relaying in terms of service failure rate and channel utilization in each AP. The service failure rate is defined as the percentage of users whose requirement can not be met. The channel utilization is measured at each AP individually.

In the lightly congested network, the mean user inter-arrival times in seconds for the three APs are $\tau_1 = 20$, $\tau_2 = 30$ and $\tau_3 = 60$, respectively. The average holding time $H = 10$ minutes and the bandwidth requirements of the users are set to $B = 200$ Kbps such that the hot-spot (AP 1) is lightly congested while the other two are under utilized. In order to make a heavily congested network, we adjust the system parameters to be $\tau_1 = 15$, $\tau_2 = 20$ and $\tau_3 = 60$, respectively, such that AP 1 an AP 2 are congested while AP 3 is under utilized. Other parameters remain the same as above.

We show in Fig. 4 the service failure rate in the hot-spot network with and without ad hoc relaying connections. From the figures, we see that for both lightly and heavily congested networks, the ad hoc relaying scheme achieves substantially lower failure rate. In the lightly congested network (Fig. 4 a.), since the traffic load is not very heavy, only around five percent of the users cannot get their service in the hot-spot AP. With ad hoc relaying connections, most users redirected by AP 1 may get their service from AP 2 or AP 3. Hence the failure rate is reduced by the ad hoc relaying scheme. However, some of them may not be able to establish ad hoc connections to the neighboring AP because they are too far away from the neighbor. Therefore, even with ad hoc relaying scheme, there are still some users which cannot get their service. In the heavy congested network (Fig. 4 b.), since the traffic load is heavier, both AP 1 and AP 2 are congested. More users, which cannot get their service from the congested APs, need to be redirected to the only AP (AP 3) that is under utilized. The users are less likely to establish connections to it. Hence the service failure rate in the heavily congested network does not decrease as much as in the lightly congested network.

In Fig. 5 we compare the channel utilization in each AP. As expected, the congested APs are fully utilized in both cases. For the other APs, which are under utilized by their own users, the channel utilization is increased with the ad hoc relaying users. Because there are only a few users need to be redirected in the

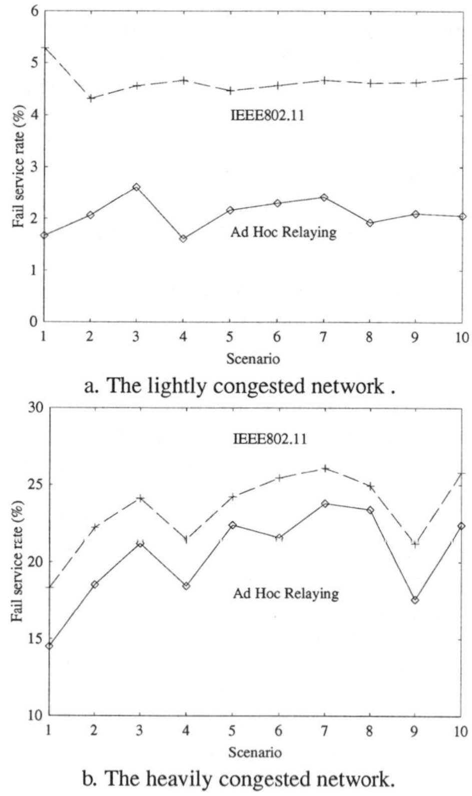

a. The lightly congested network .

b. The heavily congested network.

Fig. 4. The failure rate of proving service to the users.

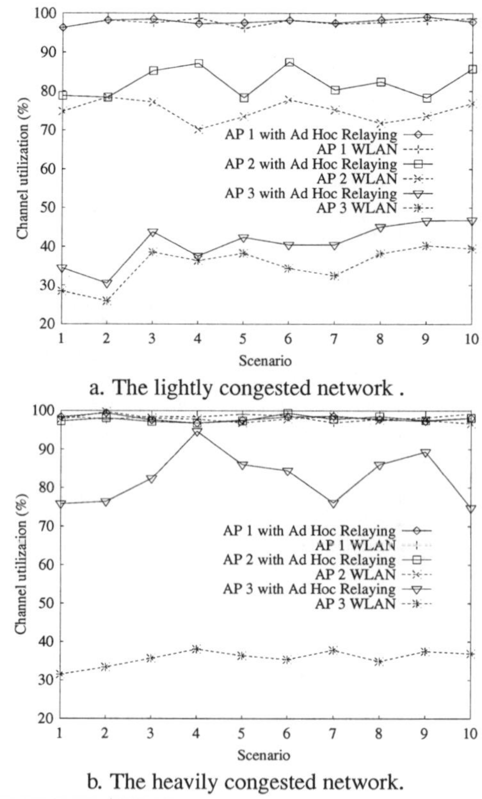

a. The lightly congested network .

b. The heavily congested network.

Fig. 5. The channel utilization in each AP of the networks.

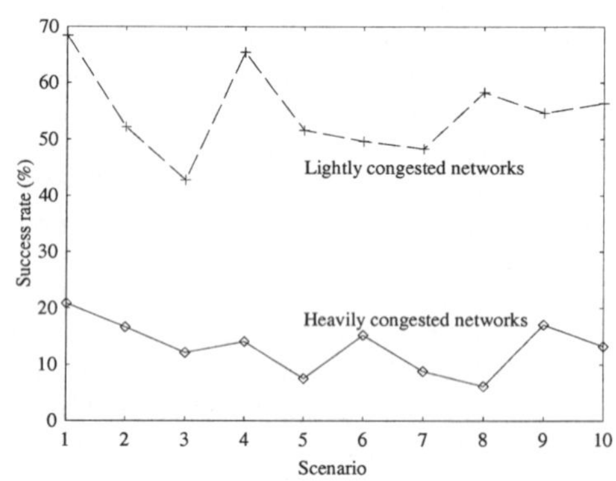

Fig. 6. The success rate of ad hoc connection establishment.

light congested network, the channel utilization is increased by only a little (as shown in Fig. 5 a.). As traffic becomes heavier, the number of users being redirected increases. The channel bandwidth of all APs is highly utilized even in the AP which is under utilized without ad hoc relaying. Therefore, with ad hoc relaying scheme, the channel bandwidth is used more efficiently and the total system throughput is hence improved.

Figure 6 shows the success rate of ad hoc connection establishment in both light and heavy traffic networks. As we can see, around half of the users being redirected can connect to their neighboring APs in the light traffic network. The failures, as mentioned before, are mainly due to the coverage limitation of the agents in the neighboring APs. In the heavy congested network, the success rate is even lower. This is because there are more users being redirected and a small fraction of them can nearly use up the channel bandwidth of the neighboring AP.

In order to further increase the success rate, we are developing a new scheme which can select a proper user (maybe an already admitted one) to redirect instead of just redirecting the new user. In this way, the location (or distance) effect can be eliminated and the selected user is more likely to connect to the neighboring AP.

IV. Conclusion

Due to the dynamic and unevenly distributed user load among the APs in public areas, some "hot-spots" may be congested with many users and cannot admit more requests, while their neighboring ones are serving few users and their bandwidths are under-utilized. In this paper, we propose an ad hoc relaying architecture to serve a mobile user in a hot spot by a neighboring AP, thus alleviating the congestion at the hot-spot. Simulation results show that using the scheme the failure rate is significantly reduced as well as the bandwidth utilization is improved in such hot-spot networks. In our future work, we will further investigate different schemes for implementing ad hoc relaying wireless LANs.

References

[1] J. Chen, S.-H. G. Chan, and J. He, "Mixed-Mode WLAN: The integration of ad hoc mode with wireless LAN infrastructure." to appear in IEEE Globecom 2003.

[2] A. Balachandran, P. Bahl, and G. M. Voelker, "Hot-spot congestion relief in public-area wireless networks," in *Proc. IEEE Workshop on Mobile Computing Systems and Applications*, pp. 70–80, 2002.

[3] H. Wu, C. Qiao, S. De, and O. Tonguz, "Integrated cellular and ad hoc relaying systems: iCAR," *IEEE Journal on Selected Areas in Communications*, vol. 19, no. 10, pp. 2105–2115, 2001.

Mobile ad-hoc networks and QoS demanding applications

Leila Boukhalfa
ESIGETEL
1 rue du port de Valvins
77215 Avon-Fontainebleau

Pascale Minet
INRIA Rocquencourt
78153 Le Chesnay Cedex
France

Laurent George
Université Paris 12, LIIA
120 rue Paul Armangot,
94400 Vitry, France

Serge Midonnet
ESIGETEL
1 rue du port de Valvins
77215 Avon-Fontainebleau

Abstract: The maturity of wireless technologies on the one hand, and the evolution of applications on the other hand, justify the introduction of Quality of Service (QoS) in ad-hoc networks. Much research has been done to date on QoS in ad-hoc networks. In this paper, we present our view of QoS architecture and define its different components. Among these components we will emphasize on QoS model which defines the achieved QoS and the needed mechanisms. Four models are comparatively evaluated. A brief state-of-the-art for QoS signaling, QoS routing and QoS MAC is given. We characterize QoS demanding applications through seven criteria and show how these criteria influence the choice of QoS architecture. These choices are illustrated by some QoS demanding applications.

Key Words: mobile ad-hoc networks, Quality of Service (QoS), QoS architecture.

1. INTRODUCTION

A mobile ad-hoc network, called MANET (Mobile Ad-hoc NETwork), consists of mobile nodes connected by wireless links forming arbitrary time-varying topologies. It does not require any existing network fixed infrastructure or centralized administration. Although the ad-hoc networks were initially put forward for military applications, now they are opening up to civilian applications having mobility requirements and generating multimedia traffic. Such applications are high demanding in term of quality of transmission. Hence, achieving Quality of Service (QoS) in an ad-hoc network corresponds to a real need. We will see why the solutions put forward to support QoS in wired networks, cannot be directly used in wireless networks. In fact, MANET networks are faced whith specific constraints: a) the limited bandwidth because of the reduced available radio resources, b) the highly dynamic topology because of versatile radio propagation and the node's mobility, c) the power constraints because network nodes rely on battery power for energy. These MANET specificities make difficult to achieve QoS in these networks.

The goal of this paper is to present our view of a QoS architecture as well as a synthesis of the various work carried out on QoS for ad-hoc networks. This paper is organized as follows. After having presented a general QoS architecture in section 2, we define the most relevant components of QoS. These components are described in sections 3, 4 and 5. Section 6 presents different QoS demanding applications characterized by seven criteria we have defined. We then show for each of them which QoS components are needed to achieve the required QoS. Finally, we conclude this paper.

2. QOS COMPONENTS

The widely accepted definition of QoS[6] is given by CCITT recommendation E. 800 as the collective effect of service performance that determines the satisfaction degree of a service user. QoS can be represented by a puzzle illustrated by figure 1, whose components must cooperate to provide the required QoS. According to [15], these components are: QoS model, QoS signaling, QoS routing, and QoS Medium Access Control (MAC). Besides these components, we introduce other components such as admission control, provisioning, shaping, scheduling and buffer management.

- *QoS model* specifies the architecture in which services can be provided as well as the necessary mechanisms such as classification. The QoS model directly influences the functionality of the other components.
- *QoS signaling* is used to propagate QoS control information in the network, to reserve and release resources, as well as to generate the QoS reports that indicate the effectively measured QoS. QoS signaling can be associated with routing with or without QoS.
- *QoS routing* aims to find routes with sufficient resources to meet the application requirements but does not reserve resources.
- *QoS MAC* protocol allows sharing a resource radio according to the QoS required. The efficiency of signaling or routing protocols directly depends on the MAC protocol efficiency. Moreover, if the MAC protocol does not support QoS, then it is not possible to ensure that the message being transmitted is the one with the highest priority among the waiting messages. We recommend the use of a QoS MAC protocol.
- *Admission control* is the mechanism that allows admission or not of a new flow in the network according to (i) the available resources on the path taken by this flow and (ii) the QoS requirements of this flow. The admission control is carried out after signaling. In that case, signaling can be a resource reservation (e.g.: INSIGNIA see section 3.1) or a QoS report that can be solicited as in SWAN for the real time traffic (see section 3.2).
- *Provisioning* [4] governs how traffic is marked and conditioned when entering a network, and how that traffic is

forwarded within that network. For instance, provisioning enables the allocation of resources to a given class via the configuration of traffic shapers.

• *Shaping* enables traffic regulation by acting on the throughput or on the jitter of the traffic. It can make traffic conform to its profile. The most used shaper is the leaky bucket.

• *Scheduling and buffer management* make it possible to implement QoS. The scheduler determines the message transmission order according to the QoS model. QoS signaling interacts with the buffer management at the time of resource reservation.

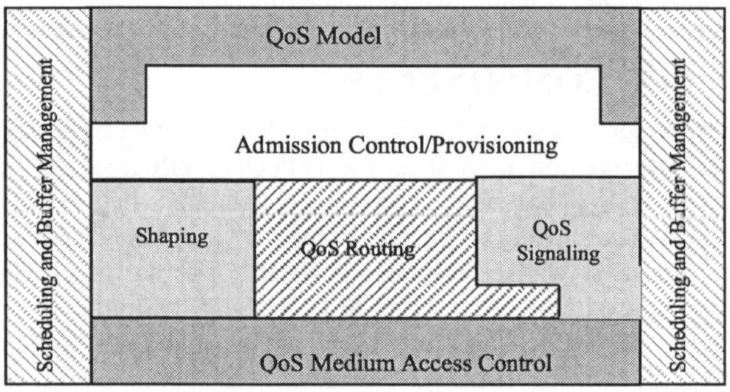

Figure 1 QoS Components.

3. QoS MODELS

As in wired networks, there are several QoS models for the ad-hoc networks. Four models are presented in this section. None of them give hard QoS guarantee, because of node mobility and versatility of radio links.

3.1. INSIGNIA

The INSIGNIA model [9] can be considered as an adaptation of the IntServ (Integrated Services) model to the ad-hoc networks. Indeed, INSIGNIA offers per flow granularity. It retains bandwidth as QoS parameter and it supports applications that adapt their flows in response to bandwidth variations. This QoS model proposes two types of QoS: base QoS that only ensures a minimum bandwidth and enhanced QoS that ensures a maximum bandwidth. These guarantees require the implementation of the signaling protocol described in section 4. This protocol constitutes the key component of this model. Like IntServ, INSIGNIA does not allow good scalability because of individual flow management. Resource reservation per flow allows a better QoS guarantee. In conclusion, this model can be adopted when the number of flows is not high and these flows require per flow guarantee.

3.2. SWAN

SWAN (Service differentiation in stateless Wireless Ad-hoc Network) [1] is an adaptation of the DiffServ (Differentiated Services) model to the ad-hoc networks. SWAN supports soft real-time traffic (i.e., traffic able to support small-scale violations of QoS) and best effort traffic. It uses a feedback-based control mechanism to support soft real-time services and service differentiation, without requiring a QoS MAC. To achieve that, it uses explicit congestion notification to dynamically regulate admitted real-time traffic in case of traffic overload due to the high dynamics of the mobile ad-hoc network. Each node is in charge of (i) regulating best effort traffic by means of a shaper and (ii) periodically estimating local bandwidth availability. Two types of measurements are used: packet transmission delay for the regulation of best effort traffic, and the end-to-end available bandwidth for the real time traffic admission and regulation. Furthermore, the admission control test is performed only at the source node according to the results of a request/response probe giving the minimum available bandwidth on the path.

3.3. FQMM

Flexible QoS Model for MANET (FQMM) [16], constitutes a hybrid approach combining per-flow granularity of IntServ and per-class granularity of DiffServ. In FQMM, traffic of highest priority, which constitutes a small portion of traffic, is given per-flow provisioning, while other priority classes are given per-class provisioning. Moreover, each node performs the conditioning of the only traffic it generates. This model has been introduced to offer a better QoS guarantee to a restricted number of flows and class guarantee to the other flows. In order to keep the differentiation between classes consistent under the dynamics of the network, FQMM uses an adaptive traffic profile. It is defined by the relative percentage of the effective link capacity requested by this traffic.

3.4. iMAQ

The iMAQ (Integrated Mobile Ad-hoc QoS) [10] framework consists of different layers able to adapt to changes in a dynamic environment and to support multimedia traffic. In this model, no hard QoS guarantee is provided and no resource reservation is done. Each traffic expresses its requirements in terms of minimum acceptable and maximum requested bandwidth. QoS routing is based on measurements of the available bandwidth. Several paths are maintained for a given destination. Only nodes having sufficient resources to support the new traffic are used. Mobility is accounted for by predicting the route breakage, replacing the current route by a new one before the breakage. Thus, QoS routing is adaptive to QoS fluctuations and topology changes.

4. QoS SIGNALING

QoS signaling protocols aim at providing a means for propagating control information related to QoS. This information can be a request for resource reservation or a QoS report. Two types of protocols can be distinguished: (i) in-band signaling protocols allow control information to be carried along in the IP packet header, (ii) out-of-band signaling protocols use explicit control packets.

RSVP (Resource reSerVation Protocol), the in-band signaling protocol of Internet is not adapted to MANET (heavy signaling overhead and no fast adaptation to frequent topology changes). The in-band signaling protocol INSIGNIA was designed to be lightweight in terms of the amount of bandwidth consumed and to be capable of quick reaction in case of network changes such as rapid host mobility and wireless link degradation. INSIGNIA supports fast reservation, restoration and adaptation algorithms. QoS reporting is periodically generated by the destination to inform the source of the ongoing status of flows. Hence the source can adapt its flow to the network state. In SWAN, QoS signaling is out-of-band (e.g., request probe before admitting real time traffic).

5. QoS ROUTING

Routing protocols in phase of IETF standardization for ad-hoc networks are:

• AODV (Ad-hoc On Demand Distance Vector) [12] and DSR (Dynamic Source Routing) [7] for *reactive protocols.* Such protocols build routes only when requested by source nodes. These routes are maintained as long as they are needed by the sources and are deleted when they are no longer used. Thus, the traffic control is reduced in the network. However, as there is no route immediately available towards the destination, the time needed to transmit the first packets is increased by the route discovery delay. Moreover, because of flooding, route discovery is bandwidth consuming.

• OLSR (Optimized Link State Routing protocol) [2] and TBRPF (Topology dissemination Based on Reverse Path Forwarding) [11] for *proactive protocols.* Such protocols maintain routing table towards every possible destination in the network. They require periodic information exchange in order to maintain a consistent network view. Thus routes are immediately available when they are demanded.

AODV and DSR use the two mechanisms that are route discovery and route maintenance. OLSR and TBRPF are based on the link state principle. OLSR uses the multipoint relays to reduce the size of control packets and to minimize their number of retransmissions. In TBRPF, each node maintains its tree of the shortest paths towards all possible destinations in the network. To achieve that, each node broadcasts to its neighbors a subset of its tree. In a comparative study [8], it has been shown that proactive protocols outperform reactive ones in case of a dense network or when traffic is not restricted to a small subset of nodes.

QoS routing consists in establishing routes that satisfy the QoS constraints. Ad-hoc network characteristics necessitate reassessment of solutions adopted by wired networks. Among MANET QoS routing protocols, AODV takes into account the QoS parameters which are bandwidth and end-to-end delay. The basic idea [6] consists in adding extensions to the routing tables maintained on every node and to the messages RREQ (Route Request), RREP (Route Reply) during the route discovery. A node that receives a RREQ with a QoS extension must be able to meet the service requirement in order either to broadcast the RREQ (if it does not have an up-to-date route to the destination) or to unicast a RREP to the source. This mecanism is used also by DSR in a similar way to account for the bandwidth constraint [3].

Providing QoS in term of bandwidth requires the evaluation of the bandwidth available on each node. This evaluation is difficult because of the time varying throughput and the interference between radio signals. A node can measure its bandwidth available at the MAC level. For instance, [13] measures the throughput obtained while transmitting a unicast packet over an IEEE 802.11 medium. These instantaneous measures are averaged. If the knowledge of the available bandwidth is required on a remote node, this information is carried out by the signaling protocol (e.g., probe for the admission of real time traffic in SWAN model) or the QoS routing protocol (e.g., AODV with QoS). Other QoS routing protocols have been designed outside of MANET. We can cite for instance CEDAR [14] and Ticket-based Probing [5].

In conclusion, QoS routing is a hard problem in mobile ad-hoc networks. In most cases, QoS routing is based on the evaluation of the available bandwidth which is the only QoS parameter taken into consideration.

For lack of space, we do not describe the QoS MAC component.

6. ADEQUACY OF QoS DEMANDING APPLICATONS AND QoS COMPONENTS

We can characterize an application by seven criteria:

• *Node mobility*: depending on the node speed, the mobility can be low, medium or high.
• *QoS adaptive*: QoS variations can be accounted for at the application level.
• *Interaction duration*: the lifetime of the communication session can be short, medium or long.
• *Message importance*: high importance messages at the application level must be privileged.
• *Number of flows*: the number of flows can be limited, medium or high.
• *Bandwidth requirement*: bandwidth requested by the application can be low, medium or high.
• *Real time constraints*: bounds on the end-to-end delay and end-to-end jitter.

6.1. Types of application to be considered

In order to clarify how the above criteria can influence the choice of the QoS components, we consider five different applications and describe them according to the seven previous criteria.
• *NVoD (Near Video on Demand).* A client can select a channel to view a video content sent periodically. Generally, mobility is required for clients embedded on mobile nodes (e.g., children watching TV in a car). Such an application is QoS adaptive and has a long interaction. All messages can be considered with the same importance. The number of flows is limited. The bandwidth requirement is high and there is no real-time constraints.
• *VoD (Video on Demand).* Same as NVoD but the client interacts with the video server by means of VCR commands (e.g., start, stop, play, pause, forward,...). VCR commands are of an importance higher than multimedia flows. Real-time constraints are imposed to VCR command reply.
• *Video conferencing.* People in separate location exchange in real-time text, audio and video. There is generally no mobility. The interaction duration, the number of flows and the bandwidth requirement are medium. The real-time

constraints are high due to synchronisation constraints between the participants.

• *Web browsing*. The mobility is low or medium. The interaction duration is short. The number of flows and the bandwidth requirements are weak. The real-time requirements are medium.

• *VoIP (Voice over IP)*. The mobility and the interaction duration are medium. The bandwidth requirement is weak and the real-time constraints are high. Some messages are privileged (e.g. urgency calls).

6.2. How QoS can be offered to them

Now we analyse the QoS components with regard to the criteria defined previously. It is clear that no QoS model is adapted to all types of mobility. It should be efficient to tune parameters of QoS components according to mobility, in such a way that the overhead is directly proportional to the degree of mobility. Node mobility also influences the shaping and the scheduling. The tuning of shaping parameters is more difficult with high mobility. A node mobility can generate a significative overhead if scheduling parameters depend on the flow number.

Moreover, all QoS models support QoS adaptive applications. Applications with soft real-time traffic tolerating small-scale QoS violations can also be supported. As concerns interaction durations, short ones are inefficient with INSIGNIA and with FQMM for high priority traffic, and they are expensive with SWAN and iMAQ.

Message importance can be reflected by means of priority or traffic class. iMAQ does not account for message importance. For a high importance message, SWAN proposes soft real-time class and FQMM assigns it the highest priority. In INSIGNIA the importance of a message is directly reflected by its requested bandwidth with enhanced QoS.

The number of flows is limited with INSIGNIA and FQMM for the highest priority traffic because of their per flow management. The choice of the scheduling policy is all the more important when the number of flows is high.

After this analysis, we conclude that NVoD application can be supported by any QoS model. The existence of high importance messages is problematic for iMAQ and in a less degree for INSIGNIA. That is why we do not recommend those models for real-time applications such as VoD and VoIP. Soft real-time traffic needs (e.g., bounded delay and bounded jitter) can be met at the price of a bandwidth overdimensionning. For short interactions such as in web browsing, INSIGNIA is inefficient, because of resource reservation.

7. CONCLUSION

In this paper, we have presented our view of a QoS architecture and described its different components. We have presented briefly the various researches dealing with QoS in ad-hoc networks. We have also studied the adequacy of QoS components to various QoS demanding applications through some application examples. We have concluded that the choice of QoS component depends on application criteria such as mobility, interaction duration or number of flows …

The proposed QoS models use mechanisms enabling them to adapt to ad-hoc networks characteristics and to offer a soft QoS guarantee. Hard QoS guarantee is very difficult to achieve in ad-hoc networks because of the high dynamics of topology. All QoS models consider the requested bandwidth as QoS parameters. In the best case, they also consider the importance of the message at the application level. They all use the measure of the available bandwidth to perform QoS management. QoS MAC is of prime importance to achieve a strong differentiation between services.

REFERENCES

[1] G.S. Ahn, A.T. Campbell, A. Veres, L.H. Sun, 'SWAN: Service Differentiation in Stateless Wireless Ad-hoc Networks', IEEE INFOCOM'2002, New York, June 2002.

[2] C. Adjih, T. Clausen, P. Jacquet, A. Laouiti, P. Minet, P. Muhlethaler, A. Qayyum, L. Viennot, 'Optimized Link State Routing Protocol for Ad-hoc Networks', draft-ietf-manet-olsr-07.txt, IETF, November 2002.

[3] V. Aarthi, A. Siromoney,'Quality of Service Routing in DSR' Trusted Internet Workshop 2002, International Conference on High Performance Computing (HiPC 2002), Bangalore, India, December 2002.

[4] S. Blake D. Black, M. Carlson, E. Davies, Z. Wang, W. Weiss, 'An Architecture for Differentiated Services', Internet RFC2475, IETF, December 1998.

[5] S. Chen, K. Nahrstedt,'Distributed Quality-of-Service Routing in Ad-Hoc Networks', IEEE Journal on Special Areas in Communications, Vol. 17, N° 8, August 1999.

[6] Z.Y. Demetrios, 'A glance at Quality of Service in Mobile Ad-Hoc Networks' Research report for CS260-seminar in Mobile Ad-Hoc Networks, 2001. http://www.cs.ucr.edu/~csyiazti/cs260.html

[7] D.B. Johnson, D.A. Maltz, J. Broch, 'DSR: the Dynamic Source Routing Protocol for Multi-Hop Wireless Ad-hoc networks', in Ad-hoc networking, edited by C. Perkins, Addison-Wesley, 2001.

[8] P. Jacquet, L.Viennot, 'Overhead in Mobile Ad-hoc Network Protocols', Technical Report 3965, INRIA, June 2000.

[9] S.B. Lee, G.S. Ahn, X. Zhang, A.T. Campbell, 'INSIGNIA: An IP-Based Quality of Service Framework for Mobile ad-hoc Network', Journal of Parallel and Distributed Computing, N° 60, pp. 374-406, 2000.

[10] MONET research group, 'iMAQ: An Integrated Mobile Ad-hoc QoS framework', http://cairo.cs.uiuc.edu/adhoc

[11] R. Ogier, M. Lewis, F. Templin, B. Bellur 'Topology dissemination Based on Reverse Path Forwarding (TBRPF)', draft-ietf-manet-tbrpf-06.txt, IETF, November 2002.

[12] C.E. Perkins, E.M. Royer, 'Ad-hoc On Demand Distance Vector Routing', In proceedings of the 2nd IEEE Workshop on mobile Computing Systems and Applications (WMCSA'99), pp. 90-100, New Orleans, Louisiana February 1999.

[13] S. Shah, K. Chen, K. Nahrstedt, 'Dynamic Bandwidth Management in Single-hop Ad-hoc Wireless Networks', Proc. of IEEE Int. Conf. on Pervasive Computing and Communications, Percom'2003, Dallas, Texas, March 2003.

[14] P. Sinha, R. Sivakumar, V. Bharghavan, 'CEDAR: a Core-Extraction Distributed Ad-hoc Routing algorithm', IEEE Infocom'99, New York, March 1999.

[15] K. Wu, J. Harms, 'QoS Support in Mobile Ad-hoc Networks', Crossing Boundaries, GSA Journal of University of Alberta, vol 1, N° 1, November 2001, pp. 92-106.

[16] H. Xiao, W.K.G. Seah, A. Lo, K.C. Chua, 'A Flexible Quality of Service Model for Mobile Ad-Hoc Networks', IEEE VTC2000-spring, Tokyo, Japan, May 2000.

QoS for Ad hoc Networking Based on Multiple Metrics: Bandwidth and Delay

Hakim Badis[1,2] , Anelise Munaretto [3], Khaldoun Al Agha[1,2] and Guy Pujolle[3]
[1]LRI Laboratory, University of Paris XI, Orsay, France {badis, alagha}@lri.fr
[2]INRIA Laboratory, Rocquencourt, France
[3]LIP6 Laboratory, University of Paris VI, Paris, France {anelise.munaretto, guy.pujolle}@lip6.fr

Abstract: A link state routing approach makes available detailed information about the connectivity and the condition found in the network. OLSR protocol is an optimization over the classical link state protocol, tailored for mobile ad hoc networks. In this article, we design a QoS routing scheme over OLSR protocol, called QOLSR. In our proposal, we introduce more appropriate metrics than the hop distance used in OLSR. In order to improve quality requirements in routing information, delay and bandwidth measurements are applied. The implications of routing metrics on path computation are examined and the relational behind the selection of bandwidth and delay metrics are discussed. We first consider algorithms for single-metric approach, and then present a distributed algorithm for multiple metrics approach. We also present a scalable simulation model close to real operations in Ad Hoc Networks. The performance of our protocol are extensively investigated by simulation. Our results indicate that the attained gain by our proposal represent an important improvement in such mobile wireless networks.

I. Introduction

The problem of QoS routing for mobile Ad hoc networks is studied. Most routing protocols for the mobile Ad hoc networks (MANETs) [1], such as OLSR [2], AODV [3], DSR [4], are designed without explicitly considering QoS of the routes they generate. The number of hops is the most common criterion adopted by such proposed routing protocols. It is becoming increasingly clear that such routing protocols are inadequate for multimedia application, such as video conferencing, which often require guaranteed QoS. QoS routing requires not only finding a route from a source to a destination, but a route that satisfies the end-to-end QoS requirement, often given in terms of bandwidth or delay. QoS is more difficult to guarantee in Ad hoc networks than in most other type of networks, because the wireless bandwidth is shared among adjacent nodes and the network topology changes as the nodes move [5].

The link state routing approach makes available detailed information about the connectivity and the topology found in the network. Moreover, it increases the chances that a node will be able to generate a route that meets a specified set of requirement constraints. OLSR protocol is an optimization over the classical link state protocol for the mobile Ad hoc networks. It performs hop by hop routing, i.e. each node uses its most recent information to route a packet. This paper aims at specifying a QoS-enhanced OLSR [2] routing in Mobile Ad Hoc Networks, the QOLSR. We propose to implement QoS functionality to deal with limited available resources in a dynamic environment.

This paper is organized as follows. Section II depicts the metrics measurements proposed by the QOLSR protocol. The routing table calculation of QOLSR is presented in Section III. Therefore, we validate the proposal by means of performance evaluation (Section IV). Finally, we present our conclusions (Section V).

II. Metrics Measurements

The delay and bandwidth metrics are taking into account as QoS constraints for the proposed QOLSR protocol. Such metrics are included on each routing table entry corresponding to each destination. We use the IEEE 802.11b as the medium access control to achieve the bandwidth measurement.

A. Delay Metric

Each node includes in the *Hello* message, during the neighbor discovery performed by the QOLSR, the creation time of this message. When a neighbor node receives this message, it calculates the difference between such a time and the current time, which represents the $\mathrm{measured\_delay}$. Such one-way $\mathrm{measured\_delay}$ includes the queuing time, the transmission time, the collision avoidance time and the control overhead time. Otherwise, the measurement of the one-way delay avoids the increase of traffic load at adding acknowledgment messages to the QOLSR protocol. Therefore, we suppose a synchronized network in our works.

Due to the characteristics of sparse ad hoc networks, classical clock synchronization algorithms are not applicable. Meanwhile, the aim of this paper is not to solve the synchronization issue. Time synchronization in Ad Hoc Networks is a wide subject of research, e.g. the work presented in [6].

The proposed QOLSR calculates the average delay with one of the two proposed methods: AV, RTT.

In a first work [7], we use Average delay and Variance method (AV) during the routing calculation, as show in Figure 3.

In this paper, we propose a new approach, the RTT method. The average delay is calculated through the following formula: $\mathrm{Average\_delay} = \alpha \times \mathrm{Average\_delay} + (1 - \alpha) \times \mathrm{measured\_delay}$. Then, the first results of such implementation. Figure 4 depicts the best performance achieved using $\alpha = 0.4$.

B. Bandwidth Metric

A remarkable work is presented in [8], considering the acknowledgement time from the data packets. For our approach, the bandwidth will be calculated between a node and its neighbors having direct and symmetric link. We consider for our analysis data packets and signaling traffic that also use the available bandwidth and then must be taken into account.

Let be a node i and j its neighbor, then we define our available bandwidth for i to j through the following formula:

$\mathrm{Bw}_{(i,j)} = (1-u) \times \mathrm{Throughput}_{(i,j)}$, where u is the link utilization.

The throughput seen by one packet of S bits can be calculated as:

$$\mathrm{Throughput}_{packet} = \frac{S}{t_q + (t_S + t_{CA} + t_{Overhead}) \times R + \sum_{r=1}^{R} B_T}$$

where: t_q is the Mac queuing time, t_S the Transmission on time of S bits, t_{CA} the Collision Avoidance phase time, $t_{Overhead}$ the Control Overhead time (e.g. RTS, CTS, etc), R the Number of necessary transmissions, B_T the Backoff time for retransmission r. This formula reveals some undesirable characteristics such as packet size dependence and high variance due to random per packet effects. To increase statistical robustness of the measurements, a packet window of 16 or 32 samples (packets) is adequate.

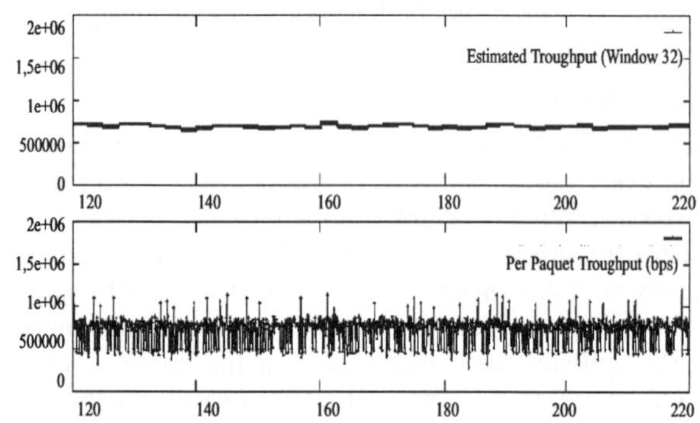

Fig. 1. Window operation

To illustrate, Figure 1 shows where the high variance per packets measurements is aggregated on a window of 32 packets.

The idle time and window duration are calculated to produce the link utilization factor and the permissible throughput measurement as:

$$\frac{\mathrm{idle\_time\_in\_window}}{\mathrm{window\_duration}} \times \mathrm{Throughput\_measured}$$

III. Routing table calculation

Let $G = (V, E)$ be the network with $|V|$ nodes and $|E|$ arcs and $p = (i, j, k, ..., q, r)$ a directed path.

The first approach presented uses one single metric in route decisions such as hop count or delay or available bandwidth.

For delay metric, each arc (i, j) in the path p is assigned a real number del_{ij}. When the arc (i, j) is inexistent or j is not a MPR of i (Referring to the OLSR routing mechanism), then $\mathrm{del}_{ij} = \infty$. Let $\mathrm{del}(p) = \mathrm{del}_{ij} + \mathrm{del}_{jk} + ... + \mathrm{del}_{qr}$. The routing problem is to find a path $p*$ between i and r so that $\mathrm{del}(p*)$ is the minimum. In such a case, we use the well-known Dijkstra routing algorithm.

For bandwidth metric, each arc (i, j) in the path is assigned a real number Bw_{ij}. When the arc (i, j) is inexistent or j is not a MPR of i, $\mathrm{Bw}_{ij} = 0$. Let $\mathrm{Bw}(p) = \min\{\mathrm{Bw}_{ij}, \mathrm{Bw}_{jk}, ..., \mathrm{Bw}_{qr}\}$. The routing problem is to find a path $p*$ between i and r that maximizes $\mathrm{Bw}(p*)$. In order to implement such a metric, we propose a variant-Dijkstra algorithm.

The second approach treats each metric individually. Such approach is not feasible due to the algorithm complexity. The problem of finding a path with n additive and m multiplicative metrics is *NP-complete* if $n + m \geq 2$ [9]. Including a single metric, the best path can be easily defined. Otherwise including multiple metrics, the best path with all parameters at their optimal values may not exist. For example, a path with both maximum bandwidth and minimum delay may not necessarily exist. Thus, we must define precedence between bandwidth and delay. Queuing delay is more dynamic, thus bandwidth is probably often considered as more important. If the bandwidth requirement cannot be met, the chance is that queuing delay will also be high. Our strategy is to find a path with maximum bandwidth (a widest path), and when there is more than one widest path, we choose the one with the shortest delay. We refer to such a path as the shortest-widest path. In our proposed work, we will introduce an algorithm of shortest-widest path. The widest path problem is to find a path $p*$ between i and r that maximizes $\mathrm{Bw}(p*)$. The widest path problem can be solved with similar technique used in shortest path algorithms such as Dijkstra's routing algorithm or Bellman-Ford algorithm.

IV. Performance evaluation

In this section, we evaluate the performance of QOLSR applying multiple metrics. We have carried out simulations to analyze OLSR and QOLSR in different configurations and scenarios. We use the OPNET simulator for our evaluation.

A. Simulation Model

The simulation model introduced in [10] aims the flexibility and hence it implements several parameters:

- The network parameters: number of nodes, region, inter-arrival, etc;
- The OLSR parameters: $Hello\ interval$, $TC\ interval$, use of MPRs (on/off), etc;
- The QoS parameters: variation threshold, RTT (on/off), routing calculation technique, etc;
- The CSMA/CA parameters: radio range, noise ratio, RTS/CTS, signal decay, etc;
- The mobility parameters: speed-min, speed-max, length mobility intervals, etc.

The simulation model is very close to a real Ad-Hoc network operations. At each time, we can detect the position of mobiles by our mobility model. Each node is represented by a subqueue and placed in the region by randomly selecting its x and y co-ordinates. With our method, the simulation model is very optimized that enables to reduce the CPU time and consequently to increase the time of simulation.

The random mobility model proposed is a continuous-time stochastic process. Each node's movement consists of a sequence of random length intervals, during which a node moves in a constant direction at a constant speed. A detailed description can be found in [10].

B. Implemented Algorithm

In order to achieve the performance evaluation including delay and bandwidth metrics in the QOLSR proposed protocol, we implement the algorithm as represented in Figure 2.

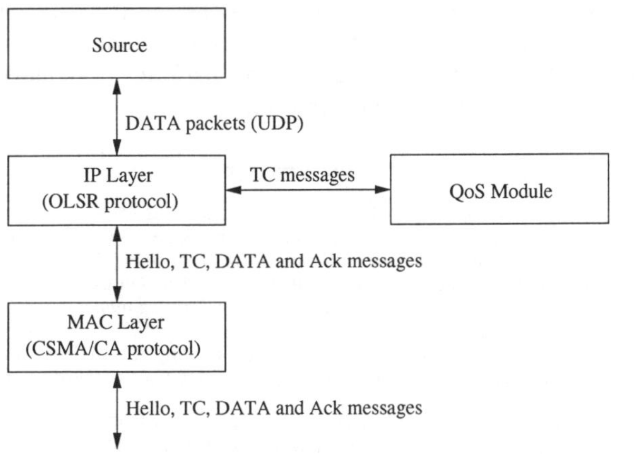

Fig. 2. The generic QOLSR scheme

The source layer sends a DATA packet (for instance, an UDP packet) to the OLSR layer with a specified *interarrival* time. Such DATA packet contains the address of the node, which has originally generated this message. This field should not be confused with the source address from the UDP header, which is changed each time to the address of the intermediate node witch is retransmitting such message. It contains also the destination address and the packet length.

The OLSR layer includes the next hop in DATA packet. The next hop router is identified by the entry of the destination in the host routing table. The OLSR layer sends Hello messages each $Hello\_interval$, TC messages each $TC\_interval$ and DATA messages to the MAC layer. The QoS module maintains for each neighbor a window of 32 packets. It calculates the idle time and window duration to produce the link utilization factor. Before that when OLSR layer sends a TC message, QoS module includes for each MPR selectors, the permissible throughput measurement and the average delay.

The MAC layer transmits the packet using CSMA/CA protocol to its neighbors and forwards such packet to the OLSR layer. Such layer uses an acknowledgment for a point to point packet, as Data packet. However, broadcast packets, as TC messages are not acknowledged.

C. Simulations

The configured parameters in the simulated scenarios are: threshold variance is fixed in 10%, the range transmission is 60 meters, the threshold of the reception is 0,00027 watts and the interarrival time is 60s.

In the first scenario, we perform the evaluation analyzing the average transmission time with AV method, varying the number of mobiles in a region of 100^2m^2. We suppose the bandwidth equal to 10kb/s. The results found in such a scenario are shown in the following Figure:

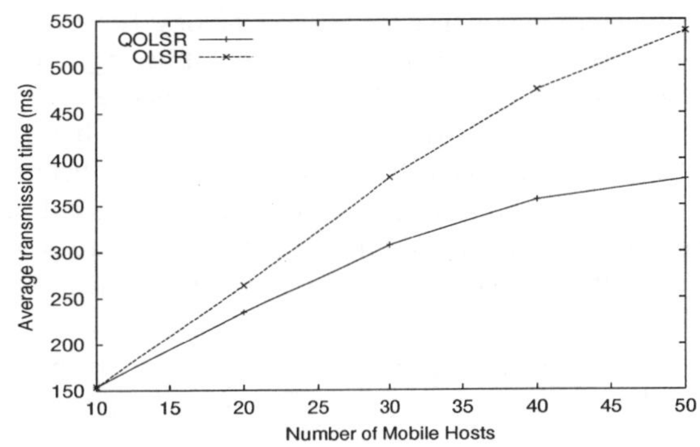

Fig. 3. Average transmission time with AV method

Figures 3 depicts an improvement comparing with the standard OLSR protocol to the average transmission time, i.e., the complete time to a data packet leaves the sender node and arrives in the destination node. As expected, the obtained gain increases (can reach 17.9%) with increasing number of mobiles. The presented results are very interesting, because of in a real wireless network, usually we have an important number of the collision due to the interferences of the radio transmission, and then, the obtained gain can improve the transmission time of such wireless networks.

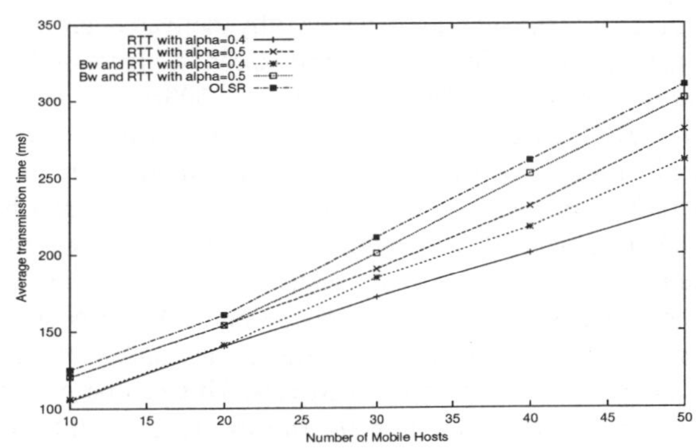

Fig. 4. Average transmission time with RTT method

Figure 4 depicts the average transmission time versus number of mobile nodes comparing four different routing table calculation with the standard OLSR protocol in a 200^2m^2 area network and 1Mb/s of bandwidth. The results plotted are:

- RTT with $\alpha = 0.4$: QOLSR with the RTT method, where $\alpha = 0.4$ using the delay as a single metric approach.
- RTT with $\alpha = 0.5$: QOLSR with the RTT method, where $\alpha = 0.5$ using the delay as a single metric approach.
- Bw and RTT with $\alpha = 0.4$: QOLSR with multiple metrics (bandwidth and delay), where $\alpha = 0.4$.
- Bw and RTT with $\alpha = 0.5$: QOLSR with multiple metrics (bandwidth and delay), where $\alpha = 0.5$.

The best performance is achieved using $\alpha = 0.4$. RTT with $\alpha = 0.4$ has the less average time because the routes are chosen with delay which is an additive metric. In this case ($\alpha = 0.4$), we increase the dominance of the last measured_delay in the formula: $\text{Average\_delay} = \alpha \times \text{Average\_delay} + (1 - \alpha) \times \text{measured\_delay}$ because of the node's movement. So, we take with more importance the actual topology. However, with $\alpha = 0.5$, the Average_delay is the classical average. Bw and RTT with $\alpha = 0.4$ ($\alpha = 0.5$) has less performance than RTT with $\alpha = 0.4$ ($\alpha = 0.5$), because the routes are chosen first with bandwidth and second with delay metric. The bandwidth metric follows the concave composition rule. Then, the distance between the source and destination is not taken into account. Consequently, the end-to-end delay is more important.

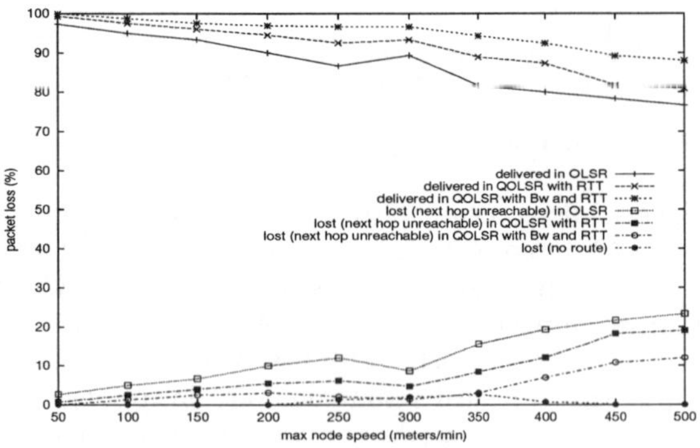

Fig. 5. Data transmitted in varying node speeds

Figure 5 shows the results of our simulation in which the data packets sent and successfully delivered are plotted against the increasing speed. The speed is increased from $50 meters/minute$ ($3Km/hr$) up to $500 meters/minute$ ($30Km/hr$). In this simulation, 50 nodes constitute the network in a region of $1000^2 m^2$, and all the 50 nodes are packet-generating sources. We also keep the movement probability as 0.3, i.e., only 20% of nodes are mobile and the rest are stationary. Each mobile node selects its speed and direction which remains valid for next 60 seconds. We can see that when the mobility (or speed) increases, the number of packets delivered to the destinations decreases. This can be explained by the fact that when a node moves, it goes out of the neighborhood of a node which may be sending it the data packets. There are about 99.92% of packets delivered for QOLSR with Bandwidth (Bw) and delay (del) as metrics at a mobility of 2 meters/minute (99.3% for QOLSR with del and 97.3% for OLSR). At a mobility of 500 meters/minute, 88% of packets delivered for QOLSR with Bw and del as metrics (80.9 % for QOLSR with del and 76.6 % for OLSR). QOLSR with Bw and del as metrics has the highest packets delivered because the routes are chosen with minimal interferences and the next-hops are valid about more than window of 32 packets. The data packets are lost because the next-hop node is unreachable. A node keeps an entry about its neighbor in its neighbor table for about 6 seconds. If a neighbor moves which is the next-hop node in a route, the node continues to forward it the data packets considering it as a neighbor. Also, the next-hop is unreachable if there are interferences. Few of packets are also lost because of unavailability of route and it is the same for OLSR with or without QoS. This happens when a node movement causes the node to be disconnected from the network temporarily, until it re-joins the network again.

V. Conclusions

This paper presented a Link state QoS routing protocol for Ad hoc networks. In order to improve quality requirements in routing information, delay and bandwidth measurements are applied. The implications of routing metrics on path computation are examined and the rationales behind the selection of bandwidth and delay metrics are discussed. In our simulations, the QOLSR protocol produces better performance comparing with the best effort OLSR protocol.

References

[1] "http://www.ietf.org/html.charters/manet-charter.html."

[2] T. Clausen and P. Jacquet, "Optimized Link State Routing Protocol," *In IETF Internet Draft, draft-ietf-manet-olsr-11.txt*, July 2003.

[3] C. Perkins, E. M. Royer and S. R. Das, "Ad Hoc On-Demand Distance Vector routing," *In IETF Internet Draft, draft-ietf-manet-AODV-12.txt*, November 2002.

[4] D. Johnson and Al, "Dynamic Source Routing in Ad Hoc Wireless Networks," *In IETF Internet Draft, draft-ietf-manet-DSR-8.txt*, march 2003.

[5] S. Chakrabarti and A. Mishra, "QoS Issues in Ad Hoc Wireless Networks," *IEEE Communications Magazine*, Febrary 2001.

[6] Kay Römer, "Time Synchronization in Ad Hoc Networks," *In Proceedings of ACM MobiHoc 2001*, October 2001.

[7] A. Munaretto, H. Badis, K. Al Agha and G. Pujolle, "A Link-state QoS Routing Protocol for Ad Hoc Networks," *In the proceedings of IEEE MWCN2002, Stockholm, Sweden*, September 2002.

[8] Kazantzidis and M. Gerla, "End-to-end versus Explicit Feedback Measurement in 802.11 Networks," *In Seventh IEEE Symposium on Computers and Communications*, 2002.

[9] F. Kuipers, P. Van Mieghem, T. Korkmaz and M. Krunz, "An Overview of Constraint-Based Path Selection Algorithms for QoS Routing," *IEEE Communications Magazine*, vol. Vol.40, No.12, December 2002.

[10] H. Badis and K. Al Agha, " Scalable model for the simulation of OLSR and Fast-OLSR protocols ," *Med-hoc-Net*, June 2003.

SESSION 1B:
BEYOND 3G AND 4G

SERVICE PROVISION AND RECONFIGURATION MANAGEMENT IN 4G MOBILE NETWORKS

VANGELIS GAZIS, NIKOS HOUSSOS, NANCY ALONISTIOTI, LAZAROS MERAKOS

Communication Networks Laboratory, University of Athens, 157 84, Athens, Greece

Abstract: **Global vision consensus on the next generation of wireless mobile communications, broadly termed 4G, sketches a heterogeneous infrastructure, comprising different wireless access systems in a complementary manner and vested with reconfiguration capabilities that will facilitate a more flexible and dynamic adaptation of the wireless infrastructure to meet the ever-changing service requirements. We present the generic stages of reconfiguration procedures and identify the management requirements in the context of these stages. Next, we introduce a generic, platform-independent reconfiguration management framework to be hosted on end systems and elaborate on its main components.**

Key Words: Wireless, reconfiguration management, 4G.

1. Introduction

Fourth generation wireless means different things to different people: for some it is just a higher-capacity (e.g., 100 Mb/s) new radio interface, while for others it is an interworking of cellular and WLAN technologies that employs a variant of Mobile IPv6 mobility management (e.g., Hierarchical Mobile IPv6) for inter-system handoff and IETF AAA technologies for seamless roaming. Most importantly, however, 4G will be characterized by a horizontal communication model, where multiple different access technologies such as cellular, cordless, wireless LAN, short range connectivity and even wired systems will interface to a common platform over the IP protocol, complementing each other in an optimal way for different service requirements and radio environments.

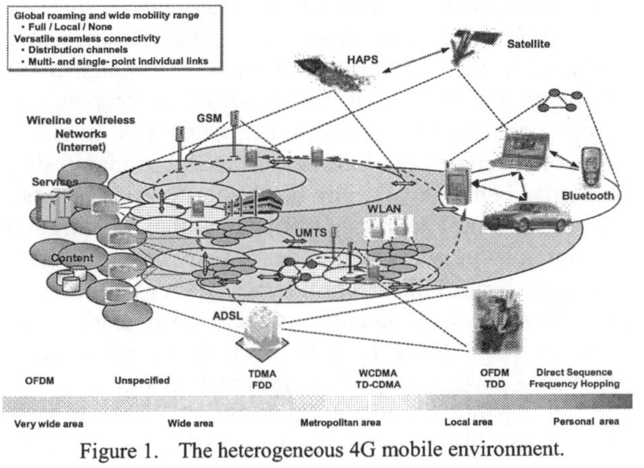

Figure 1. The heterogeneous 4G mobile environment.

Supported by their personal intelligent agent(s), users who roam in the 4G 'technological mosaic' will enjoy "always best" connectivity as well as seamless access to value-added services, third-party applications and content over the most efficient combination of systems available [1]. Considering that the observed proliferation of wireless access technologies is likely to persist and that future mobile devices will have to support multiple dissimilar wireless standards, the mobile communication industry has been focusing on reconfigurability and the dynamic instantiation of – previously inexistent – wireless standards within a mobile communication device as a technological enabler of future (multi-standard) mobile systems and flexible joint radio resource management across different wireless standards. The rest of the paper is structured as follows: Section 2 highlights the fundamental concepts of reconfigurability, providing key definitions. Section 3 discusses reconfigurability-related 4G service provision issues, identifies functional requirements and introduces a staged reconfiguration procedure. Section 4 introduces the proposed reconfiguration management framework and its main functional components. Finally, Section 5 concludes the paper and provides directions for future work.

2. Reconfiguration concepts & issues

2.1. *Definitions*

Primarily, the term reconfiguration refers to the dynamic instantiation, parameterization and inter-connection of protocols (i.e., communication-related functional entities) within the user, control and management planes of a collection of operating communication systems in a manageable, consistency-preserving and – preferably – transparent fashion. Reconfigurability facilitates a dynamic and flexible adaptation of the (mobile) communication equipment's behavior by exploiting over-the-air download of in-situ installation of appropriate communication protocols and/or middleware components. Thus, for the rest of the present paper, the term reconfiguration will refer to dynamic adaptations of implementation mappings of internal mobile communication equipment components [2] that do not compromise their consistency or their ability to continue to provide their services. We propose that network equipment be enhanced with a *generic capability for reconfiguration*, i.e., the ability to change the implementation mappings of their constituent (functional) components. Basically, that requires research and standardization in two key thematic areas:

a) Object-oriented information models [3] to capture and express the internal organization and structure of network equipment in an abstract, implementation neutral way that effectively provides the unified view necessary to start specifying the generic change capability. Through object orientation and inheritance, common structural parts can be factored out and reused as an abstract information model from which wholly different structures will inherit their common parts, thus allowing a fine-grain mix of standardized behavior with innovative, performance-focused, proprietary instrumentations.

b) Reconfiguration-related behavioral semantics recorded within a platform-independent reconfiguration management framework that specifies reconfiguration-enabling functional interfaces and the blueprints of their interactions (e.g., message sequence charts, state machines diagrams, etc) so as to support generic reconfiguration procedures.

2.2. *Related work*

SDR Forum pioneered in producing a Software Radio Architecture (SRA) specification [4] providing a framework for building, configuring, connecting and tearing down distributed, embedded (radio) applications within a radio device. SRA specifies software (OMG IDL) interfaces for the installation and use of distributed (radio) applications within a single device to support reprogrammable communication capabilities. It uses a variant of the CORBA Components specification to describe the hardware and software components of an SDR system, their properties and their interconnections. The OMG Software Radio (SWRADIO) Special Interest Group (SIG) is primarily focusing on the development of an architecture that will allow applications to be independent from the operating environment within a radio, thereby being portable across a variety of hardware configurations. Notably, it recognizes [6] that such portability requires hardware interfaces to be abstracted and application management interfaces to be standardized along with common metadata and loading processes.

In brief, all existing work on architectures for reconfigurable communications focuses mostly on the structural view and does not address other, equally important concerns, such as decision support for optimal reconfigurations, or what the generic phases of reconfiguration procedures are. The focus of the present paper in on these issues and its contribution is a staged reconfiguration procedure and an open reconfiguration management framework to support the particular stages of that procedure.

3. Service provision and reconfiguration in 4G

3.1. *Service provision issues*

The efficient provision of applications and/or value-added services over the dynamically reconfigurable networking and computing infrastructures that will be commonplace in the 4G era, requires a common set of support mechanisms that, based on each application's requirements, identify and trigger the appropriate reconfiguration actions on the underlying network infrastructure, including the mobile devices themselves, in a manner that is transparent to the user. Besides structural concerns dealing with the internal organization of affected communication systems, these mechanisms must also address issues related to context capture and discovery, capability negotiation procedures and representation formats, decision support regarding the optimal communication mode, over-the-air deployment of software implementations (e.g., MAC protocols, codecs, modulation components, etc), in-situ installation and validation checks, as well as remote service management (e.g., activation/deactivation of protocol instances within a mobile device). Treating reconfiguration as the selective adaptation of implementation mappings for modular (i.e., component-based) subsystems, we consider the following classification for reconfiguration-related mechanisms:

a) Mechanisms that realize functionality prerequisite to the reconfiguration action per se and are orthogonal to it, e.g., context-capture functionality, capability negotiation, intelligent decision making, over-the-air download functionality, etceteras.

b) Mechanism that support the reconfiguration action per se, i.e., functionality related to instantiation, inter-connection, activation/deactivation of communication components within their execution environment (e.g., communication subsystem).

The latter category of reconfiguration-related functionality will be typically available on each communication system that may be subjected to reconfiguration, while the former can be factored out to a service provision platform that undertakes the bulk complexity of pre-reconfiguration actions, including necessary interactions with any affected (proximal) network elements [5].

3.2. *A generic staged reconfiguration procedure*

Using the aforementioned classification of reconfiguration-related mechanisms, we postulate that reconfiguration procedures progress sequentially through these stages:

Context discovery and identification: As reconfigurability deals with the dynamic adaptation of implementation mappings for functional communication components, it requires contextual awareness, i.e., being able to discover and exploit contextual information useful in manifesting a particular behavior [7]. For illustration purpose, consider the case of a mobile device equipped with a reconfigurable communication subsystem under WLAN and UMTS radio coverage. From the reconfiguration viewpoint, context identification will identify which particular network elements (e.g., Node B, RNC, WLAN access point) are capable of participating in reconfiguration signaling, i.e., a sort of capability negotiation exchange. This discovery process considers the immutable capabilities of network equipment (e.g., hardware and/or software revision levels, execution environment features, supported communication standards, etc) and identifies the mobile network elements that can be reconfigured in the current context. To achieve this, standardized information models that express the functional features and mechanisms of intra-equipment components in a manner independent of any implementation artifacts are necessary. Standardized flexible representation formats (e.g., RDF, XML) can be employed to encode such information models so as to facilitate the realization of intelligent capability negotiation procedures required by reconfiguration signaling.

Discovery of feasible communication personalities: The set of candidate network elements is used to identify the initial set of possible communication personalities (i.e., standards) on the basis of necessary hardware and software capabilities. This set is further refined by ruling out combinations of communication personalities for which equipment resources do not suffice. Thus, this phase interacts with resource management to consider dynamic equipment capabilities (e.g., buffering capabilities, traffic load, processor utilization level, etc) as filtering criteria.

Decision of optimal communication personalities: The need for personalization through the 4G mobile service provision process requires that user preferences must be accounted for when choosing the optimal personalities for personal mobile devices. For instance, users may wish to maximize the available communication options of their mobile device (i.e., the set of communication standards it supports) or prefer a particular wireless access standard (e.g., WLAN) to others. To support arbitrary selection criteria and decision-making algorithms, reconfiguration management must furnish open interfaces to input the decision algorithm's data, decision variables and criteria values, to access its output results and to support switching between different decision algorithms during runtime [8].

Deployment of selected communication personalities: Having identified the set of optimal communication personalities (i.e., sets of protocols), reconfiguration signaling proceeds to download the appropriate implementations from online software repositories. Once download is complete, each particular protocol is provisioned based on the identified contextual information. Subsequently, intra-system inter-protocol communication links (e.g., IPC, message queues, shared memory, etc) are established and the protocol graphs of the chosen communication personalities are built. Finally, protocol instances are activated (i.e., given their own thread of control) and register their Service Access Points (SAP) for use by dependent intra-system protocols. During this stage, besides injecting and activating additional protocols within the communication subsystem, it may also be necessary to deactivate and purge existing ones (e.g., to economize system resources).

The last (i.e., the deployment) stage encompasses generic functionality to download, install, activate – and possibly deactivate, uninstall and remove – functional components within communication subsystems; functionality which must be resident in all reconfigurable systems. The rest of the functionality may be outsourced to a service provision platform in the wireline infrastructure to take advantage of its abundant resources and to facilitate capture of much broader contextual information that will, on average, lead to more accurate decisions about which is the most suitable communication standards and/or the best reconfiguration strategy for a given mobile device.

4. Reconfiguration management framework

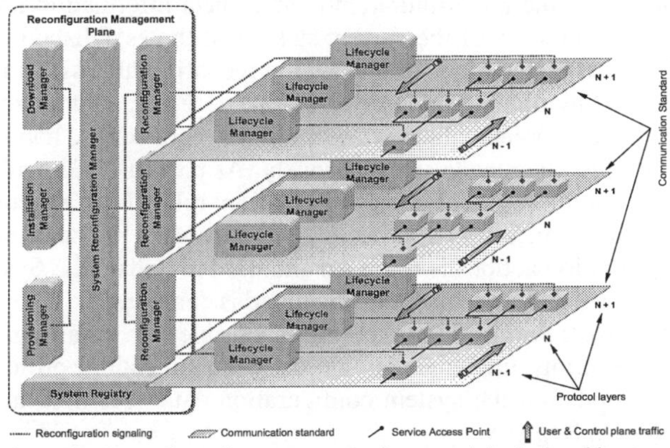

Figure 2. Reconfiguration management framework logical architecture.

The framework is founded upon a set of management interfaces that support asynchronous command dispatching to managed objects and asynchronous notifications from those objects towards the respective managing entity. Reconfiguration is specified on top of this basic set, following a 'separation of concerns' principle that discriminates between orthogonal concerns, thereby leading to modular and extensible design. Reconfiguration management functionality results through the composition – rather than the inheritance – of all identified concerns, facilitating parallelism in the design of reconfiguration signaling. Simply put, that means that signaling exchanges related to different concerns may take place concurrently, even if the dealing with the same subject. For example, deactivation of an existing protocol and download of a new implementation for that particular protocol may progress simultaneously, because the inherited basic management functionality provides for the asynchronous interactions needed in concurrent operations and composition allows multiple instances that employ asynchronous interactions.

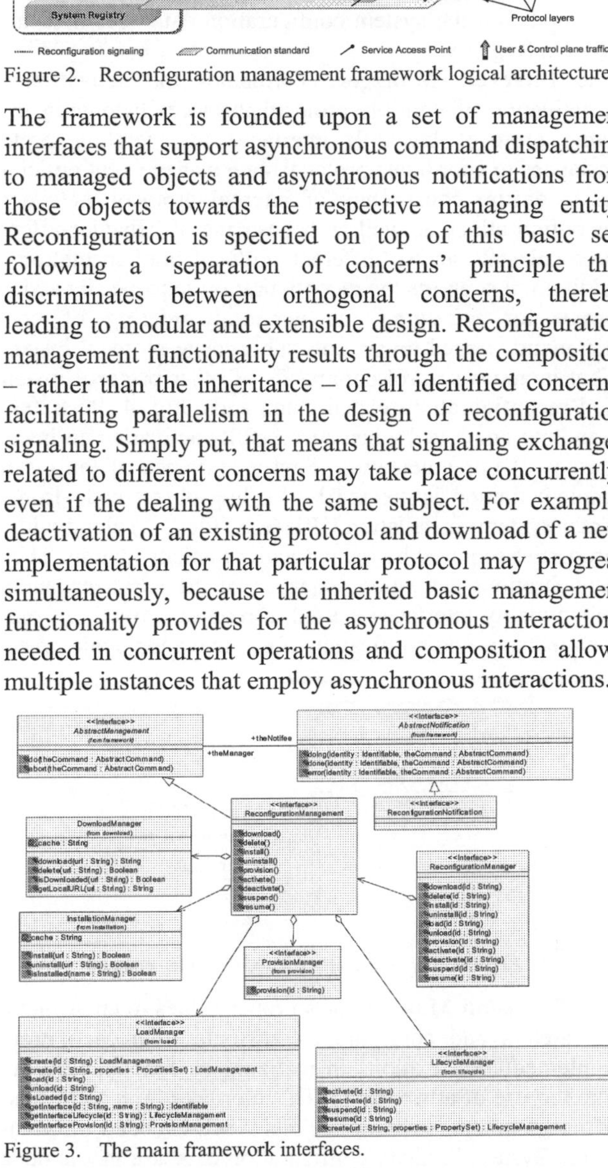

Figure 3. The main framework interfaces.

The **System Registry** is the core of the reconfiguration management framework by providing a single reference point for management and discovery of globally available reconfigurability meta-information, such as reconfigurable equipment hardware versions and revisions, communication subsystem capabilities, mobile execution environment properties, protocol implementation versions, etc. For each reconfigurable system, the System Registry provides unified access to the reconfiguration-related meta-information and a consistent view of the current status and the established (i.e., installed) capabilities of the hosting reconfigurable system.

The **Download Manager** undertakes secure download of protocol implementations from online repositories, handling multiple downloads simultaneously. As part of the download procedure, protocol meta-information related to execution environment settings as well as to necessary pre- and post-installation actions is retrieved and used to update the System Registry database once the download is complete.

The **Installation Manager** is responsible for installing and uninstalling implementations of communication protocols and updating the system configuration databases held in the System Registry.

The **Lifecycle Manager** coordinates the activation and deactivation of multiple protocol instances belonging in the same protocol layer. It interfaces to operating protocol instances to coordinate protocol state management policies so as to ensure consistency of internal protocol state across reconfigurations as well as persistence of behavior-affecting protocol state across different protocol implementations. In addition, it supports the reconfiguration sequence by acting as proxy for managing the temporal ordering of control commands issued towards its subordinate protocol instances. That includes control commands of other protocol instances and reconfiguration-related commands. Fig. 4 illustrates the signaling employed when successfully activating an installed (protocol) implementation. First an instance of the installed implementation is created by loading object code from the (inter-system) installation URL (e.g., file://...) along with any properties required by the particular implementation. Then an asynchronous activation sequence of the basic management functionality interfaces is used to request activation from the new instance and notification of the activation's outcome.

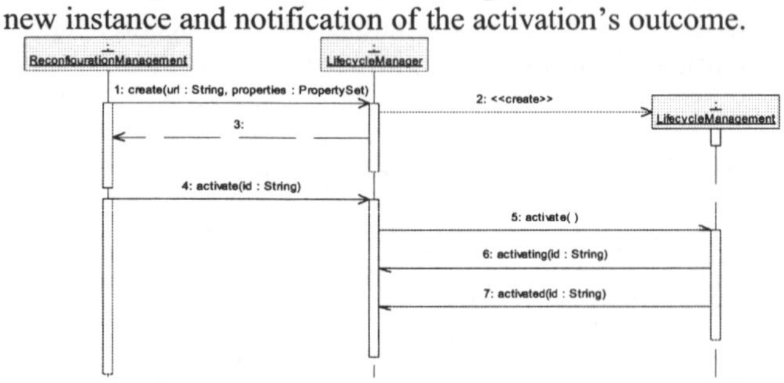

Figure 4. Activation of an installed (protocol) implementation.

The **Provision Manager** sets proper values to environmental parameters and configuration attributes that each protocol implementation exploits in realizing its functionality, much like the use of attributes in IDL interfaces. The identification of necessary parameters exploits XML-encoded information in the System Registry configuration database that is updated by the installation manager after a successful installation.

The **Reconfiguration Manager** deals with reconfiguration of a single access standard that may comprise multiple protocol layers. It guarantees the structural consistency of the communication standard and coordinates reconfiguration signaling between protocol instances and the Download, Provisioning and Lifecyle manager objects. It also provides a consistent, up-to-date structural view of its subordinate communication standard to the System Reconfiguration Manager.

The **System Reconfiguration Manager** embodies all the reconfiguration-related intelligence needed to schedule and coordinate reconfigurations for multiple communication standards without jeopardizing system integrity. It acts as a central control point for managing reconfigurations within the realm of a single system, e.g., mobile terminal.

To achieve platform-independence, the entire framework has been modeled in (pure) UML – as opposed to using a UML profile for, say, CORBA – so that portability to different platform-specific models (PSM) such as .NET or Java is possible by using OMG-standardized mappings that can preserve formality properties. This modeling approach is also endorsed by the OMG SWRADIO SIG.

5. Conclusions

In 4G, performing and managing reconfiguration actions over mobile systems and networks will constitute a major part of management activities. We introduced a generic staged procedure and a platform-independent framework for reconfiguration management that can be formally mapped to appropriate implementation technologies and/or deployment instrumentations using OMG approaches. We intend to investigate the effectiveness of these mappings in a prototypical implementation of the framework.

References

[1] V. Gazis, N. Housos, A. Alonistioti, L. Merakos, "Evolving perspectives of 4G mobile communication systems", Proc. 13th IEEE Personal Indoor Mobile Radio Communications, Lisbon, Portugal, 2002.

[2] Z. Tang, "Dynamic reconfiguration of component-based applications in Java", M.Sc. thesis, MIT, September 2000.

[3] Common Information Model (CIM), Distributed Management Task Force, http://www.dmtf.org/.

[4] J. Bickle, "Software Radio Architecture (SRA) 2.0 technical overview", OMG TC meeting, December 11, 2000, Orlando, Florida.

[5] *Software Defined Radio: Architectures, Systems and Functions*, Wiley, ISBN 0-470-85164-3.

[6] OMG SWRADIO SIG document 01-03-03, available from http://www.omg.org/.

[7] A. Dey, "Providing architectural support for building context-aware applications", PhD thesis, College of Computing, Georgia Institute of Technology, December 2000.

[8] Nikos Housos, Vangelis Gazis, Athanassia Alonistioti, "Application-transparent adaptation in Wireless Systems beyond 3G", m-Business 2003, 23–24 June 2003, Vienna, Austria, pp.43-54.

[9] SDR Forum R&D Working Group, 2002 Summary Report, available from http://www.sdrforum.org/.

GMSME: An Architecture for Heterogeneous Collaboration with Mobile Devices

Sangyoon Oh[1, 2], Sunghoon Ko[1], Geoffrey C. Fox[1, 2]
[1] Community Grids Computing Laboratory, Indiana University, IN, USA
501 N. Morton Suites 224, Bloomington, IN 47404
[2] Computer Science Department, Indiana University, IN, USA
{ohsangy, suko, gcf}@indiana.edu
Tel: 812-856-0751
Fax: 812-856-7972

Abstract:
The importance of network-ready personal devices in collaborative systems is becoming apparent. The participants of a collaborative environment use mobile devices to overcome the physical constraint of using a conventional PC. However, the integration of disparate devices like mobile devices and PCs is difficult, because of the connectivity and the computing power limitations of a mobile device. We believe a dedicated middleware layer that supports universal access, event mapping, user management, and content adaptation is a great help to integrate mobile devices into the collaborative system. In this paper, we propose a ubiquitous computing environment architecture for a collaborative system called Garnet Message System Micro Edition (GMSME). This architecture provides the environment to resolve those issues and is designed efficiently considering the computing resource limitations of mobile devices.

Key Words: collaboration, mobile, universal access, filtering, middleware

1. INTRODUCTION

Collaborative systems enable people to communicate and cooperate remotely. The Garnet Collaborative system [1] [2] is a universal access collaboration environment, which uses the NaradaBrokering Event system [3] as a messaging system.

In recent years the use of network-ready mobile devices like Personal Digital Assistants (PDAs) and Smart phones as client devices has emerged as a major focus in collaborative systems. The combination of personal devices with conventional PCs in the collaborative system makes a great impact on the degree of collaboration. For instance, the universal access model of the Garnet system is extendable to more general universal access through personal devices. But it is not an easy goal to achieve. In many projects aiming at such integration, the limitation of the personal devices in terms of the connectivity and computing power has been an unavoidable issue.

Even though there are many improvements recently, personal devices still lack in computing resources. We claim that the middleware layer between the collaborative system and the mobile devices resolves this drawback by efficiently mapping event messages and adapting content of messages. The middleware layer is a NaradaBrokering node that filters shared events and reposts them to a different stream [4].

In this paper, we propose an extensible and scalable architecture that provides universal accessibility and efficient server side computing. To make the architecture concrete, we have implemented a version, called Garnet Message System Micro Edition (GMSME) on the Garnet Collaborative System. The basic design issues and requirements for GMSME as a middleware layer are discussed in detail in Section 3. We present the architecture overview in Section 4.

2. GARNET AND NARADABROKERING

The Garnet collaboration system [5] supports distance education, collaborative computing and building electric communities. It provides rich features of collaboration such as a video conferencing with SIP VoIP [6] and H.323 and a shared export with SVG. With the experience of developing TANGO [7] and using commercial collaboration system, for instance WebEx [8], Centra [9], the Garnet system integrates synchronous and asynchronous collaboration systems that are based on the publish/subscribe mechanism. For publish/subscribe semantics, we build on the NaradaBrokering system, which is a Java Message Service (JMS) [10] compliant message brokering system. NaradaBrokering is an event brokering system designed to run on a large network of cooperating broker nodes. It was originally designed to provide uniform software multicast to support real-time collaboration linked by publish/subscribe. As noted above, it is JMS compliant and provides supports not only for JMS clients, but also peer-to-peer communication providing JXTA interactions. NaradaBrokering provides a transport framework that supports the various choices of protocols between nodes.

GMSME adds the universal connectivity to the Garnet collaboration system, which primarily focuses on the traditional desktop client. GMSME supports heterogeneous protocols for mobile client, message conversion between NaradaBrokering messages and HHMS, and customizing contents of the message to make them appropriate to PDAs or smart phones.

3. DESIGN ISSUES

In the collaborative system, personal devices like PDAs and Smart phones are drawing more interest for two major reasons. One is that the devices are performing better than before. They are now sufficiently capable to be a collaborative client device. The other reason is better network availability. There are more buildings and rooms that have WLAN connection for your PDA, and more products and service plans for cellular phones by which people can connect to the Internet.

However, there is a gap between PC and personal device, because of the nature of the latter, which is portable and convenient to use. The middleware, GMSME, fills the gap. The overview of GMSME architecture is depicted in Figure 1.

The challenges we worked on are:

Heterogeneity:
Heterogeneous device support is a core concern to the middleware layer in the ubiquitous or pervasive computing environment. The issues include universal connection support, event mapping and content adaptation. GMSME provides connection management and user management that are encapsulated in a universal interface to the personal mobile device. The implementation is presented in Section 4.1.

The Garnet collaborative system and GMSME have different event models. The Personal Server of GMSME, which is illustrated in Section 3, maps the event between the asynchronous JMS based NaradaBrokering [10] [11] and synchronous Hand Held Message Service (HHMS).

Content adaptation has been explored by many research groups [12] [13]. For collaborative systems, it is a vital design issue to adapt those devices to collaborate together with the conventional PC, which has more computing and display strength. We use a Garnet XML Object Specification (GXOS) based User Profile. GXOS [14] is the primary specification to define objects in the Garnet collaborative system. Depending on the user profile and device type of his current machine, GMSME adapts the content of messages appropriate to the user device. Shared display applications and Scalable Vector Graphics (SVG) [15] applications on the mobile device heavily rely on this idea, since using those graphic native applications on a small display device and processing large images costs heavily in computing resources. For more discussion of how a SVG document, say, can be adapted for a PDA's display see for example [16]

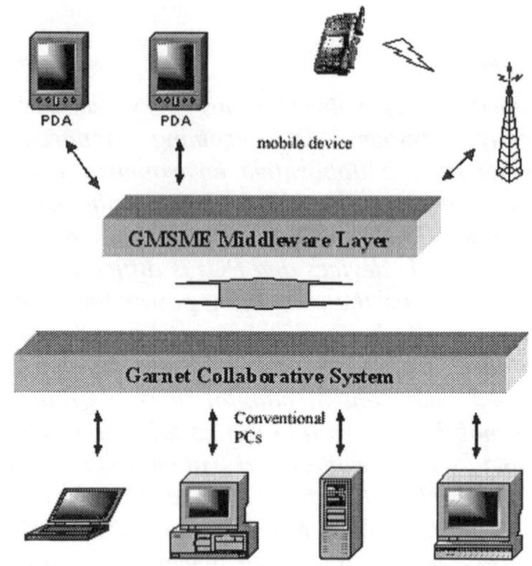

Figure 1. *Overview architecture of GMSME with Garnet Collaboration system and mobile devices*

Scalability:
Our approach to support the user is to provide a Content Adaptor and a Connection Manager per user. Since Connection Manager is multithreaded and non-blocking, both receiving messages and sending messages do not interrupt each other; this ensures a large number of users can share a GMSME middle layer. The NaradaBrokering event system as a primary event broker on the Garnet collaborative system guarantees that multiple GMSME middleware layer communicate, thus each GMSME instance can serve a large number of mobile device users at the same time.

Efficiency:
We have two detailed design issues about efficiency. The first is server side computing. We move computation that is required to support the heterogeneous device to the middleware layer to reduce the burden on the personal device. When a Content Adaptor gets the event message from the collaborative server, each corresponding application module processes the message and delivers only vital parts of the message to the Connection Manager. The other is the communication protocol. Since we are stripping down the original message from the collaborative

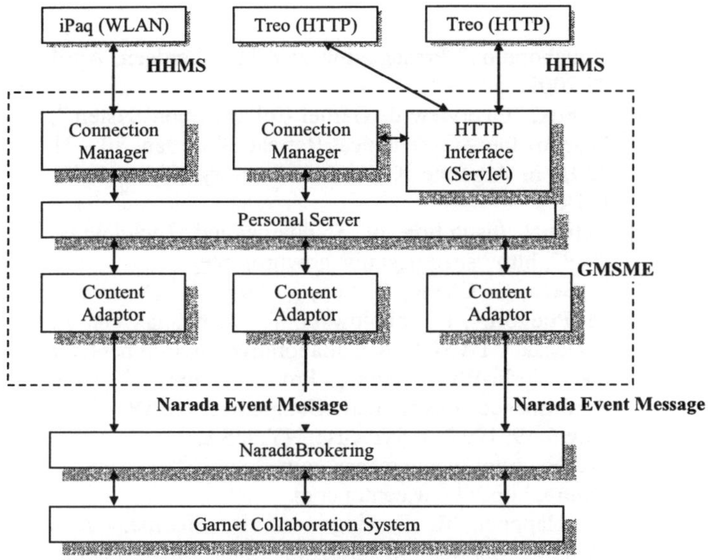

Figure 2. *Internal Component of GMSME middleware layer and protocols between components*

server, as well as other message servers, such as Jabber [17], we defined a new efficient mobile device specific protocol, HHMS. It is a natural modification of the event service of GMSME considering the constraints of the personal device. HHMS is byte-oriented messaging protocol, which includes information about the machine type, application type, event type, and event body.

4. EVENT MAPPING AND INTEGRATION WITH CONVENTIONAL COLLABORATIVE ENVIROMENT

This section describes the implementation of architecture, GMSME. The components and data flows are depicted in Figure 2. A dashed-box wraps GMSME components, which are integrated in a single application. GMSME communicates via messages to outside the box. To PDAs, it is linked with HHMS and to the Garnet system, it linked with NaradaBrokering messages.

4.1. CONNECTION MANAGER

Connection Manager has network interfaces to the protocols that personal mobile devices use most: HTTP and 802.11b. HTTP support is crucial in GMSME and the Garnet system, since most cellular phones access internet through it. A dilemma with HTTP is that it was designed specifically as a stateless protocol. This is a mismatch with the design of synchronous collaborative systems. To settle this problem, we apply a polling technique to the communication of Connection Manager and collaboration application on the personal device. A Java Servlet is added as a mediator to route HTTP message to the general Connection Manager, described in [16]. In general, the Connection Manager provides the same communication interface to the various protocols. There is a demon server component per protocol. It creates a Connection Manager when a mobile device attempts to join the collaborative session. We introduced the job-queue to balance the job-load of mobile devices in a previous paper [16]. The job-queue is well matched with the polling technique. When the mobile application polls, the Connection Manager checks the job-queue and delivers a ready message individually to the application.

In a different project, we have a web service implementation of the collaborative system. Initially, the SVG interface is implemented to have more choice of participation from the user [18].

4.2. PERSONAL SERVER (ADAPTER)

This is the main component of GMSME middleware. It sits between the applications on the mobile device and Garnet collaborative system. It looks like a typical Garnet client to the Garnet collaborative server. And it manages mobile users by a User Table. With a new connection from a mobile device, Personal Server creates a Content Adaptor module and assigns it to the user. The message routing is another key function of Personal Server. Because it routes messages internally and the entire user list is in memory, plainly Personal Server links the matching Connection Manager instance with the user message to route them.

4.3. CONTENT ADAPTOR

Content Adaptor is responsible for processing the collaborative event message. It listens to the entire collaborative message on behalf of the mobile participants and maps the message to HHMS. Personal Server routes a HHMS message to corresponding Connection Manager. Each application on the mobile device has a process module in the Content Adaptor. A collaborative application developer appends an additional module by following the GMSME application API. There are four application modules at this time: Instant Messenger, Textchat, Shared Display, and SVG. The implementations of applications are described in detail in [19].

4.4. APPLICATIONS

The instant messenger and textchat transport module in the personal server connects to the Jabber open server to deliver a message as well as check presence of other users when it gets request from the client device through the personal server. Textchat client on PDAs and Smart phone

works in a similar fashion but it publishes messages to the NaradaBroker instead of connecting the Jabber server.
Shared exports are one of the core applications of the collaboration system. The SVG module in the personal server gets the URL of a SVG file and processes the image based on the user profile. And the personal server sends the filtered image to PDAs client through communication manager over HHMS. The Shared display module works the same way, except it captures the raw image of a shared image and process image filtering.

5. CONCLUSION AND FUTURE WORK

To validate the GMSME architecture, we've implemented it for the desktop PC, local wireless (802.11b) and a cellular phone that has internet access. On the client side, the software is MyProfessor for Windows CE iPaq and Palm OS Cell-PDA combination device. This integrates client side collaboration applications (shared display, SVG, Text chat, Instant Messenger using Jabber). Through the implementation, we illustrated the importance of a middleware layer supporting heterogeneous devices in the collaborative system. We believe our approach — to push down more computing on the middleware side and use a simple and efficient protocol between a mobile application and the middleware — is suitable for most pervasive environments, as well as to the Garnet collaborative system.

We will extend this architecture so it can be integrated into the general NaradaBrokering framework. As noted in Section 2, NaradaBrokering has transport framework to choose appropriate protocols separately on an each hops between nodes. After this integration, HHMS will become the choice for the message to PDA.

REFERENCES

[1] G. C. Fox. "Architecture and Implementation of a Collaborative Computing and Education Portal", ERDC Technical Report, May 2001.

[2] G. Fox, W. Wu, A. Uyar, H. Bulut, and S. Pallickara. "Global Multimedia Collaboration System", Proc. of the 1st International Workshop on Middleware for Grid Computing co-located with Middleware 2003 (MGC2003). June 17, 2003 Rio de Janeiro, Brazil.

[3] G. C. Fox and S. Pallickara. "The Narada Event Brokering System: Overview and Extensions", Proc. of the 2002 International Conference on Parallel and Distributed Processing Techniques and Applications (PDPTA`02). CSREA Press (2002) edited by H.R. Arabnia Volume I pages 353-359.

[4] G. Fox, "Solving Some of the Problems in Collaboration" Presentation at ARL Aberdeen, April 22 2003.

[5] G. Fox, "Overview of Garnet Collaboration System", Proc. of the AG Technical Retreat 2001, Jan. 30 - 31 2001, in Argonne National Laboratory, Chicago, IL, USA.

[6] National Institution of Standards and Technology, "SIP", http://snad.ncsl.nist.gov/proj/iptel

[7] L. Beca, G. Cheng, G. Fox, T. Jurga, K. Olszewski, M. Podgorny, P. Sokolowski, T. Stachomiak and K. Walczak, "TANGO-A Collaborative environment for the World-Wide Web", Proc. of the 5th High Performance Distributed Computing (HPDC `96), Aug 6 - 9. 1997, in Syracuse, NY, USA.

[8] WebEx, http://www.webex.com

[9] Centra, http://www.centra.com

[10] M. Happner, R. Burridge and R. Sharma, "Java Message Service Specification" Sun Microsystems, Inc. http://java.sun.com/products/jms/

[11] R. Monson-Haefel and D. Chappell. *Java Message Service,* Sebastopol, CA: O'Reilly &Associates, Inc., 2000.

[12] R. Han, P. Bhagwat, R. LaMaire, T. Mummert, V. Perret, and J. Rubas. "Dynamic Adaptation in an Image Transcoding Proxy for Mobile Web Browsing", IEEE Personal Communications Magazine, vol. 5, no. 6, Dec. 1998, pp. 8-17.

[13] T. Phan, G. Zorpas, and R. Bagrodia, "An Extensible and Scalable Content Adaptation Pipeline Architecture to Support Heterogeneous Clients", Proc. of the 22nd International Conference on Distributed Computing Systems (ICDCS '02), July 2-5, 2002, in Vienna, Austria.

[14] O. Balsoy and G. Fox, "Garnet XML Object Specification", http://www.gxos.org/

[15] W3C. "Scalable Vector Graphics", http://www.w3.org/Graphics/SVG/Overview.htm8

[16] G. Fox, S. Ko, K. Kim, S. Oh and S. Lee, "Integration of Hand-Held Devices into Collaborative Environments", Proc. of the 2002 International Conference on Internet Computing (IC-02), June 24-27, in Las Vegas, NV, USA.

[17] Jabber, Inc. "Jabber Software Foundation", http://www.jabber.org/

[18] S. Lee, S. Ko, G. Fox, K. Kim and S. Oh, "A Web Service Approach to Universal Accessibility in Collaboration Services", Proc. of the 1st International Conference on Web Services (ICWS`03), June 2003, in Las Vegas, NV, USA.

[19] K. Kim, G. Fox, S. Ko and S. Oh, "Mobile Application Paradigm for Ubiquitous Computing: Universal Collaboration and Access" May, 2003.

3G AND WLAN INTERWORKING QOS SOLUTION

EERO WALLENIUS
Nokia Networks/ IMN, Hatanpäänvaltatie 30, FIN-33100 TAMPERE
Tel:+358 40 5004055, Fax:+358 7180 74385
eero.wallenius@nokia.com

TIMO HÄMÄLÄINEN,
University of Jyväskylä, Department of Mathematical Information Technology
timoh@cc.jyu.fi

TIMO NIHTILÄ
University of Jyväskylä, Department of Mathematical Information Technology
nihti@cc.jyu.fi

JYRKI JOUTSENSALO
University of Jyväskylä, Department of Mathematical Information Technology
jyrkij@cc.jyu.fi

HANNU KOIVISTO
Technical University of Tampere, Automation and Control Institute ACI,
hannu.koivisto@tut.fi

Abstract- The WLAN integrated mobile device is designed to extend the reach of enterprise applications and to create new collaboration environments. This means that interconnection with WLAN 802.11e and 3G technologies are needed. Supporting quality of service is one of the most important issues in 3G. In this paper end-to-end QoS issues of the co-operation of the WLAN and 3G are studied. We analyze how well 802.11e's QoS properties perform under the different kind of network scenarios when packet sizes and channel error rates are varied.

1. INTRODUCTION

WLAN connections [2,6] where they are available, can offer faster connections than 3G (2Mb/s) with cheaper prices. For example the new 802.11a version can achieve data rates of up to *54 Mbit/s* at the wireless medium using the OFDM modulation technique in the unlicensed *5 GHz* band [7].

The problem with WLAN is the high error rate probability. It can rise up to 50% causing trouble especially to streaming type of applications. 802.11e is trying to correct the situation by enabling the use maximum of eight separate priority queues for prioritizing higher priority traffic compared to other traffic [5]. This paper studies effects of the WLAN error to the 3G traffic classes. We have built an extensive simulation environment where different kind of traffic scenarios can be simulated and analyzed. Related research work has been focused mainly to QoS issues in 3G or WLAN and beyond networks [8, 9].

The rest of the paper is organized as follows: Chapter 2 presents 3G traffic classes, their characteristics and mapping to 802.11e priorities. In Chapter 3 we present our simulation environment and obtained results. Finally we conclude at Chapter with most remarking results and the future work.

2. 3G and 802.11E QoS support

The 3GPP has taken approach to the QoS matter by defining four separate traffic (QoS) classes that have their own and separate QoS attributes. All traffic in the 3/4G network will be handled according to the operator's requirements on the each of the traffic classes. The proposed n QoS method for 3G has is DiffServ [7] but also RSVP [10] could be used optionally.

2.1. *802.11e QoS mapping to 3G*

The Enhanced Distributed Coordination Function (EDCF) is the basis for the Hybrid Coordination Function (HCF) [4]. The QoS support is done with the Traffic Categories (TC). MAC service data units will be delivered through multiple back off instances within one station, each back off instance parameterized with TC-specific parameters. A single station can have up to eight transmission queues realized as virtual stations inside a station, with QoS parameters that determine their priorities. If the counters of two or more parallel TCs in a single station reach zero at the same time, a scheduler inside the station avoids the *virtual collision.*

The scheduler grants the Transmission Opportunity (TXOP) to the TC with highest priority, out of the TCs that virtually collided within the station.

Table 1. 3G Traffic class mappings into DSCP

Traffic class	PHB actions (mapped into DSCP)
Conversational	EF
Streaming	AF 1
Interactive (THP 1)	AF 21
Interactive (THP 2)	AF22
Interactive (THP 3)	AF 23
Background	AF 3

There is then still a possibility that the transmitted frame collides at the wireless medium with a frame transmitted by other stations [4]. Table 1 shows the PHB actions with DSCP mappings.

3. Simulation Model and simulation results

The goal for the simulations is to study how much the throughput of the traffic decreases with various error rates while changing the traffic mix and average packet sizes of the individual traffic class.

We used network simulator ns-2 with IEEE 802.11 EDCF functionality implemented by Ni Qiang et al. of the Planete Project-INRIA[10]. To emulate the process of packet transmission errors we extended the simulator by implementing a two-state Markov model in the air interface.

MMPP is a doubly stochastic process where the intensity of a Poisson process is defined by the state of a Markov chain (see Fig. 1). The transition matrix Q of the modulating Markov chain is defined by

$$Q = \begin{bmatrix} -\omega_1 & \omega_1 \\ \omega_2 & -\omega_2 \end{bmatrix}$$

In our error scenario, the channel switches between a "good state" and a "bad state", G and B respectively: Packets are transmitted correctly when the channel is in state G, and errors occur when the channel is in state B. When the channel is in state G, it can either remain in this state, with probability ω_1 or make the transition to state B, with probability $1-\omega_1$.

Likewise, if the channel is in state B, it remains in this state with probability ω_2 and changes state with probability $1-\omega_2$. Table 1 shows the transition probabilities we used to achieve different packet error rates in the air interface.

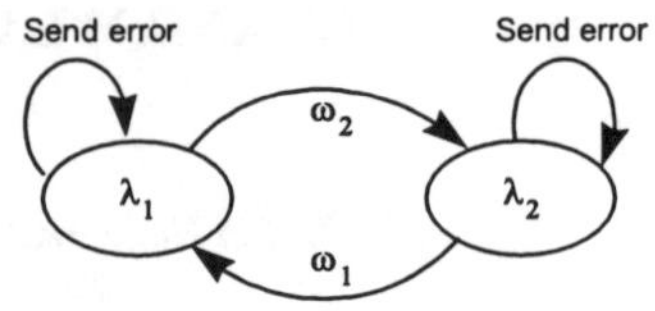

Figure 1. State machine for a 2-state MMPP.

Table 2. Transition probabilities for 2- state MMPP.

Error rate	ω_1	ω_2
30%	0.27	0.62
60%	0.90	0.60

The simulation scenario is illustrated in figure 2. The simulation scenario is illustrated in figure X. It includes eight WLAN stations. Stations 1, 2, 3 and 4 are in contact with access point AP1. Station 1 is the highest and station 4 is the lowest in priority. Stations 5, 6, 7 and 8 are connected to access point AP2. The link between AP2 and the core network is a 10 Mbit/s link with 5 ms delay.

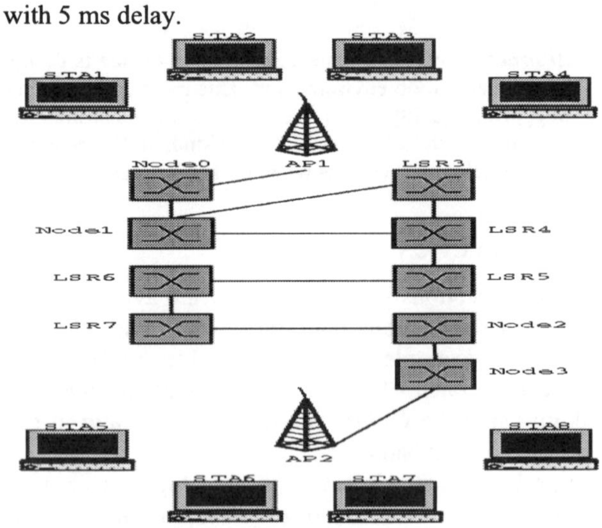

Figure 2. Used network configuration.

All other wired links in core network are 10 Mbit/s links with 0 ms delay except between LSR5 and LSR6 routers there is a 15 Mbit/s link with 13 ms delay and between LSR6 and LSR7 the delay is 14 ms.

DiffServ is used in core network. LSR4 is a DiffServ Edge router and LSR5 is a DiffServ Core router. In LSR4 there is four physical DropTail RED queues with two precedence levels in each queue. One level for packets that are in profile and one for packets that are out of profile. The queue sizes are:

Table 3. Queue sizes

Traffic class	Conversational	Streaming	Interactive	Background
In profile	30	60	30	30
Out of profile	0	30	30	0

A Token bucket policer is used. The parameters of this policer are as shown:

Table 4. Policer parameters.

Traffic class	Conversational	Streaming	Interactive	Background
CIR	3.0 Mbit/s	2.5 Mbit/s	2 Mbit/s	1.5 Mbit/s
MBS	3 KB	2 KB	1.5 KB	1.5 KB

In our simulations, stations 1, 2, 3 and 4 are sending nodes and stations 5, 6, 7 and 8 are receiving nodes. Station 1 sends to station 5, and so forth. Each station generates the same amount of traffic. The total traffic load is 2.5 Mbit/s per station. Station 1 and station 3 have both five flows all generating 500 Kbit/s constant bit rate (CBR) traffic. Stations 2 and 4 have also five flows each generating an average of 500 Kbit/s variable bit rate (VBR) traffic. VBR parameters of both of these stations are as shown:

Table 5. WLAN station traffic properties

Rate	500 Kbit/s
Rate deviation	0.25
Rate time	2.0
Burst time	1.0
No. of changes	10
Max. rate	648 Kbit/s

We used four priority levels in our simulations. EDCF parameters of different TCs are shown in the following table.

Table 6. Used EDCF Parameters

Traffic class	Conversational	Streaming	Interactive	Background
CWMin	7	10	15	127
AIFS (CWOffset)	2	4	7	15
CWMax	7	31	255	1023

In our simulations we varied the packet sizes of the traffic classes iterating through all the packet size combinations starting from 500 bytes to 1500 bytes with 500 byte steps. The possible packet sizes were thus 500, 1000 and 1500 bytes. The iteration goes as follows: at iteration 0 packet size of all the classes is 500 bytes. The lowest priority traffic class packet size is increased by 500 bytes every iteration. If the packet size is 1500 bytes, it is reset to 500 bytes and the next lowest traffic class packet size is increased by 500 bytes. This way the lowest priority traffic class iterates through all the packet size possibilities in 3 iterations and the next lowest priority traffic class in 3^2=9 iterations. The second highest traffic class goes through all the packet sizes in 3^3=27 iterations and the highest in 3^4=81 iterations. The radio channel packet error rate was varied also from 0% to 60% with 30% steps.

Figure 3 shows obtained throughput when channel error rate is 0% and 30% of totally lost packets. Figure 4 depicts the delays of these same error rate scenarios, and figure 5 shows dropped packets. As can be seen from the graphs EDCF works quite well with different channel error rates. Highest traffic class will get significantly more capacity than lower priority classes in all our simulation scenarios. This shows also the fact that EDCF is suitable for using with 3G to support QoS at the situations when 3G devices will use WLAN access to the core IP network. Interesting issue arises with the fact that highest priority traffic class will get more bandwidth when the packet size and the channel error rate are increasing. With the proposed 3G traffic classification specifications this can lead to the situation that highest class use all the capacity under certain network conditions. Our simulation results proved this issue clearly. Based on that it is important to find out the limits where packet size and channel error rate can change. If we think for example the situation where we have small size high priority packets (voice traffic) under heavily loaded network i.e. channel error rate is high, we will loose also those small size high priority packets.

Another interesting fact is that the packet size that gives the best throughput to the highest class is dependent on the error rate.

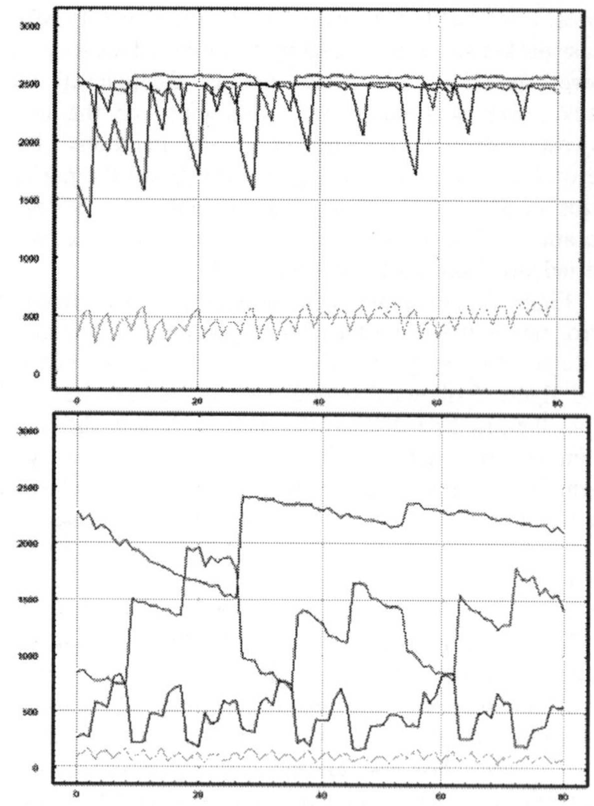

Figure 3. Throughput of the 4 traffic classes when the channel error rate is 0% and 30 % (figure on the bottom).

For example, the maximum throughput of the highest class in 0% and 30% error scenario is achieved when the highest-class packet size is about 700 bytes but as the error rate increases, the larger packet size will give the best results, when the lower priority packet sizes are small. When the error rate is very high (we tested with 60%), the maximum throughput is achieved with largest 1500 byte packet size. This is quite obvious, but it should be carefully though when building up WLAN hot spots. By using larger packet sizes the network delay (due to queuing) also increases, and that can cause several problems with real time applications.
The results depict also that if suitable packet and error rate combination can be fixed the different QoS requirements can be met. It is up to the network configuration how to tune these parameters suitable for the applications.

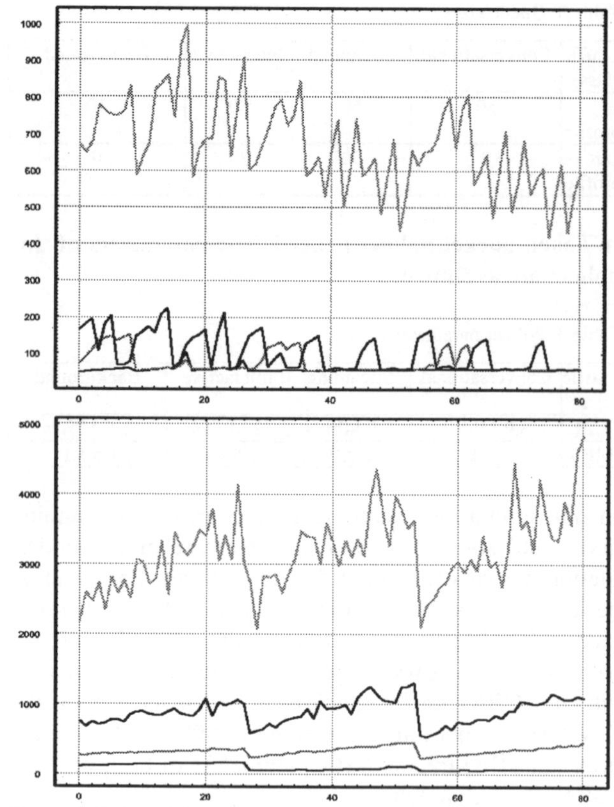

Figure 4. Delay of the 4 traffic classes when the channel error rate is 0% and 30 % (figure on the bottom).

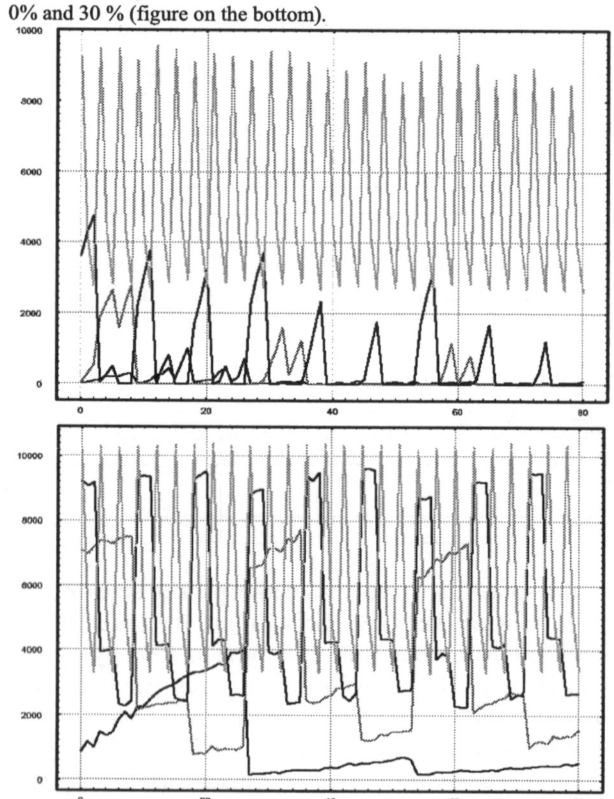

Figure 5. Dropped packets of the 4 traffic classes when the channel error rate is 0% (figure on the bottom).

4. CONCLUSIONS AND FUTURE Work

4.1. *Achievements*

In this study we have presented the behavior of 802.11e WLAN QoS used for 3GPP type services and service environment. As the WLAN is probably the first 3G complementary access solution, these kinds of feasibility studies are important and their purpose is to show methods how to use different 3G WLAN accesses.

Our simulation results showed that a throughput optimizing packet size can always be found within different channel error rate scenarios. Presented results depicted also that the mobile stations operating in EDCF can guarantee throughput even the channel error rate are high. We can also see from the results that EDCF can be implemented with 3G to support QoS at the situations when 3G devices will use fast WLAN access to the core network. Based on the obtained results the packet sizes can be changed according to the priority class and the channel error rate to achieve optimal throughput for the 3G traffic classes.

4.2. *Future Studies*

Our future work concentrates to seamless interworking between the 3GPP and WLAN, and defining the end-to-end QoS functionalities. Next we will expand the simulations to cover large number of network nodes and study how the parameters can be tuned to find optimal usage of the network resources. We will also study IP v6 possibilities, because it is a key factor in extending corporate applications to wireless handsets.

REFERENCES

[1] S. Mangold, S. Choi, P.May, O. Klein, G. Hiertz, L. Stibor, "IEEE 802.11e Wireless Lan for Quality of Service"

[2] 3GPP TS22.934, "Feasibility study on 3GPP system to Wireless Local Area Network (WLAN) interworking", Release 6.0

[3] 3GPP TS23.107, "QoS Concept and Architecture", Release 5.7

[4] 3GPP TS23.207, "End-to-End QoS concept and Architecture", Release 5.0

[5] IEEE 802.11 WG, Draft Supplement to STANDARD FOR Telecommunications and Information Exchange Between Systems - LAN/MAN Specific Requirements - Part 11: wireless Medium Access Control (MAC) and physical layer (PHY) specifications: Medium Access Control (MAC) Enhancements for Quality of Service (QoS), IEEE 802.11e/D2.0, Nov. 2001.

[6] 3GPP DRAFT TS 23.234: "3GPP System to Wireless Local Area Network (WLAN) Interworking: System Description", v1.0.0, September 2002. (Release 6)

[7] Mangold, S. and Choi, S. and Esseling, N., "An Error Model for Radio Transmissions of Wireless LANs at 5GHz," in *Proc. Aachen Symposium'2001,* Aachen, Germany, pp. 209-214, Sept. 2001.

[8] Kumar Balachandran,, "Convergence of 3G and WLANs", *IEEE ICC2002*

[9] Wolfgand Keller, Mathias Wagner, Robert Hirschfeld, Josef Noll, Stein Sveat, Jaime Ferreira, Olavi Karasti, Teddy Hudginson, Rafaelle Giaffreda, Stewart Fallis, John Charles Farancis, Christian Fischer, "Systems beyond 3G – Operator's vision", Eurescom Project P1203, December 2002

[10] Ni Qiang,<http://www-sop.inria.fr/planete/qni/Research.html>

RECONFIGURABLE EQUIPMENT MANAGEMENT: AN AGENT-BASED FRAMEWORK

Karim El-Khazen, Guillaume Vivier, Nicolas Lhuillier
Motorola Labs (CRM) – European Communications Research Lab
Parc des Algorithmes Saint-Aubin, 91193 Gif-sur-Yvette, France
{elkhazen,vivier,lhuillier}@crm.mot.com

Abstract:
Recent trends on wireless technologies as well as on software progress make the reconfiguration of wireless terminals not only an interesting feature but also an achievable one. This paper presents the motivation leading to the need of reconfigurable equipment management. A simple architecture of such equipment is given to allow the reconfiguration and its management. This architecture is based on the use of communicating agents, which provides a way for negotiating the appropriate configuration between the network and the terminal as well as easy context discovery or autonomous decision handling. Experiments based on a prototype shows that this agent-based approach is a good candidate for managing fully reconfigurable systems.

Key Words:
Reconfigurability, equipment management, agent

1. INTRODUCTION

The management of reconfigurable equipments is becoming a topic of great interest in the wireless industry as indicated by the following trends:

- New band allocated for mobile wireless services. A terminal can fully benefit from the new bands as long as it becomes agile.
- Novel application and air interface. Future applications, protocols, waveforms, cannot be anticipated now. Therefore, it is quite impossible to provision at the design time all the features that will be needed in the device terminal. Hence reconfigurable equipments will become able to download and run new functionality. Managing such equipment is of great interest.
- From military to public application of Software Definable Radio (SDR). While SDR was primarily devoted to military applications, it is nowadays moving to the public area, especially in the Beyond 3G trend.
- Beyond 3G trend. The evolution of telecommunication systems is going towards the convergence of different radio access technology towards an IP-based core network. To benefit from these heterogeneous radio systems, devices should be able to communicate through all of the radio access systems and to run all of the possible protocol stacks, applications etc.

This latter vision deserves to be further developed. Actually, telecommunication standard traditionally followed a 10-year regular life cycle. Now, the standards, whether they come from the telecommunication industry such as cellular phone systems or from the computer industry as the WLAN IEEE 802.11x standards, or from the broadcast one, converge towards the delivery of IP content. The B3G trend consists in an open, smooth and flexible integration of these different radio access technology standards to make them complementing instead of competing.

This trend assumes that future terminal will be multi-technology enable. Such ability could be provided efficiently by the reconfiguration of the terminal radio access layer parts, also known as Reconfigurable Radio, inherited from the SDR framework.

To handle the growing need of flexibility brought by the B3G vision and by the Reconfigurable Radio terminals, specific attention should be given to equipment management, a growing activity as it can be witnessed in several standardization bodies such as 3GPP, Telecom Management Forum, SyncML initiative, SDRForum which all followed different approaches.

This paper investigates another approach, amenable and open, based on the use of software conversational agents. Section 2 presents the equipment management aspects within the reconfigurability context, while Section 3 introduces and motivates the use of software agents. Then, the system architecture encompassing end-to-end and equipment architecture is presented in Section 4 and the agent platform JADE-LEAP in Section 5. Finally, lessons from the experimental testbed we developed are given in Section 6. Conclusions about agent-based reconfiguration management are drawn in the last section.

2. EQUIPMENT MANAGEMENT

Managing agile terminals is indeed a greater challenge than managing monolithic systems developed by a single vendor and communicating over proprietary interfaces and protocols. Management and reconfiguration of multi-technology (RATs) and multi-vendor systems, where software components are open and upgradeable, are of great interest to network operators and third-party service providers.

These service/network providers are looking at terminal management in order, for example, to improve their customer support, improve their network performance, and offer enhanced value-added services. The increase of device capabilities and the new expending services are raising the support costs in an exponential way. The advent of terminal management over the air will help the service/network providers.

With this perspective, complete management architecture is needed for such reconfiguration of terminals and network equipments.

3. SOFTWARE AGENT

Software agent technology encountered a renewal over the last decade, due to its foreseen suitability to the Internet ecosystem (e.g. search engines). However, the agent paradigm in itself is rather old and is sometimes dated from 1960. Because of its age, various approaches and uses of software agents have been made, so that it is now quite difficult to give a consensually agreed definition of an agent. However all agents share some common properties, which are of interest to management of reconfigurable equipments:

- Autonomy: agents are supposed to act and behave without the direct intervention of humans. Therefore, they must perform some control over their own actions and internal state;
- Sociability: agents form a community, which communicate and interact through semantically rich messages expressed in a structured language;
- Pro-activity: agents do not simply react in response to their environment; they are able to take initiatives when they consider to be appropriate;
- Adaptability: agents are able to learn on their experiences and improve their behavior accordingly.

Some elements suggest the evaluation of agent technology for terminal reconfiguration: Agent technology has today gained in maturity, as some commercial products include proprietary agents. Additionally, agents have recently moved from the Internet to the wireless device and can be executed on constrained mobile phones [1].

In this perspective, an overall management architecture was proposed, as described in next section.

4. SYSTEM ARCHITECTURE

The figure 1 depicts the overall system architecture, including a reconfigurable device. This device is built around various components. The terminal includes a reconfigurable modem that allows changing some radio parameters such as forward error correction coding rate, data rate, or even waveform type. Software radio modems [2] are typical examples of reconfigurable modems where the new configuration simply consists of changing the execution code. Such modems are appearing, see for instance [3] [4] or [5] but obviously, all the existing radio transceivers are not available today in full software version.

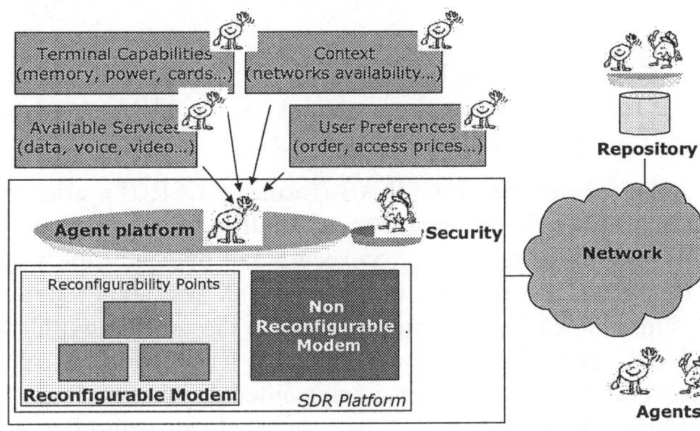

Figure 1: High level management architecture

An agent platform runs on top of the operating system. This platform provides the intelligence and the communication capability for triggering the appropriate configuration actions. It includes the following capabilities:

- Monitoring and discovery of the environment, in order to gather the information related to the available radio access technologies, the capability of the device, in terms of memory, processing power.
- Selection and negotiation of reconfiguration actions. Actually, based on its own context analysis, the device may want to change its own radio parameters but this choice may be invalidated by the radio operator, which has a larger view of the system and other priorities.

This latter capability yields to the need of having a distributed intelligence for managing the reconfiguration. Basically, the reconfiguration involves the equipment as well as supporting functions in the network. In our approach, we have chosen to have a limited network support since we originally wanted to contribute to the larger context of peer-to-peer services. The figure 1 shows a repository where all the known radio parameters and waveforms are stored, controlled by another agent.

This agent provides the basic security functions (download authorizations, subscription restriction information...). On the other hand, this agent could also prevent the user from downloading uncertified waveforms or even virus.

The monitoring operations described earlier could be performed in two ways, either retrieving the information directly or through the use of dedicated agents. For instance, an agent could be responsible for providing the most updated information on the terminal capabilities, and another one for providing the current environment characteristics in terms of networks availability...

The proposed management architecture could be used over any access technology, such as GSM/GPRS, WLAN and even Bluetooth or Ethernet...

5. JADE-LEAP AGENT PLATFORM

JADE (Java Agent DEveloper Framework) [6] is one of the major public implementations of the FIPA standard [8] for interoperable software agents.

JADE, developed by TILAB (formerly CSELT), aims at simplifying the development of multi-agent applications, by providing a runtime to execute and manage life-cycle of agents, as well as monitoring and debugging tools. While appearing as a single entity to the outside world, a JADE agent platform can actually be distributed over several hosts. JADE is implemented in version 1.2 of JAVA; hence agents can migrate or clone themselves to the different hosts of the platform, regardless of the OS. The communication architecture includes transparently to the agents an efficient intra-platform protocol based on RMI and a flexible inter-platform messaging relying on FIPA-compliant Message Transport Protocols (MTP) that are activated at run time. On the other hand, the conformance with the latest FIPA specifications ensures interoperability between agents at the syntax semantic levels.

As an outcome of the LEAP project (Lightweight Extensible Agent Platform, IST-1999-10211, [7]), the JADE platform has been enhanced with a set of libraries, which enable to deploy it over very constrained wireless devices, such as mobile phones, running Java 2 Micro Edition. The resulting platform, often refer to as JADE-LEAP, is the precursor of the second-generation FIPA compliant platforms. It represents a major technical challenge, as it brings the intelligence directly on the end-brings the intelligence directly onto the end-user device. JADE-LEAP benefits both from the development features of JADE and the lightweight properties of LEAP. It runs on a large variety of devices (computers, PDA, mobile phones, etc) and both the wired and wireless environments.

6. PRACTICAL EXPERIMENTATION

We implemented the architecture depicted in Figure 1 on classical Linux-based laptops running JADE-LEAP agent platform and demonstrated basic configuration action such as parameterizations of a WLAN networks, as described in Figure 2.

The intelligence in the terminal consists only in monitoring, filtering and managing the terminal capabilities (memory, power, cards...), the context (available networks...) and the available services (voice, video-conference...). Moreover, the user preferences rules are limited to a preference list, taking into account the requested services and the prices of the access. The repository in the network contains the WLAN connectivity information.

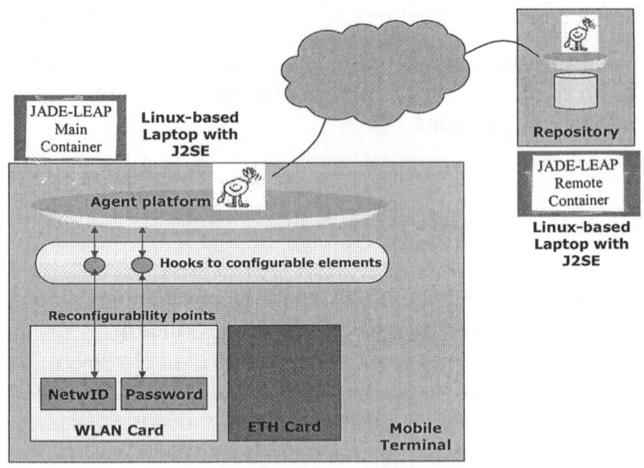

Figure 2: Demonstration Set-up

A JADE-LEAP Agent Platform (AP) is running on the terminal and contains the Terminal Agent (TA). This agent communicates with its corresponding Network Agent (NA) located in the repository. The NA belongs to the unique AP. Since both TA and NA belong to the same platform, they use the JICP protocol (Jade Internal Communication Protocol), the internal platform communications mechanism of JADE-LEAP.

Another solution could be to consider two different agent platforms running in the terminal and in the repository. The first one would contain the TA and the second platform would contain the NA. The difference resides in the communication protocol used between the two agents. The benefit of the proposed architecture with the use of only one agent platform resides in the use of the basic and efficient JICP protocol that reconciles the conflicting requirements of easy processing and reduced data transmission over wireless links.

Both TA and NA are running on Linux-based laptops with the standard version of Java (J2SE). The two agents communicate through either an Ethernet or a Bluetooth connection established between the devices. The mobile edition of Java (J2ME) could have been used instead of

J2SE. Moreover, the demonstration assumes that a WLAN network is available in the area, and the terminal has the possibility to use a WLAN PC Card.

By entering the PC Card in the laptop, the TA detects the WLAN capability since the terminal capabilities are monitored, but no connectivity information is available for the corresponding WLAN network, such as authorized user and password. The TA notifies the NA and requests the WLAN connectivity information. The NA analyses the request and responds to the TA with the needed information for the WLAN connectivity. Following the reception, the TA reconfigures its network configuration by launching a script that uses the received parameters. In this way, the terminal gets access to the WLAN network.

Though it does not include a real negotiation between yhe agents, this simple scenario demonstrates how the management architecture could be used, for instance, for getting access to the WLAN network. In the studied scenario, only parameters were retrieved from the repository in the network. However, pieces of code could be exchanged between the terminal and the network repository and executed locally. For instance, the TA could retrieve the necessary drivers for a new network card, updated waveform software code and install them on the terminal.

7. CONCLUSION

Mobile agent technology is the subject of a lot of attention in the last years, and is often considered as the next step after the client/server and object-oriented programming paradigms.

Even if reconfigurability has been for a long time a topic of interest with outcomes mostly limited to military applications, the convergence between heterogeneous radio systems provides a field of great application to configuration management of flexible equipments.

This developed setup depicts the promising future of the use of software agents in the convergence of the heterogeneous systems.

Future work might consist in addressing more complex scenarios of use, and developing the intelligence aspects as well as modelling the information exchanged between the agents.

REFERENCES

[1] G. Caire, N. Lhuillier, G. Rimassa, “A communication protocol for agents on handheld devices”, AAMAS Conference, Workshop on Ubiquitous Agents on embedded, wearable and mobile devices, Bologna, July 2002

[2] Joseph Mitola III “Cognitive Radio: An Integrated Agent Architecture for Software Defined Radio”, Dissertation Thesis, Royal Institute of Technology (KTH), Sweden.

[3] J. Chapin, V. Bose, “The Vanu Software Radio System”, 2002 Software Defined Radio Technical Conference, San Diego, November 2002, http://www.vanu.com

[4] C. Bonnet, L. Gauthier, P.A. Humblet, R. Knopp, A. Menouni-Hayar, Y. Moret, A. Nordio, D. Nussbaum, M. Wetterwald, “An all-IP Software Radio Architecture under RTLinux”, Special Issue on Software Radio In “Annales des Telecommunications”, Volume 57 N°7-8, July-August 2002

[5] GNU Software Defined Radio, http://www.gnu.org/software/gnuradio

[6] JADE, http://sharon.cselt.it/projects/jade/

[7] LEAP, http://leap.crm-paris.com

[8] FIPA, http://www.fipa.org

SESSION 2A:
WIRELESS NETWORKS — IEEE 802.11

IMPROVING THE PERFORMANCE OF IEEE 802.11 IN WIRELESS MULTI-HOP NETWORKS

GANG LU AND CAULIGI S. RAGHAVENDRA
Department of Electrical Engineering
University of Southern California
3740 South McClintock Avenue, Los Angeles, CA 90034

The IEEE 802.11 MAC Protocol and on demand ad hoc routing protocols (DSR, AODV) have been widely used in experiments and simulations in wireless ad hoc networks. However, previous researches have shown that IEEE 802.11 MAC protocol does not perform well in wireless multi-hop networks because of the interaction between MAC and routing layers. Hidden node problem makes a node fails to reach its adjacent neighbor, thus triggers false route failures in routing layer which greatly degrades the performance of the network. This paper proposes a scheme, named Adaptive Retransmission Control (ARC), to reduce the probability of false route failures. ARC is completely distributed and does not introduce any additional communication overhead. Simulation experiments are conducted to evaluate the performance with and without ARC. Results show the TCP throughput improvement by 5% to 57% under various simulated topology and different mobility.

1. Introduction

A Wireless ad hoc network is a group of nodes that communicate with each other using wireless links without a fixed infrastructure. In an ad hoc network every node could become a router and the communication may be through multi-hop where all links are wireless. Wireless network differs greatly from wired network because of its high bit error links, bandwidth limit, broadcasting nature of the radio and frequent route changes.

The IEEE 802.11 access scheme has been widely used in research of wireless multi-hop networks. However, IEEE 802.11 does not perform well which is discussed in [1]. It is pointed out that the exposed station problem and collisions prevent the intermediate node from reaching its next hop. If it fails after several retries (in IEEE 802.11, the retransmission limit is set to 7), MAC will assume a link breakage and report it to the routing layer. Both DSR [2] and AODV [3] will drop all queued packets to that adjacent node. DSR will report a route failure to the source which will begin a new route discovery process. As the link actually is not broken, we call this kind of route failure as false route failure. False route failure happens frequently in multi-hop network. For a TCP flow, a false route failure could cause out of order delivery of packets, fast retransmit, even a timeout. This can greatly reduce the performance of the network. However, [1] stopped short of providing a method to solve or alleviate this problem.

In this paper, we study in detail about the effects of the false route failures on the TCP performance in wireless multi-hop networks. We then propose a completely distributed mechanism, ARC (Adaptive Retransmission Control), to help increase the accuracy to distinguish false route failures from true route failures with zero communication overhead. We observe 5% to 50% throughput improvement in various simulation scenarios.

2. The impact of False Route Failure

We now explain the effect of route failure in detail using DSR as the underlying routing protocol. In the IEEE 802.11 protocol, when a node x fails to reach its adjacent node y after seven retransmissions, node x will notify its routing layer which will report a route error message to the source to prevent it from sending more packets on this route. DSR cache and new salvage method can help reduce the cost of a false route failure caused by contention. However, under contention and mobility, DSR cache and salvage process fail frequently. When it receives the route error message, the source will begin the route discovery process. The packets from the TCP will be stored into a send buffer. If the buffer is overflowed before a route reply message is received, packets are dropped which could incur fast retransmission or time out in TCP layer. When a new route is found, the packets in the send buffer will be released. Meantime, new packets from the TCP will often be delivered before the packets in the send buffer. This incurs out-of-order packet arrivals at the receiver, which may make the source invoke fast retransmit. What is worse, since the network is suffering contention, the

DSR control messages have a great chance to be lost again which will invoke exponential backoff of DSR. TCP normally will timeout under such a condition, which degrades the performance of the network significantly.

We examine the impact of False Route Failures on TCP throughput under different topologies by simulations. All simulations are conducted in *ns*-2, version 2.26, averaged by results from 10 seeds.

In a chain topology, all nodes are fixed and there will be only one path from a source to a destination. Successive packets from a single connection will interfere with each other. The acknowledge packets will also contend for bandwidth with data packets. The simplest and most effective solution to eliminate false route failures in chain topology is to prevent the MAC layer from reporting link breakage as there is not. We conducted simulation for different retransmission limit and hop length for a single TCP flow. Figure 1a shows that TCP throughput can achieve 90% improvement.

In a fixed grid topology, there is no real link breakage either. However, reporting a false route failure triggers the source to launch a route discovery process which may find a new route which suffers fewer contentions. For example, ack packets can find a different route rather than the reversed path of the data packets, which can reduce the contention. So there is a trade off between reducing false route failure and allowing a false route failure. However, both DSR and AODV do not intensely search for a better route to balance the traffic load in the network. From our simulations, the gain achieved by reducing number of false route failures still overweigh the gain of a possible new route.

In a random network with node mobility there are both true route failures and false route failures. Simply increasing retransmission limit could not only increase delay in detecting a true route failure, but also increase contention because of useless retransmissions when a true route failure occurs. So there is a tradeoff between reducing the number of false route failures by increasing retransmission limit and detecting a true route failure promptly.

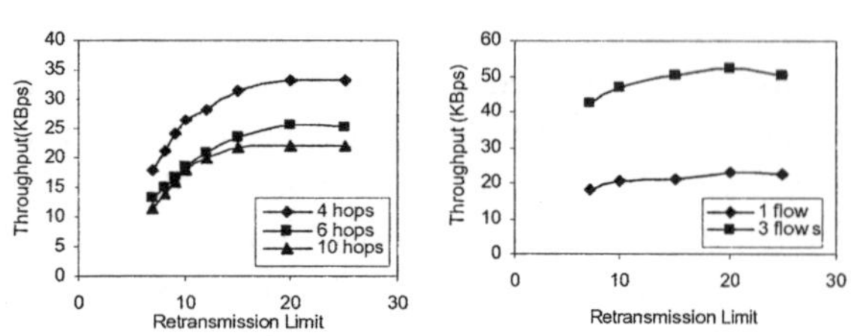

(a) Chain Topology (b) Grid Topology

Figure 1. Throughput versus MAC layer Retransmission Limit for fixed chain and grid topology

3. Improving MAC Layer Link Failure Detection

From previous section, we point out that reducing false route failures by increasing retransmission limit can improve throughput significantly when nodes are fixed since there is no true route failure. However under mobility, there are true route failures, in which we need to detect as soon as possible. This section describes the ARC mechanism which adaptively adjusts the number of retransmissions aimed at reducing both false route failures and the cost enduced by delay in detecting true route failures.

A possible solution is to explicitly discover the neighbor nodes periodically. However, this is at the cost of more control packets. If the neighbor discovery interval is too large, the topology change can not be traced promptly. But if the neighbor discovery interval is too small, the overhead of control packets will be too large.

3.1. *The ARC Mechanism*

The ARC mechanism we propose is simple with zero communication overhead. All information required can be collected locally, leading to a completely distributed scheme.

Because of the broadcast nature of the wireless radio, a node can overhear its neighbor nodes' transmissions even if it is not the data destination. Suppose a simple chain topology with three nodes A, B, C. Node A tries to reach its adjacent node B but fails in the previous ssrc_ (ssrc_ is the STA Short Retry Count in IEEE 802.11 specification.) transmission attempts, thus backoffed a relatively long period. During A's

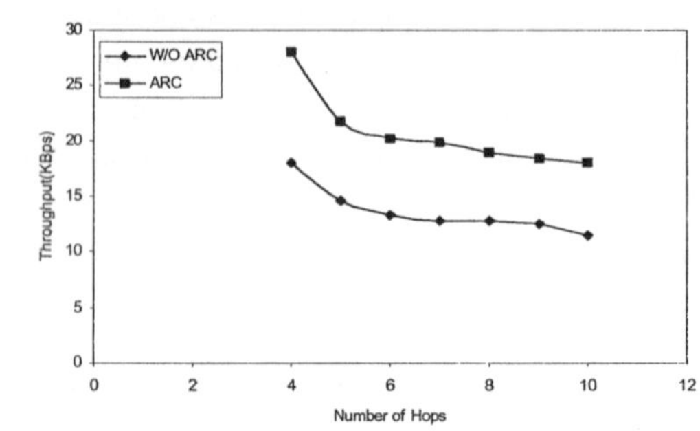

Figure 2. Throughput comparison versus number of hops.

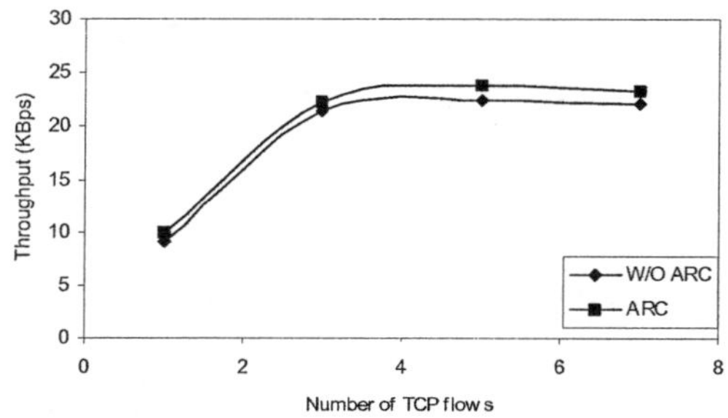

Figure 3. Throughput comparison versus number of TCP flows.

backoff period, B and C compete for the channel. If B wins the channel, B will transmit data or control packets. A can overhear B's transmission though it is in backoff stage. A then knows that B is still in its neighborhood area and then reset its ssrc_ to zero, which means it can still retry seven times before reporting a link breakage. The reason is that although a node fails to reach its adjacent node in previous ssrc_ transmission attempts, the overhearing of a transmission by its adjacent node shows that all previous failures are most probably due to contention, not because of link breakage. Here we assume symmetric links.

3.2. *Simulation Results*

We implemented ARC prototype in *ns*-2 with CMU wireless extension. We choose the aggregated throughput to evaluate the performance of ARC.

3.2.1. *Fixed Topology*

Figure 2 shows the throughput of a single TCP flow versus number of hops in a chain topology. With ARC, the throughput increase is about 50% compared to throughput without ARC. Since in such a fixed chain topology, there is no true route failure and only one route is available, ARC achieves less throughput increases than simply increasing the retransmission limit to a large number.

Figure 3 shows the aggregated throughput of multiple TCP flows in grid topology. With ARC, the throughput increase is about 5% compared to throughput without ARC. The increase is less than that achieved in a chain topology. This is because DSR can find a new route which suffers fewer contentions in grid topology.

3.2.2. *Mobility*

In this topology, 100 nodes move randomly in a 1000x500 meters area. The source is fixed on the top left corner while destination is fixed at right bottom corner so that the number of hops between them is larger than 4. We turn off the DSR cache according to [5].

Figure 4 and 6 shows the throughput of a single TCP flow under different mobility speed. In figure 6, the throughput is normalized with the throughput without ARC. In most mobility patterns, ARC achieves a higher throughput, but as mean speed increases, the improvement ratio has a higher variance. Sometimes, ARC even has a less throughput. This is reasonable as node moving speed increases, even if a node hears its neighbor's transmission in its backoff period, because of the high mobility speed, the neighbor could move out of range in next time slot. Figure 4 shows the average throughput increase is about 11% to 30% compare to that without ARC.

Figure 5 shows the ratio of false route failure and false neighbor (ARC decides to retransmit when the destination actually is out of radio range). ARC always achieves much less false route failure than without ARC. However, ARC may have a higher false neighbor ratio. We found that reducing false route failure benefits more than reducing false neighbor, especially when mobility is high. Increasing the retransmission limit has more benefit in throughput than the harm of delaying the detection of true route failures. However the decision should be made smart.

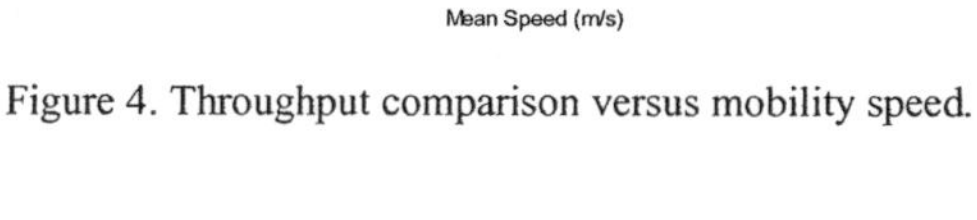

Figure 4. Throughput comparison versus mobility speed.

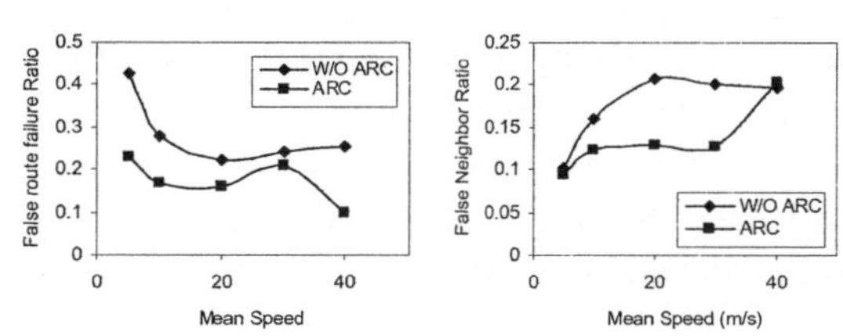

Figure 5. False route failure ratio and false neighbor ratio.

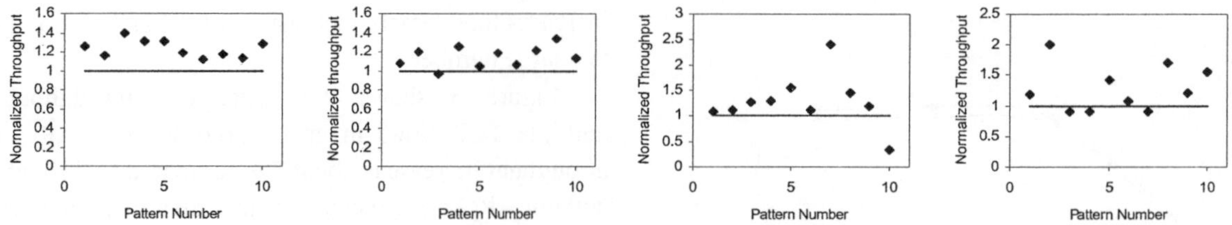

(a) mean speed = 5 m/s (b) mean speed = 10 m/s (c) mean speed = 30 m/s (d) mean speed = 40 m/s

Figure 6 . Normalized throughput for 10 mobility patterns at different mean speed with 20s pause period.

4. Related Work

Reference [1] shows that the IEEE 802.11 MAC protocol does not work well in Multi-hop Wireless Ad hoc networks. When one node fails to reach its adjacent node after it reaches its retransmission limit, it will report a route failure to the upper layer. Sometimes this is due to the contention of the transmission, not a real link breakage. The effect of retry limit on TCP performance is also discussed in [7]. However [7] simply proposed to increase the retry limit which will not perform well under mobility, while [1] does not propose a solution.

There are many papers [4, 5, 8] about TCP performance in wireless ad hoc networks. However, these papers focus on adapting TCP to the wireless communication environments. The effects of link breakage due to mobility on TCP performance in mobile Ad Hoc Networks is investigated in [5]. They show that TCP throughput decrease significantly when nodes are mobile, for TCP can't distinguish between a link failure and congestion. They propose the Explicit Link Failure Notification (ELFN) to improve TCP performance. The TCP source disables the retransmission timer and enters a "standby" mode when route failure happens. During the standby mode, TCP periodically send a packet to the destination. When an ACK is received, TCP leaves the standby mode, restores its retransmission timers and continues as normal.

5. Conclusion and Future Work

This paper proposes ARC, Adaptive Retry Control, to reduce the probability of false route failures. ARC does not introduce any additional overhead, and is completely distributed. Simulation results show the TCP throughput improvement by 5% to 57% under various simulated topology and different mobility with ARC.

In our future work, we are thinking about using the contention level in the network to adjust the retransmission limit. The effect of channel error should also be evaluated.

References

1. Shudong Xu andTarek Saadawi, "Does the IEEE 802.11 MAC Protocol Work Well in Multihop Wireless Ad Hoc Networks", in *IEEE Communications Magazine,* June, 2001.
2. D. B. Johnson and D. A. Maltz, Dynamic source routing in ad hoc wireless networks, in *Mobile Computing*, chapter 5, 1996
3. Charles Perkins and Elizabeth Royer, "Ad hoc on-demand distance vector routing", in *Proceddings of 2nd IEEE Sorkshop on Mobile Computing Systems and Applications*, Feb 1999.
4. K. Chandran, S. Raghunathan, S. Venkatesan, and R. Prakash, "A feedback based scheme for improving TCP Performance in ad hoc wireless networks", in *Proc. ICDCS,* May 1998
5. G. Holland and N. H. Vaidya, Analysis of TCP Performance over Mobile Ad Hoc Networks, in *Proc. IEEE MOBICOM*, Seattle, Aug. 1999
6. Jian Liu and Suresh Singh, ATCP: TCP for Mobile Ad Hoc Networks, in *IEEE Journal on Selected Areas in Communication*, vol. 19 No. 7 July 2001
7. Rui Jiang, Gupta V. Rvish ankar, C.V., "Interactions between TCP and the IEEE 802.11 MAC protocol", in *DARPA Information Survivability Conference and Exposition*, 2003
8. Jian Liu and Suresh Singh, "ATCP: TCP for Mobile Ad Hoc Networks", in *IEEE Journal on Selected Areas in Communication*, vol. 19 No. 7 July 2001

AN ENHANCED SUPERPOLL PROTOCOL FOR IEEE 802.11 PCF WITH RATE DIFFERENTIATION

JUKI WIRAWAN TANTRA, TECK MENG LIM, CHUAN HENG FOH AND BU SUNG LEE

Centre of Multimedia and Network Technology
School of Computer Engineering
Nanyang Technological University
Singapore

In recent years, the issue of quality of service (QoS) in wireless LANs (WLANs) has been the focus of researches. Previous works usually use different backoff scheme and inter-frame delay to achieve service differentiation. In this paper, we present a protocol that makes use of SuperPoll to achieve better utilization and Deficit Round Robin queuing technique to achieve fairness and service differentiation. Computer simulations show that the protocol can achieve better utilization and service differentiation on the WLANs.

1 Introduction

Recent technology advancement has allowed the development of high data rate WLANs with the IEEE 802.11a and 802.11g standards in addition to the popular IEEE 802.11b standard. Promoted by the widespread use of WLANs, there is a high demand of QoS for various applications (e.g. multimedia streaming).

The IEEE 802.11 specifies two mechanisms for transmission, which are the distributed coordination function (DCF) and the point coordination function (PCF). DCF makes use of CSMA/CA (Carrier Sense Multiple Access with Collision Avoidance) with binary exponential backoff for contention purpose. DCF uses two techniques for packet transmission. The default is a two-way handshaking method called basic access method. The other technique uses a four-way handshaking method, known as Request-To-Send/Clear-To-Send (RTS/CTS). PCF is an optional contention-free service that makes use of polling by a centralized point coordinator (PC).

The current IEEE 802.11 standards do not address the issue of QoS on WLAN. The QoS issue will be addressed in IEEE 802.11e standard, which proposes the enhanced distributed coordination function (EDCF) to provide service differentiation.

Many priority schemes for DCF access protocol have been proposed. Most of these schemes suggested different backoff time computation to achieve service differentiation [1]. Ranasinghe et al. [2] proposed using Deficit Round Robin (DRR) queuing technique on the PCF protocol. DRR is an efficient frame-based fair queuing technique that approximates the fluid flow model. It provides near-perfect throughput fairness with $O(1)$ packet processing.

The polling mechanism of the PCF protocol introduces excessive overheads. Ganz et al. [3] introduced a SuperPoll with chaining protocol to address this issue. Stations request to be polled by the PC during DCF frame. The PC broadcasts a SuperPoll message to a group of stations thus saving the overhead created by individual polling. This poll contains a list of stations that will be polled during the current period. A chaining mechanism is deployed to improve the reliability of this protocol in a noisy environment. This mechanism appends a list of the remaining stations to be polled in each transmitted data packet, which increases the probability that all the stations in the poll-list will receive the SuperPoll. The SuperPoll protocol achieves great improvement on the throughput (which is measured as a function of the channel noise, the SuperPoll overhead per data packet and the packet size) compared to the PCF.

In this paper, we introduce a protocol that improves the utilization of the SuperPoll protocol. We also incorporate DRR queuing technique to provide fairness and service differentiation. Our protocol is simple and has little transmission and computation overheads. The transmission overhead is due to the SuperPoll frame and the chaining mechanism employed. The computation overhead is due to the DRR queuing technique that imposes packet processing complexity of $O(1)$.

This paper is organized as follows. Section 2 explains the mechanism of the proposed enhanced SuperPoll protocol with rate differentiation. Section

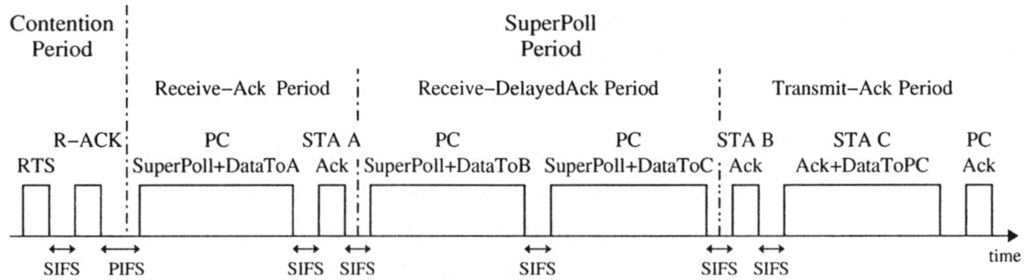

Figure 1. Enhanced SuperPoll.

3 provides some analytical results to support our scheme. Section 4 provides the performance evaluation of the scheme.

2 Enhanced SuperPoll with Deficit Round Robin

Our protocol is briefly described as follows. It makes use of the SuperPoll protocol described earlier with an addition of a request-to-be-queued indicator. This indicator enables a station currently being polled to rejoin the poll-list if it has more data to send, thus reducing the contention that happens on the DCF period as the station does not need to contend for transmission again. Our protocol reduces the overheads of extra polling cycle compared to a mechanism that polls a station until it sends an empty packet to signify that it has no more data to send. This indicator also helps the PC to schedule transmissions. We propose to use More Data field in the IEEE 802.11 frame format for the request-to-be-queued indicator. We use DRR queuing technique as the scheduling scheme to achieve fairness and service differentiation for our protocol.

In addition, our protocol merges the upstream and downstream transmission periods, hence reducing the overheads of using separate DCF and PCF periods for upstream and downstream transmissions respectively.

We divide a SuperPoll message into three segments (see Fig. 1). Each segment represents a task for stations in the poll-list to perform. These segments are the receive-ack, receive-delayed-ack-transmit and transmit-ack. In the receive-ack segment, listed stations will monitor the channel for their data packets. An acknowledgment will be sent for each received data packet. In the receive-delayed-ack-transmit segment, listed stations will monitor the channel for their data packets but acknowledgments of these data packets will be delayed. After all down-link traffic are sent, stations will acknowledge the received data packets and transmit data packets according to the polling order. The acknowledgments are piggy-backed in the MAC frames. In the transmit-ack segment, stations in the list will transmit data packets orderly and an acknowledgment is sent immediately after each packet. Owing to the possible lost of delayed acknowledgments or lost of data packets in a noisy environment, retransmissions of down-link traffic may commence in the receive-ack segment of the next frame.

When the PC has packets to send, it embeds the SuperPoll to all the packets that the PC will send in the round. Hence there is no need of another SuperPoll frame for polling purpose.

At the PC, two deficit counters (DCs) are kept for each station. One DC is used for down-link traffic control and the other is used for up-link traffic control. The quanta of the DCs are predetermined based on a prior bilateral agreement between the user and the service provider. Each DC is allocated a quantum worth of bits in each round of service using DRR. DRR serves each DC in a round robin manner. A DRR round and a SuperPoll round could be different. Therefore, a DC may not be exhausted in a single SuperPoll period. A DRR round ends only when there is a deficit of quantum in every DC or there is no more traffic to transmit. We allow negative DC in our scheme to simplify the design as the PC does not need to transmit the DC value to the station.

In the DCF period, stations only need to transmit small RTS frames hence reducing the collision period, which is determined by the length of the frames involved in a collision. Moreover, upon making a request successfully, a station will stop contending for the channel thus increasing the probabilities of the other stations successfully transmit the RTS frames.

3 Analytical Results

3.1 Fairness

Ranasinghe et. al. [2] has proved the fairness of DRR scheme that allows negative DC value. Our scheme is slightly different because a SuperPoll round may be different from a DRR round. To prove the fairness of our scheme, we prove that the SuperPoll round is independent of the DRR round.

Hypothesis *The SuperPoll round and the DRR round are not independent.*

Assume that two busy users (U1 and U2) successfully register the transmission requests on a contention round and used all of their DCs allowance on the following SuperPoll round. U1 and U2 are requeued to the poll-list via the request-to-be-queued indicators as they have more data to transmit. In the next contention round, two more users (U3 and U4) successfully register the transmission requests. In the next SuperPoll round only U3 and U4 may transmit the data as U1 and U2 have finished their DCs and the DRR round ends only if all the stations registered could not transmit any packet (i.e. all the stations have negative DCs). Hence, the hypothesis is incorrect. The SuperPoll round and the DRR round are independent of each other.

3.2 Throughput Comparison

Theorem 1 *The throughput of the original SuperPoll protocol differs from the throughput of DCF protocol by a few inter-frame spaces and the SuperPoll headers.*

Proof Assume that m packets are transmitted in a time interval using DCF protocol and SuperPoll protocol. Using DCF protocol, each packet requires $(RTS + SIFS + CTS + SIFS + DATA + SIFS + ACK + DIFS)s$ of transmission time. For m packets, the total transmission time required will be $m \times (RTS + SIFS + CTS + SIFS + DATA + SIFS + ACK + DIFS)s$.

Using SuperPoll protocol, m packets require $(m \times (RTS + SIFS + CTS + DIFS))s$ in the contention period (CP) and $(PIFS + SUPERPOLL + SIFS + (m \times (SUPERPOLL\_HEADER + DATA + SIFS + ACK + SIFS) + DIFS)s$ in the contention free period (CFP). By a simple elimination, the SuperPoll protocol requires an extra of $(PIFS + SUPERPOLL + SIFS + m \times SUPERPOLL\_HEADER + DIFS)s$ in transmitting m packets. □

Theorem 2 *The enhanced SuperPoll achieves better throughput as compared to the SuperPoll protocol.*

Proof Bianchi [4] shows that the arrival rate (λ_n) of the CP in a network is a function of the number of stations (n) contending for the channel.

Assume that all stations are busy and contending for transmissions in a WLAN. In the CP of SuperPoll, n stations will always contend for a transmission where n is the number of active stations. Therefore, the arrival rate for the channel is λ_n. Additionally, we assume that n is large such that the minimum arrival rate of the CP of enhanced SuperPoll is λ_n, which equals to the minimum of λ_i where i ranges from 1 to n.

In our enhanced SuperPoll, $(n - m)$ stations will contend for a transmission during any transient period before all stations are registered in the SuperPoll list where m refers to the number of stations in the SuperPoll list. So the arrival rate is $\lambda_{n-m} + \mu$, where μ is the arrival rate from the request-to-be-queued indicator. The value μ is equal to the service rate of the PC in CFP because after a packet is serviced, the sender is re-queued to the SuperPoll list.

When n is large and m ranges from 0 to $n - 1$, $\lambda_{n-m} \geq \lambda_n$ hence $(\lambda_{n-m} + \mu) \geq \lambda_n$. A higher arrival rate will result in a higher channel utilization, so our enhanced SuperPoll protocol achieves higher throughput compared to SuperPoll. □

Corollary 1 *At saturation, the arrival rate of the network is equal to the service rate of the PC.*

Proof When all stations have registered to the SuperPoll list ($m = n$), there will be no stations contending for transmission in the CP ($\lambda_{n-m} = 0$) as stations re-register themselves using the request-to-be-queued indicators. Therefore, the arrival rate of the channel is equal to the service rate of the PC. □

4 Performance Evaluation

Figures 2-4 show our simulation results. In the simulation, stations transmit fixed size packets of 500, 1000, and 1500 bytes distributed uniformly with exponential inter-arrival time. We use 300 μs for the length of the CP in SuperPoll scheme.

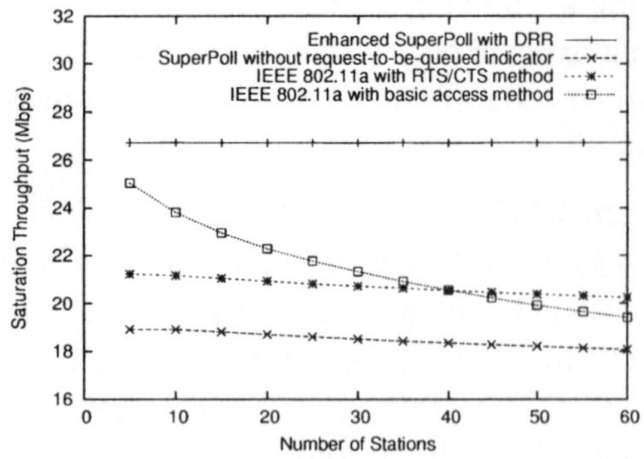

Figure 2. Saturation throughput of the different protocols with varying number of stations.

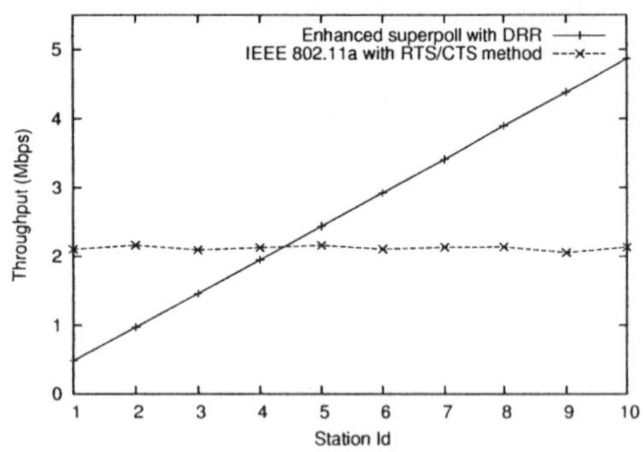

Figure 3. Throughput of each station with different rate allocated.

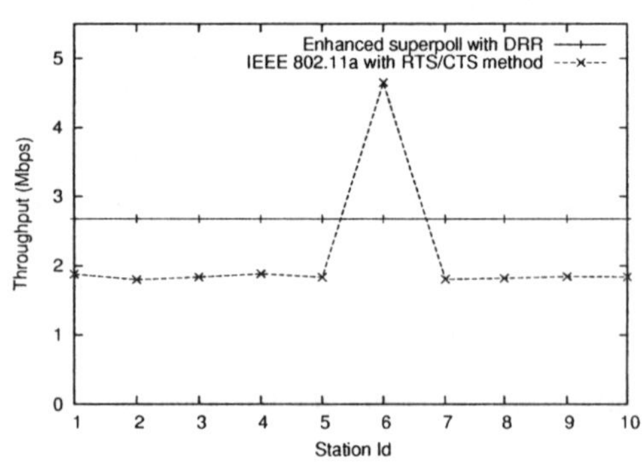

Figure 4. Throughput of each station in the existence of one rogue station.

Figure 2 shows the saturation throughput of the proposed protocol compared to the previous protocols. As we are comparing our protocol to the DCF protocol, we use only upstream traffic (i.e. PC has no packet to send). The performance of the SuperPoll protocol without using the request-to-be-queued indicator (i.e. PC will only poll a station on request) will only be slightly lower compared to the RTS/CTS method (Theorem 1). The reason is that the SuperPoll protocol only rearranges the order of the transmission with extra SuperPoll frames and extra bytes in each data packet for the chaining mechanism. Our protocol counters this problem by using the request-to-be-queued indicator, hence it achieves higher throughput (Theorem 2).

Figure 3 shows the weighted rate differentiation achieved using the DRR queuing technique. Each station makes a different bandwidth reservation in ascending order from 0.5 Mbps to 5 Mbps. The result shows that our scheme is able to provide service differentiation compared to the current standard.

Figure 4 shows the fairness achieved using the DRR queuing technique in the existence of rogue station (Station 6). We make fair bandwidth reservations for all stations. Station 6 is simulated without deploying the exponential backoff scheme, hence the station could take huge chunk of the bandwidth. The result shows that our scheme is able to ensure the reserved bandwidth distribution in the existence of any misbehaving station.

5 Conclusion

In this paper, we have introduced a scheme that improves the SuperPoll protocol, allowing better throughput, merging of upstream and downstream transmissions, and incorporation of the DRR queuing technique. The protocol also achieves fairness and rate differentiation, which are difficult to achieve using the current DCF or PCF protocols.

References

1. W. Pattara-atikom, P. Krishnamurthy, and S. Banerjee, "Distributed mechanisms for quality of service in wireless LANs," *IEEE Wireless Commun. Mag.*, vol. 4, no. 3, pp. 26–34, June 2003.
2. R. S. Ranasinghe, L. L. H. Andrew, and D. Everitt, "Distributed contention-free traffic scheduling in IEEE 802.11 multimedia networks," *10th IEEE Workshop on Local and Metropolitan Area Networks*, pp. 18–28, 1999.
3. A. Ganz, A. Phonphoem, and Z. Ganz, "Robust superpoll with chaining protocol for IEEE 802.11 wireless LANs in support of multimedia applications," *Wireless Networks*, vol. 7, pp. 65–73, 2001.
4. G. Bianchi, "Performance analysis of the IEEE 802.11 distributed coordination function," *IEEE J. Select. Areas Commun.*, vol. 18, no. 3, pp. 535–547, Mar. 2000.

Fast Handover Support in Wireless LAN based Networks

Norbert Jordan, Reinhard Fleck, Christian Ploninger
Institute of Communication Networks, Vienna University of Technology
Favoritenstrasse 9/388, A-1040 Vienna, Austria
norbert.jordan@ieee.org, reinhard.fleck@tuwien.ac.at, christian.ploninger@siemens.com

Abstract: This paper addresses the optimization of handovers in a Wireless LAN/Mobile IPv4/v6 environment by controlling the link layer handover process directly at the mobile node's side. One successful way to improve system performance of mobile clients in Wireless LAN multimedia networks is to use efficient and fast handover mechanisms when users move between different access points. We implemented some quality parameters of the wireless link provided by Network Interface Card (NIC) firmware into a network-management software running on the mobile node. This concept allows a mobile user to automatically decide about when to start a handover and to which access point to connect to. The experience shows that our multimedia-enabled solution provides the necessary guidance to monitor and control the handover process without adding new extensions to the already existing Mobile IP standards.

Key Words: · Wireless LAN, Adhoc, Fast Handover, Multimedia, Mobile IP.

1. INTRODUCTION

A WLAN handoff occurs when a mobile station moves beyond the radio range of the currently associated access point (AP), and enters another Basic Service Set (BSS) at the MAC layer. During the handoff process, management frames are exchanged between the station and the AP. Also the APs involved may exchange certain context information (credentials) specific to the station. Consequently, there is latency involved in the handoff process during which the mobile node is unable to send or receive any kind of traffic. Because of the mobility-enabling nature of wireless networks, there is opportunity for many promising multimedia and peer-to-peer applications (such as VoIP, 802.11 phones, mobile video conferencing and chat). Also, many believe that WLANs may become or supplement via hot spots the next generation 4G wireless networks. While the wireless LAN technology spreads more widely, available Mobile IP implementations do not yet handle handover of multimedia-traffic satisfactorily. The design of Mobile IPv4 is restricted by the abilities of IPv4 and leaves a lot of room for improvements. Especially the reduction of the handover latency is a topic widely discussed. Although IPv6 provides better prerequisites (in terms of flexibility and extensibility) for the implementation of a mobility supporting protocol, Mobile IPv6 still has some flaws. So one of the major concerns is, that while the wireless LAN technology and Mobile IP operate at different layers, there is no definition for the inter-layer communication between link and network layer. In practice, this is reflected in the Mobile IP implementations that are just following the behavior of the wireless LAN hardware.

Of course, this leads to delays or unwanted and unpredictable results. But especially for multimedia applications relying on QoS, the well timed transfer of network management information to the new point of attachment is very important. This can only be guaranteed, if control over the link layer handover process is provided to these applications.

2. PROPOSED HANDOVER-APPROACH

The need to rely on link layer information to allow certain means of synchronization between the handover processes of the link and the network layer is well understood [1]. The most obvious parameters that influence the wireless LAN hardware's behavior are indicators of the signal quality. These quality parameters are usually provided by the driver of the wireless LAN hardware and can be gathered if appropriate access to drivers of the NIC is given. In our case we got really kind support from Intersil[1] to make driver-development [8] possible for all PRISM2 based chipsets. These chips are implemented in a lot of WLAN products available today.

The idea is to monitor continuously the signal quality of all access points in range. In this way a client-based tool makes decisions based on signal-levels received from all access points. If any available AP provides a significant better signal quality or the actual associated base-station has too weak signal-level to serve the client with a specific link quality, it is necessary to initiate a handover-process. The whole handover-process is under control of the mobile node that is capable of performing fast handovers. Our implementation requires access to the wireless LAN hardware, as well as a communication interface to the IO-controller part. Intersil's PRISM2-DLL's are compatible with all win32 platforms and the tool was developed with assistance of Visual C++. The big advantage of this implementation is that the wireless card will stay associated with the present AP, while scanning the wireless interface for other APs in range.

[1] http://www.intersil.com

2.1 Active-Scan Mode

Active Probing is described in the IEEE 802.11 specifications [7], the active probing method uses probe request frames on each channel where it is able to detect wireless activity (to avoid blocking the discovery process on frequencies that are not in use). When an AP comes within range of the client and receives a probe request frame it will typically respond with a probe response frame containing the network SSID (see Fig. 1). This method has the little disadvantage of being unable to discover wireless networks that are configured not to advertise their Extended Service Set Identifier (ESSID) using a cloaked ESSID configuration.

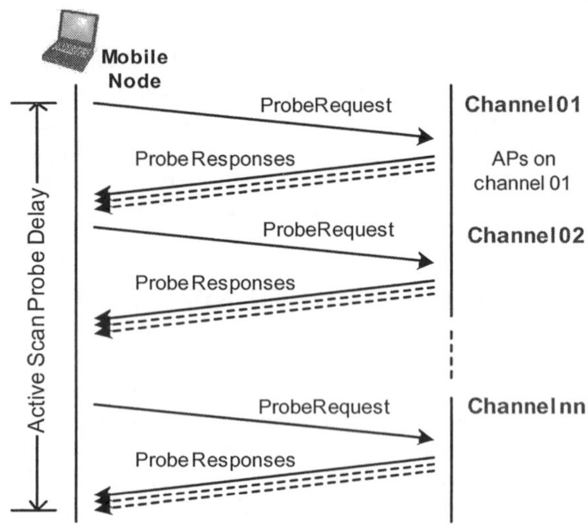

Fig.1: The Messages in an Active Scan.

2.2 PRISM2 Driver Development

The deployed PRISM2 DLL's provide easy integration into various application development environments. It is designed to provide the presentation layer described in the OSI Network architecture model. Fig. 2 illustrates the architectual overview of our implementation in an Windows32 Network Driver Interface Specification (NDIS) environment. On the left hand side there is presented the typical implementation of the Windows NDIS layers. The right side illustrates how the ``ptamacctl.dll" fits into the NDIS architectural concept. This NDIS stack now represents a proprietary protocol layer which dynamically binds to the NDIS adapter driver, it also includes a private API to communicate to the protocol driver in user mode. With this approach an adapter can be completely unbound from the rest of the protocols. So this implementation allows for a more controlled environment including firmware based functionalities.

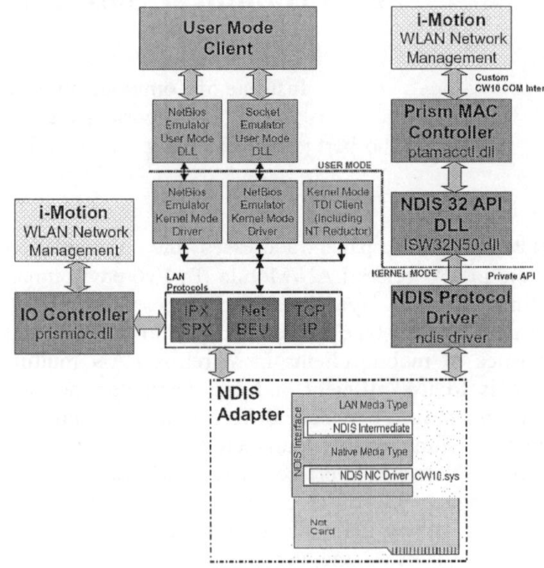

Fig. 2: NDIS32 Architecture including i-Motion.

3. CLIENT BASED HANDOVER

The operational design is based on a program which has direct access to the hardware of the pc-card in touch with 802.11b. This allows to run firmware based functions and thus directly supervise the handover process.

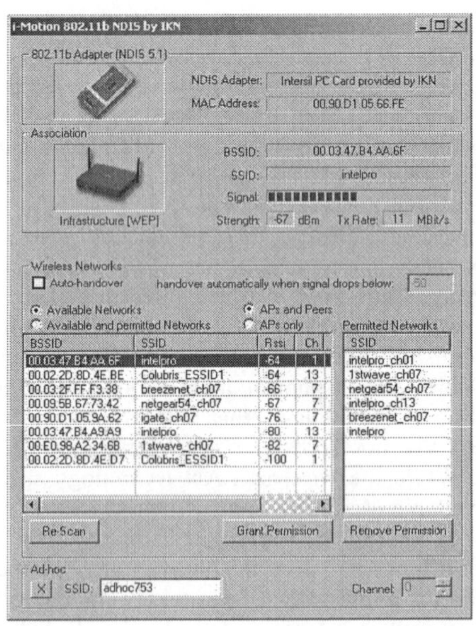

Fig. 3: Graphical User Interface of i-Motion.

There are three different mechanisms integrated in our handover-tool. Every t_{UPDATE} seconds (in our implementation this interval is set to 250ms) a signal-level measurement of the actual associated base-station is performed. In case that the level is below a configurable (-90 dBm is default) threshold, a hostscan will be activated automatically. This hostscan initiates an active scan and tells the firmware that the host desires a "non-destructive" sweep for all available access-points. The sequencing sweep is non-destructive in that the firmware will maintain its current association with the BSS while it performs the sweep. When the sweep is completed, the HostScanTable is returned to the host in a HostScanResults information packet. This information is used to detect alternative base-stations with better signaling-levels than the actual one. In case that the wireless card was not able to find any access-point the mobile node will switch to adhoc-mode defined in IEEE 802.11 standards. So clients that do not notice any AP, will build up an adhoc network in peer-to-peer mode and talk to each other in a direct way. To be able to leave the adhoc-mode, every t_{ADHOC} seconds another hostscan is performed, to see if infrastructure mode (AP in range) is possible again. The other possibility is that the PC-card finds immediately alternative APs with at least 15dB better signal-level. If this condition is met, the client will automatically start a handover-process and handoff to the AP with best signal-level. A more detailed description with all possible events, is presented in Fig. 4.

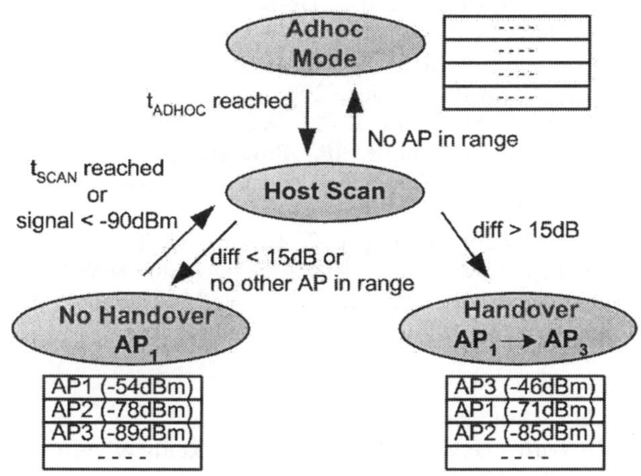

Fig. 4: Application Flow for Handover.

4. IMPLEMENTATION AND EVALUATION

The first evaluation step is to determine and validate the improved handover behavior. In normal case when there is no controlling background monitor installed at the client, the only way to detect lack of radio connectivity are failed frame transmissions. This will happen at the earliest if signal strength is about -100dBm. In contrast to this, i-Motion will promptly recognize that the power level drops below our pre-defined threshold (e.g. -90dBm) and will start searching for alternative base stations in range immediately. Results from such a client-controlled handover between two access-points can be seen in Fig. 5.

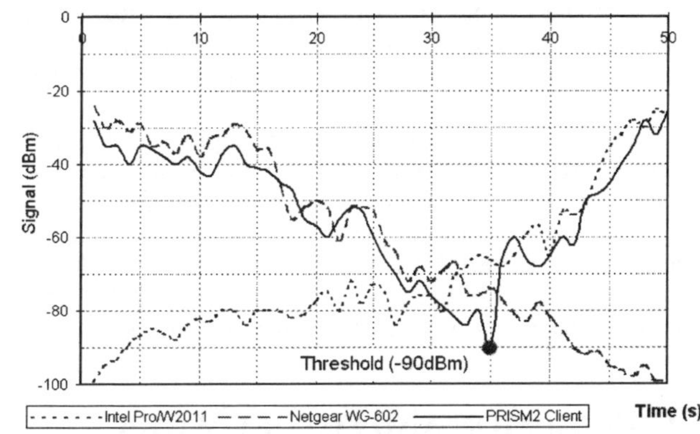

Fig. 5: Monitored Handover and AP Signal Strength at Client.

As presented in [2] the handover process can be split into three phases. The first phase, called detection, is the discovery of the need for the handover. The second phase, search, covers the acquisition of the information needed to perform the handover. Finally, the handover is performed during the third phase, execution. Detection is the longest phase (~800ms), while execution could be neglected. Additionally, detection and search times widely vary among different client card models [3]. Main problem during handover is that stations have to detect the lack of radio connectivity based on unsuccessful frame transmissions. The difficulty is to determine the reason for the failure among collision, radio signal fading or the station being out of range.

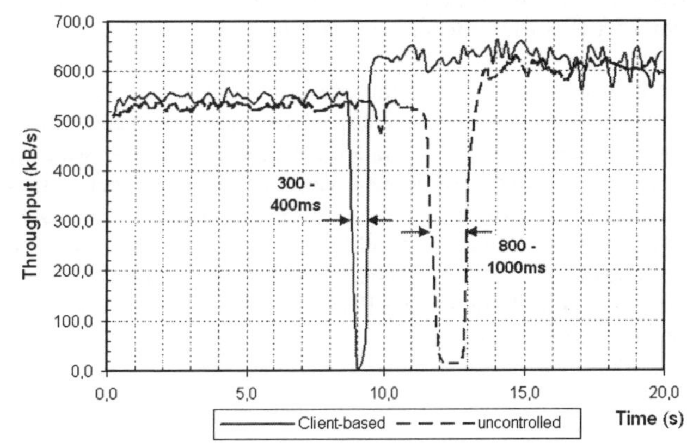

Fig. 6: Handover Latency in Comparision.

In our implementation the signal strength is monitored continuously. In case that the power level falls below a pre-defined threshold, the tool automatically tries to handoff to APs in range that provide much better connectivity. So the long phase of detection can be saved and handover is carried out much faster. Figure 6 shows comparison between handover from Intel Pro/Wireless 2011 to a Netgear WG-602 access point with and without the tool presented in this paper. At the client side a PRISM2-based PC-card was in use and Chariot from NetIQ[2] was taken for realtime networking measurements.

What can be seen is that our WLAN implementation takes advantage of the information provided by the physical layer and can completely skip the detection phase. Such stations start the search phase when the quality of radio signal degrades below the pre-defined threshold. In that case, the search starts before any frame has been lost. This has the favorable effect that overall handover-time can be reduced to about 350ms.

5. CONCLUSION AND FUTURE WORK

The primary contribution of this work is a detailed evaluation of our handover-application, which shows that it is possible to skip detection phase during link-layer handover in IEEE 802.11 networks. This is done by detecting the decrease of connection quality precociously. However in this state the available signal is strong enough and no packets are lost before initiating the deassociation process. All available wireless base stations are scanned in background all the time and so decisions can be made to associate with other access points in range. As our current release only works on PRISM2 based client cards, we now implement an improved version that will be based on Microsoft's NDIS 5.1 interface. Thus in future all IEEE 802.11a/b/g client cards with NDIS 5.1 compatible command-set can be controlled automatically during handoff. Another feature of the upcoming version will be handover between APs with same SSID by using Basic Service Set Identifier (BSSID) informations instead.

We believe that tools like the one presented in this paper will improve mobility in wireless LAN networks arbitrative. Handover-algorithms that implement prediction of the way the mobile node moves can be deployed to assist new mechanisms in wireless networks. In this way communication is enabled between MAC layer and others above. Examples of such mechanisms are "IEEE 802.1x pre-authentication", "Low Latency Handoffs in Mobile IPv4" [4, 5] and "Fast Handovers for Mobile IPv6" [6].

[2] http://www.netiq.com

ACKNOWLEDGEMENTS

Part of this work has been performed within the project "WISQY - Wireless InterSystem Quality-of-Service" at the Telecommunications Research Center Vienna (ftw.[3]) and has been funded in the framework of the Austrian Kplus Competence Center Programme. Also we would like to thank Thomas F. Divine from PCAUSA[4] and Kim Shorb from NetIQ for making development and evaluation possible.

REFERENCES

Proceedings Papers:

[1] A. Weyland, G. Stattenberger and T. Braun, Mobile-Controlled Handover in Wireless LANs, IEEE Workshop on Local and Metropolitan Area Networks 2002 (LANMAN 2002), Sweden, August 2002.

[2] Héctor Velayos and Gunnar Karlsson, Techniques to Reduce IEEE 802.11b MAC Layer Handover Time, KTH Technical Report, ISSN 1651-7717, Stockholm, Sweden. April 2003.

[3] Arunesh Mishra, Minho Shin and William Arbaugh, An Empirical Analysis of the IEEE 802.11 MAC Layer, Technical Report CS-TR-4395, University of Maryland, Department of Computer Science, September 2002.

[4] Mobile IPv4 Handoffs Design Team, Low Latency Handoffs in Mobile IPv4, Internet Draft, IETF Mobile IP Working Group, <draft-ietf-mobileip-lowlatency-handoffs-v4-05.txt>, June 2003.

[5] C. E. Perkins and K.-Y. Wang, Optimized Smooth Handoffs in Mobile IP, Proceedings of the IEEE Symposium on Computers and Communications, Egypt, July 1999.

[6] G. Dommety, A. Yegin, C. Perkins, G. Tsirtsis, K. El-Malki and M. Khalil, Fast Handovers for Mobile IPv6, Internet-Draft, IETF Mobile IP Working Group, <draft-ietf-mobileip-fast-mipv6-06.txt >, March 2003.

Books:

[7] Gast, Matthew S., *802.11 Wireless Networks, The Definitive Guide*. Sebastopol, CA: O'Reilly & Associates, Inc., April 2002.

[8] Intersil Corporation, *PRISM Driver Programmer's Manual*, Development Information, Version 2.10, August 2001, (For Distribution Under NDA Only).

[3] http://www.ftw.at

[4] http://www.rawether.net

Study of IEEE 802.11 Fairness and its Interaction with Routing Mechanism

Zhifei Li, Sukumar Nandi, and Anil K. Gupta
School of Computer Engineering, Nanyang Technological University, Singapore-639798
Email: {pg03802331, assukumar, asgupta}@ntu.edu.sg

Abstract: The MAC protocol IEEE 802.11, used in the mobile ad hoc networks (MANETs), suffers from the well-known fairness problem. In this paper, using static routing, we first study the MAC fairness by considering three factors including the hidden-terminal, capture, and imprecise EIFS, which are all related to the signal attenuation property of the wireless networks. Since the MAC protocol and the routing protocol interact with each other, we then study the interaction between MAC fairness and routing mechanism by replacing static routing with AODV and DSDV. From the two studies, we have learnt many useful lessons.

Key Words: IEEE 802.11, Fairness, Carrier Sensing, EIFS

1. INTRODUCTION

IEEE 802.11 [3] has been the most popular MAC protocol adopted in the research-work on the Mobile Ad-hoc Networks (MANETs). The Distributed Coordination Function (DCF) defined in the IEEE 802.11 is a CSMA/CA based protocol, which has two main components: Carrier Sensing (CS) and Collision Avoidance (CA). Since CS and CA work in a distributed manner without having precise information, they cause unfairness as pointed out in [4,5]. While [5] mainly focuses on the unfairness caused by the CA algorithms, [4] points out that the imprecise EIFS value, which is related to the CS, is also a major reason for unfairness. In [4], an enhanced CS (ECS) is proposed, which solves the imprecise EIFS problem largely and thus improves the fairness substantially. In addition to CA algorithms and imprecise EIFS problem, the hidden-terminal problem and signal capture, which are the two most important issues in wireless networks, also affect the MAC fairness. Therefore, it is very important to simultaneously consider the effects of hidden-terminal, capture, and imprecise EIFS in studying the MAC fairness. Moreover, since the MAC protocol and the routing protocol interact with each other, which in turn affect the overall performance of an MANET, it is very important to characterize the interaction between the MAC fairness and the routing mechanism. However, to the best of our knowledge, no such work has been done. Bearing in mind these observations, we carry out our research work with the following two objectives: *(i)* using static routing, we illustrate how hidden-terminal problem, capture, and imprecise EIFS problem affect the fairness of IEEE 802.11; *(ii)* replacing static routing with Ad-hoc On-demand Distance Vector (AODV) and Destination-Sequenced Distance-Vector (DSDV), we study the interaction between MAC fairness and routing mechanisms.

The rest of the paper is organized as follows. In Section 2, we present the factors affecting MAC fairness. In Section 3, we study the MAC fairness. The interaction between MAC fairness and routing mechanism is studied in Section 4. The paper is concluded in Section 5.

2. FACTORS AFFECTING MAC FAIRNESS

Now we discuss three factors that affect the MAC fairness.

2.1 Capture Effect

Based on the attenuation of the transmission power, two ranges are defined: the transmission range (TR) and the sensing range (SR). Normally, the SR is greater than the TR. Correspondingly, we call a frame detected by a node within TR as the TR frame, and a frame detected by a node out of TR but within SR as the SR frame. If there is no collision or any other error, a node can receive a TR frame correctly. In contrast, a SR frame can be detected by the carrier sensing but it cannot be received correctly. If capture is not allowed, a collision occurs whenever there are multiple frames around a receiving node and all the frames will be destroyed. On the contrary, if capture is allowed, one of the multiple frames may be captured (i.e., received correctly) if both the following conditions are satisfied: *(i)* the Signal to Noise Ratio (SNR) of the captured frame is greater than a threshold; *(ii)* the frame satisfying the first condition arrives at the receiving node earlier than any other frames.

2.2 Hidden Terminal Problem

In addition to the capture issue, there exists the well-known hidden-terminal problem in wireless networks. Specifically, when two nodes are out of the SR range of each other and they simultaneously try to transmit to nodes, which are in the overlapping area of the two transmitting nodes'SR, the two transmitting nodes are hidden from each other. Note that our definition of hidden-terminal problem is different from the traditional one because we assume that the SR is greater than TR. The example shown in Figure 1 illustrates this problem. There are two single-hop flows: flow from A to B and flow from D to C. The distance between any two neighboring nodes is about one hop. In other words, two neighboring nodes are within the TR of each other. The SR is assumed to be about two times of TR. Therefore, nodes A and D are hidden from each other and they may initiate transmission simultaneously, resulting in collision or capture.

Figure 1: 4-nodes with Two Single-hop Flows

2.3 Imprecise EIFS Problem

To combat with the hidden-terminal problem, IEEE 802.11 defines a four-way handshaking technique, where a sequence of Request To Send (RTS), Clear To Send (CTS), Data, and Acknowledgement (ACK) frames, is transmitted for a data packet. For the convenience, we call the exchange of RTS/CTS/Data/ACK frames as a frame exchange sequence (FES). FES (A, B) represents a FES between nodes A and B, initiated by node A.

In the IEEE 802.11, whenever a node detects an SR frame, which is treated as an error, this node defers its transmission by a duration indicated by the Extended Inter Frame Space (EIFS) constant. The idea, that a node defers its transmission with EIFS when the node gets an SR frame, is to allow the next frame in a FES between two other nodes to be completed. Since the next frame can be of any type (i.e., CTS, Data, or ACK), the transmission time of this frame may differ substantially. Therefore, corresponding to different SR frames received, the EIFS value should be different. However, the IEEE 802.11 does not distinguish among different SR frames and uses the same constant EIFS value in all the cases, which results in *large-* and *small-EIFS* problems and thus exhibits unfairness. For convenience, we refer to these two problems as the *imprecise EIFS* problem.

To illustrate the large-EIFS problem, let us consider the scenario in Figure 1 and a situation where nodes C and D are deferring their transmission, while FES (A, B) is in progress (see Figure 2). After node B sends a CTS to A, since node D is in the sensing range of node B, node D gets a SR frame and defers its transmission by EIFS. Then, node A begins to send its Data frame to B. However, since node D is out of the SR of A, node D cannot hear this frame. Moreover, since the EIFS is equal to SIFS + TxTime (ACK) + DIFS [3] which is much smaller than the transmission time of node A's Data frame, after deferring for the EIFS duration, node D starts to contend for the medium, and may send out a frame even though node A is still transmitting the Data frame to B. The frame sent by D is definitely discarded by C. We call this small-EIFS problem because the EIFS value is smaller than it should be to reflect the state of the medium. In this situation, the EIFS value should be large enough to allow for the complete transmission of the A's Data frame.

Now let us consider that node B sends an ACK back to node A. Corresponding to this ACK, node D gets an SR frame and defers its transmission with EIFS. We can see that, after the completion of the ACK, before node D backs-off by a random period, it has to defer by EIFS duration rather than by DIFS as done by other nodes (e.g., nodes A and B). We call this large-EIFS problem because the EIFS value is larger than it should be to reflect the medium state. In fact, the EIFS value should be equal to DIFS in this case.

In [4], we have shown that the imprecise-EIFS problem results in substantial unfairness. To solve the problem, Enhanced Carrier Sensing (ECS) has been proposed [4]. In ECS, whenever a node gets a SR frame, the node should identify the type of this SR frame based on its length, which is made different for different frames. Then the node adopts different EIFS values correspondingly, rather than adopting a fixed EIFS value.

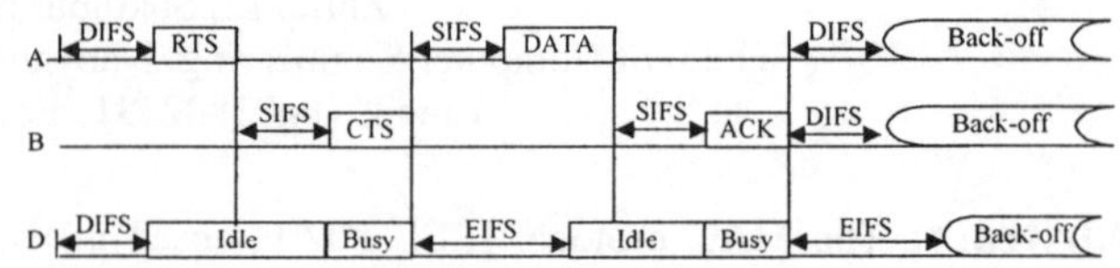

Figure 2: Time Diagram when FES(A,B) is in Progress while node D defers

3. IEEE 802.11 FAIRNESS UNDER STATIC ROUTING

Since the example of Figure 1 incorporates all the factors under consideration, we conduct an extensive study of this topology. Specifically, under a given routing protocol (e.g., static routing), we conduct simulations under four different scenarios, called S1 to S4, which are described in Table 1.

Table 1: Four Different Scenarios

Scenario	IEEE 802.11/ECS	Allow capture
S1	IEEE 802.11	no
S2	IEEE 802.11	yes
S3	ECS	no
S4	ECS	yes

The simulation environment is as follows. NS-2 with CMU wireless extensions [1] is used. For each single-hop flow, a Constant Bit Rate (CBR) traffic generates 200 packets per second. Each packet is 1000-bytes long, resulting in a traffic source rate of 1.6 Mbps. Due to the MAC overhead, the maximum throughput using a raw bandwidth of 2 Mbps is about 1.4 Mbps. Mobility and transmission errors are not explicitly considered in the simulation.

3.1 Throughput Comparison

Table 2 presents the average throughputs under the four scenarios. From the table, we notice that the throughput of each flow under static-S1 is much smaller than that under the remaining three scenarios, while the throughputs under the other three scenarios are very close to each other.

Table 2: Average Throughputs under Static Routing

Scenario	Flow	Average Throughput(M bps)
static-S1	A to B	0.314
	D to C	0.307
static-S2	A to B	0.683
	D to C	0.641
static-S3	A to B	0.662
	D to C	0.672
static-S4	A to B	0.665
	D to C	0.670

Let us first analyze the average throughputs under static-S1. When a node (say node A) is transmitting a Data frame, as node D suffers from the small-EIFS problem if the IEEE 802.11 is used, node D may transmit its RTS. Since the capture is not allowed in this scenario, node B will drop the Data frame, resulting in wastage of bandwidth. This explains why the throughputs under static-S1 degrade a lot.

The dropping of the Data frame at node B can be avoided in two ways. One is to prevent node D from sending out the RTS when node A is transmitting its Data frame. The ECS achieves this by preventing the small-EIFS problem and thus the throughputs under S3 and S4 do not degrade. Another way to avoid frame dropping is to allow the capture of the Data frames as being done in the static-S2.

In static-S4, the capture is used with ECS, but the throughput does not increase compared to that under static-S3. The reason is that in the S4, as explained later, the capture occurs only of the RTS frames and this kind of capture does not enhance the performance.

3.2 Jain's Index Comparison

The average throughput can only reflect the *long-term* fairness. In order to study fairness on a *short-term* basis, we need to make use of the popular Jain's fairness index [2]. The value (say F_J) of the Jain's index depends on the number (say w) of packets transmission time constituting an observation. Generally, the F_J value increases with w, and the complete fairness is achieved when $F_J = 1$.

Figure 3 presents the Jain's index with reference to w. The index has been averaged over all sliding windows of w packets, which occur in the simulation run. We notice that all the scenarios exhibit short-term unfairness to a certain extent. Since the common problem existing in all the four scenarios is the hidden-terminal problem, we can conclude that the hidden-terminal problem results in short-term unfairness. Interestingly, though the static-S1 attains the worst bandwidth utilization, it achieves the best fairness. This shows the conflict between optimizing throughput and achieving fairness. Compared to the static-S2, which shows the worst fairness, the ECS (S3 and S4) improves the fairness while the throughput does not degrade. Moreover, it seems that capture dose not play much role in the fairness under ECS as S3 and S4 almost show the same fairness.

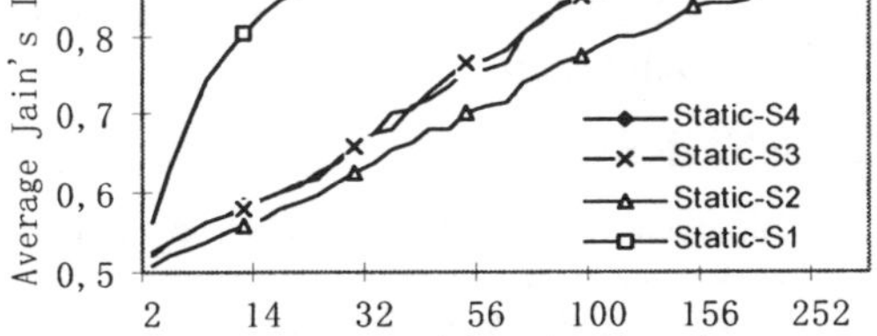

Figure 3: Fairness Comparison under Static Routing

Let us first consider the situations under which the frames sent by nodes A and D overlap. One such situation is, node D (or A) starts transmitting its RTS frame even though node A (or D) is transmitting a Data frame. This situation occurs whenever the imprecise EIFS problem exists. As analyzed in Section 3.1, when capture is allowed in such situations, the Data frame can be received correctly. Otherwise, both the frames are dropped. Another overlapping situation is that both nodes A and D transmit their RTSs as the two nodes are hidden from each other. Consider the situation that the RTS of A is captured by node B. However, after completely receiving the captured RTS, carrier sensing at node B indicates a busy medium since node D is still transmitting its RTS. This forbids node B from sending a CTS back to node A, and thus the four-way handshake initiated by node A fails even though its RTS is captured. Therefore, both the RTS frames are dropped irrespective of whether the capture is allowed. Note that, if a Data frame is captured, the four-way handshake can be successful because the node capturing the Data frame always sends back an ACK without considering whether the medium is busy [3].

Now we analyze the results of the Jain's index shown in Figure 3. In static-S1 and S2, the imprecise-EIFS problem leads to the overlapping between the Data and RTS frames. When the capture is not allowed (static-S1), both the frames are dropped. On the other hand, if capture is allowed (static-S2), the Data frame will be captured. Then, the node that sends the Data frame will get an ACK and reset its CW while the node that sends the RTS frame will retransmit its RTS in vain, leading to a larger CW. This obviously results in unfairness and explains why the static-S2 shows worse fairness than that under static-S1. In scenarios static-S3 and S4, the overlapping of frames occurs only when both nodes A and D transmit their RTSs, and the difference between these two scenarios is that the capture is not allowed in S3. However, a capture of a RTS does not lead to any advantage as explained above, and as a result, the two scenarios show similar fairness. Moreover, they show better fairness than that under static-S2 as the ECS solves the imprecise EIFS problem, which is a major reason of unfairness [4].

From the study, we deduce that: *(i)* hidden-terminal problem results in substantial short-term unfairness; *(ii)* capture, if effective, increases the capacity at the cost of fairness, showing that there is a conflict between achieving fairness and maximizing throughput. On the other hand, if the capture is not effective in the handshake, it does not have any effects on the fairness as well as the throughput; *(iii)* imprecise EIFS results in unfairness, and the ECS improves the fairness without degrading the throughput performance.

4. INTERACTION BETWEEN MAC FAIRNESS AND ROUTING MECHANISMS

To study the effects of the interaction, we replace static routing with AODV and DSDV, one at a time.

4.1 Throughput Comparison

Table 3 presents the average throughputs under different routing protocols. We notice that the average throughputs under DSDV are almost the same as those under the static routing. On the other hand, under the AODV-S1 to S3, the flow from node D to C completely starves while the other flow gets the entire bandwidth. This is because the AODV at node D cannot even find a route to node C as explained later. Note that the results presented here are obtained under the situation where the flow from A to B starts first. However, if we start the flow from D to C first, exactly opposite results are obtained, i.e., the flow from node A to B starves.

In order to explain the flow starving, we first describe the route discovery process of AODV. The AODV uses two kind of messages for the route discovery: Route Request (RREQ) and Route Reply (RREP). When a node, say node S, needs a route to some destination node, say node X, node S first broadcasts a RREQ and then waits for a RREP. When a node receives the RREQ and has a fresh enough route to node X, the node generates a RREP message and then unicasts it to the next hop toward the originator of the RREQ (i.e., node S). If a RREP is not received by the time RREQ_TIMEOUT expires, to find the route, node S will broadcast another RREQ, up to a maximum of RREQ_RETRIES times. The value of RREQ_TIMEOUT follows a binary exponential back-off mechanism.

Table 3: Throughputs under Different Routing Protocols

Scenario	Flow	Average Throughput(M bps)		
		Static	DSDV	AODV
S1	A to B	0.314	0.314	1.401
	D to C	0.307	0.308	0
S2	A to B	0.683	0.684	1.402
	D to C	0.641	0.669	0
S3	A to B	0.662	0.675	1.402
	D to C	0.672	0.659	0
S4	A to B	0.665	0.650	0.660
	D to C	0.670	0.685	0.678

Now we explain why the AODV's route discovery process at node D always fails in the first three scenarios. Figure 4 demonstrates when node D will determine the medium to be idle or busy while FES (A, B) is in progress. When the medium is determined to be idle, D's MAC layer decrements its back-off timer and transmits the RREQ when the timer becomes zero. We notice from the diagram that the RREQ can be successfully transmitted only if it is sent in a very small crucial interval of duration υ. At other instances, however, the carrier sensing at node D may indicate the medium to be idle though really it is not idle, and thus the transmission of the RREQ will result in a collision. This has been shown as *collision* in the figure. If we denote the sum by λ of all *idle* durations determined by node D within a *single* FES, then the ratio υ/λ reflects the probability that node D can successfully transmit the RREQ. It is easy to verify that the υ/λ is near to zero in three scenarios, AODV-S1 to AODV-S3. Therefore, whenever node D transmits the RREQ, it will always collide with a frame (either RTS or Data) sent by A. This explains why the AODV at node D cannot start in the first three scenarios. In fact, what we have analyzed above applies not only to the AODV, but also to other on-demand protocols (e.g., DSR).

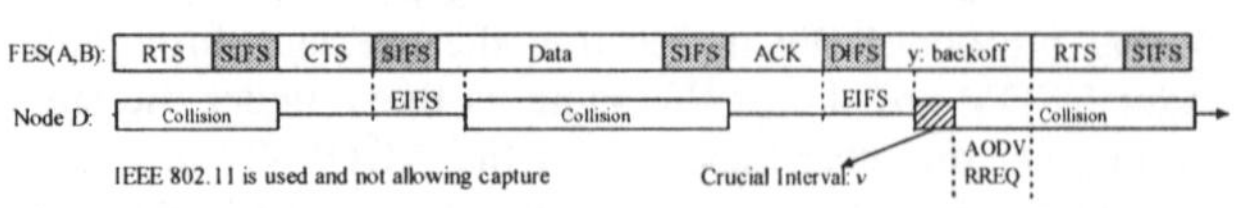

Figure 4: Medium State determined by Node D when FES (A, B) is in Progress

4.2 Jain's Index Comparison

Fairness for scenario S4 under all three routing protocols is compared in Figure 5. Since the DSDV is very similar to the static routing except that the DSDV periodically exchanges route messages, they show similar fairness. We notice that the AODV-S4 shows the worse fairness. The reason is as follows. Whenever the MAC layer at a node drops a packet after several unsuccessful attempts, the AODV views this as an indication of the link breakage, and thus will initiate a fresh route discovery process that may take a long time during which the other node gets the entire bandwidth of the medium. This situation may occur very frequently due to the hidden-terminal problem. On the contrary, the situation described above does not occur for the DSDV and static routing, explaining why the AODV shows worse fairness.

From the above discussion, we can deduce that: (i) the short-term MAC unfairness may lead to the stalling of the on-demand routing protocols (e.g., AODV and DSR); (ii) the routing mechanisms (e.g., route discovery and link breakage detection) also affect the system fairness.

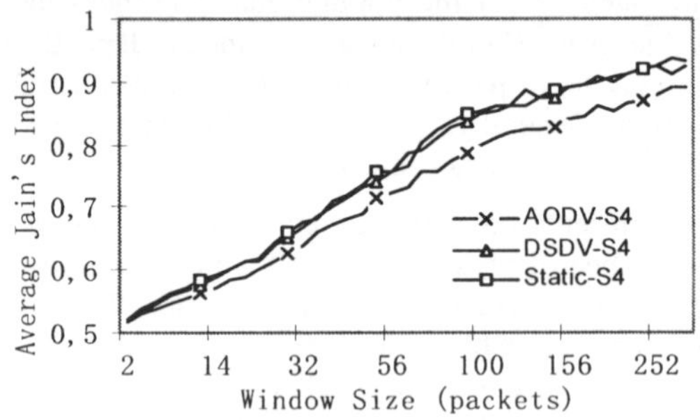

Figure 5: Fairness Comparison under S4

5. CONCLUSIONS

From this paper, the main lessons learnt are: (i) we should distinguish between the *MAC fairness* and the *system overall fairness*, since under identical MAC conditions different overall fairness performance may be obtained if different routing protocols are used; (ii) in studying the *MAC fairness*, we should isolate the factors that are not directly related to the MAC layer, and therefore, the study is to be carried out under the static routing, CBR backlogged traffic, static topology etc; (iii) to study the *overall system fairness*, in addition to the study of MAC fairness, we should also consider many other factors such as the routing mechanism, traffic pattern, mobility etc.

REFERENCES

[1] CMU Monarch Group. CMU Monarch Extensions to NS, http://www.monarch.cs.cmu.edu/.

[2] Chiu, D.-M., Jain, R. "Analysis of the Increase and Decrease Algorithms for Congestion Avoidance in Computer Networks.," Computer Networks and ISDN Systems 17 (1989),1-14.

[3] IEEE, "Wireless LAN Medium Access Control (MAC) and Physical Layer (PHY) specifications," IEEE 802.11 standards, June 1999.

[4] Zhifei Li, "Improving Performance of MAC protocols in Multi-hop Wireless Ad Hoc Networks", First year report for confirmation of Ph.D. candidate, SCE, Nanyang Technological University, June 2003.

[5] T. Nandagopal, T. Kim, X. Gao, V. Bharghavan, "Achieving MAC Layer Fairness in Wireless Packet Networks," ACM MOBICOM, 2000.

SESSION 2B:
SECURITY

MUTUAL AUTHENTICATION AND GROUP KEY AGREEMENT FOR LOW-POWER MOBILE DEVICES

EMMANUEL BRESSON

*Cryptology department – CELAR, 35170 Bruz, France *– Emmanuel.Bresson@polytechnique.org*

OLIVIER CHEVASSUT AND ABDELILAH ESSIARI

Lawrence Berkeley National Lab. † – Berkeley, CA, USA – {OChevassut,AEssiari}@lbl.gov

DAVID POINTCHEVAL

École normale supérieure – Paris, France – David.Pointcheval@ens.fr

Wireless networking has the power to fit the Internet with wings, however, it will not take off until the security technological hurdles have been overcome. In this paper we propose a *very efficient* and *provably-secure* group key agreement well suited for unbalanced networks consisting of devices with strict power consumption restrictions and wireless gateways with less stringent restrictions. Our method meets practicability, simplicity, and strong notions of security.

1 Introduction

Wireless technology has become more pervasive as Internet electronic and commerce transactions on mobile devices have gained in popularity. Institutions and industries are hankering for the power and flexibility of wireless networks, but many are postponing rollouts in strategic areas until they are convinced that their systems are not at risk. The security technologies currently deployed to protect the Internet against attacks are not fully applicable to the wireless Internet since the traditional Internet does not typically place constraints on available power consumption or bandwidth. The nodes in a wireless network are usually mobile and have computation bandwidth capabilities that place severe restrictions when designing cryptographic protocols. Storage limitation on the other hand is becoming less of an issue as many add-on memory cards are now widely available.

In the present paper we have focused on computing applications involving clusters of mobile devices [3,9,10]. The Wired Equivalent Privacy (WEP) protocol, which is part of the IEEE 802.11 standard, specifies how to protect the traffic between mobile devices and access points (i.e. gateways) using *pre-established* session keys without specifying how the keys are established. This lack of proper key-establiment scheme has opened the door to various attacks [4]. Our contribution in this paper is a *provably-secure* authenticated group key-exchange scheme based on public-key cryptography that can complement the WEP protocol. Schemes based on symmetric cryptography have obvious performance advantages over public-key cryptography, but suffer from a complex key management; they require trust in the entire network as a device moves from one domain to another. Other schemes based on public-key technology trade less computation for more communication rounds, but are still too costly to be practical for wireless networks that involve low-power computing devices [1,2,5,6].

Our key-exchange scheme allows a cluster of mobiles and one wireless gateway to dynamically agree on a session key. It shifts the computational burden to the gateway and provides mobile devices with the ability to perform most of the public-key cryptographic computations off-line. This scheme can furthermore be combined with a group Diffie-Hellman key exchange scheme [5] to cover wireless environments involving more than one gateway [8]. A mobile device would only perform one public-cryptographic computation as it moves from one wireless domain to the other.

The paper is organized as follows. In Section 3 we present a scheme that achieves "implicit" authentication and in Section 5 we discuss enhancements to achieve "explicit" authentication.

*Work done while at ENS, Paris.

†The authors are supported by the Director, Office of Science, Office of Advanced Scientific Computing Research, Mathematical Information and Computing Sciences Division, of the U.S. Department of Energy under Contract No. DE-AC03-76SF00098. This document is report LBNL-52542. Disclaimer available at http://www-library.lbl.gov/disclaimer.

2 Modeling Unbalanced Wireless Networks

Wireless Nodes. The wireless-communication system we model is a set $\mathcal{C}$, of N wireless-capable mobile devices (also called *clients*), and a wireless gateway (also called *server* or base station). We assume the clients and the server do not deviate from the algorithm they are expected to execute. We consider a nonempty subset of $\mathcal{C}$ which we call the *wireless client group* $\mathcal{G}_c$ that consists of the clients communicating with the server. The server S has the special role of adding and removing clients from the group $\mathcal{G}_c$. (In practice, this server covers an entire wireless region called a cell or domain and, thus, it never leaves, hence its special role.) Each mobile U, as well as the base station, also holds a long-lived key LL_U which is a pair of matching public/private keys.

Abstract Interface. We define the basic structure of a group key agreement scheme for unbalanced wireless networks. The scheme GKE consists of four algorithms:

- The *key generation algorithm* $\mathsf{GKE.KGen}(1^\ell)$ is a probabilistic algorithm which on input a security parameter 1^ℓ, provides each client U_i in $\mathcal{C}$ and the base station with a long-lived key.
- The *setup algorithm* $\mathsf{GKE.Setup}(\mathcal{J})$ is an interactive protocol which on input of a set of clients $\mathcal{J}$, sets the wireless client group to be $\mathcal{G}_c = \mathcal{J}$ and provides each client U in $\mathcal{G}_c$ with a secret value *sk* shared with the base station.
- The *join algorithm* $\mathsf{GKE.Join}(\mathcal{J})$ is an interactive protocol which on input of a set of clients $\mathcal{J}$, updates the wireless client group $\mathcal{G}_c$ to be $\mathcal{G}_c \cup \mathcal{J}$, and provides each client U in $\mathcal{G}_c$ with a new shared secret value *sk*.
- The *remove algorithm* $\mathsf{GKE.Remove}(\mathcal{J})$ is an interactive protocol which on input of a subset $\mathcal{J}$ of the wireless client group $\mathcal{G}_c$, updates the latter to be $\mathcal{G}_c \backslash \mathcal{J}$, and provides each client U in $\mathcal{G}_c$ with a new shared secret value *sk*.

3 Key Agreement

This section provides a method accommodating group key agreement to mobiles lacking the computational resources to perform multiple on-line computation in a finite cyclic group (such as modular exponentiation), but with enough computational resources to perform symmetric-cryptographic operations. The session-key space SK associated to this method is $\{0,1\}^\ell$ equipped with a uniform distribution, where ℓ is a security parameter. Arithmetic is in a finite cyclic group $\mathbb{G} = \langle g \rangle$ of ℓ-bit prime order q. Both g and q are publicly known. There are also three hash functions $\mathcal{H} : \{0,1\}^\star \rightarrow \{0,1\}^\ell$, $\mathcal{H}_0 : \{0,1\}^\star \rightarrow \{0,1\}^{\ell_0}$, where ℓ_0 needs not be equal to ℓ, and $\mathcal{H}_1 : \{0,1\}^{\ell_1} \times \mathbb{G} \rightarrow \{0,1\}^{\ell_0}$, where ℓ_1 is the maximal bit-length of a counter c used to prevent replay attacks.

We consider a signature scheme $\mathsf{SIGN} = (\mathsf{SIGN.KGen}, \mathsf{SIGN.Sig}, \mathsf{SIGN.Ver})$. Each client U_i holds a pair of signing private/public key (SK_i, PK_i) which are the output of the key generation signature scheme algorithm $\mathsf{SIGN.KGen}$. One would probably argue that when dealing with low-power computing mobiles, special low-cost [11] or on-line/off-line [12] signature schemes are required. However, the messages to be signed are in our setting known at pre-computation time and, thus, a mobile does not have to compute anything on-line to send its contribution.

3.1 Algorithms

Key Generation. The algorithm $\mathsf{GKE.KGen}$, on input the set of clients $\mathcal{C}$ and a security parameter ℓ, performs the following steps:

1. Run $\mathsf{SIGN.KGen}(1^\ell)$ for each client U_i in $\mathcal{C}$ to provide each client with a pair (SK_i, PK_i) of signing/verifying keys;
2. Choose $x \in_R \mathbb{Z}_q^\star$ and set the Server's private/public keys to be: $(SK_S, PK_S) = (x, g^x)$. One denotes $y = g^x$.

Setup. The algorithm $\mathsf{GKE.Setup}$, on input a set of client-devices $\mathcal{J}$, performs the following steps (see also Figure 1):

1. Set the wireless client group $\mathcal{G}_c$ to be the input set $\mathcal{J}$.
2. Each client $U_i \in \mathcal{G}_c$ chooses at random a value $x_i \in \mathbb{Z}_q$ and *precomputes* $y_i = g^{x_i}$, $\alpha_i = PK_S^{x_i} = y^{x_i}$ as well as a signature σ_i of y_i, under the private key SK_i.
3. Each client U_i sends (y_i, σ_i) to S.
4. For each $i \in \mathcal{G}_c$, the server S checks the signature σ_i using PK_i, and if they are all correct, computes the values $\alpha_i = y_i^x$.
5. The server S initializes the counter $\mathsf{c} = 0$, as a bit-string of length ℓ_1 and computes the shared secret value:
$$\mathrm{K} = \mathcal{H}_0(\mathsf{c} \| \{\alpha_i\}_{i \in \mathcal{G}_c})$$
and sends to each client U_i the values c and $K_i = \mathrm{K} \oplus \mathcal{H}_1(\mathsf{c} \| \alpha_i)$.
6. Each client U_i (and S) recovers the shared secret value K and the session key *sk* as described below, and accepts:
$$\mathrm{K} = K_i \oplus \mathcal{H}_1(\mathsf{c} \| \alpha_i) \quad \text{and} \quad sk = \mathcal{H}(\mathrm{K} \| \mathcal{G}_c \| S).$$

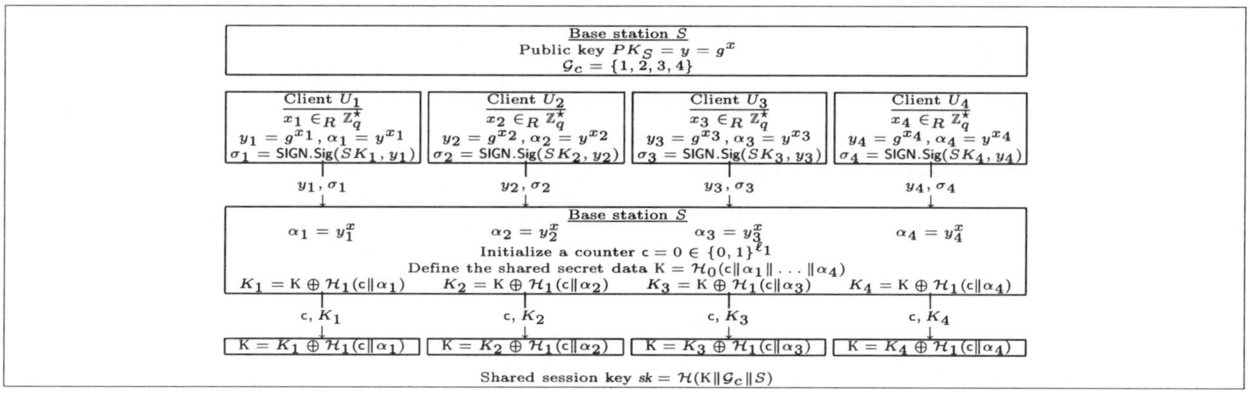

Figure 1. An execution of the Setup algorithm with the five devices U_1, U_2, U_3, U_4 and S.

Remove. The algorithm GKE.Remove, on input the set $\mathcal{J}$ of disappearing client-devices, performs the following steps :

1. Update the wireless client group $\mathcal{G}_c = \mathcal{G}_c \backslash \mathcal{J}$.
2. The server S operates as in the Setup phase. It increases the counter c and computes the shared secret value $K = \mathcal{H}_0(c\|\{\alpha_i\}_{i \in \mathcal{G}_c})$.
3. Then it sends to each client $U_i \in \mathcal{G}_c$ the values c and $K_i = K \oplus \mathcal{H}_1(c\|\alpha_i)$.
4. Each client $U_i \in \mathcal{G}_c$ already holds the value $\alpha_i = g^{xx_i}$, and the old counter value. So it first checks that the new counter is greater than the old one, and simply recovers the secret shared value K and the session key *sk* as described below, and accepts:

$$K = K_i \oplus \mathcal{H}_1(c\|\alpha_i) \quad \text{and} \quad sk = \mathcal{H}(K\|\mathcal{G}_c\|S).$$

Join. The algorithm GKE.Join, on input the set of appearing clients $\mathcal{J}$, performs similar steps to the GKE.Setup.

A complete description of the protocols can be found in the full version of this paper [7].

3.2 Efficiency

The protocol presented in this paper is very efficient, since almost everything can be precomputed off-line for the clients, while achieving a strong level of security. The amount of memory available on the clients may provide a trade-off:

- by storing many distinct triples $(y_i, \sigma_i, \alpha_i)$ one increases the security level, but one hashing and one XOR have to be performed on-line;
- by storing many $\mathcal{H}_1(c\|\alpha_i)$, for each $(y_i, \sigma_i, \alpha_i)$, for several values of the counter, one increases efficiency, since only one XOR has to be performed on-line.

4 Security Analysis

4.1 Assumptions

The security analysis of our protocol is based on the security of the underlying signature scheme that is used to authenticate messages, and on the computational Diffie-Hellman problem.

The security of the signature scheme is measured as the probability that an adversary, having access to an oracle that can sign arbitrary messages (chosen-message attack, or cma), produces a new, valid *(message, signature)* pair. Such a probability is denoted $\mathsf{Succ}^{\mathsf{cma}}_{\mathsf{SIGN}}(t, Q)$ where t is the adversary's working time and Q is the number of signing oracle queries.

The intractability of the Computational Diffie-Hellman (CDH) problem in the group $\mathbb{G}$ is measured as the probability that an attacker, on input two random elements g^a and g^b in $\mathbb{G}$, can compute g^{ab}. Such a probability is denoted $\mathsf{Succ}^{\mathsf{cdh}}_{\mathbb{G}}(t)$, where t is the adversary's working time.

4.2 Our result

The security of our protocol is measured as the probability that an adversary can get some (partial) information on the key. This probability is denoted $\mathsf{Adv}^{\mathsf{ake}}_P$ and depends on the number of messages sent on the network.

Security Theorem. *Let $\mathcal{A}$ be an adversary against the Authenticated Key Exchange (AKE) security of our protocol P, making at most q_s active requests, and asking at most q_H queries to the hash oracles ($\mathcal{H}_0$ and $\mathcal{H}_1$). Let N denote the total number of low-power devices. Then we have:*

$$\mathsf{Adv}^{\mathsf{ake}}_P(\mathcal{A}) \le 2N \cdot \mathsf{Succ}^{\mathsf{cma}}_{\mathsf{SIGN}}(t, q_s) + 2q_s q_H \cdot \mathsf{Succ}^{\mathsf{cdh}}_{\mathbb{G}}(t).$$

The above theorem shows that the security of our protocol is based on the intractability of the well-studied computational Diffie-Hellman problem (CDH) and on the security of the signature scheme

(CMA) to prevent existential forgeries under adaptive chosen message attacks.

Sketch of Proof. Unless the adversary breaks the signature scheme for one among the N clients (so, $N \cdot \mathsf{Succ}^{\mathsf{cma}}_{\mathsf{SIGN}}(\cdot)$), any valid message is output by a regular node, and can be simulated. We then note that the simulated flows can be derived from a CDH problem (g^a, g^b), while the hash values that would involve g^{ab} are answered randomly. The adversary breaks the scheme if it is able to discover which hash value corresponds to a given flow (so, $q_s q_H \cdot \mathsf{Succ}^{\mathsf{cdh}}_{\mathbb{G}}(\cdot)$). The full proof can be found in the full version [7].

5 Mutual Authentication and (Partial) Forward Secrecy

Mutual authentication ensures each player that all other parties did actually compute the same key. Our protocol can be modified to achieve this goal. The natural modification, wherein each player sends to all the other ones an "authenticator", requires that each low-power device computes N hashings and sends one flow to the server S. This computational overhead is tolerable only if N does not get too large, but for larger values of N this overhead can also be kept to a minimum by performing mutual authentication through the server. Each client authenticates to the server which then in turn authenticates to each client only after all clients have been authenticated. This approach has the attractive advantage of being not only provably secure, in the random-oracle model, but to also add little overhead to the original protocol.

About forward-secrecy, it is clear that as soon as the long-term key x of the server is leaked, all the session keys can be recovered, since all the α_i can easily be computed from the y_i and x. Therefore, no forward-secrecy exists when the server long-term key is revealed. However, the long-term keys of the low-power devices (the signing keys) are used for implicit authentication only, and not for hiding the session key. Therefore, the leakage of clients' long-term keys do not reveal anything about previous session keys. Furthermore, strong (partial) forward-secrecy (where any internal data is revealed in case of corruption, *i.e.* the signing key, but also the x_i, y_i and α_i) is also achieved if the x_i's and α_i's are erased as soon as they are not useful (the client has left from the group). As a consequence, no information about previous session keys can be found in the memory of the low-power devices. We claim these considerations make sense since the server can be reasonably seen as more reliable that the mobile devices.

6 Conclusion

In this paper we presented an efficient key agreement protocol for heterogeneous wireless networks. Our protocol allows a set of heterogeneous mobiles devices to form a secure group and to handle the continuous disappearing and reappearing of mobiles due to communication failures. Our protocol has been proved secure in the random oracle model under the computational Diffie-Hellman assumption.

References

1. N. Asokan and P. Ginzboorg. Key Agreement in Ad-hoc Networks, February 2000. Expanded version of a talk given at the Nordsec '99 workshop.
2. G. Ateniese, M. Steiner, and G. Tsudik. New Multiparty Authentication Services and Key Agreement Protocols. *IEEE J. of Selected Areas in Communications*, Apr. 2000.
3. K. Berket, P. M. Melliar-Smith, and L. E. Moser. The InterGroup Protocols: Scalable Group Communication for the Internet. *3rd Global Internet Mini-Conference*, Nov. 1998.
4. N. Borisov, I. Goldberg, and D. Wagner. Intercepting Mobile Communications: The Insecurity of 802.11. In *ACM MobiCom'01*, 2001.
5. E. Bresson, O. Chevassut, and D. Pointcheval. Dynamic Group Diffie-Hellman Key Exchange under Standard Assumptions. In *Eurocrypt '02*, LNCS 2332, pp. 321–336. Springer, May 2002.
6. E. Bresson, O. Chevassut, and D. Pointcheval. Group Diffie-Hellman Key Exchange Secure Against Dictionary Attacks. In *Asiacrypt '02*, LNCS 2501, pp. 497-514. Springer, Dec 2002.
7. E. Bresson, O. Chevassut, A. Essiari, and D. Pointcheval. Mutual Authentication and Group Key Agreement for Low-Power Mobile Devices. In *5th IEEE MWCN*. IEEE, Oct 2003. Full version available from http://www.di.ens.fr/users/pointche.
8. G. D. Crescenzo and O. Kornievskaia. Efficient Kerberized Multicast in a Practical Distributed Setting. In *Information Security Conference (ISC'01)*. Oct 2001
9. A. Harbitter and D. A. Menace. The Performance of Public Key-Enabled Kerberos Authentication in Mobile Computing Applications. In *ACM CCS'01*, pp. 78–85.
10. T. Phan, L. Huang, and C. Dulan. Challenge: Integrating Mobile Wireless Devices Into the Computational Grid. In *MobiCom '02*, pp. 271–278, 2002.
11. G. Poupard and J. Stern. Security Analysis of a Practical "on the fly" Authentification and Signature Generation. In *Eurocrypt '98*, LNCS 1403, pp. 422–436. Springer, May 1998.
12. A. Shamir and Y. Tauman. Improved Online/Offline Signature Schemes. In *Crypto '01*, LNCS 2139, pp. 355–367. Springer, Aug 2001.

Light TLS: Extended Handshake for Wireless Communications

Mohamad BADRA, Ahmed SERHROUCHNI
Ecole Nationale Supérieure de Télécommunication
46, Rue Barrault, Paris, 75634 cedex 13, France
{badra, serhrouchni}@enst.fr

Abstract: Wireless network architectures are subject to several constrains, namely scarce radio resources, high error rates and limited computational power on client side. Available security protocols suffer from such deficiencies and should be specially tailored in order to adapt to these environments. In this paper, we suggest a new protocol based on TLS which enables both the mobile clients (WAP, GPRS, IEEE 802.11, etc.) and the server to mutually authenticate each others and securely negotiate cryptographic secrets. In fact, it efficiently optimizes the computational time needed to open secure sessions for mobile clients

Key Words: Security, TLS, Wireless communications.

1. INTRODUCTION

The increase of data exchange between wired and wireless networks during the past few years opened a door for the development of security protocols to be used in a wide range of applications, namely e-commerce and bank transactions. The major constraints in wireless environments lie in the scarce radio resources, the high error rates and the limited computational power of mobile devices. This should be taken into consideration whenever we are to implement security protocols for such environments.

In this paper, we focus on the security of data exchange between a mobile client (e.g. WAP, GPRS) and the web server. Several protocols has been proposed by IETF security and WAP Forum groups; especially TLS [2] and WTLS [7] (Wireless TLS).

TLS is the most deployed security protocol, due to its native integration in browsers and web servers. In fact, it is used to secure the exchanges between a client and a server. This protocol is situated between the application and transport layers and provides connection security that has the following basic properties: peer entity authentication, data integrity, data confidentiality and security parameter negotiation.

The majority of wireless security protocols; especially WTLS are not directly compatible with TLS. This incompatibility urges the need for a gateway to translate between WTLS and TLS messages and vice versa. traditional gateways do not have the capability of providing end-to-end security between the WAP terminal and the application server. As a result, the messages transmitted between clients and servers are unencrypted, for a brief moment, inside the gateway.

To provide an end-to-end secure connection, we propose a new *Handshake* protocol based on TLS, called Light TLS that solves the wireless security problem and prohibits any gateway or middlemen from having plain text messages.

In section 2, we describe TLS problems in wireless environments. Next, we present our protocol in section 3, which we analyze and discuss in section 4. Finally, we conclude with a synthesis of the approach.

2. TLS IN WIRELESS ENVIRONMENTS

TLS is a transaction security standard that provides connection security with the following basic properties: peer entity authentication, data confidentiality, data integrity, key generation and distribution, and security parameter negotiation. This protocol consists of several sub-protocols; specially the *Record* and the *Handshake* protocols. The *Record* protocol provides basic connection security for various higher layer protocols through encapsulation. The *Handshake* protocol is used to allow peers to agree upon security parameters for the *Record* layer, authenticate themselves, instantiate negotiated security parameters, and report error conditions to each other. This protocol is used to negotiate the secure attributes of a session [2]. Once a transport connection is authenticated and a secret shared key is established with the TLS *Handshake* protocol, data exchanged by application protocols can be protected with cryptographic methods by the *Record* layer using the keying material derived from the shared secret.

The TLS *Handshake* involves the exchange of certificates and cryptographic information to allow peers to be authenticated. This step requires several operations: First, peers must verify the integrity of certificates and should generally support certificate revocation messages. Also, the certificate should always be verified to ensure it is trusted and signed by a known Certification Authority (CA). Finally, the client should be able to view the information about the certificate and root CA. For this, the use of TLS in wireless applications introduces several problems, to name a few:

- The certificate format and the attached certificate chain are X.509 [1] compliant and hence are encoded using an ASN.1 syntax in a binary DER format. The client needs to understand such formats imperatively in order to understand server certificate.

- The server certificate status needs to be checked against its issuer's CRL [6], which is a list of revoked certificates that is updated regularly by the certification authority. A CA often revokes certificates if their corresponding private key is compromised. CRL files are often heavy in size, and having to download them to check a single certificate imposes an undesirable constraint in wireless networks where the scare bandwidth is the bottleneck.
- TLS itself does not merge seamlessly with other wireless security protocols such as WTLS. In WAP architectures [7], WTLS messages must be converted into TLS messages before being routed to the end server. To do so, the gateway must decrypt the incoming WTLS message with a client-shared master key, and then encrypt it again using a server-shared master key. Between those two phases, the messages remain unencrypted on the gateway. This by itself is a security hole which is prone to numerous "man in the middle" attacks.

To overcome such deficiencies in the integration of both protocols, some additions need to be made. It seems logical that the expanding the functionalities of TLS's *Handshake* is the best solution for this problem. For that, we suggest the introduction of a trusted third party (TTP) between the client and the server. This enormously reduces the cryptographic load on the client, which is basically used to verify the certificate and its certificate chain. The TTP acts as an intermediate trusted CA or simply a proxy known to the client. Its role is to verify the server's certificate, retrieve its public key, sign this public key and then send it to the client.

In our protocol, the client and the server share one secrete key invisible to any middlemen. Thus, the client's security partner is the web server itself instead of third party.

To apply this protocol, we suggest adding a new *Handshake* to the TLS protocol. This "*wireless Handshake*" is based on TLS and is defined in the following section.

3. WIRELESS HANDSHAKE

The *wireless Handshake* is a sub-protocol added to TLS server and replaced the TLS *Handshake* of the mobile client. Its goal is to allow mobile clients to open secure session with the server and then to share a secret key in an economic way. In order to distinguish between wired and wireless clients, we should add a simple field to the *ClientHello* message. Its aim is to enable the server (Fig.1) to choose between *Handshake* and *wireless Handshake*.

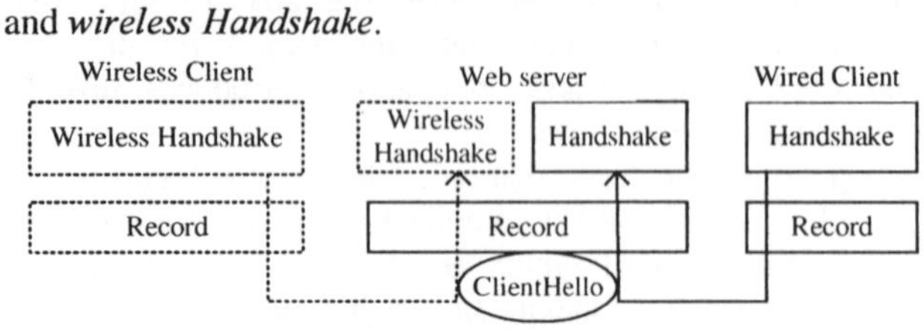

Fig.1. Phase hello

The wireless *Handshake* protocol allows the mobile client to authenticate the server through a trusted third party and to negotiate encryption algorithms and cryptographic keys. Once the wireless *Handshake* is made, the two parties share a secret key, which can be used to construct a secure channel over which application data can be exchanged.

This protocol begins with the client sending a *ClientHello* message to the TTP. This message contains a random number, the protocol version, the client certificate URL, the server's address to be connected and a list of supported *cipher_suites* algorithms. The client signs the concatenation of its certificate URL, its random and the server's address. Thus, it provides an explicit verification of its certificate in case of mutual authentication.

At the TTP, the client's signature is verified and then the *ClientHello* message is forwarded to the server.

When the server receives this message, it verifies the client signature in case of mutual authentication. Otherwise, this signature will be ignored. Then, the server sends the *ServerHello* message in response to *ClientHello* message if it is able to find an acceptable set of algorithms. It supplied also a *random* number that is used as input to the key generation process. At the same time, the server sends the certificate message containing its public key in order to allow the client to send the *pre_master_secret* key. Finally, the server sends the *ServerHelloDone* message to indicate the end of its hello.

The TTP verifies the server certificate (or the certificate chain) by verifying the server identity and checking the validity of CAs' signatures. This verification may require an on-line consultation of the CRL [1]. Afterwards, the TTP gets the server public key and transfers it with the server's random and the selected *cipher_suite* to the client using a new message called *CertificateOK*. This message contains also the Distinguished Name of the server X.509 certificate and the hash of server public key, the client's *random*, the server's *random* and a flag proving the validity of the server's certificate. The TTP then signs this hash and sends it to the client. The main of this message is to prove to the client the identity and the validity of the server's certificate and to protect applications from replay attacks since it contains, among other parameters, the signed client's *random*.

When receiving the *CertificateOK* message, the client checks the validity of TTP's signature and that the contents are correct. In fact, the client calculates the hash of the concatenation of the server's public key and the client's *random* sent in *CertificateOK* and then compares them with the value signed by the TTP. Next, it generates a random number called *pre_master_secret* and encrypts it using the server's public key. For encryption operations, peers use the algorithm which is being used for key agreement and authentication. The client then sends the result using the *ClientKeyExchange* message. It is worth noting that the *pre_master_secret* key is not visible to the TTP. In fact, it is encrypted by server's public key. This key is used to generate a *master_secret* that is shared between the client and the server. The *master_secret* is used to calculate and generate symmetric keys for encryption and message authentication.

Once received, the *ClientKeyExchange* is decrypted by the server using its private key to obtain the *pre_master_secret* and then to generate the same *master_secret* key.

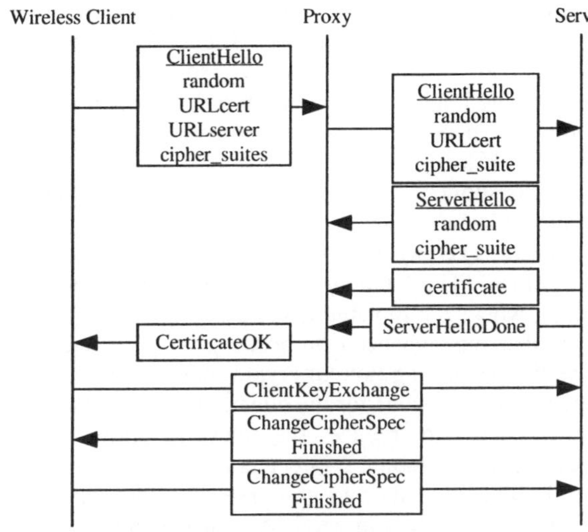

Fig.2. Wireless Handshake protocol

At this time, the client and the server have to exchange *ChangeCipherSpec* and *Finished* messages. When the client or the server receives a *ChangeCipherSpec* message, it copies the pending read state into the current read state. Whenever the client or the server sends a *ChangeCipherSpec* message, it copies the pending write state into the current write state. The *Finished* messages allow both peers to verify that the negotiation has been successful and that security has not been compromised. The *Finished* message is the first message that is protected using the negotiated *cipher_suite*. This means that it is encrypted and authenticated according to that suite. If the receiving party cannot successfully decrypt and verify the message, then clearly something has gone awry with the security negotiation.

Once one side has sent its *Finished* message and received and validated the *Finished* message from its peer, it may begin to send and receive application data over the connection.

In our *Handshake*, the client doesn't have all *handshake_messages* exchanged between the TTP and the server like the server's certificate.

For TLS, the contents of the *Finished* message are 12 bytes created by applying the PRF (Pseudo Random Function) to the master_secret, the finished_label (client or server) and the concatenation of the MD5 hash of all *handshake_messages* and the SHA-1 hash of all *handshake_messages* (i.e. the value of *handshake_messages* includes all *Handshake* messages exchanged between the TTP and the server starting at *ClientHello* up to, but not including, the *Finished* messages). This allows the TTP to send the hash of all *handshake_messages* back to the client instead of the *handshake_messages*. Since the MD5 (respectively SHA-1) algorithm produces as output a 128-bit (respectively 160-bit) "fingerprint" or "message digest" of the input, there is nothing wrong, in wireless environments, with sending back the hash (36 octets) instead of the *handshake_messages* themselves (2 to 3 kilo octets). In this case, we must add a new message to our *Handshake*. This message contains the hash of all *Handshake_messages* exchanged between the server and the TTP. It is sent to the client after receiving the *ClientKeyExchange* message. With this method, there is no change in server's TLS. The client's TLS must be modified in a simple manner in order to integrate coherently wireless security with TLS.

The principle of *Finished* calculation is the same with SSL (Secure Sockets Layer) but the PRF is applied, among others parameters, to the hash of the concatenation of all *handshake_messages* and the master_secret shared only between the client and the server. Since the TTP doesn't have the master_secret, it becomes necessary to apply the PRF to *Handshake* data, known by the client and the server, needed to be protected from any change; namely the *random* numbers, the selected *cipher_suites*, the *version protocol*, *acceptable certificate authorities* and the *URL of client's certificate*. Add to these parameters, the use of *master_secret* shared between the client and the server. In this case, we can avoid more delay in wireless link (the TTP will not send the *handshake_messages* back to the client). Thus, we can conserve the radio bandwidth, which is the bottleneck in wireless environments (i.e. the *handshake_messages* are about 2 kilo octets).

If the peers use the TLS protocol, our *wireless Handshake* can always use this method to calculate the *Finished* message by applying the PRF to some *Handshake* data instead of all *handshake_messages*.

4. ANALYSES AND DISCUSSION

Our *Handshake* Protocol allows the server and client to authenticate each other (the client authenticates the server via the third party) and to negotiate encryption algorithms and cryptographic keys before the application protocol transmits or receives data. This *Handshake* assures the same TLS1.0 functionality [2] and provides connection security that has the basic properties:

- Authentication: the peer's identity can be authenticated using asymmetric cryptography. This authentication can be made optional, but is generally required for at least one of communications' peers.
- Confidentiality: only the legitimate peers can obtain the content of a message sent across the connection.
- Negotiation of a secure shared secret: the negotiated secret is unavailable to eavesdroppers, and for any authenticated connection the secret cannot be obtained, even by an attacker who can place himself in the middle.
- Integrity: no attacker can modify the negotiation communication without being detected by peers.

Our *Handshake* can be used with any wireless networks based on Access Points (AP). In fact, Access Points will assure the services of the TTP or the CA for their clients. For example, the TTP can be the GGSN (Gateway GPRS Support Node) in GPRS, the Switch in Wi-Fi or WLAN and it is the HLR (Home Location Register) in GSM. In these cases, the authenticated client will ask its AP to verify the server certificate and to send back the server public key.

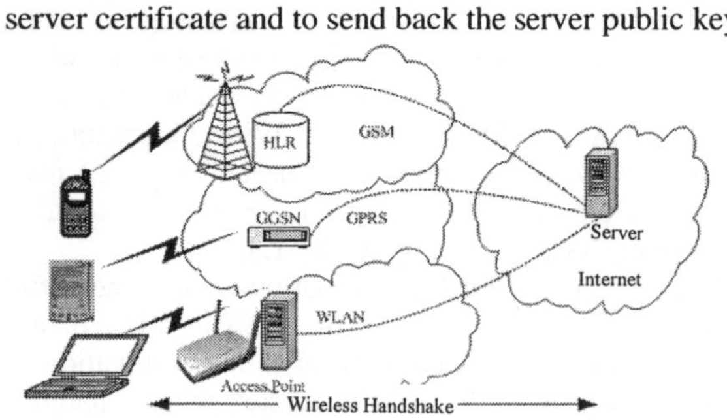

Fig.3. Wireless Handshake and mobile environments

During the *Handshake* phase, the client signs both its URL certificate and its random. This random contains a universal time in order to ensure that attackers do not replay old parameters. Furthermore, this signature can be use in order to protect the TTP (the Access Point) from *Denial Of Service* attacks. In fact, the TTP will ignore any request coming from clients don't belong to TTP's database or simply don't have the permission of access.

In TLS, the client sends the *certificate_verify* message in response to server's *certificate_request* [2]. It is used to provide explicit verification of the client's certificate. In our *Handshake*, the client signs the concatenation of its random, its URL certificate and the server's address in order to get the same *certificate_verify* services.

In this phase, the message flux exchanged over the radio link and the cryptographic loads of the client are reduced. In fact, the TTP send the server's public key (64 or 128 octets) back to the client instead of the server's certificate as in TLS1.0 (about 1 kilo octets). In addition, the client verifies the TTP's signature instead of the server's certificate. Thus, the client's load is about 8.3 ms instead of 20.36 ms. The later is quickly increased in parallel with depth increase (the number of certificates in the processed certificate chain) whereas it remains stable using our protocol.

At the *Finished* phase, the client doesn't have all *handshake_messages* (e.g. server's certificate). For that, the *verify_data* will be calculated by applying the PRF to parameters selected from *handshake_messages*. Those parameters must be carefully selected to protect sessions from active and replay attacks and to allow peers to detect any change in exchanged messages.

Before the *Finished* phase, a man-in-the-middle or an adversary attacker might try to influence the *handshake* exchange to make the peers select different encryption algorithms than they would normally choose. This attacker can change one or more *handshake* messages. If this occurs, the client and server will compute different *Finished* values. As a result, peers will not accept to open the secure session (i.e. without the *master_secret*, the attacker cannot repair the *Finished* messages and then the attack will be discovered).

It is worth informing that [3] adds new extensions to TLS1.0 for new services and applications. The parameters of these extensions should be protected against adversary attacks. For this, the selected parameters may be added to our *Finished* calculations. One of these parameters is the maximum fragment length to be sent between peers.

In order to evaluate the performance of our proposed protocol, an implementation of the cryptographic computations was carried out. In addition, we have analyzed a full *Handshake* for Light TLS and TLS protocols and we have computed the necessary cryptographic loads for the client in order to open a secure session. The benchmarks were performed on a 205 MHz iPAQ PDA [5].

The results show that the necessary cryptographic load for the client to open a secure session is about 0.76 second using Light TLS whereas it is about 1.63 second using TLS protocol. The latter increases as the certificate chain increase, whereas it remains stable using our protocol. This reveals that our enhanced proposed protocol manages to reduce the processing time of cryptographic loads in order to open secure sessions for the client. On top of that, Light TLS manages to reduce the message flow of the *Handshake* phase (about 550 octets instead of about 1400 octets in TLS) and then to conserve the radio bandwidth which is the bottleneck in wireless environments.

5. CONCLUSION

In this paper, we propose a new TLS *Handshake* that avoids the drawbacks of wireless protocols stacks (e.g. WAP). Its goal is to allow the client and the server to authenticate each other and to share a secret key in a secure way and thus would have an impact on new services and applications (e.g. E-commerce and bank transactions) independent of middlemen. Moreover, it protects applications from replay and active attacks.

In addition, Light TLS manages to save the wireless bandwidth by reducing the message flow. It also reduces the cryptographic processing time needed to open secure sessions on mobile clients.

REFERENCES

[1] R. Housley and W. Polk, D. Solo, Internet X509 PKI, certificate and CRL profile, RFC 2459, January 1999.

[2] T. Dierks and C. Allen , The TLS protocol, Version 1.0, RFC 2246, January 1999.

[3] S. Black-Wilson, et. al., Transport Layer Security (TLS) Extensions, RFC3546, June 2003.

[4] A. Freier, P Karlton, and P. Kocher, The SSL protocol, version 3.0, Internet Draft, March 1996.

[5] M. Badra, P. Godlewski, A. Serhrouchni, "Extended WTLS". ASWN IEEE workshop on Applications and Services in Wireless Networks, July 2003, Bern.

[6] S. Boeyen, et. al., Internet X.509 PKI operational protocols-LDAPv2, RFC2559, April 1999.

[7] WAP Forum, Wireless transport layer security, version April 2001.

A NOVEL SECURE HANDOFF APPROACH FOR WIRELESS LANS

S. NANDURI AND S. ZEADALLY
High Speed Networking Laboratory
Department of Computer Science
Wayne State University
Detroit, MI, 48202

Handoff is an important component of wireless networks. During handoff, a mobile station transfers connectivity from one access point to another. One of the challenges for secure wireless LANs is to enable secure handoff with minimum delays. We propose a novel scheme for the Extended Service Set (ESS) by using the TESLA (Time Efficient Stream Loss-tolerant Authentication) broadcast authentication protocol. Our proposed scheme performs re-authentication based on pure symmetric cryptographic function such as Message Integrity Check (MIC) and using Pseudo Random Function (PRF) generates the new Pairwise Master Key required between the Access Point (AP) and the mobile client while the latter is in the discovery phase (that is, searching for an AP which is nearest to it) which minimizes re-association delay during handoff.

1. Introduction

The goal of Task Group I is to enhance the security of the current 802.11 MAC , proposed 802.1X "Network Port Authentication" [9] and Temporal Key Integrity Protocol (TKIP) [1][2][3]. 802.1X authentication and TKIP can solve the deficiencies of the Wired Equivalent Privacy (WEP)[10] mechanism. However Task Group I do not address the problem of secure roaming for a mobile client among access points in a large infrastructure. That is, when a mobile station moves from one access point to another. In this case, a full re-authentication by an 802.1X server is too slow to support real-time applications like Multimedia and voice over IP traffic. To address this issue of secure handoff with minimum delay, we propose a novel scheme that exploits mutual re-authentication and generates new Pairwise Master Key on-demand. Our proposed scheme incurs lower, secure, handoff delay than 802.1X authentication

2. Background

Our novel scheme exploits the Time Efficient Stream Loss-tolerant Authentication protocol (TESLA)[4], Key hierarchy as a result of 802.1X authentication, and the Inter Access-Point Protocol (IAPP). We describe them very briefly to enable an understanding of our proposed method.

TESLA is a broadcast authentication protocol for authenticating routing messages. TESLA is efficient and adds only a single Message Integrity Check (MIC) to a message for broadcast authentication. Adding a MIC (computed with a shared key) to a message can provide secure authentication in point-to-point communication. TESLA differs from traditional asymmetric protocols such as RSA in that TESLA achieves this asymmetry from clock synchronization and delayed key disclosure.

With 802.1X authentication a hierarchy of Keys [8] is generated, a Master Key is established between the mobile client and Authenticating Server. The Authenticating Server passes the Pairwise Master Key (PMK) to the mobile client and the AP.By using Pairwise Master key, the mobile client and the AP generate the Pairwise Transient key (PTK), which eventually split into the confirmation key, the encryption key, and the temporal key. The Pairwise Master key is true as long as the mobile client is connected to that AP. When the mobile client switches to another AP the Pairwise Master Key is changed which leads to change of Pairwise Transient Key.

To support the smooth handoff of mobile (roaming) stations, a protocol is required between APs to maintain up to date, real-time information regarding the whereabouts of the mobile device. One such protocol is called IAPP whose goals include: to assure fast update of bridge tables in order not to lose traffic that is directed to a mobile station, to share authentication data regarding the mobile station to allow a fast reauthorization, and pass information about the identity of the other

APs within the same network, to be used in diagnostic tools

3. Related Work

In [6][7[[8], the authors use the discovery phase solely for searching the new access point but in our proposed scheme, we exploit the discovery phase time to generate the new Pairwise Master Key and perform re-authentication. In addition to this, their approach generates more Pairwise Master Keys whereas in our design approach Pairwise Master Keys are generated on demand (that is, only when a handoff is required). Mishra et al. [8] proactive caching is a bottleneck for the APs memory. Moreover, in [8], proactive distribution generates new Pairwise Master Keys without any authentication which could lead to the flooding of Keys. In our approach new Pairwise Master Keys are generated only when the mobile is authenticated using TESLA.

4. Design and Implementation of Novel Approach

We propose a novel design which supports secure roaming of clients among APs in wireless LANs using IAPP. Our secure roaming approach achieves mutual authentication between the mobile client and new AP and generates new Keys on demand with minimum delay. Our method assumes that the mobile client, the first AP(old) and the Authenticating server are all authenticated mutually during 802.1X authentication when the mobile client first joins the wireless LAN network. With the result of 802.1X authentication the mobile client gets Master Key, PMK and PTK. While deriving the PTK we assume that mobile and the first AP (old) derives another Key called (unique for each mobile client) the Roaming key (RK) (which is used for TESLA authentication) and is time synchronized with it.

$$RK = PRF\ (PMK\ ,\ STANonce,\ APNonce,\ STA\ MAC,\ APMAC)$$

We assume the Authenticating server and all Access points will already have secure communication among themselves.

Mishra et al. [5] found that discovering the right AP is the largest dominating factor in the handoff process. Our protocol achieves security from the start of the discovery phase. Normally the discovery phase in handoff works as follows. For each channel the mobile client sends a message (probe broadcast) for an active new AP and waits for the response (probe response) from the new AP. Based on the signal strength the mobile client receives, it selects the new AP and performs re-authentication and re-association. We enumerate the steps involved in our proposed secure handoff approach:

1. In our scheme, when a mobile client needs handoff, instead of first sending a probe request, the mobile client broadcasts a TESLA packet on each channel with a bit set indicating that it has the Roaming key and it needs re-association. Basically the TESLA packet consists of Mobile MAC address,{STANonce1, STANonce2}PMK_{old} (i.e. STANonce1 and STANonce2 encrypted under PMK_{old}) , old AP MAC address, and the Message Integrity Check (MIC), which is calculated with an undisclosed key called the TESLA Key.

2. The new APs before forwarding the TESLA packet to the old AP adds the signal strength of the client. Then after some time, the mobile client discloses the TESLA Key on all channels. When the new APs gets the disclosed TESLA Key packet, before sending to the old AP it adds the signal strength of the mobile client to the TESLA Key packet and forwards it to the old AP. It starts the discovery phase, that is, listens for active APs. Meanwhile, steps 3 and 4 will occur.

3. Old AP receives TELSA packets and TESLA Key packets from new APs, which are in the transmission range of the mobile client. Then the old AP accepts a TESLA packet, which has the highest signal strength, and performs TESLA authentication for that packet. If the client is authenticated then that new AP is selected and discards other TESLA packets and for other APs it sends a message, which says that, not selected.

4. After the mobile client is authenticated, the old AP generates new Pairwise Master Key (PMK_{new}) for the new AP and the mobile client and decrypts the STA Nonce1 and STA Nonce2.

$$PMK_{new} = PRF\ (PMK_{old},\ RK_{old},\ STAMAC,\ AP_{new}\ MAC\ )$$

The old AP forwards the packet containing the new Pairwise Master Key, STA Nonce1 and STA Nonce2 to the new AP, which records the packet and in turn forwards the packet (ACCESS REQUEST) to the Authenticating server. The Authenticating server responds with an (ACCESS ACCEPT) with encrypted message, consists of new Pairwise Master Key for the mobile client, STANonce1 and STANonce2 using the Master Key shared between the mobile client and the Authenticating server (the Master key obtained as a result of the initial 802.1X authentication).Then the new AP generates random nonces namely AP Nonce1 and AP Nonce2, and generates the new Pairwise Transient Key (PTK_{new}) and Roaming Key (RK_{new}).

5. After the discovery phase, the mobile client selects the AP with the highest signal strength and sends a packet requesting for the new Pairwise Master Key. If the new AP has received the new Pairwise Master Key, it forwards to the mobile client the packet containing the new Pairwise Master Key, STANonce1 and STANonce2 encrypted using the Master Key and AP Nonce1 and AP Nonce2 encrypted using the new Pairwise Master Key. Upon receipt of this packet, the mobile client first decrypts using the Master Key and checks the STANonce1 and STANonce2. If they are the same it accepts the new Pairwise Master Key. Using the Pairwise Master Key mobile decrypts further to get AP Nonce1 and AP Nonce2, and generates Pairwise Transient Key and Roaming Key. We present a summary of the above steps in the Figure below.The new AP and mobile generates PTK_{new} and RK_{new}

$$PTK_{new} = PRF\,(PMK_{new},\ STANonce1,\ APNonce1,\ STA\ MAC,\ AP_{new}\,MAC)$$

$$RK_{new} = PRF\,(PMK_{new},\ STANonce2,\ APNonce2,\ STAMAC,\ AP_{new}\,MAC)$$

For our implementation we have made APs on 442 MHz desktop systems running FreeBSD and mobile running on the 2.4 MHz laptop. Using our novel design the measurement for handoff is in the range of 32 and 68 ms.

5. Conclusion

In this paper we propose the design and implementation of a secure and fast handoff for wireless LANs using IAPP. Our protocol is faster than 802.1X .Our novel approach exploits pure symmetric cryptographic function for re-authentication, some Pseudo Random Function is used for generation of the new Keys, re-authentication and Handoff Keys occurs on-demand during the discovery phase and minimizes the handoff delay. Our design does not have to know the topology of the Wireless LAN. In our design 4-way handshake for generation of PTK is eliminated

Acknowledgements

This work was supported by an equipment grant from Wayne State University. The authors would like to express their gratitude to Monica Brockmeyer and Farshad Fotouhi for their support in making the wireless equipments available for us for implementing our design.

References

1 N. Ferguson, “Michael: an improved MIC for 802.11 WEP,” IEEE 802.11 doc 02-020r0, January17,2002. http://grouper.ieee.org/groups/802/11/
2 D. Stanley, “IV Sequencing Requirements Summary,” IEEE 802.11 doc 02-006r2, January18,2002. http://grouper.ieee.org/groups/802/11/
3 R. Housely and D. Whiting, “Temporal Key Hash,” IEEE 802.11 doc 01-550r1 October 31,2001. http://grouper.ieee.org/groups/802/11/
4 Adrian Perrig, Ran Canetti, Dawn Song, and J. D. Tygar. Efficient and Secure Source Authentication for Multicast. In *Network and Distributed System Security Symposium, NDSS '01*, pages 35–46, February 2001.
5 A. Mishra, M. Shin, and W. Arbaugh , “An Empirical Analysis of the IEEE 802.11 MAC Layer Handoff Process,” 2002.
6 S. Pack and Y. Choi “Fast Inter-AP Handoff Using Predictive Authentication Scheme in a Public Wireless LAN,” in Proc. of IEEE Networks 2002 (Joint ICN 2002 and ICWLHN 2002).
7 S. Pack and Y. Choi, “Pre-Authenticated Fast Handoff in a Public Wireless LAN based on

IEEE 802.1X Model," in Proc. of IFIP TC6 Personal Wireless Communications 2002, Singapore, October 2002.

8 A. Mishra et al., "Proactive Key Distribution to Support Fast and Secure Roaming," IEEE 802.11doc03084r1,January2003. http://grouper.ieee.org/groups/802/11/

9 IEEE Standards for Local and Metropolitan Area Networks: Port based Network Access Control, IEEE Std 802.1x-2001, June 2001.

10Borisov, I. Goldberg, and D. Wagner, "Intercepting mobile communication: The Insecurity of 802.11," in Proceedings of MOBICOM 2001.

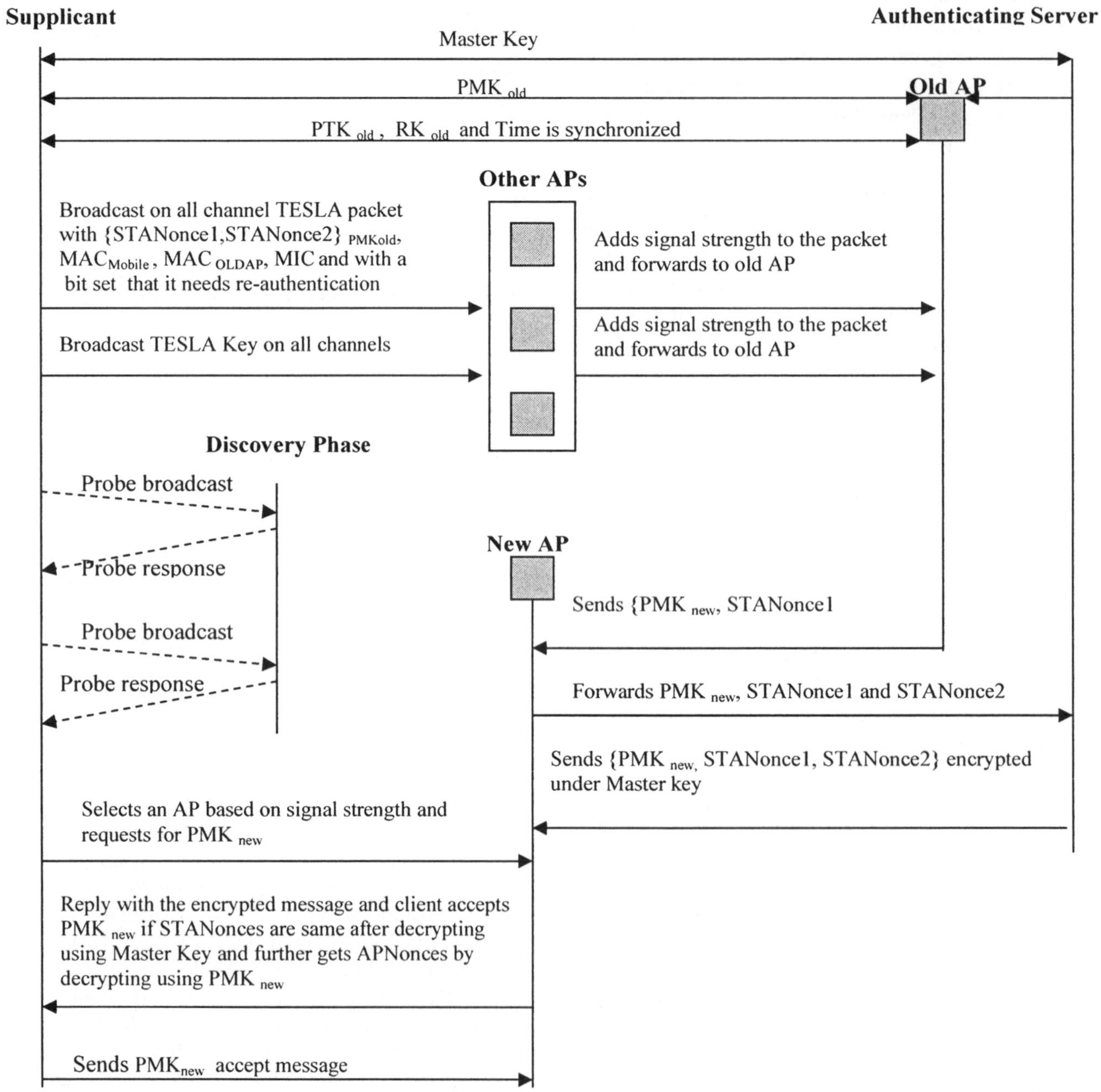

Modelling a PKI-Based Network System using Multi-Agents Systems Engineering

Gustavo A. Santana Torrellas
Leonid Sheremetov
Luis A. Villa Vargas
Instituto Mexicano del Petróleo
Programa de Matemáticas Aplicadas y Computación
Eje Central Lázaro Cárdenas N°152
CP 07730, México, D.F.
México

Abstract: **In the current networked world, it is clear that security has become a "sine qua non" condition for several public and private commercial, administrative and technical activities carried out by means of network applications. Recent developments have made it possible to interoperate complex business applications at much lower costs. To stay competitive in this scenario, it is crucial for organizations to have the flexibility to changing security factors, such as authentication, authorization, access control, new technologies, and cost of policy deployment. Such information security changes often encourage the creation of new security schemas or security improvements. Accommodating frequent systems information changes requires a PKI-based network security system be more flexible than currently prevalent systems. Consequently, there has recently been an increasing interest in flexible PKI-based network security and policy deployment systems. This article deals with the deployment of a secure IT infrastructure on a company-wide basis. This means to be able to provide secure PKI-based network services to the users, independent of their physical location: on-site (via LAN) or off-site (via WAN or an external ISP). The security solution is based on the usage of X.509 certificates to perform authentication both of the users and the network's nodes/services. The corresponding private key can be stored either on a smart-card (for mobile users) or on a secure workstation (for users that always access the services from the same workstation). The system will support security at two main levels: *application-level*, via SSL channels or S/MIME messages, and *network level*, via IPsec. Moreover, the infrastructure is conceived with flexibility as a primary goal.**

Key Words: Public Key Infrastructure, Information Security, Authentication, Autorization.

1. INTRODUCTION

Because organizations rely so heavily on their IT infrastructures to interact with customers, suppliers, partners and government agencies, they require a holistic, enterprise-wide approach to IT security—one that allows all of their constituents to participate in trusted online transactions. To this end, basic security is not enough: simple password protection is insufficient to protect the flow of information back and forth between enterprise web servers, application servers and back-end databases. To overcome this vulnerability, enhanced security is required. Public-key certification authority services represent one of the most common approaches to enhanced Internet security. Based on a public-key infrastructure (PKI), this method employs digital signing and encryption to inject trust into online transactions. Such CA services can be maintained internally or managed by a trusted third party. Public Key Infrastructures (PKIs) can speed up and simplify delivery of products and services by providing electronic approaches to processes that historically have been paper based. These electronic solutions depend on data integrity and authenticity. Both can be accomplished by binding a unique digital signature to an individual and ensuring that the digital signature cannot be forged. The individual can then digitally sign data and the recipient can verify the originator of the data and that the data has not been modified without the originator's knowledge. In addition, the PKI can provide encryption capabilities to ensure privacy. There are two important aspects to a flexible PKI-based network security system: hardware infrastructure and the corresponding planning and control software. The latter is the heart of a flexible network security system; appropriate software architecture can improve system performance significantly. In this paper, we focus mainly on the software aspect. Distributed Artificial Intelligence covers the intersection of Artificial Intelligence and Distributing Computing. Multi-Agent Systems (MAS) are commonly used in solving difficult problems in the areas of Distributed Artificial Intelligence. This paper takes the multi-agent approach in which a team of agents, each with only limited local knowledge and local information, collaborates to satisfy both local and global network security objectives. The overall behavior of the system emerges through the dynamic interactions of the agent's local behaviors.

In recent research activities author has addressed the use of MAS for the control of PKI-based network system. Santana et al. [2] consider methodological issues for designing a flexible multi agent-based network security for authentication and authorization in mobile environments; the agents use a contract-net protocol for negotiation and dynamic security authentication in inter-domain networks. In other approach Santana, Sheremetov and Contreras [3] proposes a rather original approach to security assessment task planning for multiple policies by associating a Component Agent Platform, where a set of agents collaborate to assembly an authentication and authorization schema considering different policies to be assembled. Security policies are critical for network security management; the computational complexity of finding a coherent security schema for multi-domain systems grows exponentially with the total number of nodes, host, and segments of the network system. Communication without security is dangerous because third parties can easily read, forge, or block messages. To deploy security within closed user groups, secret-key cryptography could be a solution, but when it comes to open groups of users (as in the case of Internet) public-key cryptography represents today the most used technique. It permits to easily obtain integrity and confidentiality for the data exchanged between network applications, and also authentication of the involved parties. But the deployment of public key cryptography in worldwide networks requires the so-called *public key infrastructure* (PKI) that plays the role of a trusted third party, enabling trust and confidence in network environments.

Now coming into the topic of the paper, the problem of flexible PKI-based system management is always a complex one. The solution to such problems - given the fact that common IT systems are not flexible for policy deployment - must not ignore the security aspect. The purpose of the paper is to present the design of a PKI architecture based on standards and capable to provide several security features: protecting the communication channels and securing the data traveling inside these channels. As a result, most practical approaches derives from the fact that the infrastructure will also incorporate advanced security services based on new or forthcoming protocols such as online certificate status protocol (OCSP), time stamping protocol (TSP), and to provide common framework for secure electronic documents to make the problem tractable.

2. PROBLEM DEFINITION

In order to introduce the reader into the specificity of the problem, it is necessary to explain some of the security notions that will be used in the paper.

We are developing an agent-based planning and control system for a flexible PKI-based network security system with multiple policies agents. The input of the system is the general policy model and emphasizes the importance of the PKI as a *trust enabler* into the network system to be protected. The output of the system is the final security's state PKI-based network. The general flow diagram that indicates the global operation of the system is shown in Figure 1. This flow is mainly divided in two stages: an off-line stage and an on-line stage.

The off-line stage performs PKI-based network security task decomposition. It produces a preliminary security plan that consists of a sequence of security procedures and operations and the precedence relationships among them. The input to this off-line stage is the general policy model of a system that is composed of parts. It then generates a preliminary security assessment plan based on risk analysis and vulnerabilities evaluations about accessibility and network stability. The security assessment operations that make up a preliminary information security plan are task level operations. Currently we have implemented two such operations:

- Security Requirements Assessment
- Authentication, Authorization, and Access Control Assessment

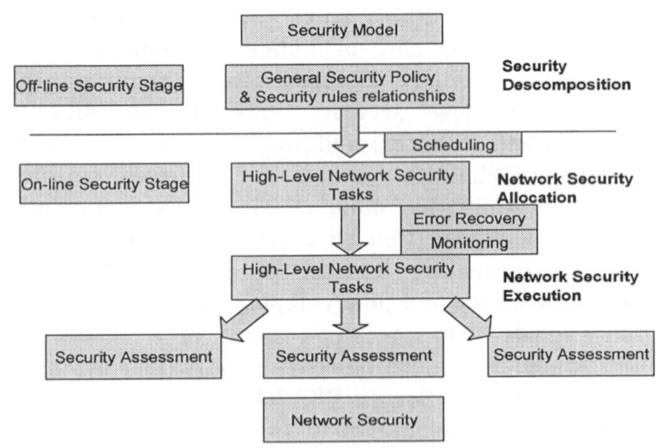

Figure 1: Global Security Flow

The preliminary information security plan is used as the input to the security assessment on-line stage. The security assessment on-line stage focuses on PKI-based network security allocation and PKI-based network security execution. In this stage, agents of a team collaborate and cooperate to achieve common security goals. The PKI-based network security allocation phase assigns security operations to appropriate resources (security assessment agents). The task level operations are also mapped into low-level operations that security assessment agents can understand and execute. The access into a secure information system must always respect a preliminary procedure. The system needs the means to verify the entity requesting login and this process is known as *authentication*. Authentication is always composed of two steps, that is the identity of the entity willing to enter the system should be acknowledged (*identification*) and verified (*verification*). During the identification step, the entity presents an identifier to the security system. Identifiers should be assigned carefully, because authenticated identities are the basis for other security services, such as access control service (i.e., authorization). During the verification step, the system generates authentication information that confirms the binding between the entity and the identifier.

Since the process of authentication is straightforward and implies only identification and verification of the identity of a certain user, a secure system must have the means for allowing or limiting the rights that user may have inside the system after being authenticated. This is called shortly *access control* or *authorization*.

Authorization denominates the process of granting rights to a user of an information system. Once you have authenticated a user, the user may be authorized to perform different types of access or activity.

Moreover the concept of single sign-on is extremely interesting for the article's topic. Typically, a user logs in just once, and then he is transparently granted access to a variety of permitted resources, with no further login being required until the user logs out. Single sign-on prevents users from choosing bad passwords, a major component of systems insecurity and is therefore highly desirable but difficult to implement.

Automatic authentication/authorization can be easily obtained by a cryptographic challenge based on public keys: one key per user, per node or per application server. Therefore such a system has the advantages of being user friendly and enabling authentication to be managed consistently across an entire enterprise, and has the disadvantage of requiring all hosts and applications to trust the same authentication mechanism. In general, authorization to computing resources is classified into two main categories: ACL-based authorization and capability-based authorization. In the second category, capability is usually represented by a *token*: an unforgeable data value (sometimes called a *ticket*) that gives the holder the right to access a system resource. Possession of the token is accepted by a system as proof that the holder has been authorized to access the resource named by the token. This type of authorization has the disadvantage of being less practical and more prone to attacks. The usual way authorization is granted is by means of the so-called *access control lists* (ACL) — these are lists of entities, together with their access rights, which are authorized to have access to a resource. A form of identity-based access control is the *role-based access control* (RBAC) where the system entities that are identified and controlled are functional positions in an organization or process. This paper focuses on the security assessment on-line stage. We present a Multi-Agent System planning and control architecture for PKI-based network security coordination. The multi-agent paradigm has been adopted as it provides the following advantages:

- **Homogenous Framework**: All system components are considered and modeled in a uniform and homogenous manner
- **Modularity and Scalability**: Different modules are developed and implemented independently. This approach can simplify agent component development/maintenance and make addition/ removal of agents much easier.
- **Dynamic Reconfigurability**: The capability to quickly reconfigure a system enables timely response to changing business process reengineering conditions.
- **Distributed Control**: Each agent has local control with the capability to coordinate its activities through messages-passing communication. Agents can reside on different computers and security assessment controllers.
- **Fault Tolerance**: Techniques for fault detection, failure recovery, and execution monitoring during network security execution is provided to make system performance more robust.

3. AGENT TYPES

The multi-agent paradigm is based on the premise that complex problems can be partitioned into simpler sub-problems that have limited mutual dependencies. The fewer dependencies the system has, the more successful the agent-based approach can be. Therefore, problem decomposition and design of individual agent become the most important software design decisions for agent-based system. We have adopted a hybrid decomposition approach [14]. Our agent architecture is shown in Figure 2.

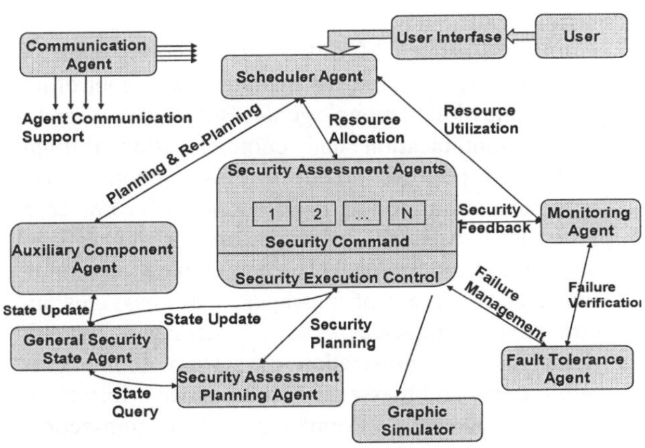

Figure 2: Agent System Architecture

We have designed the following agents in our system:

- Scheduler Agent is responsible for scheduling the detailed network security operations and assigning them to available resources (security assessment agents).
- Security Assessment Agents are responsible for the low-level control of individual security assessment agents, and for converting task level PKI-based network security operations to low-level security requirements.
- Auxiliary Component Agents are responsible for the control of the auxiliary components, i.e. CAP – Component Agent Platform in our system.

- General Security State Agent contains both static and dynamic information of all components in the system. Static information includes connections, transactions, and dynamic information of all system components. Dynamic information includes current component configuration and any dynamic state information.
- Security Assessment Planning Agent is responsible for generating the collision-free security task of the three security assessment agents on the fly.
- Fault Tolerance Agent manages and coordinates the failure recovery process.
- Communication Agent handles message passing and message dispatching in the agent system
- Monitoring Agent monitors task planning and execution to improve resource utilization. It notifies the fault tolerance agent in case of system failures.
- Graphic Simulation Agent provides visual verification of task execution.

4. AGENT MODEL

Typical agent architecture in our system is illustrated in Figure 2. Agents are involved in two types of activities: internal activities that lead the agent to achieve its own goals, and external activities through which the agent communicates and collaborates with others agents to achieve common goals. To support both internal and external activities, a typical agent contains the following modules:

- **Communication Module.-** The communication module is responsible for managing and controlling communications among agents. Both direct peer-to-peer communication and communication through PKI-based network architecture is supported. The communication module itself consists of a low-level PKI-based network interface and a higher-level communication protocol. The communication module manages two types of messages: state messages and commands messages. State messages contain dynamic state information of system components, while command messages contain other information, such as connection information, security transactions and dynamic performance, or execution commands. Command communications are implemented by priority messages, buffers and without time restrictions.
- **Security Module.-** The security module provides an agent with an information source about the agent itself and the environment (others agents). This module has a knowledge base that stores necessary data and knowledge for the agent to perform its activities. The knowledge base includes two kinds of knowledge:
- **Local security knowledge base**: It contains information related to agent capabilities and its own state information.
- **Global security knowledge base**: It contains information of other agents, with which the agent will work. The agent collaborates and cooperates with these agents during the security assessments security assessment and PKI-based network security execution stage.

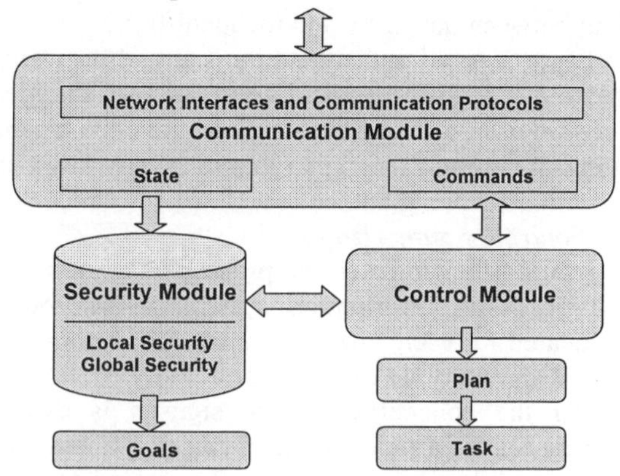

Figure 3: Agent Communication Model

- **Control Module.-** The control module determines the agent's behaviors. It operates in a goal-driven fashion. When an agent is assigned a set of tasks, every task in the set is converted into a set of goals. The agent can achieve the goals as long as its capabilities can meet the requirements. Agent goals are mapped into a collection of plan scripts, which can be achieved by the control module. The control module can run either independently or cooperatively with others agents.

5. CONCLUSION

In this paper, we present a distributed planning and control architecture for autonomous security assessment systems using a multi-agent paradigm. We have focused on the security assessment on-line stage (security assessment and PKI-based network security execution) of PKI-based network security operations. The PKI-based network security execution part is actually developed, implemented, and tested in the Mexican Oil Institute. All agents except the fault tolerance agent have been designed and are currently being implemented. We belief that our multi-agent architecture is an adequate framework for flexible PKI-based network security management. The system is modeled by a team of autonomous agents that develop cooperative strategies to achieve the common goals. We anticipated that our approach can be applied and extended to other PKI-based systems; moreover we belief in a huge field of application ranging form Continuous Security Management to e-business applications. We are currently applying and extending the planning and control strategy to a Continuous Security Management system with four security assessment agents: Security Requirements Assessment, Threat & Risk Assessment, PKI-based network Security Assessment, and Internet Security Assessment.

REFERENCES

Will be provided upon request.

SESSION 3A:
QUALITY OF SERVICE

PREDICTIVE FH CONTROL FOR FH-OFDMA SYSTEMS*

BONGSEOG JANG†

Multimedia Engineering, Mokpo National University, 61 Dorim Moon-an, Chonnam 534-729, Korea

CHANGHOI KOO, D.S. PARK, HYEONWOO LEE

Samsung Electronics, Telecom. R&D Ctr. Soowon, 442-742, Korea

In this paper, we propose a method to predict FH patterns in FH-OFDMA systems. For 4th generation wireless communication systems, OFDM is recognized as a promising modulation technique. Multi-user allocation in OFDM system can use FH that provides the spectrum-spread techniques. If one can generate more predictable FH sequence, then performance of the system can be easily improved. Current random FH and simple adaptive FH methods, however, do not consider the important of accurately predicting FH sequence. In this paper we show that the sampling of the wireless faded signal is not realized as a certain probability nature. With this regard, the proposed predictive FH allocation method is designed to embed the unknown probability in the model. Simulation study shows that the predictive FH method is more accurately predict FH sequences than the random or simple adaptive FH method. We will further improve this proposed method to apply QoS control and MAC function development in OFDM based wireless physical structure.

1. Introduction

Orthogonal frequency division multiplexing (OFDM) is one of the most promising solutions to provide broadband wireless communications. It is regarded because of its high spectral efficiency and multi-path tolerance [1][2]. OFDM allows for a high spectral efficiency as the carrier power, and modulation scheme can be differently controlled for each sub-carrier. OFDM has been also shown to be an effective technique to remove multi-path fading over wireless channels. By sending multiple narrowband signals in parallel, the allocated sub-channels are approximated to the flat faded signals. If the channels are flat faded, the equalization can be avoided to be decoded the symbols. OFDM keeps guard band to minimize inter-subcarrier interferences and to maintain orthogonal between sub-channels. In addition, adding cyclic prefix to the start of each OFDM symbol, successive symbols within a sub-carrier can be not interfered with each other [3].

Multi-user accesses in OFDM systems are accomplished thru time, frequency, and code division processes. These three techniques applied to OFDM are compared in [4]. In this comparison study, transmitter has no information on the channel state before sending data. When the transmitter has knowledge of the channel states for each of the receivers, the system performance can be better achieved. An algorithm to find the optimum sub-carrier allocation for OFDM-FDMA system is derived [5], and the algorithm is refined in [6] to lower the complexity and convergence rate. Dynamical allocating the sub-carriers based on channel conditions can utilize the channel resources more efficiently. Assuming knowledge of the instantaneous channel gains for all users, [5] proposes multi-user resource allocation in OFDM systems with fixed bit rate for each user. If adaptive modulation is used with OFDM, significant performance is achieved [5]. The employment of high modulation scheme for sub-carriers with large channel gain can increase the spectral efficiency of the overall system. OFDM-FDMA multi-user allocation technique can be easily achieved the dynamic allocation of the sub-carriers and adaptive modulation than the OFDM-TDMA and OFDM-CDMA. Frequency hopping (FH) with OFDM-FDMA (or FH-OFDMA) can allocate the clustering sub-carriers to users causing frequency diversity. If some clusters are in deep null or existed the interferences from other users, FH-OFDMA can be hopped to other clusters. In

* This work is supported by Samsung Electronics, Telecom. R&D Ctr., Korea.

† Assistant Prof., Mokpo National University, Korea. Email : jang@mokpo.ac.kr

multimedia traffic control in OFDMA systems, efficiently designed FH scheme can provide various different FH sequences in terms of required QoS demands.

In this paper, we investigate predictive FH pattern estimation method based on FH-OFDMA system over frequency selective fading channels. It maintains a monitoring of sub-channel gain then compares threshold and the gain at the predetermined OFDM symbol period. After the comparison, channel state is determined as good or bad. The channel state occurrences are averaged to estimate ensemble average of the state. This value is primarily used to sort the all the sub-channels. The highly ranked sorted sub-channels are predicted to be a good state at the next FH cluster starting time. Simulation result is provided indicating that the proposed method are better than the traditional FH schemes in terms of the accuracy of channel state prediction.

In the next section, we discuss the method proposed. The brief review of conventional FH method is provided in section 2. Section 3 is shown the simulation result. The simulation is conducted using MATLAB and C code. Finally in section 4, conclusion and further research works are mentioned.

2. FH methods

2.1. *Conventional FH methods*

When OFDMA system allocates sub-channels with symbol clusters of FH pattern, many different methods can be applicable. The simplest method is random FH. In this method, we can suppose that FH cluster is selected by uniformly distributed probability. Also random cluster allocation assumes that the prior channel state information is not available [2]. We use this random FH method for the comparison purpose in simulation study.

Recently appeared method in literature is the adaptive FH method [7]. In this method, usually SNR (signal-to-noise ratio) is measured and the FH clusters are allocated to the highest SNR among users. We find that this adaptive allocation method is becoming random FH if as more users are in the system. Simulation study on this statement, however, is not provided in this paper. In addition, when the sampling intervals of the faded signal are going to be larger, the measurement of SNR in the past short time period is not accurately reflected the sub-channel state changes. Because the sampled points can be represented as unknown random variables, it is hard to realize in the simple method. In this study, we assume that the adaptive FH method as follows,

1. Check the previous G state sub-channels.
2. Select the checked sub-channels to allocate users. In this case uniform distribution is used for the selection criteria.

Fig.1 shows that hopping scheme in FH-OFDMA system. Such as Flarion FH-OFDMA system, hopping sequences are cyclically assigned at every T interval. Also assigned hopping sequences within T interval are distinguished to the neighboring cells. Symbol by symbol hopping is implemented in the duration T. Note here that if T is being larger, than sampled points at each end of T interval are probabilistically characterized to a random process. This fact can be revealed that random FH or simple adaptive FH methods can not be accurately predict the channel state information.

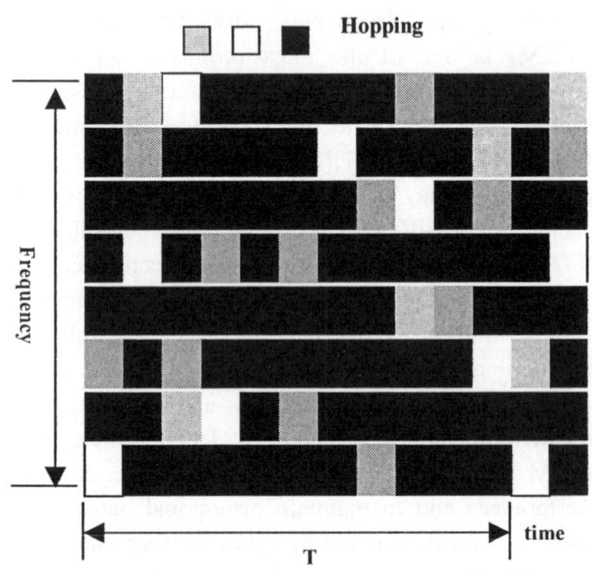

Figure 1. FH and T relation in FH-OFDMA system

2.2. *Predictive FH method*

We suppose that FH cluster is defined in T of OFDMA symbol duration. At every time T, the system measures the channel gain from receiver's feedback information. Transmitter determines the current channel state by comparing threshold R and the sub-channel gain, $z(n)$ where n represents the number of T interval occurrences. If $z(n) > R$, the sub-channel is considered to Good (G) state. Otherwise, the sub-channel is Bad (B) state. Using this comparison result, the transmitter can calculate the average length of the G states as follows,

$$w_i^{(G)}(n) = (1-\beta) \cdot w_i^{(G)}(n-1) + \beta \cdot c_i^{(G)}(n) \qquad (1)$$

where c is the number of continuing occurrence of event G before the state changes to B state, i shows the sub-channel index, β represents weighting factor ranged $0 \leq \beta \leq 1$, and w is the ensemble average of channel state length of the G state. This equation is similar to exponentially weighted moving average (EWMA) in time-series research. Using Eq.(1), we define the factor of channel state changing as follows,

$$y_i^{(G)}(n) = \beta \cdot w_i^{(G)}(n) + (1-\beta) \cdot z_i^{(G)}(n-1) \qquad (2)$$

where z represents the sub-channel gain at event $n-1$, and y is the value in which the transmitter uses to predict the state of the sub-channel i for the next event n. Eq.(2) means that the average length of the sub-channel state G is added to the most recently available sub-channel gain.

After the transmitter calculates y_i for the sub-channel i, the values of y_i for all sub-channels are sorted by descending order. The highest y_i is probabilistically being a G state at the next event n.

3. Simulation

In this section, the simulation results are discussed. Each sub-channel generates OFDM based transmitted signals over the wireless fading characteristics. Using MATLAB, we generate this faded OFDM signals. In this signal generation, we assume that each sub-carrier has a bandwidth much smaller than the coherence bandwidth of the channel and the channel is frequency selective Rayleigh fading for a user. We also assume the system with a fixed total power constraint, and assume that modulation scheme has a fixed bit rates for simplicity of simulation. In Table 1, the simulation key parameters are shown. We only use one user in this simulation. The user requires different numbers of sub-channel in the unit symbol time, and simulation runs separately for each case.

Fig.2 shows that sub-channel gain generated. This faded channel gain is used to test random, simple adaptive, and the proposed predictive FH methods.

Table 1. Key parameters for simulation

Number of sub-channels	64
OFDM symbol size	0.1 ms
Channel Characterization	Frequency selective fading
Delay spread for fading channel	8
Sub-channel modulation	BPSK
Number of user	1
User's required sub-channels	{ 1, 5, 10, 20 }

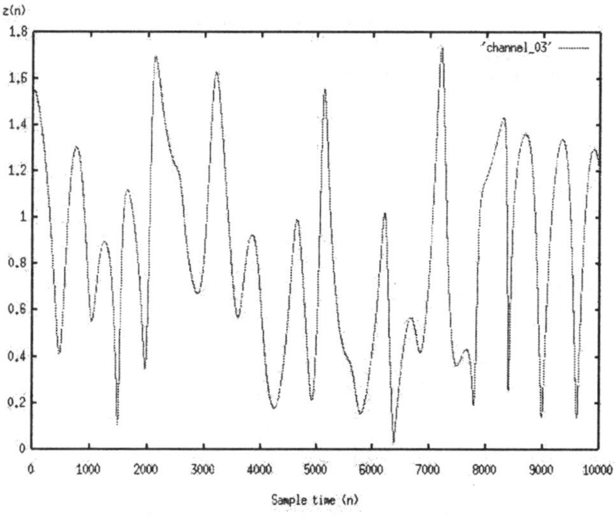

Figure 2. Generated sub-channel gain

The value of T is set to 100, 200 and 300 times the one symbol time, which means that the all generated symbol is sampled at every T interval to transmit the signal. Since the symbol time is set to 0.1ms, if $T = 100$ then 100 symbols are sent at every 10ms. In Fig. 3 and 4, simulation results are presented to show the prediction accuracy for each mentioned FH methods. The proposed FH outperforms the random FH and performs better than the simple adaptive FH. In this simulation, we set $R = 0.6$ for all FH cases and $\beta = 0.1$ for the proposed FH. Fig.3 shows that all methods are not affected the number of sub-channels. Fig.4 reveals that T interval is increased then the simple adaptive and predictive methods decrease the prediction accuracy rate. The random FH method, however, maintains the same success rate. It tells us that randomness is increased with T intervals.

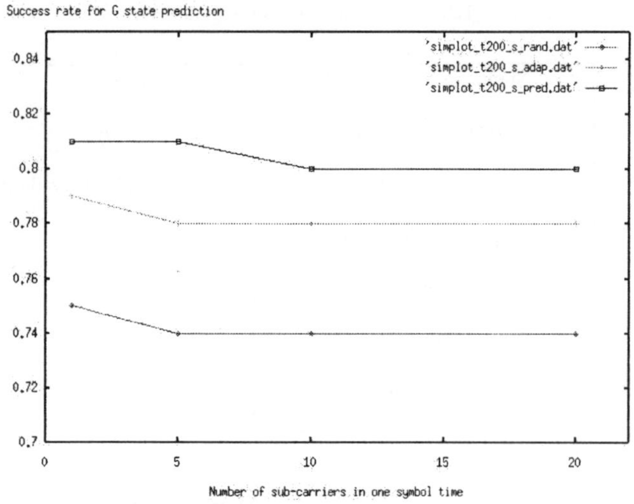

Figure 3. Prediction success rate comparison with T=200 and all sub-carrier cases

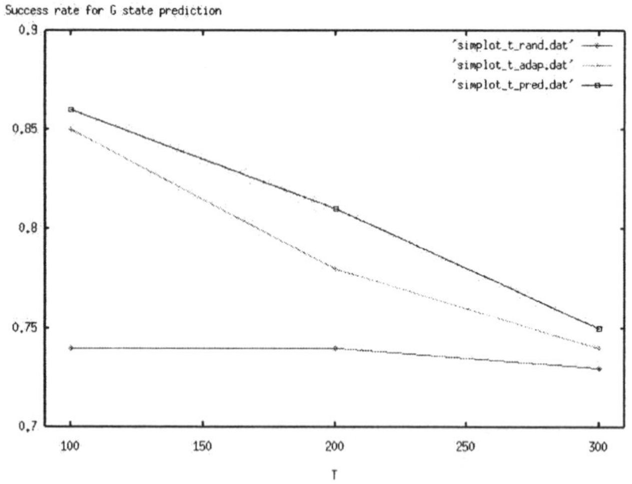

Figure 4. Prediction success rate comparison with different T cases

4. Conclusion

In this paper, we propose the method to predict the future channel state in FH-OFDMA system. Simulation results show that the proposed method is better predictable the channel state than the random or simple adaptive FH methods. This is due to the sampling points are characterized to the unknown random process in terms of *T* interval lengths. Other FH methods could not be easily acknowledged the unknown characteristics however the proposed FH can embed the probability during operation. We should further investigate the proposed one in more dynamic and realistic wireless fading and structural environment.

This work result is to prepare the development of QoS control algorithm in OFDM wireless physical structure. More predictable channel state information can be provided better fairness and throughput for multimedia traffic in such system. We will continue work to develop the scheduling algorithm and MAC control functions using the proposed FH method as a base model.

Acknowledgments

First author would like to thank Kwang-Bo Chung at Multimedia Detp. of Mokpo Natl. University who did programming for the simulation.

References

1. R.V. Nee and R. Prasad, "OFDM For Wireless Multimedia Communications", Artech House Publishers, 2000.
2. E. Lawrey, "Multiuser OFDM", The 5th Inter. Natl. Sym. on Signal Processing and its Appl. (ISSPA'99), Australia, pp. 761-764, August, 1999.
3. Ye Li, et al., "Robust Channel Estimation for OFDM systems with Rapid Dispersive Fading Channels", IEEE Trans. on Comm., vol 36, no 7, pp. 902-915, July 1998.
4. S. Hara and R. Prasad, "Design and Performance of Multicarrier CDMA System in Frequency-Selective Rayleigh Fading Channels", IEEE Trans. on Vehicular Tech., vol 48, no 5, pp.1584-1595, Sep. 1999.
5. C. Wong, R. Cheng, K. Letaief, and R. Murch, "Mutiuser OFDM with Adaptive Subcarrier, Bit, and Power Allocation", IEEE J. of Selected Areas in Comm., vol 17, no 10, pp. 1747-1758, Oct. 1999.
6. H. Yin, and H. Liu, "An Efficient Multiuser Loading Algorithm for OFDM-based Broadband Wireless Systems", IEEE Global Telecomm. Conf., vol 1, pp. 103-107, 2000.
7. W. Rhee, and J. Cioffi, "Increase in Capacity of Multiuser OFDM System Using Dynamic Subchannel Allocation", Proc. of 51st IEEE Vehicle Tech. Conf. Proc., Tokyo, Japan, pp.1085-1089, 2000.

A CLASS-BASED ADMISSION CONTROL ALGORITHM FOR SHARED WIRELESS CHANNELS SUPPORTING QOS SERVICES

PATRICK HOSEIN

Ericsson Wireless Communications, Inc., 6455 Lusk Blvd, San Diego, CA 92121, USA
E-mail: patrick.hosein@ericsson.com

Third Generation (3G) wireless networks all support a shared high-speed downlink channel for efficient transport of data. These channels use several common techniques (fast link rate adaptation, Hybrid ARQ, fast cell selection and fast scheduling) to improve performance. In addition, future 3G networks must provide the necessary framework to allow operators to offer Quality of Service (QoS) based applications. Service differentiation will also become necessary so that operators can provide differing service levels based on user classes. Pricing will be based on these user classes so that the operators can optimize their revenue. Typically, the shared forward link channel will be the bottleneck resource and hence usage of this resource must be optimally managed. This means that admission controls must be used to ensure that admitted users receive their QoS guarantees for the duration of their connection. If Class-Based Grade of Service (GoS) is offered then the admission control algorithm must also provide class dependent connection blocking rates. In this paper we present a framework that allows an operator to provide Class-Based connection admissions.

1 Introduction

Third Generation wireless networks all support some form of a shared high-speed downlink channel for efficient transport of data (the Forward Packet Data Channel or F-PDCH in the case of 1X-EVDV and HRPD networks and the High Speed Downlink Shared Channel or HS-DSCH in the case of WCDMA networks). These channels all use several common techniques (fast link rate adaptation, Hybrid ARQ, fast cell selection and fast scheduling) to improve the throughput and delay performance for data applications. In addition, these networks must provide a flexible framework to allow operators to offer QoS based services (e.g., streaming, gaming, Voice over IP). Service differentiation will also become necessary so that operators can provide differentiated levels of service based on the subscription plan of the user and seek to maximize revenue. This all implies that a flexible framework is needed to allow for sophisticated pricing plans.

In practice, the shared forward link channel will be the bottleneck resource. This must therefore be carefully managed through (a) optimal scheduling to ensure that users attain their subscribed service levels and the corresponding QoS level guarantees, (b) optimal congestion control to ensure that system performance does not degrade under congestion and that during these overload periods resources are distributed based on class levels and (c) optimal admission control to ensure that sufficient resources are reserved for already admitted users so that their QoS guarantees can be maintained and also ensure that blocking and dropping rates are class dependent. In a wireless environment, the stochastic capacity available to users together with the bursty nature of the user applications complicates the scheduling, congestion control and admission control algorithms.

These three resource management algorithms must be jointly designed for optimal system performance. In this paper we present an admission control algorithm for a generic shared high-speed forward packet data channel. This algorithm tries to maintain the QoS guarantees of the admitted users and also provide connection blocking and dropping rates based on user classes. Since the admission control algorithm depends on the optimization framework used by the scheduler, we first provide an overview of the corresponding scheduler. We then present the admission control algorithm together with a simple illustrative example. Although congestion controls are not addressed in this paper it is important that such controls also be implemented.

Many papers have been written on call admission control algorithms for both TDMA and CDMA based wireless networks where a constant rate, dedicated channel is provided to the user. In such cases the management of resources for QoS services is well understood since it is quite similar to the wired network environment (e.g., see Choi[1], Chou[2], Hou[3], Iraqi[4] and Kwon[5]). However, providing QoS services over a shared channel with stochastic capacity is quite more challenging.

2 The Scheduling Algorithm

We follow the problem formulation of Kelly[6] as follows:

$$\begin{aligned} \text{maximize} \quad & F(\vec{r}) \equiv \sum_{i=1}^{K} U_i(r_i) \\ \text{subject to} \quad & \sum_{i=1}^{K} r_i < C \\ \text{over} \quad & r_i \geq 0, \;\; 1 \leq i \leq K. \end{aligned}$$

where,
K = the number of active (transmitting) users,
r_i = the average throughput of user i,
C = the channel capacity,
$U_i(r_i)$ = the utility function of user i.

If we assume the utility function of each user is strictly concave and differentiable, then the same also holds for the objective function F. Since the feasible region is compact then an optimal solution exists, is unique and can be found by Lagrangian Methods. We assume that the downlink rate that can be supported by each mobile (denoted by $d_i(n)$ for the nth decision period) is periodically reported to the base station. If a maximum gradient ascent algorithm is applied to the above optimization problem then one can show (see Hosein[7]) that at each scheduling decision point the optimal choice is the user for which $d_j(n)U_i'(r_j(n))$ is maximum.

For example, assume that the provider charges based on the number of bits delivered to the user. In this case the utility function, U, is a linear function of the average throughout of a user and hence $U(r) = \alpha r$ for some constant α. The resulting scheduler will pick the user for which:

$$j* = \arg\max_j \{d_j(n)\}$$

which essentially means pick the user with the highest achievable bit rate. This results in maximum sector throughput and hence maximum revenue.

In order to provide QoS guarantees as well as take advantage of multi-user diversity gains, we introduce the notion of a barrier function. The utility function now consists of a base function (i.e. the desired optimization criterion when all QoS guarantees are satisfied) plus a "barrier" function that is used to maintain a user's QoS levels by introducing a penalty based on the degree of violation. The resulting scheduling problem is again solved using a maximum gradient ascent algorithm. In this paper we focus on throughput constraints but the approach can also be applied to other types of QoS constraints. We use the maximum throughput function $U(r) = r$ as the base function though others (such as the proportional fair function $U(r) = \log(r)$) can also be used.

Assume that a minimum throughput must be maintained for a particular connection. We introduce an exponential barrier function to incur a penalty for dropping below the minimum throughput r_{min}. The resulting utility function (base plus barrier) is given by:

$$U(r) = r + (1 - e^{-\beta(r - r_{\min})}).$$

As long as the throughput exceeds the minimum value then this is approximately the maximum throughput objective function. When r drops below the minimum value then the utility function rapidly decreases. Hence the scheduler would be forced to serve users with throughputs below the minimum (since the gradient rapidly increases). The parameter β determines the rate at which the penalty for violating the constraint increases. This value should be based on the Service Class of the user (e.g., Gold, Silver and Bronze services).

Figure 1 provides an example of this utility function for different values of β. Each user requires a minimum throughput of 60 kbps. As long as bandwidth is available then this requirement can be satisfied for all users. If not, a small penalty is incurred for violating the throughput constraints of Bronze users but a much larger penalty is incurred for Gold users. Hence Gold users continue to be given higher scheduling priority over the Silver users even when insufficient resources are available. In this case the gradient ascent algorithm picks the user for which $d\,U'(r)$ is maximal where

$$U'(r) = 1 + \beta e^{-\beta(r - r_{\min})}.$$

We can similarly determine appropriate barrier functions for delay constraints, jitter constraints and queue length (buffer occupancy) constraints. The sum of these barrier functions plus the base function will constitute the user's overall utility function. The user is referred to Hosein[7] for further details.

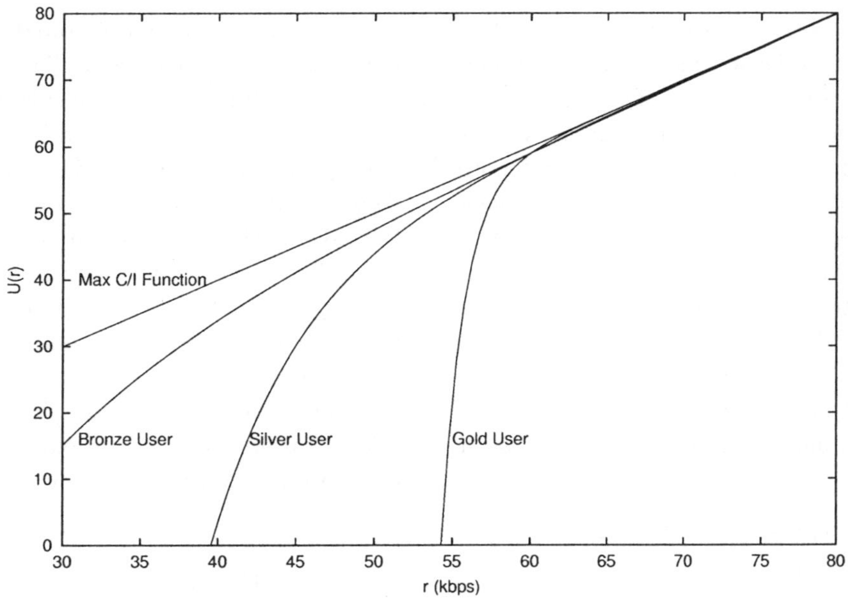

Figure 1. Barrier Functions for Different User Classes

3 The Admission Control Algorithm

In this section we use the framework in the previous section to develop the proposed admission control algorithm. Let $U_i(r)$ represent the utility function of user i. Denote the base and barrier functions of user i by $G_i(r)$ and $B_i(r)$ respectively. The barrier function imposes a penalty which we denote by the penalty function $P_i(r) \equiv -B_i(r)$. As long as resources are sufficient then all users will lie outside of the regions affected by their barrier function (this is enforced by the scheduler) resulting in $P_i(r) \approx 0$ for all users. When radio resources become exhausted then all users incur penalties. The degree of the penalty incurred will depend on the user's barrier function which in turn depends on the application and class of the corresponding user. Therefore, the total user penalty given by,

$$P(\vec{r}) = \sum_i^K P_i(r_i),$$

reflects the degree by which users' QoS constraints are being violated. Note that this metric takes into account the fact that high priority users (e.g., Gold users) are more sensitive to congestion than the low priority users. We therefore use this metric to (a) detect when forward link capacity is exhausted and (b) determine which new users should be allowed to enter the system.

When a new user requests a connection[a] we must first estimate the resource needs of the connection. We assume that, as part of the connection request, the throughput requirements of the connection are provided through the QoS request message. Note that we are focusing on throughput (as opposed to other QoS metrics) since it is the most resource intensive. Given the throughput requirement we then estimate the penalty that would be experienced if the connection was accepted. We assume that the total sector capacity remains the same even if the new user is accepted. In reality the capacity may go up because of user diversity gains or it may go down because of congestion. If we have any geometry information for the user at the connection request stage then we can use this to obtain an estimate of the request rate of the user otherwise the user's request rate is taken as the average request rate over all admitted users $\hat{d} = \sum_{i=1}^K d_i/K$. where d_i is the most recent request rate of user i. Assuming that the system is in steady state then we have

$$d_i(G_i'(r_i) - P_i'(r_i)) = \Gamma \quad \text{for} \quad i \in \{1, \ldots K\},$$

[a] Note that the Admission Controller must take into account many other system resources. The algorithm described in this paper is performed in addition to the other resource usage controls.

for some constant Γ since all user gradients tend to some constant value. If the new user is admitted then the rates for all users will decrease. The decrease for a particular user will depend on his channel conditions and his priority. For the new user let $\hat{d}$ denote his expected request rate and let $\hat{r}$ denote his "guaranteed"[b] throughput. Note that the user's barrier function (and hence penalty function) will be a function of $\hat{r}$. We denote the new user's base and penalty functions by $\hat{G}$ and $\hat{P}$ respectively. Denote the (expected) steady state rates if the user is accepted by r_i^* for the presently admitted users and $\hat{r}^*$ for the new user. Then we must compute $\{r_i^*\}$ and $\hat{r}^*$ such that

$$\begin{aligned} d_i(G_i'(r_i^*) - P_i'(r_i^*)) &= \Gamma^* \quad \text{for} \quad i \in \{1, \ldots K\} \\ \hat{d}(\hat{G}(\hat{r}^*) - \hat{P}(\hat{r}^*)) &= \Gamma^* \\ \text{such that} \quad \hat{r}^* + \sum_{i=1}^{K} r_i^* &= \sum_{i=1}^{K} r_i. \end{aligned}$$

The constraint is due to the fixed sector capacity assumption. Once these equations are solved we can now obtain the expected penalty

$$\tilde{P} = \sum_{i=1}^{K} P_i(r_i^*) + \hat{P}(\hat{r}^*).$$

if the new user is admitted. We can now compare this to a threshold to determine admittance. In order to provide differentiated blocking probabilities based on user class, we make these thresholds class dependent. For each User Class j (or GoS specification) we associate a penalty threshold γ_j. This threshold will be less stringent (higher) for the high priority users (or those needing a lower blocking probability). The new connection will then be accepted if $\tilde{P} < \gamma_j$ where j is the class of the owner of the requested connection.

4 An Illustrative Example

In this section we illustrate the admission control algorithm with a simple example. Assume that an operator provides three service levels (Gold, Silver and Bronze) and that the blocking probability of Gold users must be less than that of Silver users which in turn must be less than that of the Bronze users. If possible, all users are provided with a minimum rate of 64 kbps. If this is not possible then the Bronze users are the first ones to have their rate limit constraint violated followed by the Silver users and finally the Gold users. The primary objective of the operator is to provide this minimum rate but, once this constraint is satisfied for all users, a secondary objective is to maximize sector throughput. This would correspond to an operator who charges a fixed monthly rate (which is highest for the Gold then Silver then Bronze users) for a certain amount of transfered data (e.g. 20MB per month) and also a fixed rate per MB (for all users) above this minimum allocation.

In this case the base function is given by $G(r) = r$ for all users. We use an exponential barrier function so that all users have the same utility function given by,

$$U(r) = r + (1 - e^{-\beta_j(r-64)}),$$

where the subscript j is used to denote the class of the user.

Suppose that the system is presently running with 8 Gold, 8 Silver and 8 Bronze users and that the sector throughput is 1560 kbps. For simplicity, we also assume that all users are experiencing similar channel conditions so that they all have the same request rates $d_i = d$. Furthermore, assume that β is 0.040, 0.035, 0.030 for the Gold, Silver and Bronze users respectively. For admission, we use penalty thresholds of 1.5, 1.0 and 0.5 for the Gold, Silver and Bronze users respectively. In this way the blocking rate of Gold users will be lower than that of Silver users which will be lower than that of Bronze users, as required. Recall that the maximum gradient ascent algorithm attempts to equalize the gradients for all users. We can therefore find the steady state rates by setting all gradients to be equal and solving for the resulting rates. This would result in average throughputs of 68.7, 65.6 and 60.7 kbps for the Gold, Silver and Bronze users respectively. Note that the Gold and Silver users all achieve their minimum rates but not the Bronze users. The associated penalty cost is given by

$$\begin{aligned} P = & -8(1 - e^{-0.040(68.7-64)}) \\ & -8(1 - e^{-0.035(65.6-64)}) \\ & -8(1 - e^{-0.030(60.7-64)}) \\ = & -0.95 \end{aligned}$$

[b]In practice strict guarantees cannot be provided because of the elastic capacity and user demands.

Note that the penalty is negative because of the fact that the Gold and Silver users far exceed their minimum throughput constraints.

Now assume that a new Gold user requests a 64 kbps connection. We can estimate the penalty that would be incurred if this user was admitted by using the approach outlined in the previous section. If we do this we find that the throughputs for the Gold, Silver and Bronze users now becomes 66.3, 62.8 and 57.5 kbps respectively resulting in a penalty

$$\begin{aligned}\tilde{P} &= -9(1 - e^{-0.040(66.3-64)}) \\ &\quad -8(1 - e^{-0.035(62.8-64)}) \\ &\quad -8(1 - e^{-0.030(57.5-64)}) \\ &= 1.24.\end{aligned}$$

Since the penalty is below the 1.5 threshold set for Gold users then the user is admitted.

Suppose instead that a Bronze user had requested a 64 kbps connection then, if this user is accepted the resulting throughputs are 66.6, 63.2 and 57.9 for the Gold, Silver and Bronze users respectively and the resulting penalty is given by

$$\begin{aligned}\tilde{P} &= -8(1 - e^{-0.040(66.6-64)}) \\ &\quad -8(1 - e^{-0.035(63.2-64)}) \\ &\quad -9(1 - e^{-0.030(57.9-64)}) \\ &= 1.23.\end{aligned}$$

Since this exceeds the penalty threshold set for Bronze users then the connection is rejected.

Note that this framework allows for pre-emptive service. For example, suppose that a Gold user requests a new connection and we find that the penalty exceeds the threshold set for Gold users. Since the potential revenue from this user is higher than that of a Bronze user, one can determine the expected penalty if a Bronze user was dropped and the new Gold user was added. If this penalty is within the specified threshold then a swap is made instead. Note that this decreases the blocking probability of the Gold users at the expense of an increased dropping probability for Bronze users.

5 Summary and Conclusions

Future 3G networks will be required to support QoS services over their shared forward high speed packet data channel. In this paper we presented a flexible framework within which a wireless operator can provide these QoS services and at the same time give users priority (in terms of adherence to QoS guarantees as well as blocking and dropping probabilities) based on their subscription plans. This framework allows the operator to fine tune their pricing scheme in order to maximize revenue. We also presented a simple example to illustrate the approach. The issue of congestion control (e.g., see Hosein[8]), although not addressed in this paper, must also be properly addressed for efficient operation of the network.

References

1. S. Choi and K. Shin, "Adaptive Bandwidth Reservation and Admission Control in QoS-Sensitive Cellular Networks," *IEEE Transactions on Parallel and Distributed Systems*, vol. 13, pp 882-897, Sept. 2002.
2. C. Chou and K. Shin, "Analysis of Combined Adaptive Bandwidth Allocation and Admission Control in Wireless Networks", in *Proceedings of IEEE Infocom*, New York, USA, June 2002
3. J. Hou, J. Yang and S. Papavassiliou, "Integration of Pricing with Call Admission Control to Meet QoS Requirements in Cellular Networks", *IEEE Transactions on Parallel and Distributed Systems*, vol. 13, pp 898-910, Sept. 2002.
4. Y. Iraqi and R. Boutaba, "An Adaptive Distributed Call Admission Control for QoS-Sensitive Wireless Mobile Networks", in *Proceedings of the IEEE Wireless Communications and Networking Conference*, Chicago, USA, Sept. 2000.
5. T. Kwon, Y. Choi and C. Bisdikian, "QoS Provisioning in Wireless/Mobile Multimedia Networks Using an Adaptive Framework", *Wireless Networks*, vol. 9, pp. 51-59, 2003.
6. F. Kelly, "Charging and rate control for elastic traffic", *European Transactions on Telecommunications* no. 8, pp 33–37, 1997.
7. P. Hosein, "QoS Control for WCDMA High Speed Packet Data", in *Proceedings of the Fourth IEEE Conference on Mobile and Wireless Communications Networks*, Stockholm, Sweden, Sept. 2002.
8. P. Hosein, "A TCP-Friendly Congestion Control Algorithm for 1XEV-DV Forward Link Packet Data", in *Proceedings of the 57th IEEE Semi-Annual Vehicular Technology Conference*, Jeju, Korea, April, 2003.

A Framework for End-to-End QoS Provisioning in Mobile Internet Environment

Salem Lepaja[1], Arben Lila[1], Natalia Kryvinska[1], Hoang M. Nguyen[2]

[1]Vienna University of Technology
Favoritenstrasse 9–388, A1040, Vienna, Austria

[2]Networking Department, Communications and Device Division,
Institute for Infocomm Research (Member of A*STAR)
21, Heng Mui Keng Terrace. Singapore 119613

E-mail: {salem.lepaja, arben.lila, natalia.kryvinska}@tuwien.ac.at; hoangnm@i2r.a-star.edu.sg

Abstract: In this paper, a framework for end-to-end QoS provisioning to the mobile Internet users, is discussed. In order to meet requirements of mobile user's real-time applications, a new protocol for QoS provisioning is proposed. The proposed protocol is based on an existing flow transparency Mobile IP and RSVP interworking scheme and can be considered as its complimentary for meshed access network topologies. Performances of the protocol are evaluated in terms of the RSVP signalling delay and the maximum allowed delay for QoS provisioning.

....

Key Words: MIPv6, QoS, IntServ, RSV, MPLS.

1. INTRODUCTION

During the last decade, we have seen an exponential growth of Internet and mobile communications. A further rapid growth of these two main communication technologies, particularly towards real-time multimedia service provisioning, is expected in the next few years. Furthermore, it is expected that mobile users will request the same real-time services as fixed Internet users. It is obvious that, in mobile Internet environment, these services require both mobility and QoS support. MIPv4 and MIPv6 protocols provide transparency of host mobility to transport and higher layer protocols but do not provide QoS. On the other hand the Internet QoS techniques: IntServ/RSVP, DiffServ, and MPLS do not consider mobile environment. Hereby, several solutions addressing QoS provision to mobile hosts have been proposed. Related to the issue, in this paper we propose a new framework for end-to-end QoS provisioning to mobile users applications, encountering heterogeneous access networks. In order to meet requirements of mobile user's real-time applications, also a new protocol for QoS provisioning is proposed and evaluated.

In the next section the basic principles of MIPv6 are presented. In Section 3 brief discussions of RSVP is given. Section 4 brings a short overview of the state of the art related to end-to-end QoS provision to mobile hosts. In Section 5, the new proposed framework and protocol for end-to-end QoS provision to mobile users is given. In Section 6, performance evaluation results are presented. Conclusions are given in the last section of the paper.

2. MOBILE IPV6 PRINCIPLES

The mobile IPv6 protocol [1] enables a Mobile Host (MH) to communicate with other hosts, Corresponding Host (CH), after changing its point of attachment to the Internet without changing its IP address. The goal behind this feature is to maintain MH's transport and higher layer protocol sessions, which depend on a permanent IP address. This is achieved by assigning two IP addresses to a mobile host: a permanent IP address, which is called IP home address, and a temporary IP address, named Care-of-Address (CoA), which is assigned each time the mobile host is visiting a new foreign subnet. Home address is used for transport and higher layer sessions whereas CoA is needed to route packets to the actual point of attachment. Every time that MH acquires a new CoA, it has to register it to the Home Agent (HA) and CH, by sending Binding Update (BU) messages. An HA is a router in the MH's home network, handling the mobility of the MH. Packets of new calls are intercepted by the home agent and then tunnelled to the MH's CoA by using IPv6 encapsulation, whereas packets belonging to active sessions are sent directly to CoA. Packets from the MH are sent to the CH through the Internet as usual.

3. RESOURCE RESERVATION PROTOCOL

Resource Reservation Protocol – RSVP, is a receiver-oriented signalling protocol for resource reservation [2]. RSVP signalling configures QoS flow handling mechanism in RSVP capable routers along the path of the traffic flow. There are several RSVP messages among which Path and Resv are the most important ones. The Path message is sent by the sender to set up the *path state* at the routers along the same path as data from the sender to the receiver. The Resv messages are sent by each receiver along the reverse path from the receiver to the sender hop by hop to ask for resources. They create and maintain the *reservation state* in each node along the path. The RSVP protocol contains features such as *soft state*, *local repair* and *merging*. RSVP uses *soft state* to manage the reservation in routers and hosts. *Local repair* provides fast adaptation to routing changes. *Merging* is used to reduce signalling in forwarding reservation requests.

4. RELATED WORK FOR QOS PROVISION TO MOBILE NODES

In mobile Internet environment real-time services require both, mobility and QoS support. Several solutions addressing QoS provision to mobile hosts have been proposed, which may be grouped into the following schemes:

- Advance reservation,
- RSVP and Mobile IPv6 interworking.

4.1. Advance Reservation of Resources

There have been many proposals to perform reservation in advance. The challenging issue for these schemes is how to predict the MH's movement so that reservation in advance is done only in some of the potentially new cells. In [3], a Mobile RSVP (MRSVP) protocol is proposed, which allows a MH to make advance reservation along the traffic flow paths to and from the cells it may visit during the lifetime of the connection.

4.2. Interworking of RSVP and Mobile IPv6 Protocols

Interworking of Mobile IP and RSVP protocols is another approach for QoS provisioning to mobile users. Based on this approach, several schemes are proposed.

In [4], a combined scheme of RSVP and Mobile IPv6 is given. The basic feature of this scheme is that flow identification (source and destination addresses) is based on the MH's CoA. Each time that the MH changes the CoA during a session, an end-to-end RSVP signalling between the MH and the CH is needed. This results in high QoS signalling delay during handover. The QoS signalling delay from the time packets using the new CoA are released into the network until the time QoS handling mechanisms are configured along the new path is one round trip delay when the MH is a sender and one and a half round trips when the MH is a receiver.

To alleviate the above-mentioned problems a more advanced scheme, named Flow Transparency (FT) [5] is proposed. The idea of the FT concept is to keep host mobility transparent to the network layer flow-handling mechanism. This is achieved by keeping the flow identity constant, regardless of the change of the MH's CoA. Hereby, end-to-end resource reservation signalling, is avoided.. However, this has an impact on both the RSVP and Mobile IPv6 protocols. In order to keep the flow identity constant, Path and Resv messages use MH's home address to identify the flow. Concerning MIPv6 modifications, when MH is a sender, new header options called *the flow transparent route alert option* is carried in the hop-by-hop option header, whereas the MH's CoA and home address remain as they are in MIPv6. When the MH is a receiver, the flow handling mechanism can process packets according to the flow source and flow label parameter in MIPv6, without examining the packet's flow destination.

5. A FRAMEWORK FOR END-TO-END QOS PROVISIONING

5.1. Network architecture

The proposed network architecture (Fig.1.) for end-to-end QoS provisioning to mobile users, consists of IPv6 based access and core networks. Due to lack of scalability IntServ/RSVP model is applied on the access networks, whereas core network is assumed as pure MPLS domain, i.e. MPLS is used for both control and data traffic [6].

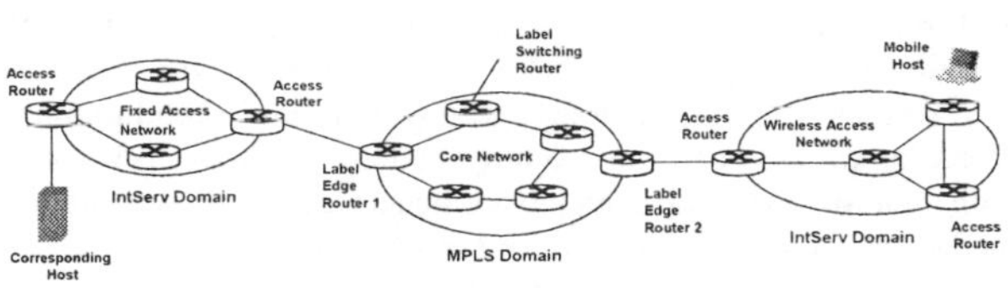

Fig .1: Network architecture for end-to-end QoS provisioning to mobile users

We assume that RSVP and MIPv6 messages are transparently transmitted through the MPLS network. For the establishment of Label Switched Paths (LSPs) in the MPLS domain RSVP-TE signalling protocol is assumed. Consistent with RSVP protocol, admission control functions for requested access network resources are performed at the routers, hop by hop, along the path form the receiver to the sender. For the MPLS domain, admission control is performed at the ingress Label Edge Router (LER). Also at the LER of the MPLS domain, service mapping between service classes of IntServ and MPLS is performed. A proposal of possible scenario for service mapping from IntServ classes to MPLS service classes is given in the Table 1.

Table 1. Service mapping from IntServ to MPLS

IntServ Service	MPLS EXP Field	MPLS Service Class
GS	111	MGS
CL_1	110	MCL_1
CL_2	101	MCL_2
CL_3	100	MCL_3
CL_4	011	MCL_4
CL_5	010	MCL_5
CL_6	001	MCL_6
BE	000	BE

In the proposed model, the following QoS classes for IntServ are defined: **guaranteed service** class with the highest priority; **control-load services**, divided into six subclasses, from CL_1 to CL_6, with different priorities; and Best Effort (BE) service, with lowest priority. Based on the experimental bits on the MPLS header up to eight MPLS service classes can be defined. In order to match to the defined IntServ QoS classes, we have also defined eight appropriate MPLS service classes as follows: **MGS** (MPLS guaranteed service) for guaranteed service class, six **MCL** (MPLS controlled-load service) classes, from MCL_1 to MCL_6, and the **BE** service class.

5.2. Protocol description

For simplicity, we will consider only the case when mobile host is a receiver in a wireless IntServ access network, whereas the correspondent host is a sender in the fixed IntServ access network. We also assume that RSVP signalling messages travel end-to-end between CH and MH to support RSVP/IntServ reservations outside the MPLS network domain and are tunneled across the MPLS domain.

Signalling message flow for new connections

Related to the defined architecture (Fig.1), the following message sequences illustrate the process by which an application obtains resource reservation for end-to-end QoS provision [7].

1. The QoS process on the sending host CH generates an RSVP Path message that describes the traffic offered by the sending application. This message is send towards the MH.
2. At the LER 1, the Path message is subjected to standard RSVP processing and *path state* is installed in the router. The Path message is ignored by Label Switching Routers (LSR) in the MPLS network domain and then processed at LER 2 and send onward to access network towards the MH.
3. When the Path message reaches Mobile Host, a Resv message indicating the offered traffic of a certain IntServ service type, is generated.
4. The Resv message is carried back towards the MPLS network domain and the sending host. Consistent with standard RSVP/IntServ processing, it may be rejected at any RSVP capable node in the path if resources are insufficient to carry the traffic requested.
5. At the LER 2, the Resv message may be rejected if resources on the downstream interface of the LER 2 are insufficient. If it is not rejected, it will be carried transparently through the MPLS network domain, arriving at the LER 1.
6. In the LER 1, the Resv message triggers admission control processing. The LER 1 compares the resources requested in the RSVP/IntServ request to the resources available in the MPLS network at the corresponding MPLS service level.
7. If the LER 1 approves the request, the Resv message is send upstream through the network domain to which the CH is attached. Any RSVP node in this network domain may reject the reservation request due to inadequate resources or policy. If the request is not rejected, the Resv message will arrive at the sending host, CH.

8. At CH, the QoS process interprets receipt of the Resv message as indication that the specified traffic flow has been admitted for the specified IntServ service type.

Signalling message flow when handover occurs

In this section, we present a new protocol for QoS provisioning to mobile users when handover occurs [8].

The proposed protocol is based on an existing flow transparency Mobile IP and RSVP interworking scheme and can be considered as its complimentary for meshed access network topologies. The basic idea of our proposal is to use the previous access router(s) as a Nearest Crossover Router (NCR) for the old and the new added flow paths. We have added two new RSVP messages: PathState_discovery and PathState_reply, to optimise the use of already reserved resources. As far as MIPv6 is concerned, for address mismatch avoidance, we propose the solution given in Section 4.2.

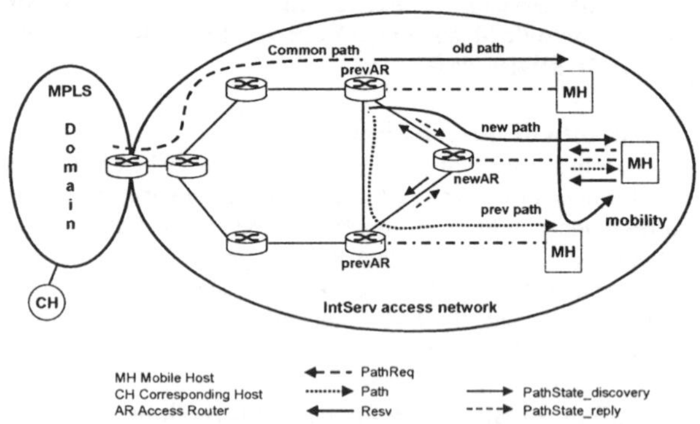

Figure 2. Handover within access network

When the mobile host acts only as a receiver the mobility information, which contains the home address and the CoA of the MH, is carried in the PathReq message. When the access RSVP router receives the PathReq message from MH, it checks first if it is the DNCR (Downlink NCR), by comparing the MH's home address with the existing one in the path state information. The following two cases may be considered:

1. If the new access router is not a DNCR, then the router sends a PathState_discovery message, containing the flow address, to all adjacent routers. All routers that receive this message with reply with the PathState_reply message for the indicated flow, which besides other needed information contains a one-bit flag to indicate whether the router has already a related *path state*.

1.a. If two or more adjacent routers have a *path state*, than the access router selects the next hop router, based on the routing table, and forwards the PathReq message (Fig. 2). Upon receiving the PathReq message, the selected router will respond with a Path message towards the MH's new CoA and will also send a PathTear message to the old MH's CoA to trigger the release of the old reserved resources. After receiving the Path message the MH sends Resv message to reserve resources along the new path.

1.b. If none of the adjacent routers has a Path State then the access router selects the next hop router, and sends the PathReq message to that router. The selected router will forward the PathReq message to the next hop router. Each next hop router will check for being a DNCR. If it is, it will reply with a Path message towards MH CoA, if it is not, it will forward the PathReq message to the next hop router.

2. If the new access router is a DNCR, then the protocol procedures are the same as in the existing protocol [5].

6. SIMULATION AND PERFORMANCE EVALUATION

Performance evaluation of the proposed protocol is done through software simulations.

6.1. Simulation environment

The basic topology of the simulated network is hierarchical with the lowest level being meshed (Fig.3), Nodes **1** to **7** are access routers and we assume that each router is responsible for one wireless subnetwork. Nodes **8** to **10** are intermediate routers in the access network, whereas node **11** is the gateway to the core network. In the core network only the correspondent host and the home agent are shown, nodes **12** and **13** respectively. There are four types of links with different delays and capacities. Links **a** are used to connect mobile hosts to access routers and they are assigned a constant delay of 1 ms and a constant capacity of 2Mb/s. Links between the access routers are denoted with **b** and they have a delay of 2 ms. All other links in the access network, denoted with **c**, have a delay of 4 ms. To the links **d**, in the core network, a delay of 20 ms is assigned.

By assigning different capacities to the links, we have simulated two network congestion scenarios. In the first scenario, we assumed that there is congestion in the access network, modelled by assigning a capacity of 2 Mb/s to the links in the access network and 10 Mb/s to the links in the core network. In scenario 2, congestion is assumed in the core network by assigning a capacity of 2 Mb/s to the links **d** and a capacity of 10 Mb/s to the links **b** and **c**.

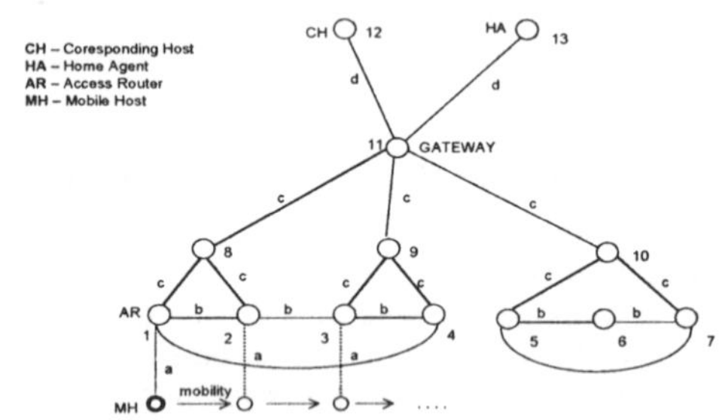

Figure 3. Access network with meshed topology

Two mobility scenarios are simulated considering one MH in the network. In the first scenario, we assume that the MH is moving in the pattern 1-2-3-4-5-6-7, which is typical for the case when a MH moves along a highway. In the second mobility scenario, we assumed the mobility pattern 1-2-3-2-1-4-5-6-7. In both mobility scenarios, handover occurs every 10 seconds i.e. the MH stays in each subnetwork for 10 seconds.

Two types of traffic are encountered in simulations: the traffic from the CH to the MH and the background traffic. Traffic from the CH to the MH is a 500 kb/s exponentially distributed with constant packet size. The background traffic consists of 7x 64 kb/s streams transmitted from the CH to the access routers. This traffic is also exponential distribution with a constant packet size.

In order to demonstrate some of the features of the proposed protocol, we have also simulated the existing protocol presented in the first paragraph of the Section 4.2. Both protocols are simulated for the case when the MH is a receiver.

6.2. Performance evaluation

Performances of the proposed protocol are evaluated in terms of the RSVP signalling delay and the maximum allowed delay for QoS provisioning. The RSVP signalling delay is defined as the amount of time elapsed since a MH acquires a new CoA until resources in the new added path are reserved. The maximum allowed delay for QoS provisioning is defined as the time from the instant when the BU is sent to the CH until the first packet with the new CoA arrives at the NCR.

The simulation results for the proposed and the conventional protocol are shown in Fig. 4. First it can be seen that, for the proposed protocol, the RSVP signalling time is smaller than the maximum allowed delay for QoS provisioning at all handover instants, whereas in the case of the conventional protocol it is the other way round. This means that during all handover instants, the proposed protocol ensures QoS provisioning for all packets that use the new care-of address. In the conventional protocol such packets lack QoS after the BU is processed in the CH and before QoS for the new path is ensured.

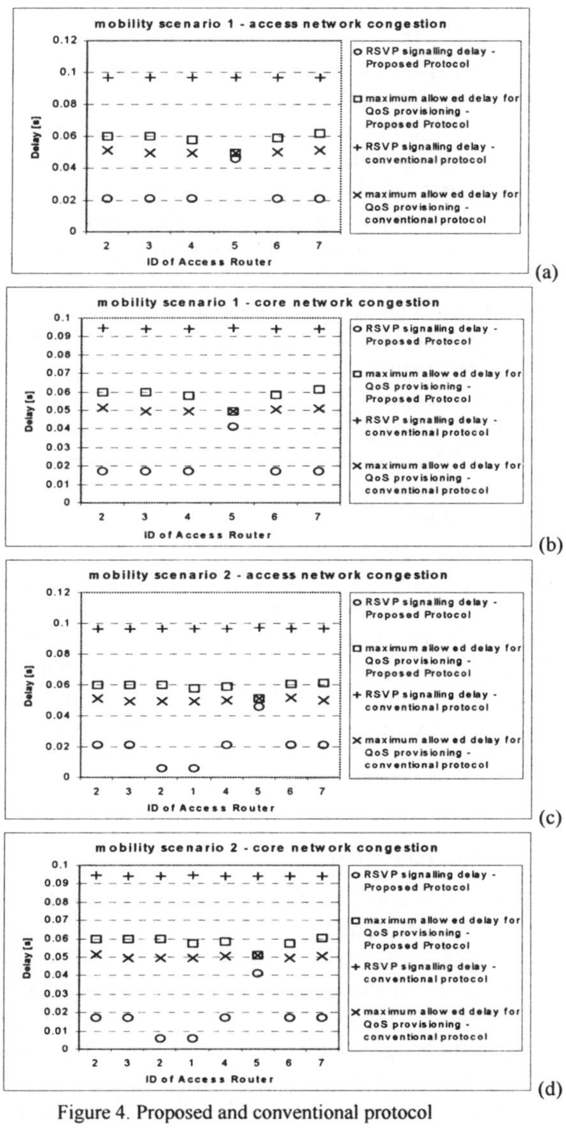

Figure 4. Proposed and conventional protocol

Further on, simulation results show that for both congestion and mobility scenarios, the proposed protocol obviously results in much lower RSVP delays. During the handover from subnetwork 4 to subnetwork 5 (where there is no direct link in between), the RSVP delay is higher than in other handover cases, due to the fact that in this case the NCR is node **11** (gateway) and the *path state* discovery procedure applied in our protocol (PathState_req and PathState_reply signals) results in longer delays. From the plots also it can be seen that the maximum allowed delay for QoS provisioning is longer in the case of the proposed protocol, because the NCR is located closer to the MH than in the case of conventional protocol. The handover from subnetwork 4 to 5 is an exception, where the values are equal because in this case node **11** (gateway) serves as NCR.

Concerning the congestion scenarios, the simulation results show that the congestion in the access network results in slightly longer RSVP delays than congestion in the core network for all handover cases. This is obvious as more congested links are involved in the path connection. The maximum allowed delays for QoS provisioning are also slightly longer for the access network congestion scenario. In general, in our simulations there is not a big difference between the two congested scenarios. There are two reasons for this: First we have generated very low background traffic and second we have assigned the highest priority to the signalling messages.

Comparing the results between mobility scenario 1 and 2, it can be seen that the proposed protocol results in much lower RSVP signalling delay, which shows the advantage of the proposed protocol for the meshed access networks.

7. CONCLUSION

In mobile Internet environment, real-time services require both mobility and QoS support. Hence, in this paper a framework for end-to-end QoS provisioning to the mobile Internet users, encountering heterogeneous access networks, has been presented. In order to meet mobile user's QoS requirements for real-time services, a new protocol has been proposed. The proposed protocol is based on an existing flow transparency Mobile IP and RSVP interworking scheme and can be considered as its complimentary for meshed access network topologies. The basic idea of our proposal is to use previous access router(s) as a NCR for the old and the new added flow paths. The proposed protocol requires minimum extensions to the RSVP and Mobile IPv6 protocols, keeping the basic principles of the protocols.

The simulation results, in terms of RSVP signalling delay and the maximum allowed delay for QoS provisioning, show that the proposed protocol leads to significant improvements in comparison to the conventional protocol, in particular for meshed access networks.

REFERENCES

[1] D. Johnson and C. Perkins, "*Mobility support in IPv6*", IETF Internet draft, draft-ietf-mobileip-ipv6-12.txt, April 2000.

[2] IETF RFC 2205, "*Resource reservation Protocol (RSVP)* Version 1, Functional Specification", September 1997.

[3] A. K. Talukdar, et al "MRSVP: A Resource reservation Protocol for an Integrated Services Network with Mobile Hosts", Wireless Networks, Vol. 7 No.1, Jan/Feb 2001, pp. 5-19.

[4] G. Chiruvolu, A. Agrawal, and M. Vandenhoute "Mobility and QoS support for IPv6-based real-time wireless Internet traffic", 1999 IEEE Int. Conf. on Communications, Vancouver, BC, Canada, June 1999, Vol.1, pp.334-8.

[5] Shen et al, "An interoperation framework for using RSVP in Mobile IPv6 Networks", IETF Internet draft <draft-shen-rsvp-mobileipv6-interoip-00.txt>, July 2001.

[6] E. Horlait and N. Rouhana, "*Differentiated Services and Integrates services Use of MPLS*", http://www-rp.lip6.fr/~eh/Files/mpls.pdf

[7] Y. Bernet at al., "*A Framework for Integrated Services Operation over DiffServ Networks*", NWG RFC 2998, November 2000.

[8] S. Lepaja, R. Fleck, R. Donner and N. Hoang, "*QoS Provisioning to Mobile Internet Users*", In the Proceedings of the ECUMH 2002, April 2002, Colmar, France.

Adaptive Quality of Service for a Mobile Ad Hoc Network

Antonis Dimakis, Linhai He, John Musacchio, Hoi-Sheung Wilson So, Teresa Tung, Jean Walrand
Department of Electrical Engineering and Computer Sciences
University of California, Berkeley, CA 94720
{dimakis, linhai, musacchj, so, teresat, wlr}@eecs.berkeley.edu

Abstract— **This paper presents a QoS routing system for MANET supporting multiple traffic classes. The system takes into consideration clustering and channel allocation. Simulation experiments show that our algorithms are convergent. The system also yields a higher total throughput compared to the case in which every interface uses the same channel.**

I. Introduction

A. *Motivating Scenario*

Cellular phone systems have been tremendously successful but they rely on a fixed infrastructure. In some cases, such as after an earthquake, it is desirable to set up a network in an area without any infrastructure as quickly as possible. In this paper, we present the design and simulation results of a QoS routing system that can provide different classes of service in a mobile ad hoc network. We believe the entire QoS routing system should be considered as a whole because optimizing different components of a system separately may lead to poor performance, or worse, an unstable system. In this paper, we take a first step towards an integrated system by considering clustering and channel allocation in addition to routing.

B. *Related Work*

Traditionally, QoS routing applies to only fixed wireline networks, see e.g., [2] and the references therein. There have also been many ad hoc routing protocols proposed for MANETs without QoS support as in [12]. Recently, there have been proposals to provide QoS in MANETs as in [3], [9]. In [3] the authors propose an algorithm to minimize utilization that does not take into account the effects of the coupling of interference between neighboring nodes. The work in [10] considers coupling, but the authors use a centralized integer programming approach that does not scale well as the number of nodes becomes large.

Previous work on clustering focuses on metrics such as cluster diameter ([8],[1]) and stability [11]. Our work tries to respect the natural clusters by focusing on the number of inter-cluster links. Although such links are necessary for intercluster communication, only a subset of them are used as gateway links. By minimizing the number of intercluster links, one also minimizes the number of unused links. This metric is considered in [6] using a centralized solution for a stationary network. We propose a distributed algorithm designed for a dynamic environment.

The frequency assignment problem for cellular networks has been studied extensively. [5] provides a comprehensive survey on different approaches to this problem. However, those solutions are not suitable for our purpose as they are designed for fixed infrastructure networks. Furthermore, our problem requires each node maintain connectivity to the rest of the network in addition to reducing interference. We solve the channel allocation problem using simulated annealing which has been widely applied to many combinatorial optimization problems. [14] provides a good coverage of its theory and some of the typical applications. However, most simulated annealing work assumes a central agent has all the state information and fully controls the adaptation process. Our work applies simulated annealing technique in a completely distributed way.

C. *Assumptions*

In this paper, we will use the terms *node* and *router* interchangeably. We assume each node of the ad hoc network is mobile and has one or more wireless network interfaces. Each interface can operate at one of the many independent channels (i.e., frequencies) chosen from a predetermined set. Equipping nodes with multiple interfaces can increase the capacity of the network. These interfaces can have different maximum transmission ranges and link rates. The rate of a link between two nodes depends on the channel quality. We assume that the interfaces use a contention-based MAC protocol suitable for a multihop ad hoc network but the design of such MAC protocol is outside the scope of this paper.

The network supports unicast applications with different priorities and bandwidth requirements. Our system supports the following classes of application in decreasing order of priority: (1) Urgent Messages (UMs) are connectionless and they demand minimum end-to-end delays. (2) Urgent Flows (UFs) are urgent calls with low to medium bandwidth requirements of up to about 100Kbps. They require fast call set-up, fast rerouting in case of failures, but are relatively infrequent. (3) Regular Flows (RFs) are non-urgent voice and video calls with potentially high bandwidth requirements. They are frequent and demand a low blocking probability. (4) Best Effort (BE) flows are connectionless and do not require any bandwidth guarantees.

II. System Description

A. *System Architecture*

Our system consists of three major components: clustering, channel allocation, and route computation. An efficient routing system must support load balancing of traffic. This capability requires some knowledge of the network topology beyond the local neighbors. However, without any hierarchal structure, each node would have to gather global topology information and incur a high control overhead. Our system partitions the network into clusters of nearby nodes and perform hierarchical routing at two levels (within a cluster and across clusters) to reduce control overhead. Another advantage is when a link fails in a transit cluster, local rerouting of flows within the cluster may suffice to contain the fault without bothering the end nodes. Although hierarchical routing is sub-optimal because some links are not used, we believe the differences are small and acceptable.

The clustering algorithm dynamically organizes the nodes

into a two-level hierarchy. Each cluster is connected to nearby clusters via one or more *gateways* which are nodes at the boundary of two clusters. These gateways, also determined by the clustering algorithm, are responsible for finding routes for flows crossing clusters. The goal of the channel allocation algorithm is to assign, in a distributed fashion, suitable channels to different interfaces to maximize the network throughput while keeping the network connected. The channel allocation algorithm implicitly defines the links in the network.

Within each cluster, each node participates in a link state protocol to exchange information on the topology within the cluster. In addition, all gateways in the network participate in another link state protocol exchanging connectivity information at the cluster level. Based on the link information within a cluster and the cluster summaries exchanged among the gateways, a distributed routing algorithm finds the best routes for different flows and traffic classes. The QoS routing algorithm is also responsible for rerouting flows after a failure. We describe each of these three components in the following sections in details.

B. Clustering Algorithm

Our objective is to form clusters near size n while minimizing the resulting number of intercluster links. We choose the target cluster size n to be the square root of the total number of nodes in order to minimize the sizes of the routing tables. In our clustering algorithm, each node maintains two types of clustering: prospective and actual. The following distributed algorithm computes the prospective clustering. A leader node is associated with each prospective cluster. Initially, every node starts as its own prospective cluster and as the leader of that cluster. At fixed intervals, all nodes broadcast a vector, (node ID, cluster ID, path to the leader, number of nodes in cluster), to their one-hop neighbors. Each node also sends a hello message to the leader node of its cluster. By counting the hello messages, the leader determines the current size of the cluster. A node updates this size from a neighbor with the shortest path to the leader. If no such neighbor exists, the field is set to infinity.

Nodes dynamically update their cluster ID based on their neighbors' broadcast. After waiting for a random period of time, each node decides whether to update its cluster ID with probability P_{switch}. P_{switch} is lower when the number of nodes in a cluster is near n. The updated cluster ID of a node is a random variable that takes on values from the set of cluster IDs of neighboring nodes and a new cluster ID. The value is chosen in a way that favors the more popular IDs and better cluster sizes.

The prospective clustering may vary frequently and take on suboptimal clustering configurations. However this behavior is necessary to discover better clustering configurations. The actual clustering is updated from the prospective one at a slower rate. The more stable actual clustering is used for channel allocation and routing. The algorithm also chooses gateway nodes so exactly one link exists between any two neighboring actual clusters.

C. Channel Allocation

One may formulate channel allocation as an integer program. However, this approach is unsuitable for distributed implementations because its solution requires global network knowledge. We propose to use a simulated annealing approach to solve our combinatorial optimization problem as in [7] [13].

By an intra-cluster link state protocol, each node can determine the set of reachable nodes within its cluster and the routing distances. Whenever a node becomes idle, it changes the channel on one of its interfaces and tests how good the new channel is. This goodness, called potential in simulated annealing literature, is defined by the number of neighbors and how well it is connected to the rest of the network. If its potential is reduced, it adopts this new channel; otherwise, it reverts to the original one with a probability depending on the magnitude of the increase. As the adaptation progresses, this probability is gradually discounted by a parameter called temperature to achieve convergence.

A node stays on a channel for a random period before switching. The length of this period depends on the potential and the temperature. The higher they are, the faster a node tends to switch. This design stabilizes the distributed algorithm and speeds convergence. When the network topology changes suddenly, a surge in the potential of affected nodes triggers the temperature to be *reheated* and the adaptation process restarted.

D. Routing

The presence of different priority levels in many disaster relief and military applications prompts us to use different routing algorithms for flows of different levels:

1. Urgent Messages (UM) are sent over shortest paths in terms of end-to-end transmission delay. Currently, a static link cost of $1/C$ is used, where C is the link speed, but we plan to incorporate queueing delays too.
2. Urgent Flows (UF) use shortest paths with link cost given by $1/C + \lambda I$, where C is the link speed, I the amount of bandwidth used by interfering links, and λ a tuning parameter. I is calculated by information obtained through a link state protocol (see below) and is the rate at which a link is busy either sending or receiving data. Note that its value is not obtained by measurements, but by information contained in flow setup and teardown requests. The parameter λ captures the tradeoff between resource availability and delay optimality.
3. Regular Flows (RF) uses the least loaded routes. Because interference constraints are coupled, the resulting program is NP-hard [10] and hence a *myopic* algorithm is unlikely to work. For these reasons, upon arrival of a new RF call, our algorithm solves a linear program that maximizes available capacity to find the route for the flow. solved. The solution of a linar program splits a single flow over multiple paths in general. We discretize the solution by selecting the least loaded link on each hop, starting from the source node[1].
4. Best Effort(BE) use shortest hop count paths (unit link costs), as is the case for signaling traffic.

Using shortest path algorithms for UMs and UFs avoids the high route computation delay associated with RFs. The use of an additive interference link metric for UF is justified because these flows account for only a small fraction of the traffic. On the other hand, this is certainly not the case for RFs.

UF packets have priority over RF and BE packets, while UM packets have priority over UF. When the interference metric I

[1] Loops are prevented by appropriately perturbing the objective function

for a link is calculated, as is needed in the routing algorithm for UF, the bandwidth usage of RF and BE for *that link* is not taken into account. On the other hand, since we not assume any MAC layer QoS capabilities, *I* needs to include RF and BE bandwidth usage of interfering links.

When a node attempts to allocate link bandwidth for a UF or an RF, it first requests approval from all of its interfering neighbors. The request is granted by a neighbor, if and only if it does not violate the interference constraints on any of that neighbor's links. If the allocation request is for UF or RF traffic, then a neighbor might need to inform its neighboring nodes to throttle back their BE traffic by an appropriate amount.

For scalability reasons, we define a two level routing hierarchy. At the lower level, node clusters exchange network state information within the same cluster. This is done by using a link state protocol similar to OSPF but with additional information such as per class bandwidth usage of interfering nodes. Cluster gateways export a topology summary of the cluster to other gateways in different clusters using another link-state protocol.

Link failures due to mobility are handled at two levels. The gateways of the cluster containing the failed link try to reroute any transit flows crossing the cluster if possible. Flows that cannot be rerouted locally are handled by the gateways of the source and destination nodes. At both levels, many flows may need to be rerouted by the same node. In these cases, the new paths are computed solving a single optimization problem for each class: Urgent Flows are routed first, then Regular Flows.

III. Simulation Results

A. Simulation Framework

To evaluate the performance of our proposed system, we implement a simple time-driven flow level simulator in MATLAB [4]. The simulator models the arrival and departure of flows with different requirements, and the volatile link connectivity among the nodes due to mobility. However, individual packets are not simulated. All control information is kept in a global data structure in the simulator. To simulate the distributed nature of the system, different lgoical instances of various algorithms can only access part of the data structure.

In our mobility model, the nodes belong to static administrative groups. The members of each group roughly travel together, but the relative positions of the nodes vary over time randomly. The center of gravity of each group follows a random walk over a square area. The speed and the direction of travel of the nodes can change by a small amount in every time step. As the groups move, they may overlap. The simulator models the wireless links of various speeds. Each link can take on a speed from a set of predefined speeds based on the distance between the two interfaces.

B. Clustering Algorithm

We compare the results of our distributed clustering algorithm with an optimal solution computed by a slow, centralized, exhaustive search. We generate scenarios of 25 nodes with an ideal cluster size n of 5 with various average node degrees. The target cluster size used in this section is also 5. The network is stationary. We run our algorithm for 20 iterations.

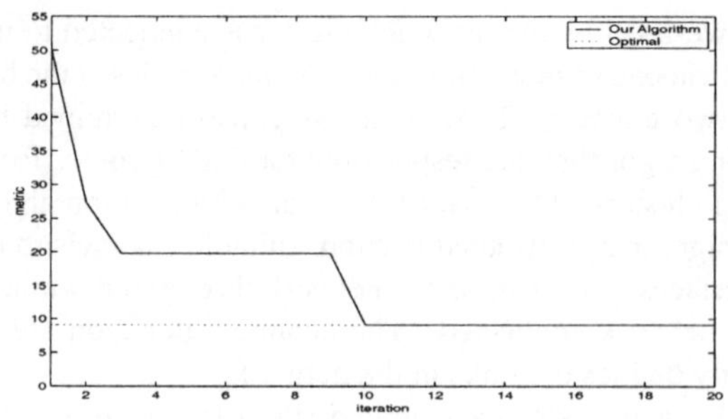

Fig. 1. Metric after each iteration.

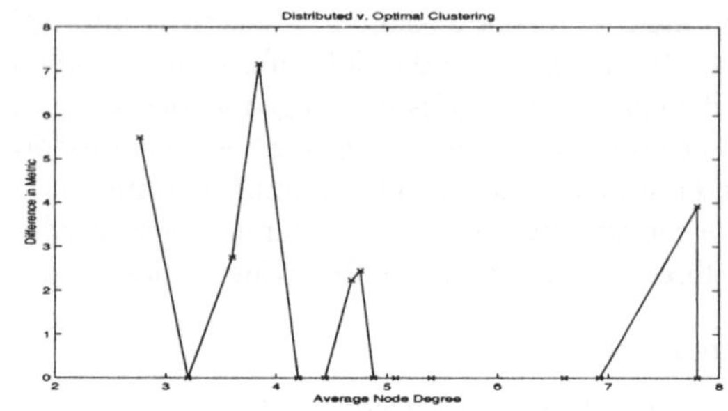

Fig. 2. Performance difference between ours and the optimal scheme for scenarios of different average node degrees.

Define the mean square error of a clustering (*MSE*) to be the sum of the squared difference between each cluster size and n, divided by the total number of clusters. We define the goodness metric (or simply the metric) of a particular clustering to be the sum of *MSE* and the total number of intercluster links. Fig.1 shows the how this metric changes after each iteration for a single scenario. The optimal metric is shown by the dotted line. Our algorithm does not always converge to the optimal. Fig.2 shows the performance difference for various scenarios denoted by average degree. For these scenarios, the difference is small even after a limited number of iterations especially when the optimal cluster size is near the target cluster size. However, our distributed algorithm computes the clustering in seconds while the centralized exhaustive search take days.

C. Clustering and Channel Allocation

To study the interaction between clustering and channel allocation algorithms, we simulat a 100-node network divided into 10 groups. Nodes move in groups as described earlier and reorganize themselves using the clustering algorithm. Once new clusters are formed, nodes execute the channel allocation algorithm to find the best way to connect with each other. Fig.3 shows the average potential of all nodes over time. In this scenario, nodes stop moving after 70 time steps (each step corresponds to 2.5 seconds). The channel allocation algorithm is able to at least maintain the performance of the network when nodes are moving. After that, it gradually "cools" down and settles into its steady-state, as indicated by the downward trend in the last part of the curve. For the same scenario we also show the gain in the average throughput per node that our algorithm can achieve in Fig.4. The curve at the bottom of the plot shows the average throughput per node when all interfaces are assigned to

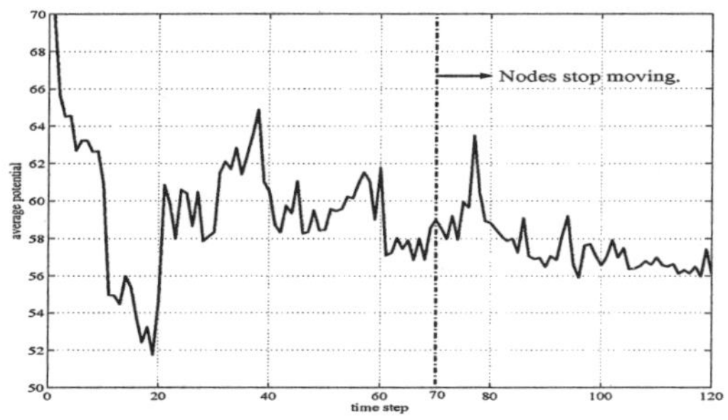

Fig. 3. Average potential vs. time.

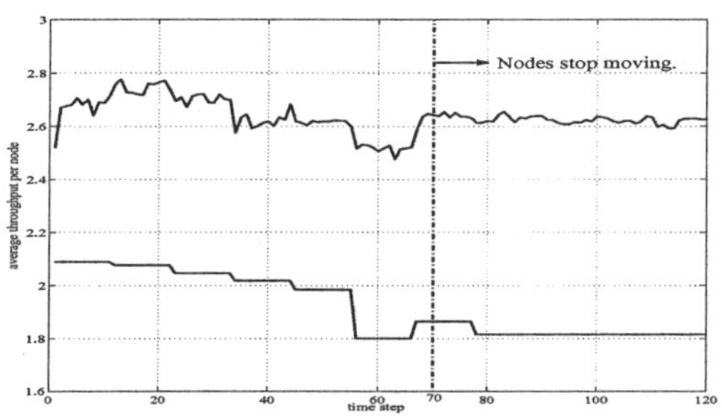

Fig. 4. Througput vs. time.

the same channel. So in this experiment our algorithm achieves a factor of 1.44 increase in average throughput per node.

D. Routing

In this section, we present the simulation results of our routing algorithm running on top of a stationary network. We compare our routing algorithms with one that does not account for the coupling between interference among neighboring links. Namely, each time a Regular Flow arrives it is routed over the path with the highest available bandwidth without considering interference between neighboring links. This algorithm can be seen as a myopic version of our RF algorithm, in that the available bandwidth of each link is independent of other links. A similar myopic algorithm has been used in [3].

The simulated network consists of a single cluster of 20 nodes. Each node has a random number of radios that varies from one to three, and each radio can use a fixed frequency out of three possible. The topology was randomly generated and is depicted in Fig.5. Regular flows arrive randomly in the network at a very high rate averaging 1 flow each second.

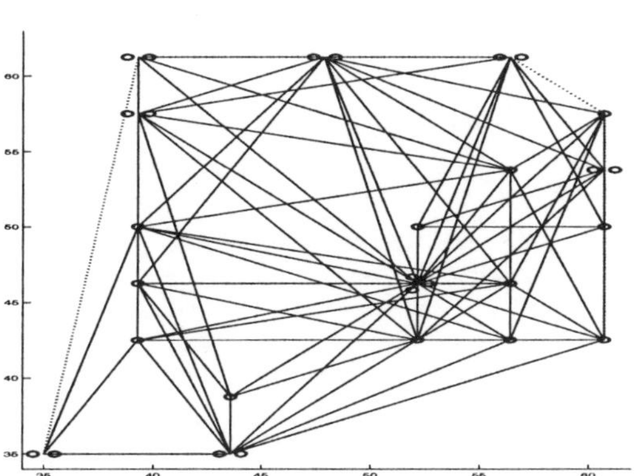

Fig. 5. Network topology used in simulation.

We ran two simulations: one using our algorithm and another one using the myopic one, with the same arrival pattern (generator seed) for 500 seconds of simulation time. Our algorithm accepted rejected 122 flows out of 613 (rejection rate 20%), while the myopic algorithm rejected 142 (rejection rate 23%).

IV. Conclusion

In this paper, we have presented a QoS routing system that considers both channel allocation and clustering. Our simulation results also shows fast convergence of our clustering algorithm producing good sized clusters after about 20 iterations. Additionally, once the clusters reach the target cluster size, the clustering remains stable when network topology is stable. In networks with mobility (when the neighbors of each node vary slightly per iteration), the algorithm is able to adapt to changes and assign clusters accordingly.

Finally, we have presented a set of fast routing algorithms that provide a diverse set of QoS guarantees to different applications. This was made possible by introducing a two level routing hierarchy that decomposes the routing computation into several lower complexity subproblems. This hierarchical approach is necessary in a large wireless network where the provision for QoS in the presence of interference constraints significantly increases the complexity of the problem.

This work is part of a joint project between Cisco and UC Berkeley. David Jaffe is the lead contributor from Cisco. The UC Berkeley team includes Venkat Anantharam, Eric Chi, Bill Hodge, Zhanfeng Jia, David Tse, and Pravin Varaiya. We would especially like to thank DARPA (contract no. N66001-00-C-8062) for their funding support for the project.

References

[1] S. Basagni, I. Chlamtac, and A. Farago. A Generalized Clustering Algorithm for Peer-to-peer Networks, July 1997.

[2] S. Chen and K. Nahrstedt. An overview of quality-of-service routing for the next generation high-speed networks: Problems and solutions. *IEEE Network Magazine*, 1998.

[3] S. Chen and K. Nahrstedt. Distributed quality of service routing in ad-hoc networks. *IEEE JSAC*, 17(8), 1999.

[4] The MathWorks Inc. Matlab 6.5 release 13, June 2002.

[5] I. Katzela and M. Naghshineh. Channel assignment schemes for cellular mobile telecommunications: A comprehensive survey. *IEEE Personal Communications*, pages 10–31, 1996.

[6] R. Krishnan, R. Ramanathan, , and M. Steenstrup. Optimization Algorithms for Large Self-Structuring Networks. In *Proc. IEEE INFOCOM, 1999*, New York, March 1999.

[7] P. Laarhoven. *Simulated Annealing: theory and applications*. Kluwer Academic Publishers, 1987.

[8] C. R. Lin and M. Gerla. Adaptive Clustering for Mobile Wireless Networks, September 1997.

[9] C. R. Lin and J-S. Liu. Qos routing in ad hoc wireless networks. *IEEE JSAC*, 17(8), 1999.

[10] E. Magana, D. Morato, H. W. So, B. Hodge, J. Walrand, and P. Varaiya. A Wireless Overlay Network with QoS Capabilities, December 2002.

[11] A. McDonald and T. Znati. A Mobility-Based Framework for Adaptive Clustering in Wireless Ad Hoc Networks, August 1999.

[12] C. E. Perkins and P. Bhagwat. Highly dynamic destination-sequenced distance vector routing (dsdv) for mobile computers. In *ACM SIGCOMM'94*, London, UK, September 1994.

[13] B. Preneel and J. Walrand. Convergence of a Quasi-Static Frequency Allocation Algorithm. In *technical report to ERL at U.C. Berkeley*, 1995.

[14] P. J. M. van Laarhoven and E. H. L. Aarts. *Simulated Annealing : Theory and Applications*. Kluwer Academic Publishers, 1987.

SESSION 3B:
ENERGY CONSUMPTION

ENERGY BALANCED COMMUNICATION IN WIRELESS SENSOR NETWORKS

AMOL BAKSHI AND VIKTOR K. PRASANNA
Dept. of Electrical Engineering-Systems, University of Southern California, 3740 McClintock Ave, EEB200C, Los Angeles, CA 90089-2562 USA
Ph: 213-740-4483, Fax: 213-740-4418
Email: {abakshi, prasanna}@usc.edu

We propose threshold randomized routing (TRR) in multi-hop wireless sensor networks as a strategy for structured communication that provides a ***tradeoff among total energy consumption, total latency, and energy balance.*** Energy balance is a special concern because the collaborative nature of sensor network applications can render an energy-constrained sensor network useless if energy imbalance leads to a network partition. We evaluate the effectiveness of this strategy for worst-case and average-case permutation routing scenarios for a range of network sizes. This work is part of our effort to define a hierarchical methodology for application development for wireless sensor networks, based on energy-efficient communication and computation primitives that hide networking details from the application developer.

1 Introduction

Wireless sensor networks (WSNs) are typically designed and optimized for a specific application that is known at design time. Due to the limited energy supplies available to individual sensor nodes, the major challenge facing the designer is to maximize the lifetime of the network through energy-efficient implementation of communication patterns known *a priori*[a]. In addition to the total energy consumed by a particular (collaborative) task, energy balance is also a critical measure of energy efficiency. Intuitively, energy balanced implementations will maximize the time to network partition, which is one of the measures of system lifetime.

This paper deals with structured communication, i.e., routing problems where the communication pattern is known in advance. Unlike the Internet, where point-to-point communication is the basic primitive, the collaborative nature of communication and computation in WSNs makes structured communication relevant. Example primitives are one-to-all, all-to-one, many-to-many, permutation, etc. We propose and evaluate *threshold randomized routing* (TRR) as a general routing strategy that provides a smooth tradeoff between latency, energy, and energy balance. Although we simulate multi-hop mesh topologies for simplicity, the routing technique itself is not restricted to a particular topology.

In 1-1 permutation routing, each node transmits one packet to some other node and expects to receive one packet. Energy-efficient permutation routing in single-hop wireless networks has been the subject of recent research[3]; here we compare energy-balanced routing mechanisms for permutation routing in multi-hop networks. *Matrix transpose* is a pathological case of permutation routing in terms of network congestion and the corresponding impact on latency. In addition, naive shortest-path routing for matrix transpose leads to a severe energy imbalance in the network. Matrix transpose is interesting from a purely theoretical perspective because its extremal behavior makes it a benchmark to evaluate efficient permutation routing schemes. In terms of sensor network applications, matrix transpose is a core operation in distributed digital signal processing such as multidimensional FFT. An efficient implementation of matrix transpose could improve the performance of *collaborative signal processing* kernels for wireless sensor networks[4].

The remainder of this paper briefly describes our system model, introduces threshold randomized routing, and evaluates its performance for two different permutation routing patterns that represent the worst-case and average-case performance respectively of greedy routing strategies.

2 System Model

We consider a homogeneous wireless sensor network, where each node is equipped with one radio with omnidirectional antenna. The radio can be in the Transmit, Receive, or ShutDown state. Energy consump-

[a] Computation issues are not addressed in this paper

tion is same for the Transmit and Receive states; and zero (negligible) for ShutDown. The n^2 nodes in the network are distributed uniformly over a square 2D terrain. The terrain can be modeled as n^2 cells of unit size, with one node in each cell. Each node is identified by its row and column number; node in row i and column j is labeled (i, j), where $0 \leq i, j \leq n-1$. A node can communicate only with its four neighbors. All data is exchanged in fixed size packets with unicast addressing. A CSMA/CA mechanism with RTS/CTS signaling is used for channel access. The time (energy) for transfer of a data packet is d times the time (energy) for resolving contention over the channel, i.e., for transferring an RTS/CTS packet.

3 Routing Strategies in Multi-Hop Networks

Greedy Routing: Most routing protocols in multi-hop networks implement some variation of the greedy shortest path algorithm, which we consider the de facto baseline. For our simulated mesh topology, we use the well-known *XY routing* as the greedy routing algorithm. Each packet travels along the X dimension (row) till the destination column is reached and then along the Y dimension (column) till the destination node is reached.

Greedy algorithms are attractive because they are very simple to implement, rely solely on local decisions and require negligible state[b] per node for routing. For one-to-one routing problems, greedy algorithms perform very well in the average case and very poorly in the worst case[2] in terms of total latency. Since each packet takes the shortest path, the total energy of routing (ignoring contention) is also minimum for the greedy approach. In XY routing for matrix transpose however, all data items in a row move to the corresponding column. This causes *congestion* along the diagonals, which leads to worsening latencies as the network size increases. Of greater concern in the context of energy-constrained sensor networks is that *XY routing is not energy balanced* for patterns such as matrix transpose because heavily loaded diagonal nodes could lead to a network partition.

Randomized Routing: In the basic *two-phase randomized routing* technique proposed in [5], each packet in the network is initially sent to a random destination within the network and then to its correct destination. Each phase could use greedy routing. Randomized routing is known to be useful in reducing total latency of worst-case problems such as transpose[2], because the elements along a row are not held up waiting for the preceding elements to move one by one via the diagonal element. However, since randomized routing really converts the routing problem into two sub-problems, it could take twice as much time as greedy routing on average. Since randomization disperses the paths throughout the network, *energy balance* improves. Unfortunately, dispersing the routes through the network also causes packets to take longer paths to their destinations, thereby *increasing the total energy*.

4 Threshold Randomized Routing

Threshold randomized routing (TRR) is motivated by the need to achieve a *tradeoff between latency, energy, and energy balance*. Energy overhead for a given packet under randomized routing can be defined as the ratio of the increase in path length to the shortest path between the source and destination of that packet. Intuitively, the energy overhead with randomized routing (on average) is inversely proportional to the length of the shortest path between the source and destination. If the destination is 'close' to the source, randomized routing might not be desirable because of the energy and latency overhead, even if energy balance is achieved. On the other hand, if the destination is 'far' from the source, randomized routing will not cause much overhead, while still balancing the energy dissipation and possibly reducing the latency in worst-case scenarios.

In TRR, the packet is routed greedily if the shortest path between source and destination is less than a user-specified threshold, and randomly (two-phase) otherwise. The threshold can be tuned to obtain a smooth tradeoff between total latency, total energy, and energy balance. Decreasing the threshold will typically improve the energy balance and increase the total energy. The latency could increase or decrease, depending on the routing problem. Due to space limitations, we discuss the application of TRR only for two patterns: *matrix transpose* and *row shuffle*. Matrix transpose is the worst-case (latency)

[b]We assume that each node knows its position in the mesh.

scenario for permutation routing. The row shuffle pattern is an average case scenario, where node (i, j) sends a packet to node $(i, n-1-j)$.

Figures 1(a) and (b) illustrate TRR for these two patterns. For matrix transpose, the diagonal elements are most heavily loaded, and for row shuffle, the elements in column $\frac{n}{2}$ are the most heavily loaded. Therefore, a threshold of k results in the routing zones as shown in the figures. $k = 0$ represents complete randomized routing.

TRR might not be very useful if every node in the network is both a source and a destination of packets, e.g., in matrix transpose, the total number of packets that cross the diagonal are the same, regardless of how each packet is routed. TRR does reduce latency but energy balance is achieved not by reducing the load on the diagonals, but by increasing the load on non-diagonal elements. TRR will be of greater use if energy balance can be achieved by recruiting nodes that will otherwise not participate in the routing problem at all. For example, if the only activity in the network is node $(0,0)$ periodically sending packets to node $(n-1, n-1)$, greedy routing will cause a network partition. If TRR is used, the routes will be distributed throughout the network, thereby improving energy balance. Fortunately, for sensor network deployments with low duty cycles where activity in the network depends on the spatio-temporal properties of the events to be detected, the latter scenario can be expected to occur more frequently than the former.

5 Experimental Results

We simulated the network for three different values of the threshold. $k = 0$ and $k = 2n$ effectively modeled complete randomized and complete greedy routing. The intermediate value (labeled 'Hybrid') was chosen as $k = \frac{2n}{3}$ for matrix transpose, and $k = \frac{n}{2}$ for row shuffle. Assuming an RTS/CTS packet size of about 8 bytes and typical data packets to be 32 bytes, we used $d = 4$ for simulation purposes.

Figure 2 plots the total latency and total energy consumption of the three routing strategies for matrix transpose, for networks ranging from 100 to 2500 nodes. Figure 4 shows the same for row shuffle. As expected, greedy routing has the worst latency and best total energy for matrix transpose. Also, complete randomization does reduce latency but leads to a large increase in total energy. Hybrid routing with $k = \frac{2n}{3}$ represents an intermediate point with better latency than greedy with slightly more energy consumption. For row shuffle, the important difference is that greedy routing has the best latency as well as the best total energy.

Figures 3 and 5 show the energy balance of the algorithms for a 2500-node network. Lighter regions represent greater energy consumption. Clearly, energy balance of the hybrid and complete random routing strategies is much better than greedy. Simulator details and more analysis appears in [1].

6 Remarks

Threshold randomized routing is a powerful concept that is also very simple to implement. In terms of latency and total energy, we expect TRR (with suitable choice of threshold) to be as good as greedy routing in the average case, while still performing better than greedy approaches in worst-case scenarios as discussed in this paper.

Acknowledgments

This work is supported by the DARPA Power Aware Computing and Communication Program under contract no. F33615-C-00-1633 monitored by Wright Patterson Air Force Base.

References

1. A. Bakshi and V. K. Prasanna, *Application directed routing in wireless sensor networks*, Tech. Rep. (in progress), Dept. of EE-Systems, University of Southern California, 2003.
2. T. Leighton, *Methods for message routing in parallel machines*, Proc. 24th ACM STOC, 1992.
3. K. Nakano, S. Olariu, and A. Zomaya, *Energy-efficient permutation routing protocols in radio networks*, IEEE TPDS, 12(6), 2001.
4. K. Ramachandran, *Distributed sensor networks: Opportunities and challenges in signal processing and communications*, NSF Workshop on Distributed Communications and Signal Processing, 2002.
5. L. G. Valiant and G. J. Brebner, *Universal schemes for parallel communication*, Proc. 13th ACM STOC, 1981.

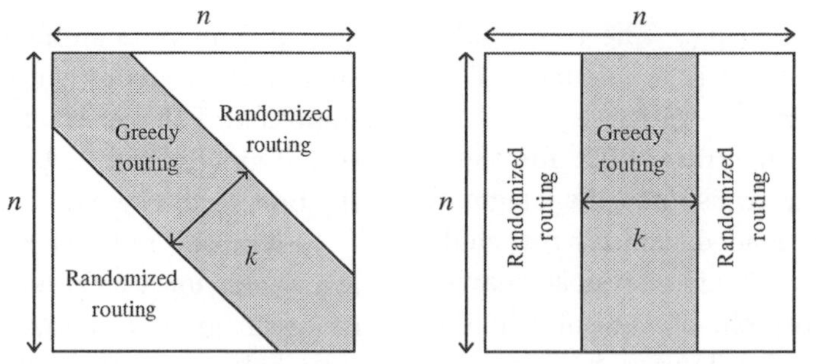

Figure 1. Threshold Randomized Routing: (a) Matrix Transpose, (b) Row Shuffle

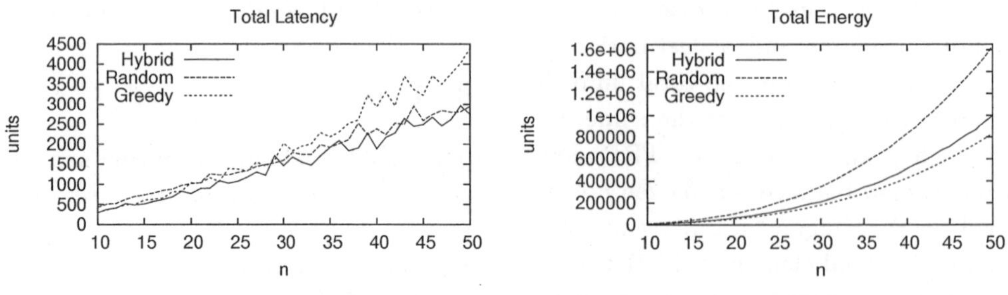

Figure 2. Matrix Transpose: Total energy consumption and total latency

Figure 3. Energy Balance for Matrix Transpose. (a) Greedy, (b) Hybrid, (c) Random

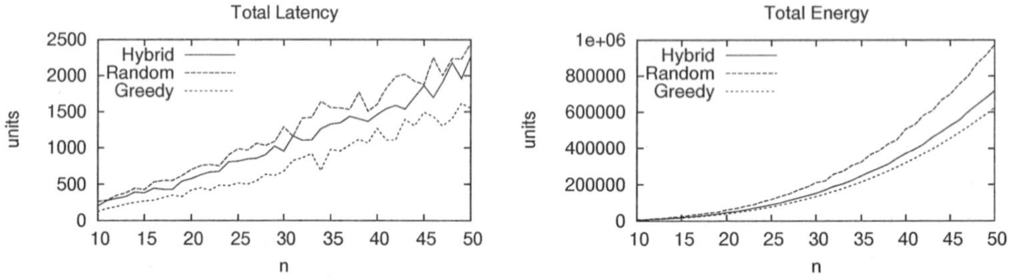

Figure 4. Row Shuffle: Total energy consumption and total latency

Figure 5. Energy Balance for Row Shuffle. (a) Greedy, (b) Hybrid, (c) Random

Adaptive Path Energy Conserving Routing in MANETs

Hasnaa Moustafa and Houda Labiod

Ecole Nationale Supérieure des Télécommunications (ENST)
GET/ENST/INFRES Department
46 rue Barrault – 75634 Paris Cedex 13 - France
Tel : +33 (0)1.45.81.74.36 Fax : +33 (0)1.45.81.31.19
Email : moustafa@enst.fr , labiod@enst.fr

Abstract

The specific features of mobile ad hoc networks (MANETs) impose new requirements for routing protocols. The aim of this paper is to present an approach that enhances routing performance by integrating ad hoc related characteristics. We propose a loop-free adaptive path energy conserving scheme which attempts to minimize both routing and storage overhead in order to efficiently provide robustness to host mobility, adaptability to wireless channel fluctuations and optimization of network resources use in large-scale networks. The strength of our scheme, based on source routing approach, named Energy Conserving Dynamic Source Routing (EC-DSR), is to handle network state related constraints such as node's energy consumption, links' availability and paths' quality. The performance of the proposed scheme is evaluated via simulation and is compared to DSR protocol. Better results are obtained with EC-DSR in terms of control overhead. Furthermore, EC-DSR shows significant improvements in delay and forwarding efficiency for large scalable networks.
Keywords:- mobile ad hoc network, source routing, energy conserving, link quality, high scalability.

I. Background and Motivation

Mobile Ad hoc Networks (MANETs) [1] refer to a collection of hosts communicating over a wireless channel. They are multihop wireless networks such that packets may traverse multiple consecutive wireless links to reach their destinations. Nodes in ad hoc networks are of limited resources and storage capacity; since they are battery operated they need to be energy conserving in order to maximize their battery life. This introduces a trade-off between link maintenance in a highly unreliable networks and power conservation for users with little battery power.
Actually, ad hoc network features pose significant challenges to solve a classical network problem that is routing. Routing protocols should be energy efficient, reacting dynamically to the changing topology and the radio environment characteristics. They should maintain and construct durable routes without generating excessive control overhead or consuming excessive energy. Until now, no standard has been adopted for a routing protocol for ad hoc networks because many critical issues remaining unsolved.

In fact, routing in mobile ad hoc networks is an important research topic where various mechanisms have been already proposed. These protocols can be generally categorized by the routing strategy, where various metrics are used to determine optimal paths. The most common metric is the shortest hop applied in DSR [2] (Dynamic Source Routing), DSDV [3] (Destination Sequenced Distance Vector), TORA [4] (Temporally-ordered Routing Algorithm), and WRP [4] (Wireless Routing Protocol). Nevertheless, not necessarily optimal routes are obtained for the possible different network configurations. Some protocols, however, can use shortest delay as the metric, although longer paths have higher probability to reach destination in an ad hoc environment. SSR [5] (Signal Stability based Adaptive Routing) has been advocated to improve routing performance, based on using link quality metrics. In this case, a shortest hop route may not be always used. However, this method could not avoid frequent rerouting due to node mobility.

In this paper, we address the problem of energy efficient routing in large-scale mobile ad hoc networks, in terms of both number of nodes and coverage area. This is a challenging environment as every node operates on limited battery resources and multihop routes are used over a changing network environment due to node mobility.
The battery energy of a transmitting node can be depleted due to: (a) processing at the node, (b) transmission attenuation due to path loss, (c) the need to maintain the transmission above a certain threshold due to signal interference [6]. In fact, users in an ad hoc network have to strike a balance between two objectives: conserving power and mitigating interference to other users on the one hand, and increasing power to maintain links on the other hand [7]. Our approach aims at extracting from dynamic, irregular topology of a MANET a topology with higher path quality, available and more stable links, and minimum energy consumption. In this context, we propose a new adaptive path energy conserving routing protocol, named Energy Conserving DSR (EC-DSR).

The following section gives a description of our proposed protocol EC-DSR. Section III, analyzes our performance results compared to DSR protocol. Finally, section IV provides concluding remarks and highlights our future work.

II. Energy Conserving Routing Protocol

In fact, in the case of high network scalability, the size of the routing tables increases significantly and the topology changing events will grow proportionally. In addition, large-scale networks introduce higher interference and weak link quality. Therefore, we should consider the trade-off between the increasing interference, and the higher connectivity of the network. Furthermore, we need to balance the trade-off between the power consumption and the link maintenance [8]. The goal of our work in this paper is to propose new characteristics in ad hoc routing to provide efficient saving in bandwidth and network resources, and to insure minimum energy consumption along the different used routes.

II.1 EC-DSR: An Overview

We propose a new adaptive path energy conserving routing protocol, named Energy Conserving DSR (EC-DSR). EC-DSR uses four proposed metrics based on battery power at each node, node's stability with respect to neighbors, quality of the link between nodes in terms of the received signal level, and availability of the link using future prediction for the link's state. Our scheme operates in a loop-free manner and attempts to minimize both routing and storage overhead providing robustness to host mobility, adaptability to wireless channel fluctuations and optimization of network resources use. To allow efficient use of network resources, route discovery process is initiated on demand avoiding channel overhead and improving scalability.

II.2 EC-DSR: Operation

EC-DSR applies a selection criteria using the previously defined four metrics to provide stable paths based on: links availability according to future prediction of links state, higher battery life paths tending to power conserving, link quality information to select one among many different routes, and location stability of nodes biasing the route selection towards routes with relatively stationary nodes.
Operation starts when a source node has data to transmit to a certain destination and it has no route to that destination. A route discovery procedure is then invoked through broadcasting a *request* packet to a selected set of neighbors searching for the destination. Neighbors' selection takes place following our selection criteria among the neighbors, through using the four proposed selection metrics. Each neighbor node receiving the *request* packet repeats the process of selection and *request* transmission; meanwhile the source route, proposed in DSR [2], accumulates in the packet header, until reaching the destination node. A destination receiving a *request* packet will take the accumulated route in its header and transmit a *reply* packet in the reverse route direction. The *reply* packet continues to be forwarded along the source route that is stored in the packet header, constituting of a set of selected nodes, until reaching the source. The source receives the packet and stores in its cache the route that is found in the packet header. This constructs a route between the source and the destination consisting of selected nodes to forward data. In fact, minimizing the broadcasting scope in the route discovery process attempts to minimize the message overhead of computing routes. A benefit of these types of selected routes also is that they are energy efficient and there will be little need to modify them frequently.

III. Performance Evaluation

We analyze the performance of our proposed EC-DSR scheme using ns2 [9]. The aim of our performance analysis is to evaluate the behavior of our proposed scheme and to compare its performance to traditional DSR protocol, as an on demand routing protocol applying the source routing concept. We applied two approaches in our analysis study. First, we study the behavior of EC-DSR with different number of nodes and different mobility types. Then a performance comparison is carried out between EC-DSR and DSR.

III.1 Simulation Model and Scenarios

Table 1, illustrates the different network configuration as well as the different simulation parameters used in our simulations.

Table 1: Network configuration and simulation parameters

Mobility	
Number of Nodes (N)	25, 50, 75, 100
Mobility Zone (Topography)	1200 x 500 m
Mobility Model	*Random Waypoint* [10]
Pause Times	0, 300, 600, 900, 1200, 1500 sec
Coverage Zone	250 m
Nodes's Speed	20 m/sec
Simulation Time	1500 sec
Traffic	
Traffic Type	CBR
Number of Connections (N/2)	13, 25, 38, 50
Transmission Rate	4 packets/sec
Routing Protocol	
Routing Protocol	EC-DSR, DSR

III.2 Analysis

The following sections illustrate our analysis for the obtained results. Several performance metrics were used in our performance evaluation and comparison, which are: (1) *delivery ratio* in terms of the number of packets correctly received with respect to the total number of packets sent, (2) *forwarding delivery* in terms of the ratio of the number of nodes forwarding packets correctly to the number of transmitted packets, (3) *forwarding control packets overhead* in terms of the ratio of the control packets to the total number of transmitted packets, and (4) *end-to-end delay*.

Firstly, Figure 1 and Figure 2 compare the performance of EC-DSR showing respectively the delay and forwarding control overhead behavior with different mobility types at different scalable network configurations. It is clear that delay increases with mobility for network topologies (25 to

75) nodes due to more probability of links' failures at high mobility cases inducing more delay. Actually, EC-DSR consumes some time to find a new route or to use another existing route. This is caused by the transmission of route error messages and routes' update in caches, in addition to route discovery in the cases where no other routes are found in the caches. On the other hand, for network topology of 100 nodes the delay behavior is better at high mobility cases and is generally better than the case of (25 to 75) nodes configuration. At this case, as the number of nodes highly increases the number of stored routes in the nodes' caches increases also. In addition, there is a higher probability for more stable routes to be found among the stored ones. Generally, we can conclude that DSR is scalable for the delay metric, as it has a weaker delay for the network configuration of huge number of nodes and very high mobility.

Considering the forwarding control packet overhead, nearly same behavior is shown at different mobility cases for network topologies (25 to 75) nodes. While at network topology of 100 nodes, control overhead increases linearly with mobility decrease starting from intermediate mobility. At this case, the probability of more stable nodes increases highly, which maximizes the broadcasting scope of the control messages during the route discovery phase.

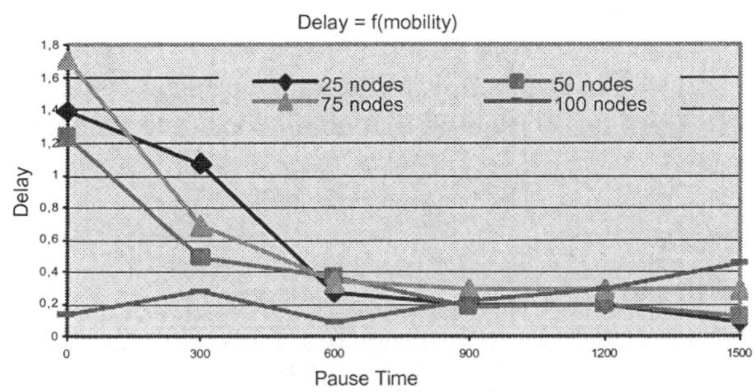

Figure 1 : EC-DSR Delay

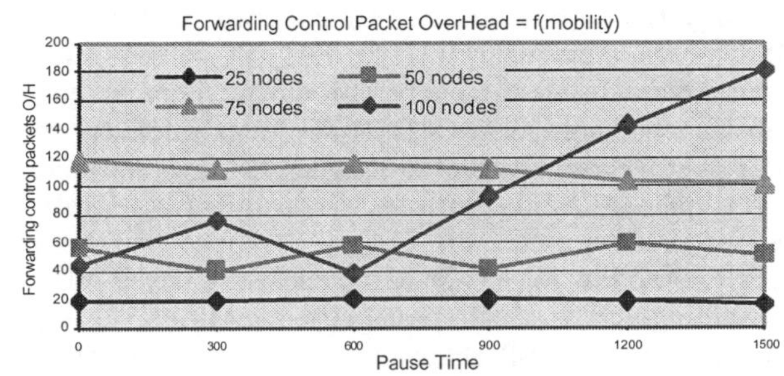

Figure 2 : EC-DSR Control O/H

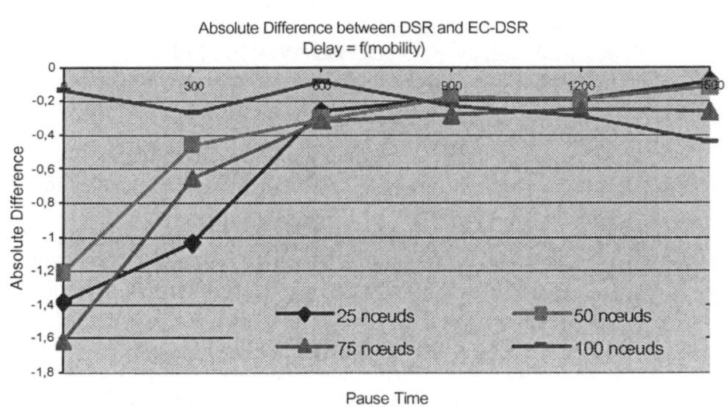

Figure 3 : DSR Vs EC-DSR absolute delay

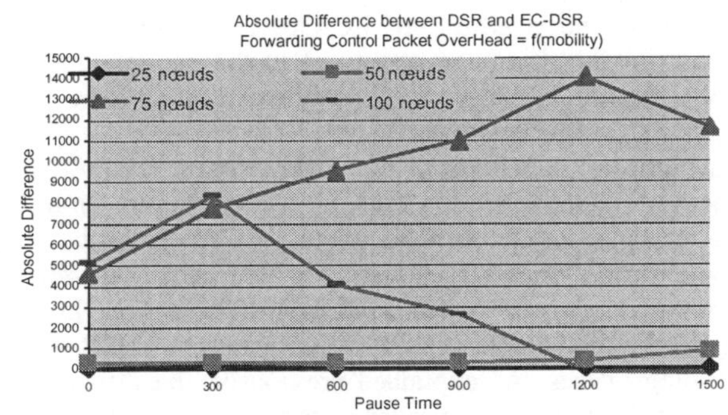

Figure 4 : DSR Vs EC-DSR absolute Control O/H

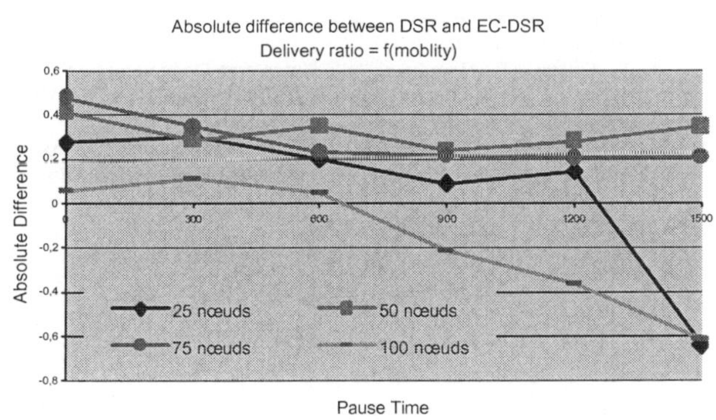

Figure 5 : DSR Vs EC-DSR absolute delivery ratio

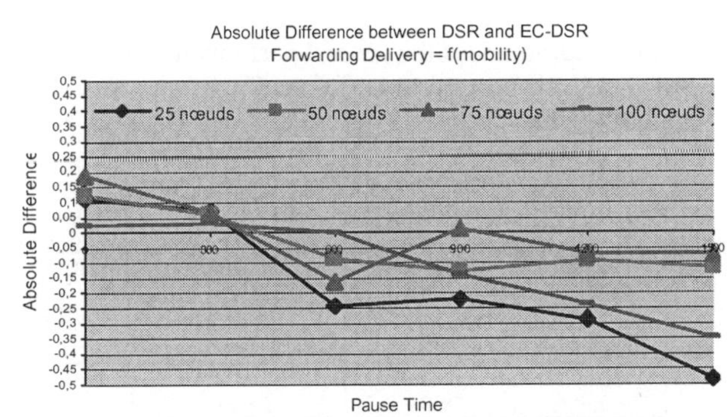

Figure 6 : DSR Vs EC-DSR absolute forwarding delivery

Secondly, we analyze the performance of EC-DSR versus DSR using the absolute difference, which we define it as (*DSR performance metric value – EC-DSR performance metric value*).

Figure 3, shows the absolute delay. For network topologies (25 to 75) nodes, DSR has weaker impact on the delay compared to EC-DSR. The delay difference increases with mobility. While for network topology of 100 nodes, EC-DSR has weaker impact on the delay compared to DSR for nearly all mobility cases. In fact, EC-DSR selection of routes allows longer route to be chosen consuming more delay, for the case of 100 nodes the shortest path criteria used by DSR in routes selection permits more probability of link failures introducing more delay.

The absolute forwarding control overhead is illustrated in Figure 4. At all mobility cases and all topologies, DSR exerts higher overhead compared to EC-DSR. This returns to the fact that EC-DSR minimizes the flooding scope during route discovery, while DSR necessitates the transmission of more control packets.

From Figure 5, it is clear that DSR outperforms EC-DSR in terms of delivery ratio at different mobilities for network topology (25 to 75) nodes. As mentioned previously, EC-DSR exerts more delay due to applying certain selection criteria during route discovery, introducing more delay. This translates the superiority of DSR in terms of delivery ratio. DSR is still more efficient even when the network tends to stable state (mobility decreases gradually). In general, the delivery ratio implies the end-to-end connection between each two nodes, for EC-DSR the packets traverse different forwarding nodes, which means different buffers, losing the efficiency of EC-DSR although its improvement in selecting more stable energy efficient routes. It is obvious that EC-DSR becomes more efficient when the topology becomes more complex (100 nodes) and mobility becomes less. At this case, the overhead of EC-DSR mechanisms has less effect on its efficiency compared to the economic energy consumption factor that allows the choice of more stable paths showing better behavior.

At higher mobility cases, DSR outperforms EC-DSR in its forwarding delivery for all mobility cases (Figure 6). In fact, more link failure probability of EC-DSR with highly dynamic networks cases translates this behavior, which is always due to the overhead in routes' selection. For less dynamic networks, EC-DSR outperforms DSR at all topology cases. As explained previously, the efficiency of EC-DSR protocol weakens the shortcoming of its mechanisms overhead leading to this improvement.

Table 2, summarizes the general performance behavior of the two protocols according to different network topologies and mobility types.

Table 2: General Performance Comparison

	Small number of nodes	Intermediate number of nodes	Huge number of nodes
Weak Mobility	EC-DSR+ DSR+	EC-DSR+ DSR+	EC-DSR++ DSR-
Intermediate Mobility	EC-DSR- DSR+	EC-DSR- DSR+	EC-DSR+ DSR+
High Mobility	EC-DSR-- DSR++	EC-DSR-- DSR++	EC-DSR- DSR+

IV. Conclusion and Future Work

In this paper, we propose the Energy Conserving DSR (EC-DSR) protocol. It uses no periodic network flood of control packets. Thanks to its selection criteria, stable paths with future links' availability and higher battery life are provided. EC-DSR may be applied on any on-demand routing protocol with source routing concept. In our performance study we applied it on DSR, and concluded that it is more efficient as the number of connectivities between nodes increases. In fact, the improvements in economic energy consumption and minimized flooding are more efficient at high scalable networks.

For future work, we intend to compare EC-DSR with other energy conserving routing protocols, considering new performance metrics such as energy-based mobility and links' stability metrics. We also intend to consider larger scalable networks in our performance analysis.

Acknowledgement

We would like to thank Erik Milhomme and Eric Municchi for their great aid in running the simulations, and gathering the results.

References

[1] Mobile Ad Hoc NETwork (MANET). (1999). URL: http://www.ietf.org/html.charters/manet-charter.html

[2] D. Johnson, D. Maltz. "Dynamic source routing in ad hoc wireless networks", in Mobile Computing, T.Imielinski and H korth, Eds. Norwell, MA: Kluwer, 1996.

[3] C. perkins, P. Bhagwat. "Highly dynamic destination-sequenced-distance-vector routing (DSDV) for mobile computers", ACM SIGCOMM, vol. 24, no. 4, October 1994, PP. 234-244.

[4] E. Royer, C. Toh. "A review of current routing protocols for ad hoc mobile wireless networks", IEEE personal communication, April 1999.

[5] R. Dube and al., "Signal Stability based Adaptive Routing (SSR) for Ad Hoc Mobile Networks", IEEE personal communication, February 1997, pp. 36-45.

[6] I. Kang and R. Poovendran, "On the Lifetime Extension of Energy-Efficient Multihop Broadcast Networks", IEEE International Joint Conference on Neural Networks 2002, IJCNN'02.

[7] M. Chiang and G. Carlsson, "Asmission Control, Power Control, and QoS Analysis for Ad Hoc Wireless Networks", IEEE International Conference on Communication ICC, 2001, volume:1, 11-14 June 2001.

[8] X. Hong, K. Xu, and M. Gerla, "Scalable Routing Protocols for Mobile Ad hoc Networks, Network, IEEE, volume:16 Issue:4, July_August. 2002.

[9] K. Fall and K. Varadhan. "NS Notes and Documentation", the VINT project, UC Berkeley, LBL, USC/ISI, and Xerox PARC, May 1998. Work in progress.

[10] C. Bettstetter, H. Hartenstein, and X. Pérez-Costa. "Stochastic Properties of the Random Waypoint Mobility Model: Epoch Length, Direction Distribution, and Cell Change Rate", ACM MSWiM'02, 2002.

A RATE-ADAPTIVE POWER CONTROL SCHEME FOR WIRELESS MICROSENSOR NETWORKS

CAIMU TANG
Department of Computer Science
University of Southern California
Los Angeles, CA 90089
Email: caimut@cs.usc.edu

CAULIGI S. RAGHAVENDRA
Departments of EE-Systems and Computer Science
University of Southern California
Los Angeles, CA 90089
Email: raghu@usc.edu

***Abstract*—In this work, we investigate how to use power controlled radio with adaptive channel coding on source coded bits to save energy in wireless microsensor network. In order to save node energy, we propose a so-called rate-adaptive power control scheme for wireless microsensor network. In the scheme, the radio uses discrete power levels to transmit data and channel codes are employed and the code rates are selected based on the power level and channel condition. To implement the scheme, we formulate an optimization problem and propose a distributed way of solving it in the context of wireless microsensor network. Simulation results on field data show that the proposed scheme can result in more than 50% energy savings.**

1 INTRODUCTION

Wireless microsensor networks are both constrained by radio link reliability and node energy. To reduce energy, lower radio power is desirable. However, lower radio transmission power means lower Signal-Noise ratio (SNR) at receiver and more retransmission is a result of it due to high bit error rate (BER). Channel code has been widely used to improve channel BER via introducing redundancy.

To achieve a reliable radio link, the rate adaptable channel code based on Convolutional Code (CC) and punctured convolutional code (PCC) could have been integrated with a radio power control scheme. In order to guarantee the reliable delivery of data packets, the reciever can use Hybrid ARQ protocol based on the ITOH algorithm [8]. We have formulated an optimization problem with the objective function directly related to the transmission energy of a given number of data bits. The objective function consists of the redundancy cost introduced by the channel coding, the retransmission cost introduced by the automatic repeat request (ARQ) protocol and the transmission energy per bit. To efficiently solve the problem while taking into account the practical implementation issues, we propose to use an iterative approach based on feedback channel information derived from the ARQ requests. Both simulation results and actual implementation have shown a significant energy savings.

2 A RATE-ADAPTIVE POWER CONTROL SCHEME

We assume that the channel BER is determined by two factors: (1) the achievable nominal BER denoted by E and (2) signal randomness at receiver radio due to various channel effects, e.g. capture and fading, and this factor is P which is for the probability when a received power is greater than the receiver's sensitivity. Therefore, the channel BER denoted by e is computed as follows,

$$e = (1 - P) + PE. \tag{1}$$

We argue this is a better approximation to the actual channel BER estimate when the radios are power controlled. Probability P depends on the radio Tx-Rx apart distance, received signal SNR. However, we take it as a constant in the analysis. The nominal BER in this analysis is estimated using the following upper bound of CC for AWGN BPSK modulated channel [3],

$$E \leq Q\left(\sqrt{\frac{2wd}{\omega_0}}\right) exp\left(\frac{wd}{\omega_0}\right) \frac{\partial \mathrm{T(N,I,D)}}{\partial I}\Bigg|_{\substack{D=exp\left(-\frac{w}{\omega_0}\right)\\ I=N=1}}, \tag{2}$$

where w is the received signal power; ω_0 is the noise power; d is the free distance which is equal to the lowest exponent of D in $T(N, I, D)$, and the transfer function $T(N, I, D)$ is as what in [7] (refer to VII-B of [7],[2] and [5] for details on the BER upper bound estimation and the free distance of CC).

For uncoded channel under the same condition as above, we also know that the channel BER E is as follows,

$$E = Q\left(\sqrt{\frac{2w}{\omega_0}}\right).$$

Let the channel code rate $R = \frac{l}{n}$ (we may use another parameter $R = \frac{1}{\lambda}$ with $\lambda \geq 1$), to save energy, the transmission power has to be reduced by a factor of R to Rw and use channel coding to enhance the channel to remedy on the BER increase. From ineq. (2), if dR is kept small, energy can be saved since the rest excluding the Q-function term in the RHS of ineq. (2) is close to 1 when the received signal SNR large enough. In other words, channel coding could be used to maintain a sustainable channel BER.

Power control is important when Tx-Rx apart distance is too large in which case the channel coding may not be enough to maintain an energy favorable channel BER. As we know, signal strength degrades inversely at least on the square of the Tx-Rx apart distance. As the signal degrades quickly on the distance, the channel BER increases. When the signal degrades to a certain level, channel coding may not be able to bring the BER to an energy favorable level since a large number of ARQ request will result. Fig. 1 shows the relationship between the channel coding and power control using $\frac{1}{2}$ CC coded channel and its derived PCCs (using the upper bound shown in eq. (2)). In Fig. 1, there are four curves of which three curves correspond to channel codes with $\frac{1}{2}$ CC and $\frac{3}{4}$, $\frac{2}{3}$ PCC, and the curve in dark corresponds to channel coded with $\frac{2}{3}$ PCC and the transmission power is raised to next level from 20 to 25 (100 scale). The extra energy dissipation from the RF is smaller than that introduced by a lower channel code rate of which the channel BER is kept at about the same. From Fig. 1, we see that it is energy efficient to raise to the higher power level instead of using lower rate channel code for certain cases. Combining the two points explained above (i.e. change of power level and usage of channel code and ARQ), in order to save energy on communication, we have to combine channel coding and power control. This could be done via selecting appropriate code rate with appropriate power level (in this study we assume the radio is able to adjust the transmission power on a fixed number of discrete levels).

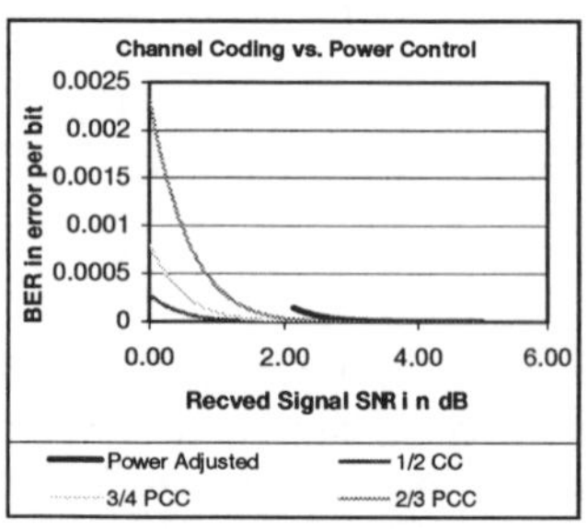

Fig. 1. Relationship between Channel Coding and Power Control

In what follows, we assume the event of any packet delivery (repeated packet or original packet) is independent, and also assume one ARQ request is replied by the receiver using ITOH algorithm if one packet has erroneous bit(s) upon channel decoding; furthermore, there is no loss of packets. Let ϵ denote the probability for a packet to be requested by an ARQ, n be the number of packet and $n+k$ be the number of packet transmission trials for the n packets. The probability mass function (pmf) $p(k, \lambda)$ for k with code rate $\frac{1}{\lambda}$ is as follows,

$$p(k,\lambda) = \begin{cases} \sum_{i=1}^{k} \binom{n}{i} P_i^k \epsilon^k (1-\epsilon)^n & k \le n \\ \sum_{i=1}^{n} \binom{n}{i} P_i^n \epsilon^k (1-\epsilon)^n & k > n. \end{cases} \tag{3}$$

where P_i^k satisfies the following recursive formula,

$$P_i^k = \sum_{j=i-1}^{k-1} P_{i-1}^j, \tag{4}$$

and $P_j^j = 1, P_1^j = 1$ for $\forall j \ge 0$. Note that $k \ge i$ in eq. (4) and the following equation can be verified so that $p(k,\lambda)$ is indeed a pmf (the proof is omitted),

$$\sum_{k=0}^{\infty} p(\lambda, k) = 1, \text{for given } \lambda \ge 1.$$

The computation of $\sum_{i=0}^{k} p(i,\lambda)$ can be very efficiently done using dynamic programming approach in time of $\mathcal{O}(k)$ (n is taken as a constant). An equivalent expression for $p(k,\lambda)$ is

$$p(k,\lambda) = \binom{n+k-1}{k} \epsilon^k (1-\epsilon)^n;$$

however, it is computationally unfavorable to use the above equation for the computation of $\sum_{i=0}^{k} p(i,\lambda)$ and the time taken is in $\mathcal{O}(k^2)$.

If we take the number of remaining packets to send as the states, there are total $n+1$ states. Figure 2 shows the state transition diagram. Equation (3) follows by noticing that each state can dwell on itself 0 to k times (Dwelling once on a state means one failed delivery of the next packet).

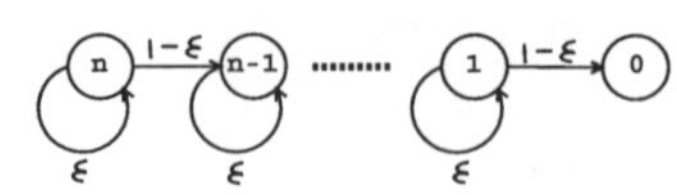

Fig. 2. Packet Delivery State Transition Diagram

From the application standpoint, the delivery of n packets has to be done in a timely fashion, the maximal allowed number of repeat request is fixed. Let k_0 be the maximal number of repeated requests for any packet, and p_0 be the minimal packet delivery probability. The rate selection based on an energy criterion is formulated as the following integer programming optimization problem (IPOP):

$$\tilde{\lambda} = \underset{\lambda,\tau}{\operatorname{argmin}}(\lceil(\lambda-1)n\rceil + k)\tau,$$

subjects to:

$$\lambda \ge 1 \text{ and is rational (implementable);} \tag{5a}$$

$$\sum_{i=0}^{k} p(i,\lambda) \ge p_0; \tag{5b}$$

$$\sum_{k=0}^{k_0} p(i,\lambda) \ge P_{sig}; \tag{5c}$$

$$\tau_0 \le \tau \le 1, \tag{5d}$$

where P_{sig} is close to 1, τ is the antenna power factor with two bounds τ_0 and 1, and it can only has discrete values according to the set power levels. The number of antenna power levels are fixed and the power on a level is also fixed. The objective function in IPOP consists of two parts, namely, $\lceil(\lambda-1)n\rceil\tau$ and $k\tau$. The first part is the redundancy cost introduced by channel coding and the second part is the repetition cost of some packets introduced by ARQ protocol. Analytically optimal solution

TABLE I

IPOP PARAMETERS AND RESULTS OF THE EXAMPLE

λ	e	ϵ	k	P_0'	τ	C_{ost}
2	1.1046e-5	0.00564	4	0.9997	0.73	75.92
$\frac{7}{4}$	1.6011e-5	0.00816	5	0.9998	0.78	62.4
$\frac{3}{2}$	4.601e-5	0.0276	8	0.999	0.8	46.4
$\frac{6}{5}$	3.7895e-4	0.1764	39	0.9991	0.85	50.15

for above problem is not always obtainable since the channel is not stationary. Due to constraint on hardware implementation, certain coding rate may be too costly and end up extra waste on circuitry energy dissipation. One solution could be to solve IPOP with the feasible solution space limited to a pre-selected set of rates and a number of available power levels (the space in question is presumably smaller). However, when the search space is too big, IPOP is intractable.

In order to help understand the above formulation, we give an example on it as follows. Let $n = 100$; $k_0 = 40$; $P_{sig} = 0.99999$; $p_0 = 0.999$; $P(\gamma, R) = 0.99999$. The common uncoded wireless channel with BPSK modulation (baseline) has BER around 0.01 per channel symbol on arrival signal with magnitude above -91dBm. Assume the packet size is 512 symbols, the baseline is not in the feasible space since $\epsilon = 0.994176$ which is too high to meet (5c) of IPOP within 40 repetitions.

Table I gives the parameters and the objective value corresponding to the above IPOP parameters for four coding rates. In Tab. I, $P_0' = \sum_{i=0}^{k} p(i, \lambda)$ is the packet successful delivery probability corresponding to k repetition, and C_{ost} is the objective value corresponding to the given parameter settings. From this simple example, we can see there is some optimal operational point on the joint space of channel codes and power levels for a given set of parameters.

In order to see how the transmission power affects the objective value, let us fix the coding rate at $\frac{1}{\lambda}$. Denote $\Delta k = k_2 - k_1$, and k_1 is the number of retransmissions after the power adjustment and k_2 is the number of retransmissions before the power adjustment, and τ is the power factor before adjustment. Suppose the power after adjustment is $\tau + \Delta\tau$, in order to save energy, the following must hold,

$$\Delta k > (\lceil(\lambda - 1)n\rceil + k_1)\frac{\Delta\tau}{\tau}.$$

Note that $k_1 < k_2$ in general if $\Delta\tau > 0$.

To estimate the radio transmission power, the formula shown in Sec. 2.1 of [6] is used, $p_t = \alpha r^{\gamma} + \beta$ for Tx-Rx apart distance r where $\gamma = 2$ and it is the propagation exponent for free space model; $\alpha = 14.8mW$ and it is the radio energy cost coefficient; $\beta = 11mW$ and it is the cost of radio transceiver electronics. We set the noise floor at -119dB. The path loss estimate uses the following formula (see Chapter 3 and Appendix B of [4]),

$$\overline{PL} = \overline{PL}(r_0) + 20\log\left(\frac{r}{r_0}\right) + X_\sigma,$$

where $\overline{PL}$ is the average path loss in dB at distance r away from the transmitter in free space, r_0 is the close-in distance, and X_σ is a random variable with zero-mean and variance σ. If the close-in distance is selected so that $\overline{PL}(r_0) = 0$, the signal strength at r can be computed in dB as

$$w = -20\log\left(\frac{d}{r_0}\right) + X_\sigma.$$

Thus, using the above formula the SNR of the received signal can be derived from the noise figure. Fig. 3 gives the relationship between symbol error rate and number of retransmissions for CC's with different rate.

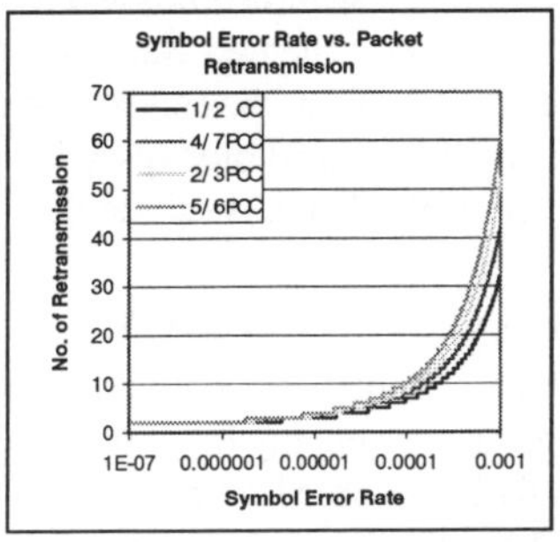

Fig. 3. Relationship between Symbol Error Rate and ARQ

Assume there are k packets transmitted successfully between two ARQ requests. This information reflects the channel condition and it is available to the sender. For n_0 packets of the same size, and $1 < k \le n$ and $n \ge n_0$, the total number of packet transmissions Λ can be calculated as follows,

$$\Lambda(\lambda) = \left\lfloor\frac{n-k}{k-1}\right\rfloor k + k + \left\lfloor\left(\frac{n-k}{k-1}\right)k - \left\lfloor\frac{n-k}{k-1}\right\rfloor k\right\rfloor + 1,$$

$$= \left\lfloor\frac{n-1}{k-1}\right\rfloor k + \left\lfloor\left(\frac{n-1}{k-1}\right)k - \left\lfloor\frac{n-1}{k-1}\right\rfloor k\right\rfloor + 1,$$

where, n and k are functions on λ. $k(\lambda)$ is an ascending monotonic function on λ and $n(\lambda)$ is a monotonic descending function on λ. When $k = 1$, $\Lambda = \infty$, and when $k = n$, $\Lambda = n + 1$. Λ reflects the energy cost for the delivery of these n_0 packets. k follows the geometric distribution when we assume the event of a ARQ request of a packet is independent of that of another packet.

In this rate adaptable power control scheme, we use one convolutional code with rate at $\frac{1}{2}$ and its derived PCCs with various puncturing matrix corresponding to different channel conditions under a given transmission power. The channel codition is measured by both the sender and receiver, since each node is both a sender and a receiver in one session in most sensor network applications, and this enables it can be done rather accurately to measure these channel parameters. We derive N number of thresholds T_i $(1 \le i \le N)$ so that a proper code is selected based on whether k is greater than T_i or not. Let P_{ARQ} be the probability of a packet to be repeatly requested.

The mean value of the geometric distribution of k is $\frac{1}{P_{ARQ}}$ and the variance is $\frac{1-P_{ARQ}}{P_{ARQ}^2}$. Based on the above analysis, we use the following formula to determine these thresholds,

$$T_i = \left\lfloor \frac{1}{P_{ARQ}} + \eta \frac{L-i}{L} \frac{\sqrt{1-P_{ARQ}}}{P_{ARQ}} \right\rfloor, \qquad (6)$$

where $0 < \eta \leq 1.0$ which reflects the fluctuation of the channel conditions in question, and it should set to 1.0 when the channel does not experience fading and is in a open space (no multi-path). In eq. (6), the second term is the tuning factor on each level and the first term is the base. It is possible to use another way to determine these thresholds and it is not critical on the usage of the proposed scheme. Based on each confirmed transmission and ARQ request, the sender can adapt the transmission power using these thresholds (pseudocode on the rate-adaptable power control scheme is omitted).

3 Experimental Evaluations

We use OPNET to simulate the network and verify how the proposed scheme can save energy for wireless microsensor networks. The simulation is based on pairs of nodes with different distances. The radio Rx-sensitivity is set at -91dBm and the propagation limit is -111dBm. The radio has 10 different power levels available to use and we assume the distances are within the propagation limit and we also assume every node has 2Kx8 source information bits in every second (field datasets [1] are used in the simulation study, and they are source coded before passing to radio model.). The network consists of 10 nodes randomly distributed in field of size 20km x 20km where 5 nodes are mobile nodes and the other five are stationary. The network used in the simulation is shown in Fig. 4 where the arrowed lines indicate the trajectories of the mobile nodes. In order to compute the transmission energy, we modified the default OPNET radio pipeline stage so that the energy profiling information can be generated.

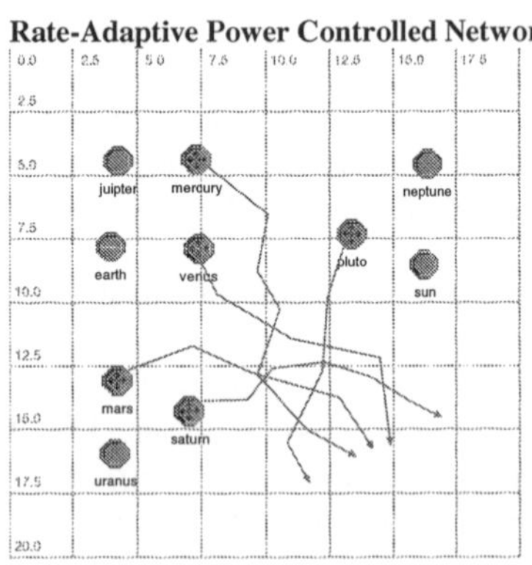

Fig. 4. Mobile Network Topology for Simulation Study

The left plot of Fig. 5 shows the energy saving using the proposed scheme compared to the fixed power case with fixed rate. The CC has constraint length of 5 and the rate is $\frac{1}{2}$. The energy reduction is at least 75% among all frames. In this measurement, we use average energy per second per node over the field. The right plot of Fig. 5 shows the energy savings using the proposed scheme compared to regular power control scheme (without rate control) where the CC used has rate $\frac{1}{2}$ and constraint length 5. The energy reduction using proposed scheme is around 50% compared to the regular power control scheme which is based purely on distance for the determination of the RF transmission power.

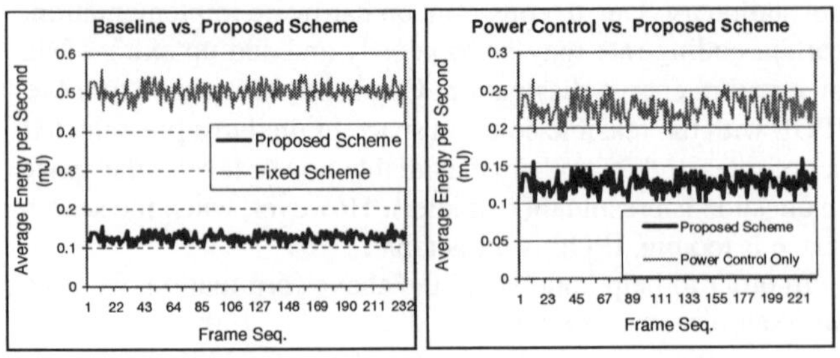

Fig. 5. Energy Comparison of Fixed, Power Controlled and Proposed Scheme

4 Conclusions

In this paper, we proposed a novel rate-adaptable power control scheme and derived a formulation to minimize communication energy. Simulation results have shown significant energy savings using this scheme compared to regular methods.

ACKNOWLEDGEMENTS: *This reserach is supported by the DARPA PAC/C project (F33615-00-C-1633), Air Force Research Laboratory, Air Force Material Command and USAF (F30602-01-2-0549). The views and conclusions contained herein are solely those of the authors.*

References

1. ATR and SensIT SITEX Datasets, DARPA and ARL 2000.
2. J. Hagenauer, "Rate-Compatible Punctured Convolutional Codes (RCPC codes) and their Applications", IEEE Transactions on Communication, vol. 36, no. 4, pp. 389-400, April 1988.
3. P. V. Kumar and H, -F. Lu, "Lecture Notes on Error-Correcting Codes [Online]", Available: http://ceng.usc.edu/~vijayk", 2002.
4. T. S. Rappaport, "Wireless Communications: Principles and Pactice", Prentice Hall PTR, 1996.
5. M. Rouanne and D.J. Costello Jr., "An Algorithm for Computing the Distance Spectrum of Trellis Codes", IEEE Journal on Selected Areas in Communications, vol.7, no.6, pp. 929-940, Aug. 1989.
6. C. Tang, C.S. Raghavendra and V.K. Prasanna, "Energy Efficient Adaptation of Multicast Protocols in Power Controlled Wireless Ad Hoc Networks", Proceedings of 2002 International Symposium on Parallel Architectures, Algorithms and Networks (ISPAN), May 22-24, 2002, Makati City, Manila, Philippines.
7. A. J. Viterbi, "Convolutional Codes and Their Performance in Communication Systems", IEEE Transactions on Communications Technology, vol. COM-19, no. 5, pp. 751-772, Oct. 1971.
8. H. Yamamoto and K. Itoh, "Viterbi Decoding Algorithm for Convolutional Codes with Repeat Request", IEEE Transactions on Information Theory, vol. IT-26, no. 5, pp. 540-547, Sept. 1980.

ENERGY AWARENESS AND AD HOC ROUTING

EROL GELENBE, RICARDO LENT

School of Electrical Engineering and Computer Science, University of Central Florida, USA
Email: {erol,rlent}@cs.ucf.edu

In this paper, we focus on a way to extend the working lifetime of mobile nodes in a wireless, ad hoc network. For this, we employ ad hoc cognitive packet networks (AHCPN), which are inspired by previous research in cognitive packet networks (CPN). These networks employ smart packets to build networked learning systems that support adaptive, quality of service-driven routing of packets.

To make smart packets seek a desired routing goal, appropriate adjustments are required in weights of a random neural network prior to use the decision making procedure. Reinforcement learning is employed to achieve this task with a new routing metric that we have introduced (path availability) to manage efficiently the low energy resources typically present in ad hoc networks. The new metric quantifies a path via the probability to find nodes and links available for routing. In this paper, we define this metric as a function of battery lifetime at nodes. This addition enables smart packets to make power-aware routing decisions, and consequently, nodes, which have low power, are able to extend their working lifetime and routes are selected based on power considerations, leading to the establishment of more robust communication paths over time.

This paper examines the AHCPN algorithm and defines path availability in relation to energy factors.

1 Introduction

Mobile ad hoc networks are autonomous systems of mobile nodes connected by wireless links that are able to dynamically create any arbitrary topology. This high level of "freedom" brings major limitations to the network, such as the amount of energy that a mobile node can carry, and restricts network reliability. As a consequence, the establishment of stable, high-quality routes for connections and an efficient use of resources are imperative.

Communication-related operations are a major source of energy consumption in mobile nodes. In particular, consumption is significantly high when ad hoc routing is employed as nodes are assigned additional operations to support the routing of packets from other nodes. This research extends the cognitive packet networks initiative to support routing of packets in wireless, mobile ad hoc networks. We explore the use of *smart packets* to enable a power management scheme, which can be directly integrated into the routing algorithm.

Cognitive packet networks offer fast, adaptive routing, which make them particularly attractive to ad hoc situations. Smart packets can be custom tailored to seek minimization of factors that produce errors within ad hoc networks. Routing decisions that prolong the battery lifetime expectancy of the nodes will reduce nodes' failure probability.

2 Routing in Mobile Ad Hoc Networks

Previous proposals for routing in ad hoc mobile networks can be classified as table driven protocols (proactive) and as source-initiated on-demand driven protocols (reactive). While proactive protocols require global information about the network to maintain routing information for every possible source-destination pair, reactive protocols initiate a route discovery only when it is needed by a node. Proactive protocols are in general more expensive in terms of use of network resources than reactive protocols, since they require a periodic transmission of routing tables by the nodes. Destination-Sequenced Distance-Vector routing (DSDV), Wireless Routing Protocol (WRP), Clusterhead Gateway Switch Routing (CGSR) are examples of proactive protocols where the control packet overhead grows as $O(n^2)$ where n is the number of nodes. Global State Routing (GSR) takes the idea of link state routing to reduce routing messages by sending them only when changes occur in the status of links. Fisheye State Routing (FSR) is an attempt to reduce the still high routing overhead of GSR by using different update frequencies depending on the distance to other nodes.

Most on-demand protocols use a variation of flooding, sending a route request packet to find a route from a source to a destination. Flooding can be restricted by requiring that each node process each request packet at most once, limiting the control packet overhead to $O(n)$. In order to accelerate route discovery, many on-demand routing proposals stop the route search process when the request packet encounters a node with knowledge of a route to the destination: DSR, AODV, LMR. However, this condition does not reduce the overall complexity because it only prevents the continuation of that particular replica of the route request. A way to reduce route discovery complexity is offered by Location Aided Routing (LAR) and geoTORA, where the propagation area of the flooding is restricted by using location information of the nodes based on Global Positioning System (GPS) information at each node.

Because the advances in battery technologies are behind the pace of computing and storage technologies, there exists an increasing interest in developing power-aware systems for wireless, mobile applications. Research in power-efficient techniques embraces all seven layers of the OSI model. Power conserving methods have been investigated in the design of hardware [13], operating systems, and applications, ranging from the design of energy-efficient schedulers for processors to power-aware applications for database transactions. In the context of power efficient, routing proposals, Singh, et al, investigated the use of power-aware metrics in the calculation of the shortest paths. The metrics described the power required for transmitting and receiving a packet on a link, so as to minimize the end-to-end power requirements for routing with a Dijkstra or Bellman-Ford algorithm. However, the proposal did not take into account the remaining energy in the nodes, which may produce a severe drain of energy in the batteries of the nodes on the least-cost route. Other proposals overcome this problem by using of battery lifetime information. Toh proposed a new metric, which calculated the summation of the inverses of the remaining battery capacities of the nodes on the path. Furthermore, the author proposed the *Min-Max* algorithm to maintain a fair use of resources by avoiding the use of nodes with the least battery capacity on the network.

The *power-aware source routing* (PSR) protocol uses a modified version of DSR that includes power computation. Like DSR, PSR sends out route requests to find new routes. All nodes but the destination calculate a link cost, which consist of the remaining battery capacity and the transmission power of the node, and include this information in the header of the request packets. The destination waits a time interval after the arrival of the first route request, so as to receive more than one possible route, and then selects the one with the minimum cost.

3 Ad Hoc CPN

Similarly to CPN, *Ad-Hoc Cognitive Packet Networks* (AHCPN) use three types of packets to achieve learning of routes and transport of data packets: *smart packets* (SP) are responsible for the discovery and maintenance of routes, *dumb packets* (DP) transport payload, and *acknowledgment packets* (AP) distribute routing information and provide confirmation of delivery to the source.

All these three types of packets have a similar structure. They contain a header, a *cognitive map* (CM) (to store routing information), and a data area. The header contains routing control information, for example, a destination address, packet length, etc. In the cognitive map, packets record routing metrics as they move on a network. The data area is used only by dumb packets for payload. Smart packets explore the network attempting to achieve their routing goals. As they travel , they record their path in their CM. Also, they collect additional characteristics about the path, for example the arrival time (timestamp) at each hop, which is used by future smart packets to make better routing decisions.

To make a routing decision in a node, a smart packet constructs a random neural network containing as many neurons as possible next-hop decisions, and adjusts the weights of the RNN with a reinforcement learning based algorithm from the information in the mailbox. The most excited neuron of the neural network in steady state determines the next-hop decision. At the destination, the cognitive maps of the packets are processed to remove possible loops before sending them to the source into acknowledgment packets. To overcome problems associated with topological changes, a small number of smart packet decisions are replaced by local broadcasts (controlled by a pre-selected broadcast-to-unicast ratio). Broadcasts are also used as a last-resource alternative whenever sufficient information is not available to make a regular RNN/RL decision.

Acknowledgments follow the reverse path that is recorded in their cognitive map. They update the mailboxes on their path with the information stored in the packet. For example, the timestamp stored in the cognitive map and the current arrival time are used to calculate the round-trip delay to the destination. Acknowledgments also store the route discovered by the corresponding smart packet in a route cache at the source node. Dumb packets use the routes in the route cache to transport payload (source routing). As in smart packets, dumb packets also collect routing metrics as they move, which are used later to maintain the information in the mailboxes.

Nodes are used as storage areas and processing units. A mailbox is a storage area where packets read or write information. Acknowledgement packets deposit in a mailbox the routing metrics acquired by a previous SP or DP. Smart packets use the information stored in a mailbox to update the weights of the RNN to make a routing decision. Nodes also store and execute the code for the RNN.

4 Tailored Routing Goal to Enable Power Awareness

Smart packets employ a routing goal to update RNN weight matrices. To enable power awareness, we introduce path availability to quantify the battery lifetime involved on a path. We define *path availability* as the probability to find all nodes and links available for routing on a path. Formally, suppose that a smart packet takes the path: $(n_1, n_2, \ldots, n_d)$, where n_i represents the i-th node on the path and (n_i, n_{i+1}) represents the link between nodes n_i and n_{i+1}.

Assume that the node n_i is available for routing with probability $P_n(n_i)$ and that the link (n_i, n_{i+1}) is available with probability $P_l(n_i, n_{i+1})$. We define the path availability $P_p(n_i, n_d)$ as:

$$P_p(n_i, n_d) = \prod_{j=i}^{d-1} P_n(n_{i+1}) P_l(n_i, n_{i+1}) \qquad (1)$$

We formulate a recursive, combined routing goal function G_i as follows which consist of round-trip delay and path availability:

$$G_i = P_p(n_i, n_d) D(n_i, n_d) + [1 - P_p(n_i, n_d)](T_o + G_i) \qquad (2)$$

where $D(n_i, n_d)$ is the round-trip delay from n_i to n_d and T_o is a predefined timeout interval.

In real systems, many factors contribute to the values of P_n and P_l. For the purposes of this paper, we assume that $P_l = 1$. Because batteries are the major source of energy in mobile nodes, they represent one major factor in determining P_n. To provide greater portability, batteries need to be small and lightweight, which unfortunately restricts the total energy that they can carry. Once the batteries exhaust their energy, they need to be replaced or recharged, which typically reduces the independence of the mobile node to a few hours of operation.

The energy consumption, in communication related tasks, depends on the communication mode of the node. A node may be transmitting, receiving, or in idle mode. Transmission naturally consumes more energy than the other two modes. From the routing perspective, our interest is in selecting routes in such a way that the transmission and reception of packets is distributed on the network so as to maximize the overall average battery lifetime of the nodes. Therefore, we are interested in getting smart packets to select—with greater frequently— the nodes with the longest remaining battery lifetime.

Let us set $P_l(n_i, n_{i+1}) = 1$ in Equation 1, for all $i \in [1, d-1]$, to simplify our discussion. Therefore:

$$P_p(n_i, n_d) = \prod_{j=i}^{d-1} P_n(n_{i+1}) \qquad (3)$$

Assume B_i is the remaining battery lifetime of node n_i. We formulate $P_n(n_i)$ in terms of B_i as:

$$P_n(n_i) = \frac{B_i}{B_m} \qquad (4)$$

where B_m is the lifetime of a fully charged battery. From Equation 3 and Equation 4 we obtain the desired path availability to destination n_d from any node on the path:

$$P_p(n_i, n_d) = \prod_{j=i}^{d-1} \frac{B_{j+1}}{B_m}$$

5 Conclusions

Both mobility and scarcity of resources are natural characteristics of wireless ad hoc networks which impede the use of the original cognitive packet networks algorithm as it requires a semi-static network topology to operate. We have discussed an extension to the CPN algorithm to enable their operation over these type of networks. Our approach was to introduce the use of broadcasts as a last-resource alternative for smart packets to move on the network. Furthermore, we enabled power awareness in the routing goal of smart packets to overcome the problems associated with power limitations in the nodes. For this, we defined *path availability* as a new routing metric, which is able to model the quality and quantity of various network resources and we employed this metric to specifically quantify the battery lifetime of the nodes.

References

1. Charles E. Perkins, *Ad Hoc networking*, Addison-Wesley, December 2000.
2. Erol Gelenbe and Ricardo Lent, "Ad-hoc wireless and wireline cognitive packet networks", in *Proceedings of the Second IEEE Workshop on Applications and Services in Wireless Networks, Paris, France*, July 2002.
3. Elizabeth Royer and C-K. Toh, "A review of current routing protocols for ad-hoc mobile wireless networks", *IEEE PERSONAL COMMUNICATIONS*, April 1999.
4. Charles E. Perkins and Pravin Bhagwat, "Highly dynamic destination-sequenced distance-vector routing (dsdv) for mobile computers", in *Proceedings of the Conference on Communications Architectures, Protocols and Applications - London England.* September 1994, pp. 234–244, ACM.
5. G. Pei, M. Gerla, and T.-W. Chen, "Fisheye state routing: A routing scheme for ad hoc wireless networks", in *Proceedings of ICC 2000, New Orleans.* June 2000, IEEE.
6. David B. Johnson, Davis A. Maltz, Yih-Chun Hu, and Jorjeta G. Jetcheva, "The dynamic source routing protocol for mobile ad hoc networks", IETF DRAFT, March 2001.
7. Charles E. Perkins and Elizabeth M. Royer, "Ad hoc on demand distance vector routing", in *2nd IEEE Workshop on Mobile Computing Systems and Applications.* February 1999, IEEE.
8. M.S. Corson and A. Ephremides, "A distributed routing algorithm for mobile wireless networks", *Baltzer Wireless Networks Journal*, February 1995.
9. Young-Bae Ko and Nitin H. Vaidya, "Location-aided routing (lar) in mobile ad hoc networks", *IEEE Mobicom*, 1998.
10. R. Dube, C. Rais, K. Wang, and S. Tripathi, "Signal stability based adaptive routing (ssa) for ad hoc mobile networks", IEEE Personal Communication, February 1997.
11. V.D. Park and M.S. Corson, "Temporally-ordered routing algorithm (tora) version 1", IETF DRAFT, November 2000.
12. Christine E. Jones, Krishna M. Sivalingam, Prathima Agrawal, and Jyh-Cheng Chen, "A survey of energy efficient network protocols for wireless networks", *Wireless Networks*, vol. 7, no. 4, pp. 343–358, 2001.
13. M. Bhardwaj, R. Min, and A. Chandrakasan, "Power-aware systems", In Proceedings of 34th Asilomar Conference on Signals, Systems and Computers, November 2000.
14. Suresh Singh, Mike Woo, and C. S. Raghavendra, "Power-aware routing in mobile ad hoc networks", in *Mobile Computing and Networking, Proceedings of The Fourth Annual ACM/IEEE International Conference on Mobile Computing and Networking*, 1998, pp. 181–190.
15. C.-K. Toh, "Maximum battery life routing to support ubiquitous mobile computing in wireless ad hoc networks", June 2001.
16. Qun Li, Javed A. Aslam, and Daniela Rus, "Online power-aware routing in wireless ad-hoc networks", in *Mobile Computing and Networking, MobiComm 2001*, 2001, pp. 97–107.
17. Morteza Maleki, Karthik Dantu, and Massoud Pedram, "Power-aware source routing protocol for mobile ad hoc networks", Proceedings of the 2002 International Symposium on Low Power Electronics and Design, 2002.
18. Ivan Stojmenovic and Xu Lin, "Power aware localized routing in wireless networks", IEEE Transactions on Parallel and Distributed Systems, November 2001.
19. Erol Gelenbe, Ricardo Lent, and Zhiguang Xu, "Design and performance of cognitive packet networks", *Elsevier Performance Evaluation 46 (2-3)*, pp. 155–176, 2001.
20. Laura Marie Feeney and Martin Nilsson, "Investigating the energy consumption of a wireless network interface in an ad hoc networking environment", in *IEEE INFOCOM*, 2001.
21. S. Udani and J. Smith, "Power management in mobile computing", Technical report, University of Pennsylvania, MS-CIS-9826, 1998.

SESSION 4A:
TOPOLOGY IN AD HOC NETWORKS

INTRODUCTION OF NODE PROMISCUITY LISTEN MODE AND OPTIMISATIONS IN LAR PROTOCOL FOR WIRELESS AD HOC NETWORKS

F. DE RANGO, G.ALOI AND S. MARANO

D.E.I.S, University of Calabria, Via P.Bucci, cubo 42C
I-87036 Rende (CS), ITALY

Wireless *ad hoc* networks offer mobile nodes the possibility of connecting to other ones. This is possible because each node operates as a router, forwarding packet towards neighbouring nodes. In this environment, routing is an essential issue and protocols with reactive or proactive nature have been proposed in literature. This work focuses on one of these protocols called Location Aided Routing (LAR). The introduction of new functionalities, such as the promiscuous listen mode and speed propagation among the mobile nodes is investigated. Promiscuity is power consuming but, its use in LAR, presents benefits of high data throughput even for high movement speed and both lower overhead and end-to-end data delay. This paper demonstrates that using optimisations of LAR protocol together with the promiscuous listen mode, it is possible to have significant improvements in protocol performances.

1. Introduction

Wireless *ad hoc* networks offer mobile nodes the possibility of connecting to other ones. This is possible because each node operates as a router, forwarding packet towards neighbouring nodes. In this environment, routing is an essential issue and protocols with reactive or proactive nature have been proposed in literature [1].

Recently, the international research community proposed reactive protocols that use location information for the route discovery phase, for packet relay and for route maintenance (i.e.LAR, Gedir, Grid [3]).

This work focuses on one of these protocols called Location Aided Routing (LAR) that can be defined as *partially location aware* protocol. This is because it uses location information only for the route discovery phase and works as a reactive protocol in the packet relay and route maintenance phase.

LAR uses the same methods as the Dynamic Source Routing (DSR) protocol [2] for the packet relay and route maintenance phase. The route discovery phase performed by LAR changes according to LAR *scheme 1* and LAR *scheme 2* as defined in [4]. Recent works [5] show how the performance of LAR, for heavy mobility scenario, give a better data throughput than DSR protocol but at the expense of a higher routing protocol overhead.

The introduction of new functionalities, such as the promiscuous listen mode and speed propagation among the mobile nodes, is investigated to discover whether they give significant improvements to LAR protocol. The authors in [5] introduced only the promiscuous mode in the DSR protocol but not in LAR and they did not investigate the impact of promiscuity on two LAR schemes.

Before explaining the LAR optimisations and the promiscuous listen mode, LAR protocol is briefly explained.

2. LAR protocol

This protocol tries to reduce the propagation of *Route Request* packet (RREQ), used in the Route Discovery Phase, adopting a *Forwarding Zone* that limits the number of nodes propagating RREQ packets. When destinations receive RREQ packets, they reply through a reverse path, sending a packet called route reply (RREP). In this way, the source will know the path towards destination included in RREP header. The forwarding zone is built according to two LAR schemes.

2.1. *LAR scheme1*

In LAR *scheme 1* (fig.1a), the request zone is determined according to the source location and the so-called *expected zone* for the destination. The *expected zone* is the region that most likely contains the destination node.

A source node S that wants to establish a route to a destination node D, computes the *expected zone* with the available location information. The *expected zone* has a circular shape around the position of D. Its radius depends both on the average velocity v of the mobile node D and on the time elapsed since the last known location of the destination is recorded. It is assumed that node S knows the average speed v of node D.

Let's assume that node S knows that node D was at location (X_d, Y_d) at time t_0, at time t_1, node S initiates a new route discovery for destination D. Using information on the average velocity v of node D, node S defines the *expected zone* at time t_1 to be the circle of radius R equal to:

$$R = v(t_1 - t_0) \qquad (1)$$

where $(t_1 - t_0)$ is the time elapsed between two successive route requests from the source node.

2.2. *LAR scheme2*

In LAR *scheme 2* (fig.1b), node S includes two additional pieces of information inside its RREQ[4]: the *distance* (DISTs) from the destination node D at some time t_0 and the last known *coordinates* (X_d,Y_d) of node D location.

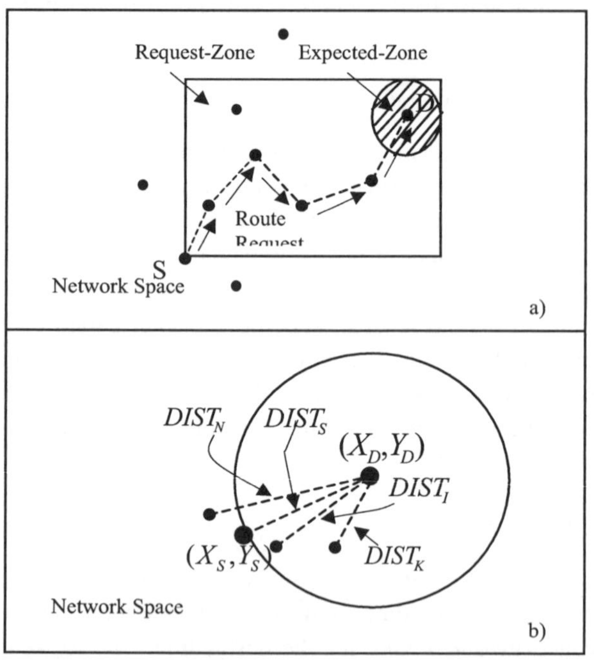

Figure.1: a) Request-Zone in LAR scheme1; b) Forwarding method for LAR scheme2.

On the reception of the route request, an intermediate node I calculates its distance ($DIST_I$) from node D location (X_d,Y_d) and forwards the request to its neighbouring nodes only if the following expression is satisfied for some parameters α and β:

$$\alpha(DIST_S) + \beta \geq DIST_I \qquad (2)$$

3. LAR optimisations

The optimisations considered in LAR are:

- Adaptation of Request Zone
- Propagation of location and speed location
- Promiscuous listen mode

In the *adaptation of request zone*, when LAR scheme 1 is used, node I may determine the *expected zone* using more recent location information for node D, and defines the adapted request zone as the smallest rectangle containing node S and the new expected zone for node D. When LAR *scheme 2* is used instead, node I may calculate the distance from the more recent location of destination D that it knows, and use this distance in the decision rule of *scheme 2*. An example is shown in fig.2a, fig.2b.

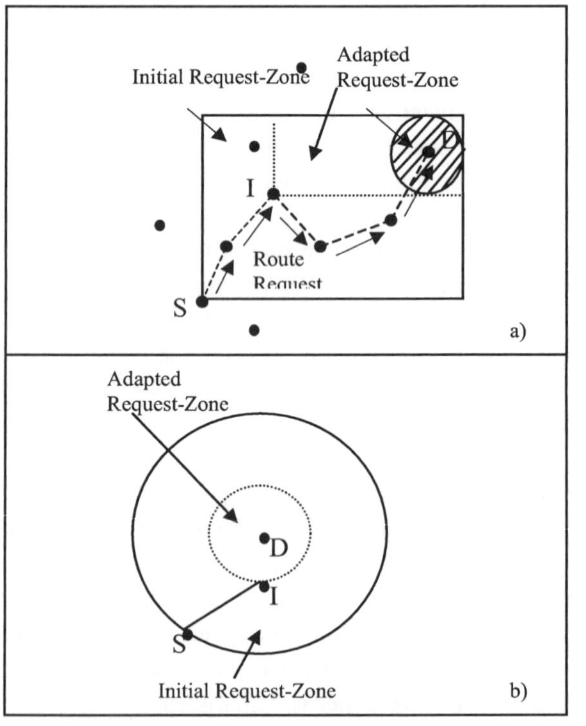

Figure. 2: a) Adaptation of Request Zone for LAR scheme1; b)Adaptation of Implicit Request Zone for LAR scheme1

In the *propagation of location* and *speed location*, instead, each node can get location as a result of its route discovery or as a result of message forwarding for another node's route discovery that includes the same node in the process. The location and speed info are inserted in the RREQ, RERR and RREP packet, so the advantage of request zone adaptation is more effective.

The last optimisation introduced is the *promiscuous listen mode* that permits a node to learn potentially useful routes by listening to packets not addressed to it. Promiscuous listening of data packets transmission significantly reduces control overhead and increases the data packets delivery ratio, especially at high node speeds, as demonstrated in [5] for DSR protocol. In LAR protocol the effect is more suitable because it improves the adaptation of the request zone and the location propagation.

The promiscuous mode used in DSR protocol reduces the stale paths stored in the route cache so cache records are more suitable giving a more effective improvement. When bad routes are stored inside the cache, the intermediate nodes cache replies could reduce the data throughput and could generate RERR and RREQ packets in the networks increasing the control overhead. Moreover, the LAR protocol, owing to its nature, can be improved by both the promiscuity and the speed propagation because it is possible to have more recent information inside to *info location cache.* More accurate location info permit a more efficient resizing of the request zone and reduces the flooding request.

4. Simulation Results

The LAR protocol was implemented in NS2 [7]. Simulations for different mobility scenarios that use the *point way model* [6] are performed. It is assumed that each node moves with average speed v and it has a pausing time of 10 seconds. The node speed is uniformly distributed in the range $[v\text{-}\alpha^* v, v+\alpha^* v]$ with α that can assume the values 0.1.for a heavy mobility scenario [5]. The total number of considered mobile node is 50.

Simulations with movement patterns generated for 5 different average speeds, 0.5, 5, 10, 15, 20 m/s are run. Scenarios with 75 different movement patterns, 15 for each value of the average speed, are generated. LAR and its modified versions run on the same 75 movement patterns. The simulated traffic sources are at a constant bit rate. The packet size is 64 bytes and the transmission rate is fixed to 4 packets/s. The considered maximum transmission distances between nodes are 100m and 250m.

A timer of $t_{timeout}$ =500ms between retransmitted route requests in the case of route discovery failure is assumed. In our implementation, a source asks its neighbouring nodes a path to destination before starting the route discovery phase. As suggested in [5], a timeout of 30ms for this first hop response is considered. The packet lifetime in buffers is 30s.

Figure 3 shows the protocol overhead for LAR *scheme 1*, LAR *scheme 2* with and without optimisations and DSR-P is shown. The introduction of speed propagation optimisation leads to better rebuilding of the Request-Zone and to the reduction of control overhead. DSR-P with route cache reply mechanism, however, leads to lower overhead as shown in [5]. The promiscuous listen mode has more suitable effects over LAR protocol than DSR-P because through it each node has more accurate information. LAR *scheme 2* has a worse performance than LAR scheme1 because it uses a larger Request-Zone than LAR *scheme 1*.

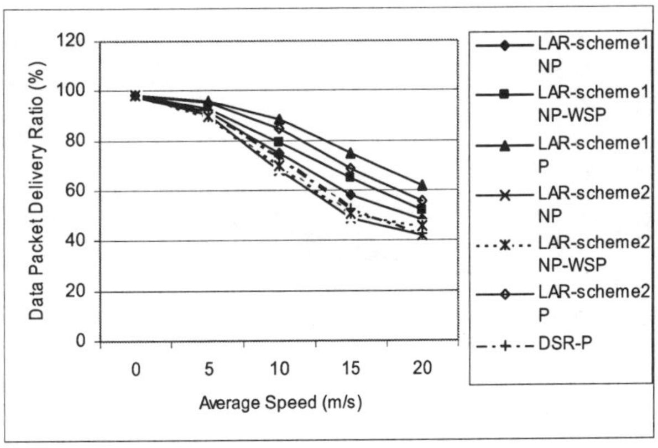

Figure3: Routing Overhead for transmission range of 100m and 300x600 grid.

Figure 4 shows how the LAR protocol *scheme 1* and *scheme 2* obtain a higher throughput trough geographical info location.

Figures 5, 6 and 7 show overhead distribution for RERR, RREP and RREQ packets. Propagation speed and promiscuous listen mode reduce the propagation of RREQ packets but increase the RREP and RERR overhead. Total protocol overhead however is reduced. It can be noted how DSR-P and LAR characteristics are extremely different.

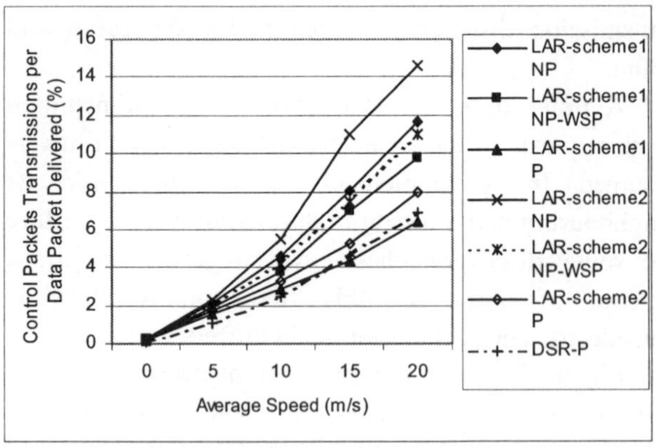

Figure4: Data Packet Delivery Ratio for transmission range of 100m and 300x600 grid.

In figure 5 the DSR-P cache reply mechanism lead to frequent route failure procedure causing a higher RERRs overhead. LAR protocol, instead, reduce route recovery trough the optimisations introduced.

The DSR RREP and RREQ overhead are lower than LAR ones because it use the route cache reply mechanism that reduce the route request propagation as showed in figure 6 and 7.

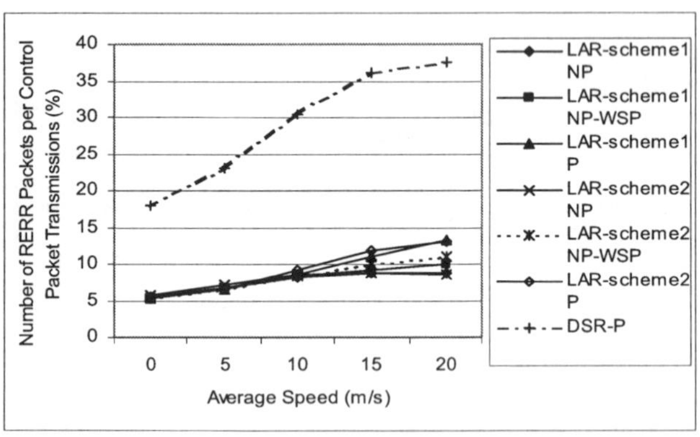

Figure5: Routing Overhead of Route Error packets for transmission range of 100m and 300x600 grid.

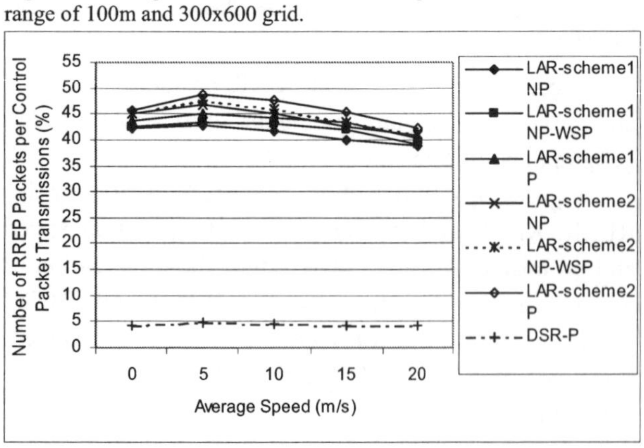

Figure6: Routing Overhead of Route Reply packets for transmission range of 100m and 300x600 grid.

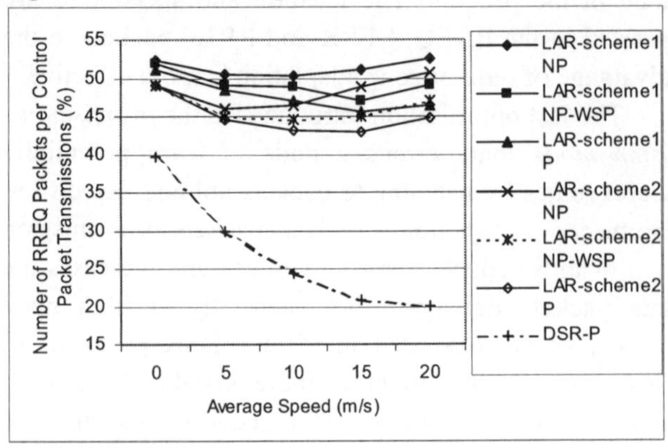

Figure7: Routing Overhead of Route Requests packets for transmission range of 100m and 300x600 grid.

Figure 8 shows how more refreshed information leads to finding the path toward destination with a better probability of success. In fact, the percentage value of the flooding requests is reduced. LAR *scheme 2* using a larger request zone than LAR *scheme 1* can obtain a lower flooding request but it has greater protocol overhead because the larger *Forwarding-Zone* includes more mobile nodes in the route discovery process.

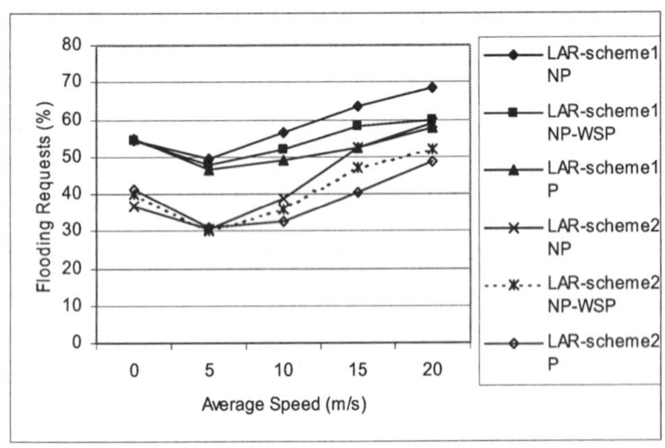

Figure8: Flooding Requests in terms of percentage value for transmission range of 100m and 300x600 grid.

Figure 9 shows the average end-to-end data packet delay. Speed propagation and promiscuous listen mode offer less delays for LAR *scheme 1* and LAR *scheme 2*. An effective building of the *Request-Zone* reduces the protocol overhead giving more space to data packets in sending buffers and reducing the latency in the tails (protocol packets have higher priority than data packets).

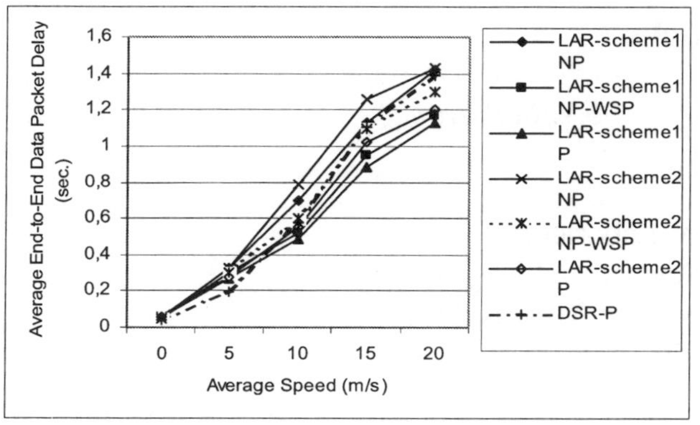

Figure9: Average end-to-end data packet delay for transmission range of 100m and 300x600 grid.

Figure10 shows the protocol overhead in the case of a transmission range of 250m. The lowering of the curve can be seen due to greater route stability. A greater transmission range guarantees greater coverage and a longer duration of the routes. Also in this case it can be noted how the speed propagation together with the promiscuous listen mode notably reduces the overhead.

All results are obtained for a mobility scenario where the nodes move with a mean speed variation of about 10% but similar considerations apply to the speeds with greater percentage change (e.g. 50%, 100%).

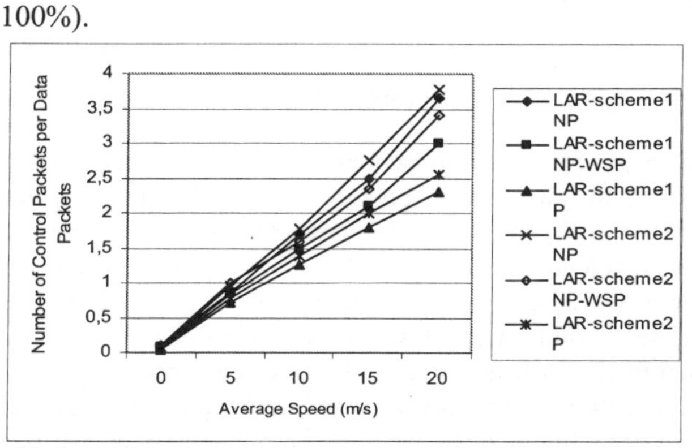

Figure10: Routing Overhead for transmission range of 250m and 750x1500 grid.

5. Conclusions

The simulation showed an improvement in LAR *scheme 1* and *scheme 2* and their performances outperform the DSR protocol. The performance parameters are evaluated in terms of flooding request, RREQ overhead, RREP overhead, RERR overhead, data throughput, average end-to-end delay. We first evaluated the performance with only speed propagation then with the promiscuous mode listen. Promiscuity is power consuming but, its use in LAR, presents benefits of high data throughput even for high movement speed and both lower overhead and end-to-end data delay.

References

1. E. M. Royer, C. K. Toh, "A Review of Current Routing Protocols for Ad Hoc Mobile Wireless Networks," IEEE Personal Comm., Apr. 1999.
2. D. B. Johnson, D. A. Maltz, Y.-C. Hu, J. G. Jetcheva, "The Dynamic Source Routing Protocol for Mobile Ad Hoc Networks (DSR)", Internet-Draft, 21 Feb. 2002.
3. M. Mauve, J. Widmer, H. Hartestein, "A Survey on Position-Based Rouitng in Mobile Ad Hoc Networks", IEEE Network, Nov. 2001.
4. Y. Ko, N.H.Vaidya, "*Location-aided routing (LAR) in mobile ad hoc networks*", Wireless Networks, vol.. 6, pp.307-321, 2000.
5. T. Camp, J. Boleng, B. Williams, L. Wilcox, and W. Navidi, "*Performance evaluation of two location based routing protocols,*" Proc. of INFOCOM, 2002.
6. T. Camp, J. Boleng, V. Davies, "*A survey of mobility models for ad hoc network research*", Wireless Communication and Mobile Computing Journal, 2002.
7. K. Fall, K. Varadhan (Editors), "Ucb/lbnl/vint Network Simulator–ns(version2), http//www-mash.cs.berkeley.edu/ns/ "

An Information Theoretic Approach to Cluster Head Selection in Mobile Ad Hoc Networks

Philips Koshy *S.V.Raghavan*

Network Systems Laboratory
Department of Computer Science and Engineering
Indian Institute of Technology, Madras, India
koshy,svr@cs.iitm.ernet.in

Abstract

Hierarchical routing is found to be more effective than flat routing in large mobile ad hoc networks. For this purpose, mobile nodes are grouped into clusters and each cluster is represented by a cluster head. Conventional methods use either the connectivity or the identifiers of the nodes to select the cluster head. We propose an activity based approach to select the cluster head. We use the entropy of the node to gauge the activity. The entropy is a measure of how confined the movements are, and this is used to classify nodes as conservative and exploratory. Simulation results show that the proposed cluster head selection causes less cluster head changes, as compared to the widely used Lowest-Id cluster head selection.

1. Introduction

Ad hoc networks consists of a collection of mobile wireless nodes with no fixed infrastructure. Ad hoc networks finds application in situations where it is very difficult or very expensive to establish fixed infrastructure based networks. Examples include battle fields, rescue and emergency applications etc. Researchers are looking into the applications of the mobile ad hoc networks in sensor networks.

This paper discusses our attempts to improve Quality of Service (QoS) for multimedia communication in mobile ad hoc networks at the network layer. It is seen that multimedia files are generally large and it takes relatively longer for their transfer, unlike other kinds of data transfer. The dynamically changing topology of mobile networks poses a serious threat to multimedia communication, as it causes frequent changes in the multimedia data transfer path. Thus a stable path is more crucial to multimedia traffic than for other data transfer applications. This is true for both flat routing as well as hierarchical routing.

Flat routing becomes very ineffective for large wireless networks, because of link and processing overhead. One solution to this scalability problem is to use hierarchical routing. Internet uses hierarchical routing and this helps in keeping the routing tables small and fewer updates. In wireless networks hierarchical routing is made possible by grouping nodes geographically close to each other into explicit clusters [1]. In each cluster one node is elected as the representative (cluster head) for performing inter-cluster communication.

There have been several proposals to elect cluster heads (CH) in the literature. To this growing mix of proposals, we are presenting a novel cluster head selection based on the activity of the mobile node. The information theoretic definition of entropy is used to measure the activity of mobile nodes. It is shown that the activity based method leads to less CH changes when compared to the traditional methods of CH selection.

2. Related Work

Researchers have proposed several techniques for cluster head selection. These techniques are broadly classified into identifier based or connectivity based CH selection. The most popular among them are the lowest-Id [2] and the maximum connectivity [3]. In the former, if a low id node ever gets highly mobile it can result in an unacceptably large number of cluster head changes. In maximum connectivity technique, the criterion used for selection keeps changing for different mobile nodes depending on the activity and distance and this can cause instability. In the clustering scheme referred to as (α, t) clustering, they assume that the link is active for a definite period of time from the time it is noted as active. This assumption is weak, in the sense that whenever it goes wrong it makes clustering ineffective. In [4] Sivavakeesar et al. predict the time and probability that a node is going to remain in a given cluster. As the cluster head selection depends on this prediction, any inaccuracy in it will make the clustering inefficient. Unlike the identifier based or the connectivity based methods, the proposed activity based method leads to more stable cluster heads.

3. Mobile Node Behavior Modeling

In this section, a framework for characterizing the mobile node behavior is developed. Such a characterization helps in classifying nodes as conservative and exploratory. The movements of a conservative node is expected to be confined to fewer clusters than that of a node in exploratory mode. Behavior modeling involves two phases - Behavior Monitoring and Behavior Characterization. Behavior monitoring involves the manner in which the location of the mobile node is sampled. This creates the movement history or movement pattern of the mobile node. The Behavior capturing phase, takes as input, the movement history developed and analyzes it and comes out with two parameters which represent the behavior of the mobile node.

Clusters are formed by dividing the region into several smaller regions or cells and a virtual cluster is associated with every cell. In this kind of clustering, the cluster boundaries are static or fixed.

3.1. Behavior Monitoring

The activity/behavior of a mobile node may be monitored in several ways. In this work, we mean by the behavior, the activity involved with the changes in position/location with respect to time. As the node keeps changing its location/position continuously, one can only take samples of the position/location. To monitor the activity of a node, its position is sampled at various instants. There are various ways in which this sampling process can be performed. We assume that all nodes are able to keep track of its position with the help of Global Positioning System (GPS) or some location information system. The behavior monitoring phase builds the movement history and it is used as input for the behavior capturing phase. We define node mobility as a stochastic process $\mathcal{V} = \{v_i\}$ where v_i forms a sequence of random variables representing the cluster i to which the mobile belongs at the instant.

3.2. Behavior Characterization

The behavior characteristics of a mobile node (MN) are captured by an incidence vector and a transition matrix. Similar descriptions were used in [5] for classifying interactive users. The number of visits to different clusters made my the mobile node in the recent past as well as their transitions from one cluster to another, are noted . The incidence vector contains the steady state probabilities and the transition vector contains the transition probabilities.

The steady state incidence vector (**S**) is of size 1xn where n is the total number of clusters in the region. Each entry in the vector gives the probability of the node being in that cluster. If all the elements of the incidence vector are equal, this means that the mobile node is equally likely to be in any of the clusters. On the other hand, if only a few elements are prominent and other entries are zero, it means that the movement of the mobile node is confined to only a small number of clusters. The transition matrix (**T**) describes the dynamics in the node behavior and is an n x n matrix. Each element P_{ij} represents the probability of visiting cluster j from cluster i.

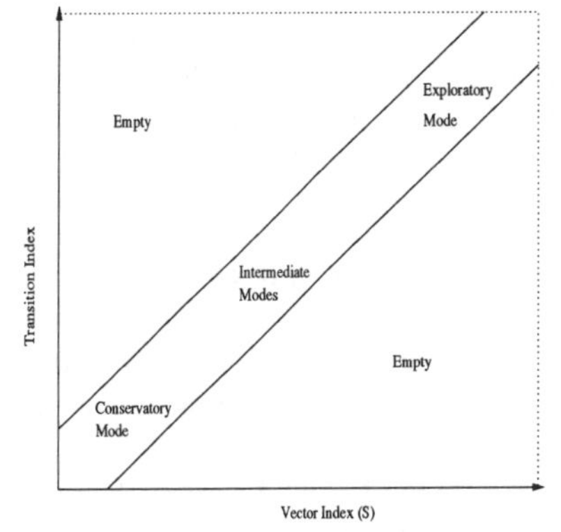

Fig. 1. Plot showing modes of mobile nodes

In order to capture the state of a mobile node (whether in exploratory mode or conservative mode) we use two indices called steady state index ($\mathcal{S}$) derived from **S** and transition index ($\mathcal{T}$) derived from **T**. The information-theoretic definitions of entropy are used for this purpose.

$$\mathcal{S} = -\sum_{i \in \mathcal{V}} p_i \log p_i$$

$$\mathcal{T} = -\sum_{i=1}^{n} \sum_{j=1}^{n} p_i \log p_{ij}$$

The tuple $< \mathcal{S}, \mathcal{T} >$ represents the current state of the mobile node. The plot of V versus $\mathcal{T}$ can be divided roughly into three regions. Figure 1 illustrates this regions. If the value of $\mathcal{S}$ and $\mathcal{T}$ are high, the node is in an exploratory mode and if the point is close to the origin it represents a node which is in the conservative mode.

4. Clustering Algorithm

We assume that the region to be covered consists of a number of static clusters associated to geographic areas. The clustering algorithm works as follows. The CH is supposed to send a HEARTBEAT message every CH-HELLO-INTERVAL. This message contains the virtual cluster id and the ids of all MNs falling within the cluster. Whenever a new MN comes into the cluster it is it's responsibility to register with the new CH.

The way the new CH is selected depends on whether a CH is presently existing or not. If a CH does exist, it is it's responsibility to initiate the CH selection process. For this purpose, at the appropriate time, it broadcasts a CH-CHANGE-OVER event. All MNs within the cluster are expected to send their $\mathcal{S}$ and $\mathcal{T}$ values to their existing CH. The CH the elects the MN with the lowest $\mathcal{S}$ and $\mathcal{T}$ values as the new CH. On the other hand if for some reason a CH does not exist, new CH election process has to be done in a distributed manner. This is done as follows. All the MNs send their $\mathcal{S}$ and $\mathcal{T}$ values to each other. If any of the received values is smaller, the MN backs out of the contest. This way they reach a consensus and the MN with the lowest values become the CH. However, this kind of CH election takes much longer than the CH selection by an outgoing CH.

5. Simulation

In this section, we present the simulation results of the proposed scheme as well as the comparison with Lowest-Id cluster head selection.

The region is divided into 25 square-shaped sub-regions, each representing a cell. We use random walk mobility model [7] for our studies. The simulator gives the location information of 500 mobile nodes at fixed intervals. We build the movement history from this location data. The movement pattern of each node is analyzed and the entropies are computed. The entropy of the node depends on the movement characteristics (such as angle and speed) and characteristics of the cell (such as size and shape). We simulate scenarios with varied speeds of nodes, varied sizes of cells and varied shapes of cells. As they all cause different entropies for nodes, we report the results of the scenario with varied speeds of the nodes. This helps us to simulate some conservative nodes and some exploratory nodes. The location of the nodes for the recent past is noted at regular intervals. Though there are more efficient ways of collecting the location information(such as activity based, distance-based etc.), we chose fixed time-interval based updating, for simplicity. We sample the position of the node at fixed time interval of 15 time units. Figure 2 shows the entropy of 500 mobile nodes. Nodes close to the origin are more conservative and their movement is confined to fewer clusters than those nodes far from origin.

For every cluster, we determine the node with the lowest entropy among the nodes present in the cluster at that point of time. This node is made the cluster head. The results of the entropy based cluster head selection is compared against lowest-Id cluster head selection. The results in figure 3 shows marked improvement in the number of cluster head changes.

To study the performance of the entropy based technique for different ranges of entropy spread, we gradually vary the

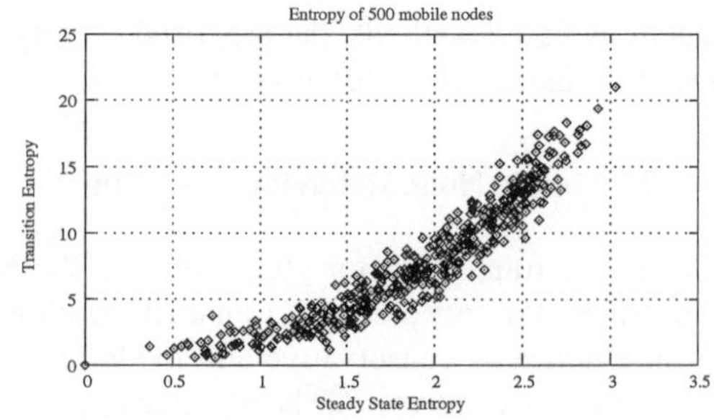

Fig. 2. Entropy of 500 mobile nodes

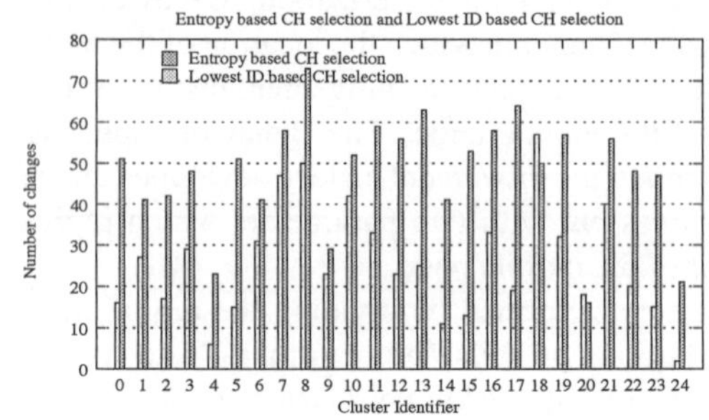

Fig. 3. Number of cluster-head changes

entropy of the mobile nodes. Figure 4 shows some such scenarios with the top left one, with entropy spread over the largest range, and bottom right with entropy spread over a narrow range. Table below show the corresponding number of cluster head changes for entropy based technique and lowest ID based technique. We find that as long as the entropy is spread over a reasonable range, entropy based technique results in less number of cluster head changes than the lowest ID technique.

Table 1. Total cluster head changes for entropy based and lowest ID based methods

Scenario	1	2	3	4
Entropy method	214	473	994	1673
Lowest ID method	1108	1157	1303	1671

6. Conclusions

In large mobile ad hoc networks, hierarchical routing is found to be more efficient than flat routing. For hierarchical routing, mobile nodes are grouped into clusters and a cluster head (CH) is selected to represent them. As all inter-cluster communication are directed through the CH, it is important

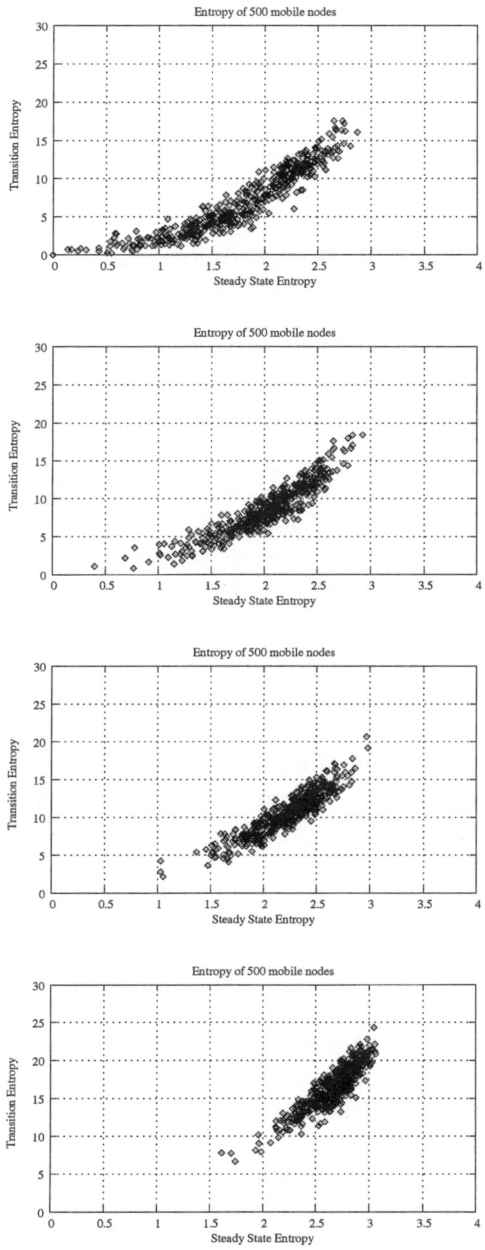

Fig. 4. Scenarios with entropy spread over different ranges

that CH does not change frequently. We have proposed an information theoretic approach to selecting a stable node as the cluster head. We observe that the entropy of the node can be used as a measure of how conservative or how exploratory a node is. The node with lowest entropy among the nodes present in the cluster, is selected as the cluster head. Our results show that entropy based cluster head selection causes less cluster head changes than Lowest-Id cluster head selection. Essentially, the entropy based characterization provides a technique to distinguish mobile nodes based on their activity. Though we have used it, to compute the stability of nodes and select them as CHs for routing, this kind of separation based on the activity/behavior can be put to use in several applications. It is also possible to compute the cell level activity and the network level activity.

References

[1] P. Krishna, N.H.Vaidya, M.Chatterjee and D.K.Pradhan, "A Cluster Based Approach for Routing in Dynamic Networks," in *Proceedings of ACM SIGCOMM Computer Communication*, April 1997, pp. 49–65.

[2] M.Jiang, J.Li and Y.C.Tay, "Cluster Based Routing Protocol (CBRP) Function Specifications," *IETF Draft*, August 1999.

[3] T.C.Hou and T.J.Tsai, "An Access Based Clustering Protocol for Multi hop Wireless Ad Hoc Networks," *IEEE Journal on Selected Areas in Communication*, vol. 19, July 2001.

[4] S.Sivavakeesar and G.Pavlou, "A Prediction-Based Clustering Algorithm to Achieve Quality of service in Multi Hop Ad Hoc Networks," in *London Communications Symposium*, 2002, pp. 676–684.

[5] S.V.Raghavan and R.Kalyanakrishnan, "On the Classification of Interactive Users Based on User Behavior Indices," in *Proceedings of the ACM SIGMETRICS Conference on Measurement and Modeling of Computer Systems*, August 26-29 1985.

[6] N.Swaminathan and S.V.raghavan, "Intelligent Pre-Fetch in WWW Using Client Behavior Characterization," in *In Proceedings of the International Symposium on Modeling, Analysis and Simulation of Computer and Telecommunication Systems, San Fransisco, California*, August 2000.

[7] Tracy Camp, Jeff Boleng and Vanessa Davies, "A Survey of Mobility Models for Ad Hoc Network Research," *Wireless Communications and Mobile Computing: Special Issue on Mobile Ad Hoc networking: Research, Trends and Applications*, vol. 2, no. 5, pp. 483–502, 2002.

A LOCALIZED TOPOLOGY CONTROL ALGORITHM FOR POWER-EFFICIENT ROUTING ON MOBILE AD HOC NETWORKS

GUO XIAOGUANG

Nanyang Technological University, BLK N4, #02-32, Nanyang Avenue, Singapore
E-mail: gxg@pmail.ntu.edu.sg

HSU WEN JING

Nanyang Technological University, BLK N4, #02-32, Nanyang Avenue, Singapore
E-mail: hsu@ntu.edu.sg

In this paper, a distributed topology control algorithm is proposed to improve the power efficiency of routing algorithms in mobile ad hoc networks. In the algorithm, each node makes local decisions about its transmission power and these local decisions collectively retain global connectivity. The power control approach not only reduces the power consumption in transmission, but also retains the global network connectivity so that the communication routes between nodes will not be disconnected due to the power control. We also apply our topology control algorithm in some geographic routing protocols to show that the routes in the multihop network are efficient in power consumption. Finally, simulation results demonstrate significant performance improvement.

1 Introduction

Mobile ad hoc networks (or MANETs for short) is a collection of wireless mobile nodes in which communications with each other are without the aid of fixed infrastructures. In such an environment, each node acts as a router and forwards packets to the next hop in order to reach the destination node hop by hop. Since in these networks, nodes operate on limited battery power. it is important to minimize power consumption to prolong the network's life time.

Topology control for ad hoc networks aims to conserve energy, reduce interference, and increase effective capacity by adjusting the transmission power on each node. Recently, some topology control schemes have been proposed for ad hoc networks. The centralized topology control algorithms [1, 2] can achieve optimality or its approximation, which are more applicable to the static networks due to the lack of adaptability to topology changes. In [1], topology control was formulated as an optimization problem to minimize the maximum power consumption subject to the connectivity constraints. In this paper, two centralized greedy algorithms were presented to guarantee connectivity of the network and bi-direction of the links that can achieve the optimal power consumption on the static networks. Another centralized topology control algorithm for static ad hoc networks was described in [2], which achieved more challenging optimization objectives than [1], such as minimizing the maximum transmission power and minimizing the total power subject to stronger constraints such as two-node connectivity.

In contrast, the distributed/localized topology control algorithms [3,4] are more suitable for mobile ad hoc networks since they are adaptive to topology changes at the cost of less optimality in reducing transmission power and maintaining network connectivity.In [3], Rodoplu and Meng proposed a distributed topology control algorithm (minimum energy mobile network protocol, MEMN for short) that guarantees connectivity of the entire network. The main idea of this protocol is that it is not required to consider all the nodes in the network to find the globally minimum power path to the desirable destination. Instead, using a very localized search, a node can eliminate some nodes in its relay regions from consideration and select only those few nodes in its immediate neighborhood to be the only potential candidates. In [4], a distributed topology control algorithm termed Cone-Based distributed Topology Control (CBTC) was proposed. CBTC attempted to limit the degree of each node by adjusting the transmission power, and it did so directly by its proposed optimizations and indirectly by increasing a node's transmission power range gradually until the condition that "a node has some neighbor in any of its small-enough pie-area" is met. This condition provably ensured the connectivity of the network, which failed to be achieved in [1].

In this paper, we propose a localized topology control algorithm that achieves power control and power-efficient routing on MANETs. The main idea of our algorithm is that, nodes can reduce their transmission power to maintain the immediate neighbors more frequently than the further neighbors. The power control decisions are made through these immediate neighbors, which can construct a power-efficient topology on the underlining ad hoc network. Our algorithm not only reduces the transmission power, but also guarantees the global network connectivity so that the communication routes between nodes will not be disconnected due to the power control. Moreover, the routing is operated on the topology so that a power-efficient end-to-end route can be formed locally.

The rest of our paper is organized as follows. Section 2 describes the network model used in the proposed algorithm. Section 3 presents the algorithm and correct it. Simulation study is presented in Section 4 to demonstrate the benefits of the proposed topology control algorithm to application performance. Finally, Section 5 concludes and suggests the directions for future research.

2 The Model

We adopt a common *path loss model* to measure the power consumption P of transmission, in which the transmit power P falls as a function of $1/d^k$, where d denotes the distance between two nodes, and k denotes a constant determined from field measurements, whose value typically lies between 2 and 4. We also assume that, in the network, each node can get its own geographical location through the support of GPS. There exists a location service system, such as GLS [6], to provide the location information of destinations for source nodes in the network. In order to achieve power control, each node has the ability to adjust its transmission power, up to a maximum value that is identical for all nodes in the network. Here, we also assume the existence of an underlying MAC layer to resolve multiple channel access and interference problems.

The following conventions will be adopted throughout the paper. $d(x, y)$ denotes the Euclidean distance between node x and node y. $P(x)$ and P_{max} denote the current transmission power and the maximum transmission power of node x. $R(x)$ and R_{max} denote the current transmission radius and the maximum transmission radius of node x respectively. $N(x)$ denotes the set of neighbors of node x. $N_{max}(x)$ denotes the maximal set of neighbors of node x with the maximum transmission power P_{max}.

Consider three nodes s, r and d in a $2D$ plane, where s denotes the sender and d denotes the destination node. If s uses r as its next hop to forward packets to d, node r is referred to as a relay node of s.

Definition 1 (*Relay Region*): *For a given sender-relay node pair (s, r), the relay region $RR(s, r)$ of the ordered pair of nodes (s, r) is defined as*

$$RR(s,r) = \{x|d^k(s,r)+d^k(r,x) \leq d^k(s,x), x \neq r, k \geq 2\}$$

The notation of *relay region* is first defined by Rodoplu in [3]. Intuitively, the relay region is the set of destination nodes where relaying through r is more power-efficient for s.

Definition 2 (*Relay Candidate*): *For a given node s, and node $r \in N(s)$. Node r is said to be a relay candidate of s if $\forall x \in N(s)$, $r \notin R(s, x)$. The set of relay candidates of node s is denoted by $RC(s)$.*

Intuitively, for each node, its relay candidates are the set of nodes that relaying through them to the other nodes is more power-efficient for it. Here, if a node j is a relay candidate of node i, and node k is in the relay region $RR(i, j)$, we say node k is *dominated* by node j for node i.

Definition 3 (*Relay Polygon*): *For a given node s, let $r_0, r_1, \cdots, r_{n-1}$ be the list of relay candidates of node s, sorted according to the clockwise direction with respect to the node s. The relay polygon $RP(s)$ is formed by linking every two consecutive relay candidates in this sequence, i.e.*

$$RP(s) = \bigcup_{0 \leq i < n} \overline{r_i r_{i+1(mod n)}}$$

There are two general cases of relay polygon. When a node is internal to its relay polygon. We say that node s is *bounded* by its relay polygon; Otherwise, when a node is external to its relay polygon, we say that the node is *unbounded* by its relay polygon.

Lemma 1 *For any arbitrary node x on the plane, the relay polygon $RP(x)$ is a convex polygon.*

Proof: We first prove that the relay polygon $RP(x)$ is a simple polygon. As shown in Figure 1 (a), by contradiction, it is easy to prove that the relay polygon $RP(x)$ is a simple polygon. So in the next

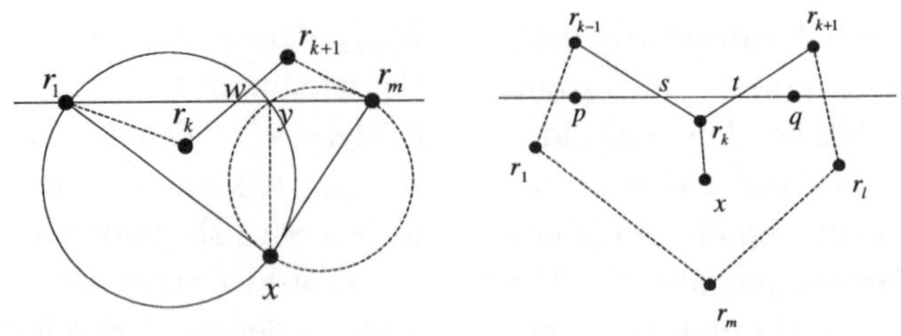

(a) RP(x) is a simple polygon (b) RP(x) is convex

Figure 1. The relay polygon of node x is a convex polygon

step, we only need to prove that the relay polygon of node x is a convex polygon. For contradiction, assume that the relay polygon is a non-convex polygon. By the definition of *nonconvex polygon* [7], there exist two points p, q interior to the polygon $RP(x)$, such that $\overline{pq}$ is not entirely contained in $RP(x)$. Assume that $\overline{pq}$ intersects $\overline{r_{k-1}r_k}$ and $\overline{r_k r_{k+1}}$ at points s and t respectively, and the segment $\overline{st}$ of line $\overline{pq}$ is not contained in the relay polygon $RP(x)$. As shown in Figure 1 (b), in the triangle $\triangle sr_kt$, the angle $\angle sr_kt$ is less than π, and hence the complementary of $\angle sr_kt$ is greater than π. So at least one of the angles $\angle r_{k-1}r_kx$ and $\angle r_{k+1}r_kx$ must be greater than $\frac{\pi}{2}$. Without loss of generality, we assume that $\angle r_{k-1}r_kx > \pi/2$. Then we know that r_{k-1} is in the relay region $RR(x, r_k)$. This contradicts to the condition that r_{k-1} is a relay candidate of node x. So combining with the fact that $RP(x)$ is a simple polygon, we conclude that the relay polygon $RP(x)$ of node x is a convex polygon. ∎

Since the *point location problem* [7] in convex polygon is simpler than in non-convex polygon, Lemma 1 can help us to simplify the decision of point location in our algorithm. It's also the reason that we only consider the closure of polygon formed by nodes in $RC(i)$, not in $N(i)$.

Definition 4 (*Relay Graph*): *Given a network $G = (V, E)$, the relay graph of this network, denoted by $RG = (V', E')$, is given by*

$$RG = \{(V', E') | V' = V, E' = \bigcup_{i \in V'} \bigcup_{j \in RC(i)} \overline{ij}\}$$

3 The Algorithm

In this section, we will describe the Relay Topology Control (RTC) algorithm in details.

3.1 *Search for Relay Candidates*

From the property of relay candidates, we know that, for a source node, its relay candidates have the power-efficient properties in contrast to its other neighbors, since its other neighbors can be reached through them power-efficiently. In this section, we propose a distributed algorithm to find the relay candidates of each node. In our relay topology control algorithm, the *Search for Relay Candidates* algorithm is the basis for other operations (power control, routing, etc.)

The pseudocode of this algorithm is shown in the following, where i denotes the node executing the algorithm, and $N(i)$, $RC(i)$ denote the set of neighbors and relay candidates of node i respectively, when using transmission power $P(i)$. This algorithm is executed in each node to find its relay candidates.

$Search\_Relay\_Candidates(i)$
{
$RC(i) = N(i);$
$for\ each\ j \in RC(i)$
$for\ each\ k \in RC(i) - \{j\}$
$if\ k \in RR(i, j)$
$RC(i) = RC(i) - \{k\};$
}

3.2 *Relay Topology Control Algorithm*

In our relay topology control algorithm (RTC), each node dynamically adjusts its transmission power in order to guarantee the network connectivity. The power control decision is based on whether a node is surrounded by its relay candidates (internal to its relay polygon).

Initially, each node i broadcasts beacons with a transmission power $P(i) = P_0$. After receiving replies from the other nodes, node i finds its relay candidates by executing the *Search for Relay Can-*

didates algorithm and calculates its relay polygon $RP(i)$. Then node i can judge whether itself lies inside of $RP(i)$. This is a *point-location problem* in computational geometry [7]. If node i is not internal to $RP(i)$, it will continue to increase the transmission power until it becomes internal to its relay polygon, or until the maximum transmission power P_{max} is reached.

4 Simulation Results and Evaluation

In this section, we simulate our algorithms to measure the impact on some performance issues, such as network connectivity, power consumption, etc.

4.1 Simulation Environment

Our simulation is implemented based on Simjava [8], a discrete event simulation package for Java. The simulation emphasizes on the topology control algorithm and packet forwarding protocol while assuming other low level parameters at reasonable values. Our simulations provide a rather wide range of variables such as node density, adjustable transmission radius as well as variations of routing methods.

4.2 Network Connectivity

In this section, the purpose of the experiment is to study the effect of varying nodes' transmission radius on network connectivity. We vary nodes' transmission radius in the range from 50 meters to 400 meters and test the network connectivity in the five cases of node density. The transmission radius is step-incremented at 50 meters. We assume that no nodes fail or disappear under any circumstances. The simulation results are shown in the Figure 2 (a).

From Figure $2(a)$, we can see that increasing transmission radius may result in increasing of the network connectivity. Moreover, the network connectivity will not further increase when certain thresholds of transmission radius is exceeded in different cases. Therefore, the networks in different cases will achieve strong connectivity when the transmission radius of nodes in this network reaches or exceeds the threshold. In our simulation, we treat the threshold of transmission radius in each node density as an *optimal* transmission radius for this case since the threshold is the smallest transmission radius which can ensure strong connectivity of the network.

4.3 Efficiency of RTC Algorithm

We will measure the performance of our algorithms under three topology control scenarios (denoted by Max, Fixed and RTC), in which their transmission radius are maximum transmission radius, optimal transmission radius and the radius by RTC respectively.

We measure the average node degree under the three kinds of topology control scenarios. The experimental results are shown in Figure $2(b)$. The node degree of the mobile ad hoc networks should not be too large because a large node degree typically implies that a node has to communicate with more distant nodes directly, which increases interference and collision. The average degree should not be too small either because that tends to increase the overall network power consumption as longer paths have to be taken. So we believe that average node degree is an important metric for mobile ad hoc network topology.

When using RTC and fixed transmission radius,

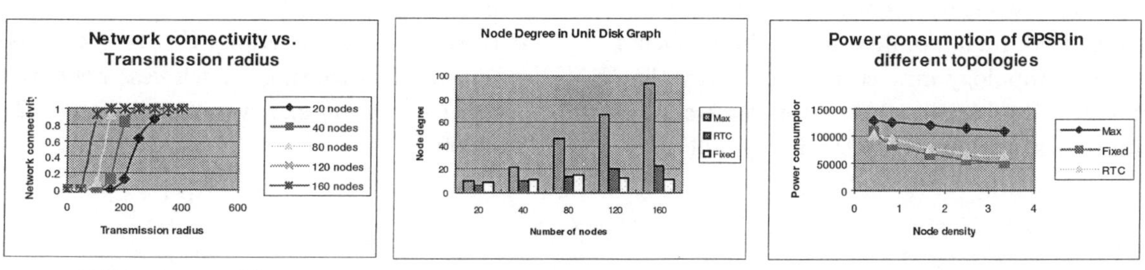

(a) Network connectivity vs. Transmission radius

(b) Node degree in Unit Disk Graph

(c) Power consumption in routing

Figure 2. Simulation results

the average node degree can be decreased significantly compared with using maximum transmission power. The average node degree in RTC is very close to the one in Fixed scenarios when the node density is not very high. However, the difference between them becomes larger when the node density increases. In RTC algorithm, the nodes that lie near the boundaries of the network contributes more to the average degree statistics. This is because most of such nodes are not internal to their relay polygons, but they are required by the RTC algorithm to increase their transmission radius to check whether they are internal to their relay polygons until the maximum transmission radius is reached. So we remark that the average node degree can be reduced if we know the boundary of the network. When a node finds itself to be very near the boundaries of the network region, it is not required to use the maximum transmission radius.

4.4 Power Consumption

Figure 2(c) shows the average power consumption for routing when using GPSR [9] routing protocols in different topology control scenarios. From the figure, we can see that our topology control algorithm reduces the power consumption significantly in routing process and tends closely to the Fixed scenario.

5 Conclusions

In this paper, we describe a distributed topology control algorithm that can not only reduce the transmission power, but also guarantee the global connectivity. In our simulation, we only verify the results in a static environment. However, when node mobility is considered, the algorithm may be less efficient, because the power-efficient graph needs to be dynamically maintained whenever nodes move. In networks where nodes have high mobility, maintaining such a power-efficient topology may cost more energy than it could save. How to design the topology control algorithm to deal with mobility efficiently will be included in our future research. In our algorithm, the topology graph maintained is sparse. Although it lowers the cost of topology maintenance, it also reduces the number of choices in setting up paths and decreases the robustness of path. As a result, a link or node may be overwhelmed by multiple requests. So how to increase the robustness of path will also be a useful item for future work.

References

1. R. Ramanathan and R. Hain, Topology Control of Multihop Wireless Networks Using Transmit Power Adjustment, *IEEE INFOCOM* , pages:404-413 (2000).
2. E. L. Lloyd and Rui Liu, *et al.* Algorithmic Aspects of Topology Control Problem for Ad Hoc Networks, *ACM MOBIHOC* , June (2002).
3. V. Rodoplu and T. Meng, Minimum energy mobile wireless networks, *IEEE Communications* , pages:1333-1344 (1999).
4. R. Wattenhofer and Li Li *et al.*, Distributed Topology Control for Power Efficient Operation in Multihop Wireless Ad Hoc Networks, *IEEE INFOCOM* , pages:1388-1397 (2001).
5. T. S. Rappaport, Wireless Communications: Principles and Practice, *Englewood Cliffs, NJ: Prentice-Hall* , pages:69-122,139-196 (1996).
6. J. Li and J. Jannotti *et al.*, A scalable location service for geographic ad-hoc routing, *Proceedings of the 6th ACM (MobiCom'00)* , pages:120-130 (2000).
7. F. P. Preparata and Michael Ian Shamos, Computational Geometry: an introduction, *Springer-Verlag, New York* , pages:41-65 (1985).
8. F. Howell and R. McNab, Simjava package (version 2), *Division of Informatics, University of Edinburgh* , (1999).
9. B. Karp and H. T. Kung, GPSR: greedy perimeter stateless routing for wireless networks, *Mobile Computing and Networking* , pages:243-254 (2000).

GROUP REGISTRATION IN WIRELESS CELLULAR SYSTEMS WITH SHORT-RANGE RADIO NETWORKS

HYUNKU JEONG, HYEON-JU YOON[1], CHEOLGI KIM, JOONGSOO MA[1], JOONWON LEE

Dept. of Computer Science, Korea Advanced Institute of Science and Technology, Daejeon, Korea
{hkjeong, cgkim, joon}@camars.kaist.ac.kr

[1] *School of Engineering, Information and Communications University, Daejeon, Korea*
{juyoon, jsma}@icu.ac.kr

In wireless cellular networks, burst registration overhead can be generated when a large number of subscribers change their zone simultaneously in mass transportation. We propose Group Registration schemes utilizing an intelligent relaying device and additional short-range radio network. Group Registration Access Point (GRAP), which is the relaying device attached to each carriage, communicates with mobile stations via short-range radio link, and carries out the registration to the wireless cellular system as a delegate of a group of mobile stations. We propose two different Group Registration procedures, Emulated Mobile Station (GR-EMS) and Virtual Moving Zone (GR-VMZ), under the condition that the short-range radio network is Bluetooth piconet. Performance evaluation results by simulation show that both schemes reduce the contentions among registration trials significantly, thus achieve much higher success rate and less delay compared with the IS-95B/2000 registration scheme.

1. Introduction

One of serious problems, which occur in the case that the wireless cellular networks subscribers use mass transportation, is a burst of simultaneous registration requests and consequent degradation of service quality. When vehicles with full of passengers who have mobile stations (MSs) cross over the border of a zone at high speed, the MSs detect it in a short period time and try to register their new location to the wireless cellular system. Because each registration trial competes with the others on a shared access channel, a burst of registration trials from hundreds of MSs increases the collisions significantly. Therefore, most MSs fail to register and the wireless cellular system loses their locations, so the paging cost increases [1].

In this paper, we propose two group registration schemes to reduce collisions among registration trials in wireless cellular networks. We introduce new system component, group registration access point (GRAP), that is placed between a base station (BS) and MSs on a vehicle, and carries out the registration requests for a group of the MSs. One or more GRAPs are attached to a vehicle and gather the information needed to the registration from MSs using short-range radio link. After MSs delegate their registration duties to a GRAP, they are free from registration while they are on board. If an MS does not find a GRAP, it operates with a standard scheme.

A GRAP does the actual registration job for a group. In brief, a GRAP performs registration for d-MSs, MSs who are delegating to the GRAP, on behalf of them. So our schemes can reduce collisions of simultaneous registration requests or the signaling load significantly between MSs and BSs. It also decreases messaging cost of the systems to track the locations of MSs [2, 3]. This is another benefit of group registration.

The outline of the paper is as follows. Section 2 shows the related work. The idea and procedures of proposed group registration schemes are described in section 3. Section 4 explains simulation environments and discusses the results, and section 5 concludes the paper.

2. Related Work

To reduce the contentions on a shared access channel, IS-95B [4] has several ways *e.g.* slotted mode, random backoff, PN randomization and so on. In our previous work [1], we worked on the performance evaluations by changing parameters of slotted mode and random backoff scheme. We observed that parameter tuning seemed to be less helpful in mass transportation, especially in the subway. On the other hand, to support PN randomization delay, BS needs the circuits per each probe. It increases the hardware cost.

Another group location tracking (GLT) scheme using relaying BS attached to a transportation system (TS) has been proposed [2, 3]. The GLT is similar to our scheme

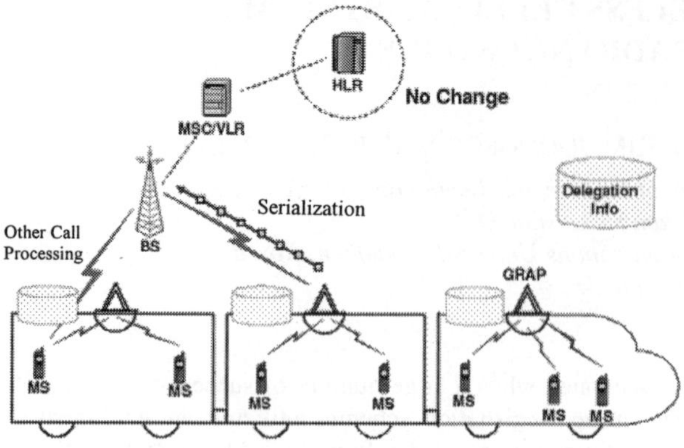

Figure 1. GR-EMS

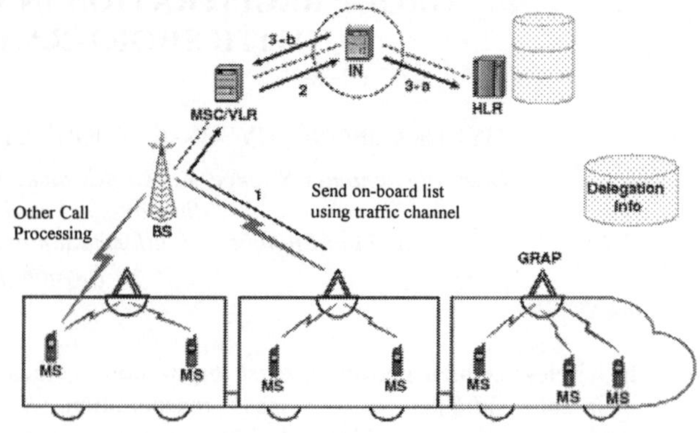

Figure 2. GR-VMZ

in the sense that the relaying BS gathers the requests of MSs and sends group location update message as a delegate. A new system component, Virtual VLR (VVLR), is added to maintain the information of TSs and the d-MSs. The GLT shows how it reduces the messaging cost of systems using group registration. The GLT, however, uses several new messages for VVLR registration, location update request, and group registration cancellation, and additional procedures for call delivery via VVLR. Also, it causes some problems to use a BS as an agent for group registration: interfering with previous fixed BSs or leading ping-pong registration for MSs not in TSs.

3. Group Registration

3.1. *Overview of the proposed schemes*

We propose two schemes, emulated mobile station (GR-EMS) and virtual moving zone (GR-VMZ), which are different in the interaction method between GRAPs and wireless cellular systems.

The Figure 1 shows the concept of GR-EMS. In GR-EMS, each GRAP maintains the delegation information, *i.e.* profiles of d-MSs. Representing each d-MS one by one, the GRAP performs the standard registration procedure as if the GRAP is an MS. It does not reduce the total number of registration trials but serializes them. So the number of concurrent registration trials is reduced to the number of GRAPs on a vehicle. The advantage of GR-EMS is that no modification of the systems is needed, but the GRAP should maintain all information and periodically update it by checking whether the MSs leave from its coverage or not.

In GR-VMZ (see Figure 2), the delegation information is managed by the systems. A GRAP transmits the delegation information to the systems via a traffic channel periodically and doesn't keep it in its memory. The systems take charge of the delegation information. Only the GRAP registers its own location to the systems on behalf of a group of d-MSs, when it enters a new zone. Using the delegation information and the location of GRAPs, the systems can track the locations of the d-MSs. When an incoming call arrives to a d-MS, the systems search the location of the GRAP which manages the d-MS and send a page signal to the location. Because the GRAP area is very smaller than the paging area, the paging signal is delivered to the d-MS with very high probability.

The most important feature of GR-VMZ is using the traffic channel and an intelligent node (IN). The traffic channel has a wider bandwidth than the access channel, so the registration is done more efficiently. Because only the GRAP registers its own location, the number of registration trials can be reduced to a minimum. The IN is in front of the home location register (HLR), filters out and processes group registration messages which are sent by GRAPs. With the help of the IN, we minimized the change of the data structure in the HLR and the call delivery procedure.

Theses schemes consist of a set of procedures: delegation, group registration, call delivery, and delegation cancellation. Delegation and delegation cancellation procedures largely depend on the short-range radio link technology. We adopted Bluetooth since Bluetooth is expected to be available for MSs and does not interfere with air interface [5]. Also, using Bluetooth, the transport service providers can give users useful value-added services. The detail procedures of GR-EMS and GR-VMZ based on Bluetooth are described in next two sub sections.

3.2. *Procedures of GR-EMS*

3.2.1. *Delegation*

A GRAP and MSs use Bluetooth *inquiry* to find each other. After the GRAP recognizes the MSs, it tries to make a connection to MSs in sequence. The connected MS transmits its profile using Bluetooth *object push profile* to delegate registration duty. After that, the GRAP changes the Bluetooth status of the MS into the *park mode*. The parked MS *i.e.* d-MS is free from zone-based registration but should keep on operating the other functions normally. When the GRAP finishes gathering all profiles from the MSs, it moves them into an on-board list, and at registration, it performs registration with this on-board list.

Aside from the delegation procedure, we also suggest the conditions for MSs to start inquiry scan because the usual all-time scanning to find GRAPs wastes much battery power. One is using a special prefix in the BS ID. After an MS detects this predefined prefix, it starts to scan a GRAP. This method is simple but requires a small change of the system and adaptable to the subway only. Measuring the cell change rate is an alternative. When a MS is moving fast, the rate becomes higher. When the rate goes beyond a threshold, the MS starts scanning. This can be used without any support from a BS and in any type of mass transportation, but it should tolerate some delay to start after the MS gets on board.

3.2.2. *Delegation Cancellation*

To prevent a false registration, the GRAP should check whether the d-MSs in the on-board list leave its coverage or not. For the exit checking, the GRAP uses Bluetooth *beacon channel*. The GRAP periodically broadcasts a beacon message to d-MSs and hears *slave-activated unparking requests* from all d-MSs through the *access window*. If the GRAP does not receive messages from a d-MS until the end of the access window, it determines that the MS was getting off and removes the related profile from the list.

In case of a d-MS, if it does not receive any beacon messages for *n* beacon trains, it determines that it exits, and then comes back to normal state. The MS does not register its location until it meets a new zone, because the previous GRAP had performed the registration already.

3.3. *Procedures of GR-VMZ*

Because GR-VMZ uses the traffic channel and new system component, IN, we need to modify the system procedures in addition to the delegation procedures. Overall GR-VMZ procedures consist of delegation, group registration, call delivery, and delegation cancellation.

3.3.1. *Delegation*

The delegation procedure of GR-VMZ is similar to that of GR-EMS. The difference is that in GR-VMZ, the GRAP periodically transmits the profiles of its on-board list to the systems via the traffic channel and clears the list.

3.3.2. *Group Registration*

Figure 2 shows the group registration procedure of GR-VMZ. 1) A GRAP periodically sends a group registration message to a BS. The message contains a GRAP ID and the profiles of an on-board list. 2) The BS and the MSC/VLR pass it to the HLR. 3) An IN placed in front of the HLR is monitoring all kinds of incoming messages to the HLR. When a group registration message arrives, the IN a) converts the delegation message into delegation relationship and save it to the HLR, and b) requests the VLR to delete all related information about the d-MSs.

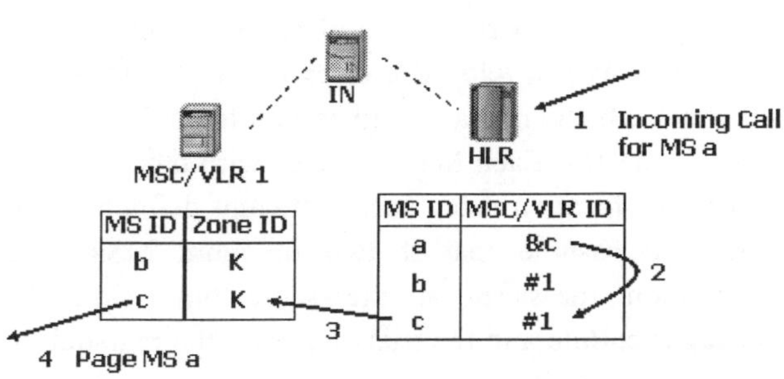

Figure 3. Example of a delegation relationship and call delivery

Figure 3 shows an example of a delegation relationship. The HLR and the VLR maintain location information based on IDs such as MS IDs or VLR IDs. To represent the delegation relationship, we used a prefix, `&`. An HLR entry starting with `&` is a GRAP ID instead of a VLR ID. We used another prefix `#` to distinguish normal location information from delegation relationship. In this example, the location of MS `b` is maintained by VLR `1` and its Zone ID is `K`, while the MS `a` delegates its duty to GRAP `c`.

When a GRAP crosses the location area, it registers its own location just as one of MSs. A successful registration of a GRAP means that all d-MSs are registered at a time.

3.3.3. *Call Delivery*

Figure 3 depicts call delivery procedure. 1) An incoming call arrives at MS `a`. 2) The HLR searches its entries and it finds that MS `a` is maintained by GRAP `c`. 3) The HLR informs VLR `1` that there is an incoming call for MS `a` and it is nearby GRAP `c`. 4) After all, the paging signal for MS `a` is broadcasted to the zone `K` where GRAP `c` is located.

3.3.4. *Delegation Cancellation*

Any kinds of registration performed by d-MS reset the delegation relationship, *i.e.* convert `&` into `#` in the HLR. It causes delegation cancellation. So d-MSs do not perform zone-based or timer-based registration while they are maintained by a GRAP. When a d-MS can not hear the beacon messages of the GRAP, it returns to normal state and registers its own location.

4. Performance Evaluations

4.1. *Simulation Environments*

We analyzed the performance of proposed schemes by simulation. We developed a discrete-time event-driven simulator which simulates registration trials from many MSs to one BS. Each MS competes with other MSs to register its location using access attempt defined in IS-95B. We assume that if two or more MSs send registration messages at the same time slot, the messages collide and the trials fail, and the registration start time is uniformly distributed from 0 to 12 sec. Finally, we assume that there are 10 GRAPs.

Varying the number of MSs and the delegation rate P, number of d-MSs over total number of MSs, we evaluated the performance: success rate and average delay of registration.

4.2. *Success Rate*

A success rate is a proportion of successfully registered MSs. Each MS tries 10 access attempts; 1 access attempt consists of 2 sequences of 6 access probes. If a MS sends a registration message to a BS without collision during 10 access attempts, the registration is successful and the MS rests. A GRAP tries access attempts as many as 10 * number of d-MSs for the equal opportunity. That a GRAP make a success means that the all of d-MSs are registered at once.

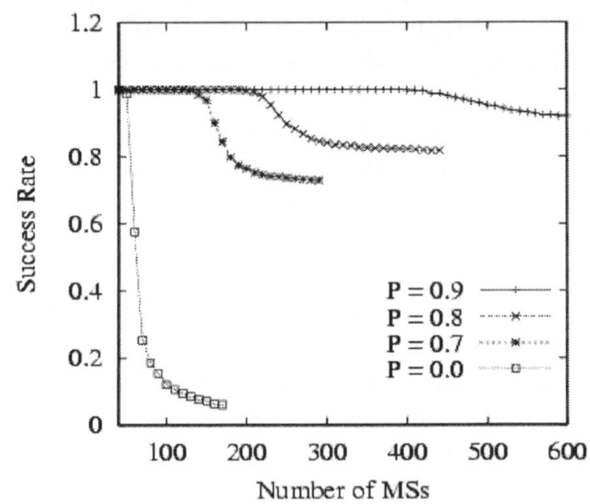

Figure 4. Success rate of GR-VMZ (P = 0.0, 0.7, 0.8, 0.9)

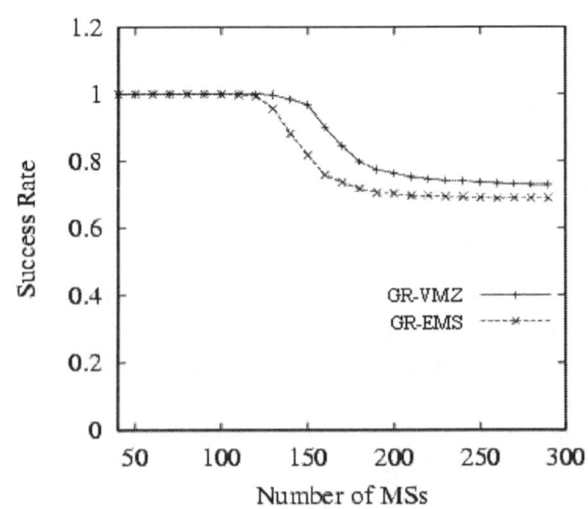

Figure 5. Success rate of GR-EMS and GR-VMZ (P = 0.7)

Compared with IS-95B/2000 registration scheme, *i.e.* P = 0, GR-EMS and GR-VMZ significantly reduce the contentions among registration trials. Figure 4 shows the success rate of GR-VMZ with various P values. At P of 0.0, the success rate falls rapidly when the number of MSs is greater than 70. As P increases, the success rate is improved significantly. We also evaluated success rate of GR-EMS with various P values, and it showed similar pattern with GR-VMZ. Figure 5 shows the comparison of our two approaches. The performance of GR-VMZ is slightly better than that of GR-EMS, however, as delegation rate increases, the difference is reduced.

4.3. *Average Delay*

We estimated the average delay which means the average time for all MSs to succeed in registration. Every registration competitor, including normal MSs and GRAPs, sends the registration message repeatedly during the simulation time, until it succeeds in the registration. The total simulation time is 10 minutes.

Figure 6 and Figure 7 show the average delay of GR-EMS and GR-VMZ, respectively. As we can see, both GR-EMS and GR-VMZ outperform the original scheme, *i.e.* P = 0. In the original scheme, when the number of MSs comes over 50, the average delay is increased exponentially. As P increases, the delay is improved significantly.

GR-VMZ shows much better performance than GR-EMS because GR-EMS has to register all of d-MSs while GR-VMZ does registration just for itself. For example, at P = 0.9, the average difference of GR-VMZ and GR-EMS is about 5 seconds.

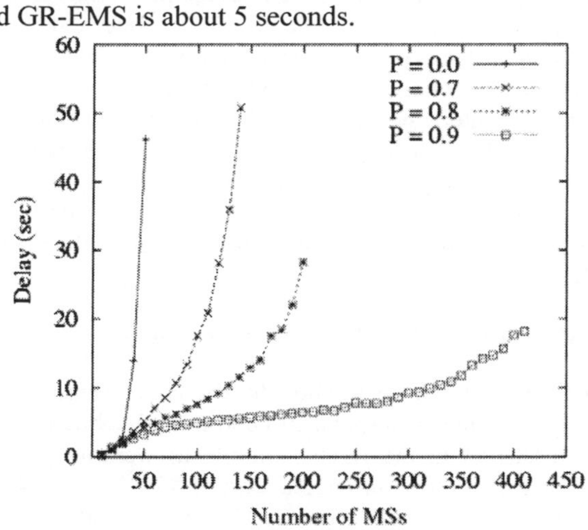

Figure 6. Average Delay of GR-EMS (P = 0.0, 0.7, 0.8, 0.9)

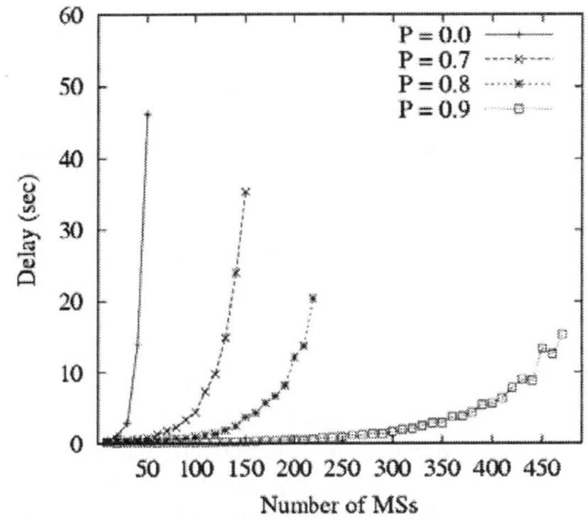

Figure 7. Average Delay of GR-VMZ (P = 0.0, 0.7, 0.8, 0.9)

5. Conclusions

In this paper, we proposed two group registration schemes, GR-EMS and GR-VMZ. Our schemes use an intelligent relaying device to carry out the registration on behalf of a group of MSs, and short-range radio link to avoid interfering with air interface. GR-EMS can achieve the group registration without the system modification, while GR-VMZ can achieve the maximum reduction of simultaneous trials with the help of additional intelligent node in the system. We also described the detail procedures of two approaches based upon the Bluetooth technology, and evaluated their performance by simulation. Both proposed schemes outperform the original scheme and GR-VMZ shows better performance than GR-EMS in two measures: success rate and average delay.

References

1. S. Cho, H. Jeong, K. Lee, and J. Ma, "Performance optimization technique of location registration in public transportation," 7th *CDMA International Conf.*, Seoul, Korea, pp. 22-29, 2002.
2. I. Han, and D. Cho, "Group Location Management for Transportation Systems in Wireless Cellular Systems," *Vehicular Technology Conference*, Vol. 4, pp. 2726-2730, 2001.
3. I. Han and D. Cho, "Hierarchical group location tracking for transportation systems in wireless cellular Networks," *Electronics Letters*, Vol. 38, No. 18, August 2002.
4. TIA/EIA/IS-95B
5. Bluetooth Specification Version 1.1

Optimizing Mobile Network Datagram Routing through Programmable Proxying

Peter Tabery, Christian Bachmeir, and Xiaohe Li
Munich University of Technology, Institute of Communication Networks, Germany
Email: tabery@ei.tum.de, bachmeir@ei.tum.de, xiaoheli@eikon.tum.de; Fax: +49-89-289-23523

Abstract: Public access to the Internet is nowadays even available in trains and airplanes. When such a network with several connected nodes changes its point of attachment to the Internet, it is called a mobile network. Unfortunately, mobile networks are not well supported by Mobile IP.
In this paper, we describe Programmable Proxying, a dynamic approach for optimizing mobile network datagram routing. Transport layer connections originating from the mobile network will be handled by the programmable proxy, bypassing the home agents. In the case of a Mobile IP handoff, the proxy forwards ongoing connections to the new point of attachment.

Key Words: Mobility Support, HTTP Proxy, Programmable Networks.

1 INTRODUCTION

Mobile networks consist of a mobile router serving nodes traveling along with that mobile router located e. g. aboard a train or aircraft. If these nodes are mobile nodes themselves, they are expected to change their mobile router infrequently compared to their overall movement.

Mobile IPv4 [1] yields support for connecting these mobile networks to the Internet and ensuring datagram delivery for the nodes attached to them, that may be configured with a local IP address or be mobile nodes as well. However, this solution yields a disadvantageous routing path for the mobile nodes' datagrams.

We avoid the unfavorable routing for mobile networks with our *Programmable Proxying* approach. By proxying outbound connections, they bypass the home agents and use the direct path to the correspondent node. After a handoff, the transport connection remains anchored at the original foreign agent and the data is forwarded to the mobile router's new foreign agent. Although this routing path is not optimal any more, in the case of *horizontal* handoffs it will still be shorter than routing the data through the home agent. However, this may not be true in the case of *vertical* handoffs (e. g. when switching to a satellite network), as the new foreign agent may be far away in network topology terms.

Programmability allows for easy deployment of customized proxies for supporting a variety of different communication protocols. We assume (Mobile) IPv4 throughout this paper, as (Mobile) IPv6 neither yields support for mobile networks [2], nor is widely deployed yet [3].

This paper is organized as follows: In Section 2, we give a short overview about Mobile IP and mobile networks. After giving an insight in Programmable Networks in Section 3, we describe the architecture for Programmable Proxying in Section 4. Finally, we present an evaluation of our approach based on a prototype implementation in Section 5.

2 MOBILITY IN IP NETWORKS

The preferred solution for dealing with mobility in the Internet is Mobile IP [1]. It limits the mobility logic to very few entities at the edge of the network, as it is impossible to distribute host-specific routes throughout significant parts of the network.

Mobile networks face a similar problem. The backbone routers' routing tables do not contain routes to all of the Internet's networks, but only aggregate routes comprising several networks and their respective subnetworks. Thus, changing the route to one single (mobile) network would imply distributing a new routing table entry for that single network only. Hence, this solution effectively poses the same scalability problems as the distribution of host-specific routes.

2.1 MOBILE NETWORKS

For supporting network layer connectivity of mobile nodes served by a mobile router, [1] suggests the use of the regular Mobile IP mechanism for the mobile nodes as well as for the mobile router's movement, resulting in *nested* Mobile IP contexts. Regarding to its set of mobility agents, the mobile router behaves as a mobile node. For the mobile nodes, the mobile router acts as a foreign agent.

In Fig. 1, the mobile network's datagram routing is shown for the case of nested Mobile IP. The mobile node's home agent (denoted HA_{MN}) intercepts and encapsulates the datagrams (*dashed line*). At the mobile router's home network, the mobile router's home agent (HA_{MR}) intercepts and encapsulates the datagrams again, yielding a double encapsulation (*dotted line*). The packets are then decapsulated by the mobile router's foreign agent (FA_{MR}) and again by the mobile router (which is also the mobile nodes' foreign agent, thus denoted $MR \,\&\, FA_{MN}$).

2.2 MOBILE IP ROUTE OPTIMIZATION

Route optimization, i. e. to allow for datagrams being delivered directly from the correspondent node to the mobile node, has been discussed in the IETF's Mobile IP working group for years. It will become an integral part of Mobile IPv6 [4], thus implemented in every IPv6 node. However, Mobile IPv6 does not support mobile networks [2].

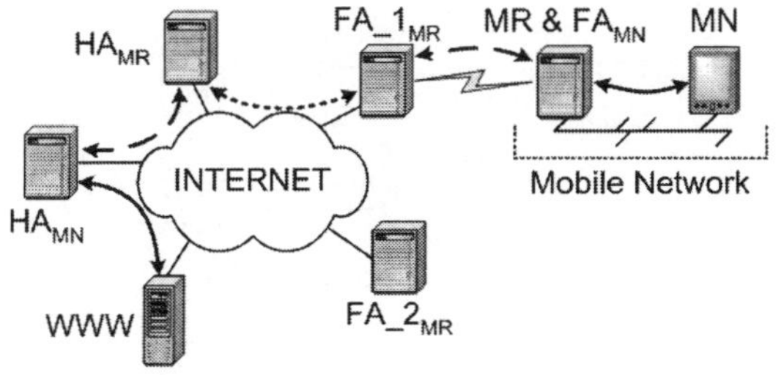

Fig. 1. Mobile network datagram routing with nested Mobile IP.

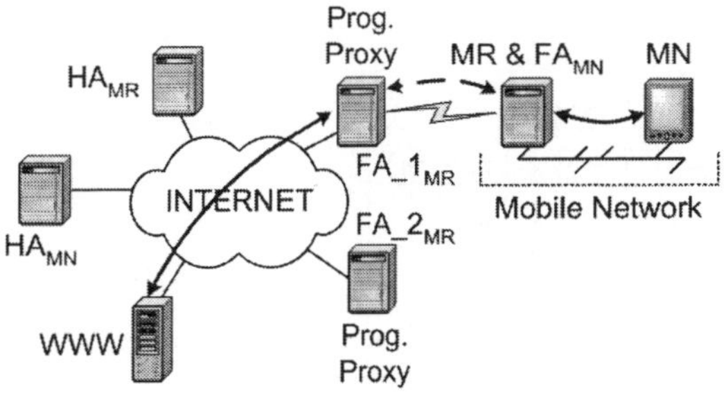

Fig. 2. Programmable Proxying for a mobile network with mobile nodes.

Adding route optimization to the current Mobile IPv4 standard as suggested in [5] has not been adopted, as this requires correspondent nodes that are able to process binding updates. In addition to this serious limitation, route optimization may not be applied across nested Mobile IP contexts, as in the case of mobile networks.

Another approach for using the direct route between correspondent node and mobile node has been suggested in [6]. The mobile node may use its topologically correct (i. e. local) care-of address for outbound connections. Although this works well for individual mobile nodes, it can not be applied to mobile networks as the mobile nodes do not possess a topologically correct IP address. In fact, the routing path would be optimized only partially this way, bypassing the mobile node's home agent (HA_{MN}).

2.3 APPLICATION LAYER ASPECTS

Furthermore, Mobile IP route optimization operates on the network layer—relying only on the destination's IP address—and does not consider higher layers. When the mobile node utilizes an HTTP proxy, the mobile node's communication with the remote server is carried over two consecutive transport layer connections. In consequence, any optimization on the network layer applies to the first transport layer connection only.

Obviously, we get a much higher gain when optimizing the entire HTTP connection at once. Therefore we suggest optimizing the mobile nodes' outbound connections on the application layer. As operating on the application layer means also dealing with a much broader diversity of demands, we base it on platforms for Programmable Networks.

3 PROGRAMMABLE NETWORKS

The aim of active and programmable networking consists in simplifying the deployment of new network services leading to networks that explicitly support dynamic service creation, deployment and management in the network infrastructure [7]. By pushing the network's functionality beyond routing, intelligence may be shifted from the hosts towards the network itself. Several proposals have been made for applying Active and Programmable Networks to Mobile IP route optimization.

The Active Delivery Scheme [8] uses *active registrations* that are evaluated by active routers in the network. The active routers then efficiently deliver binding updates to the correspondent hosts and redirect in-transit packets, that have already been encapsulated and sent to the old foreign agent. However, this redirection is being done on an active router located before the routing path to the new foreign agent forks off from the routing path to the previous foreign agent. Therefore, a considerable performance gain can only be achieved if there are many active routers throughout the Internet and if they are located close to the Internet's backbones.

The Mobile Active Overlay [9] is an approach that works well with active routers serving edge networks only, where performance is not as crucial as it is to core network routers. Here, the routing path of the mobile node's datagrams is optimized transparently by redirecting the packets through an overlay network, bypassing the home agent. Unfortunately, this approach requires not only an *Extended Home Agent* in the home network and a *Bridgehead Agent* in the visited network, but also a *Correspondent Agent* at the correspondent node's (i. e., the WWW server's) site, where it can intercept all of its packets. Therefore, this solution is not feasible for optimizing the mobile node's traffic in the Internet as it exists today.

4 PROGRAMMABLE PROXYING

Circumventing these obstacles, our approach achieves a direct routing path to the correspondent node by proxying the HTTP connection. Hence, outbound transport layer connections are terminated at the fixed foreign agent. This way, the unfavorable routing of the datagrams through the home agents is avoided.

4.1 ARCHITECTURE

The framework for Programmable Proxying is deployed in parallel to an existing Mobile IP infrastructure. As shown in Fig. 2, datagrams sent from the mobile nodes are encapsulated by the mobile router (*dashed line*) and forwarded to the fixed foreign agent, where the programmable proxy is deployed. When mobile nodes and the mobile router are using Mobile IP in co-located mode, datagrams arrive at the proxy encapsulated twice.

At the Programmable Proxy, the encapsulated datagrams are intercepted, decapsulated and redirected to the respective local proxy port. As the proxy uses its own network address for

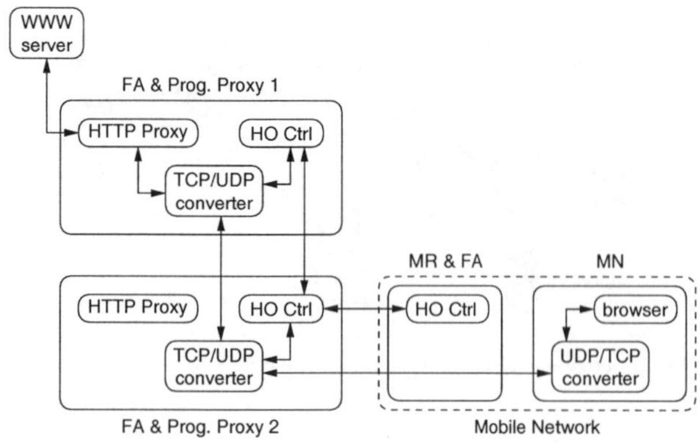

Fig. 3. Software entities for Programmable Proxying after a handoff.

the new transport layer connection to the correspondent node, this connection's packets are routed directly from the proxy to the correspondent node and vice versa, avoiding the path through the home agents. As the proxying is only applied to outbound connections of registered protocols, other protocols are not disrupted.

4.2 HANDOFF SUPPORT

To improve handoff behavior, we transmit the HTTP data between mobile node and the programmable proxy using UDP, enhanced with custom flow control and retransmission logic, that allows for active fetching rather than timer-based retransmissions like TCP. Similar to a transparent proxy, the mobile node's outbound TCP connections are redirected to a local converter that repackages the HTTP data in UDP packets along with additional overhead for assuring reliable delivery.

When the mobile router performs a Mobile IP handoff, the handoff control (HO Ctrl) at its previous foreign agent will be notified about the new location [10]. As the TCP/UDP converter keeps track of the mobile router every mobile node is attached to, it is able to identify the HTTP connections that need to be forwarded to the mobile router's new foreign agent. As shown in Fig. 3, the TCP/UDP converter at the previous Programmable Proxy starts forwarding these HTTP connections to the new one, where the UDP packets will be encapsulated and delivered to the mobile router according to the rules set by the handoff control. Acknowledgment packets will be returned to the previous proxy whereas new HTTP connections will be handled by the new proxy.

Regarding handoffs (especially *ping-pong* handoffs), using the UDP-based transmission turned out to be far more flexible than using TCP, as tearing down and re-establishing TCP connections frequently imposes significant overhead. The ability to fully control the sending buffer and scheduling of the packets contributed to the performance of our prototype.

4.3 PROGRAMMABILITY ASPECTS

Programmable network systems like AMnet [11] support the provisioning of net-centric services to end systems unaware

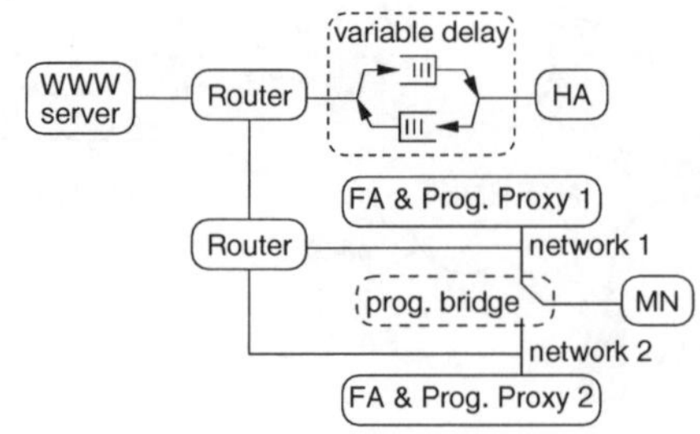

Fig. 4. Simplified testbed setup.

of the additional functionality inside the network. They allow for dynamically loading the functionality needed, e. g. for providing proxy support for new protocols or a different kind of support in addition to an existing proxy.

This is especially important, when the mobile nodes' applications pose diverse or even contradictory demands on intermediate proxies—ranging from simple *Network Address and Port Translation* to sophisticated higher layer approaches (or the explicit request that no proxy should be used). Using Programmable Proxying, every node may have a customized proxy profile containing its particular requirements.

5 EXPERIMENTAL EVALUATION

We implemented a prototype for verifying the advantages of our approach regarding throughput and delay during static as well as dynamic operation. In our test environment shown in Fig. 4, we use the Dynamics [12] Mobile IP implementation and the AMnet [11] programmable platform, both of them are mature and stable releases by now. For the performance measurements, we utilized the Squid HTTP proxy [13].

For switching the mobile node between the two networks, we utilized the bridging capabilities of a Linux computer. This way, we could move the mobile node in regular intervals.

In our measurements, we compare the standard Mobile IP mechanism with and without enabling our enhancements. First, we analyzed the *respond times* of both systems depending on the home agent's distance. We define respond time as the time from sending the first datagram towards the server until the first byte of the requested data arrives. The round trip time (RTT) stated in the graphs below is the additional delay through the home agent(s) compared to the direct routing path from the mobile node to the WWW server. In a full mobile router setup, the encapsulation, decapsulation, and forwarding delay of the mobile router would add to the measured respond times. As this delay is very small and equal for both cases, it would only yield a slight offset for both curves in the graph.

As can be seen in Fig. 5, without additional delay, our system has a slightly larger respond time than standard Mobile IP due to the TCP/UDP conversion and the HTTP proxy. When there is additional delay, i. e. when the home agent is more distant, Programmable Proxying yields a notable advantage over standard Mobile IP. The respond time of standard Mobile IP increases

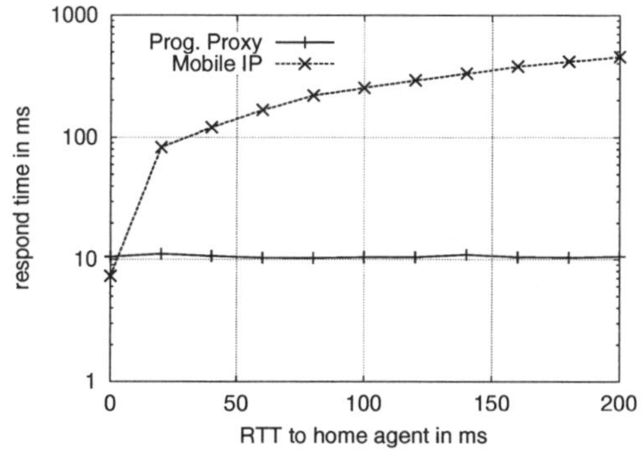

Fig. 5. Respond time depending on round trip time (RTT) to the home agent.

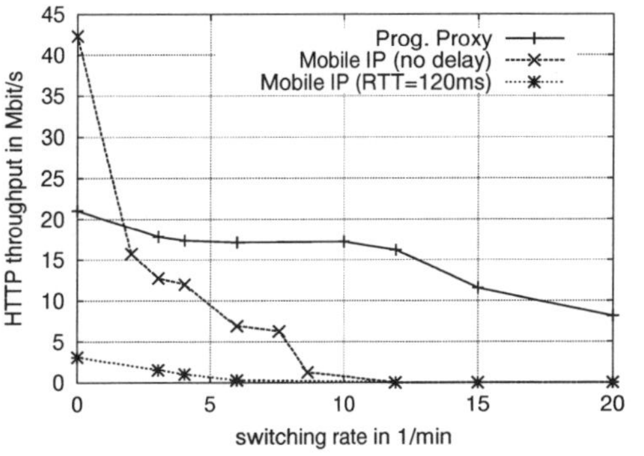

Fig. 6. Effective HTTP throughput during *ping-pong* handoffs.

linearly with the RTT, but stays rather constant in the latter case.

For comparing the HTTP throughput, we consider standard Mobile IP, first without any additional delay, second with a fixed delay of 120ms, corresponding to the round trip time from Munich to New York. In both of these cases, the HTTP data is transported within one single end-to-end TCP connection. In contrast, the Programmable Proxy approach uses a TCP connection between proxy and server, but relies on UDP-based transport for delivering the data from the proxy to the mobile node.

The behavior of these three scenarios with respect to the switching rate between the two foreign networks is shown in Fig. 6. Although our approach is outperformed by standard Mobile IP in the static case, it behaves better in the presence of so-called *ping-pong* handoffs. Mobile IP has a steep decrease in throughput, whereas our approach—although initially below Mobile IP—degrades much less. At high switching rates, our approach still yields a significant throughput whereas TCP over standard Mobile IP stalls.

6 CONCLUSIONS

In this paper, we introduce Programmable Proxying, a new approach for enhancing the network performance perceived by mobile terminals attached to a mobile network. By proxying outbound HTTP requests, the data is transported on the direct path from the remote server to the mobile network—bypassing the home agents—which results in lower end-to-end delay.

Employing a UDP-based transport protocol with custom flow control and retransmission logic significantly enhances handoff performance as well as data throughput, as confirmed by the performance measurements performed with our prototype implementation.

Our solution may be deemed robust, as it does not interrupt protocols or data flows that must not be subject to proxying. This is especially important for protocols or applications where proxying is adverse, e. g. when authentication is based on the mobile node's home IP address.

As we use Programmable Networks, the appropriate proxy code may be loaded from a remote repository when needed. This way, functionality for new protocols can be deployed easily. Furthermore, different flavors of support for existing protocols may be provided concurrently, yielding a high flexibility and alleviating the need for standardizing the proxy functionality.

ACKNOWLEDGMENTS

This work has been done as part of the FlexiNet project (`http://www.flexinet.de/`), funded by the German Federal Ministry of Education and Research.

REFERENCES

[1] C. Perkins, "IP mobility support for IPv4," IETF, RFC 3344, Aug. 2002.

[2] T. Ernst, C. Castelluccia, and H.-Y. Lach, "Extending Mobile IPv6 with multicast to support mobile networks in IPv6," in *European Conference on Universal Multiservice Networks, ECUMN'00, Colmar, France*, Oct. 2000.

[3] S. J. Vaughan-Nichols, "Mobile IPv6 and the future of wireless Internet access," *IEEE Computer*, Feb. 2003.

[4] D. Johnson, C. Perkins, and J. Arkko, "Mobility support in IPv6," *Work in progress, <draft-ietf-mobileip-ipv6-24.txt>*, June 2003.

[5] C. Perkins and D. B. Johnson, "Route optimization in Mobile IP," *Internet Draft (expired), <draft-ietf-mobileip-optim-11.txt>*, Sept. 2001.

[6] X. Zhao, C. Castelluccia, and M. Baker, "Flexible network support for mobile hosts," *Mobile Networks and Applications*, vol. 6, no. 2, 2001.

[7] A. T. Campbell, M. E. Kounavis, and J. B. Vicente, "Programmable Networks," in *Informatics - 10 Years Back, 10 Years Ahead*. Springer Berlin Heidelberg, 2001.

[8] K.-W. Chin, M. Kumar, and C. Farrell, "Enhancements to Mobile IP with active networks," in *Active Networks International Working Conference IWAN, Tokyo, Japan*, Oct. 2000.

[9] A. Juhola, K. Ahola, J. Mölsä, and S. Lehtonen, "Mobile active overlay," in *Active Networks International Working Conference IWAN, Tokyo, Japan*, Oct. 2000.

[10] P. Tabery and C. Bachmeir, "Advanced network services using programmable networks," in *IFIP Workshop and EUNICE Summer School on Adaptable Networks and Teleservices, Trondheim, Norway*, Sept. 2002.

[11] T. Fuhrmann, T. Harbaum, M. Schöller, and M. Zitterbart, "AMnet 2.0: An improved architecture for Programmable Networks," in *4th Annual International Working Conference on Active Networks, IWAN 2002, Zurich, Switzerland*, Dec. 2002.

[12] "Dynamics - HUT Mobile IP," www.cs.hut.fi/Research/Dynamics/.

[13] "Squid web proxy cache," http://www.squid-cache.org/.

SESSION 4B:
ROUTING

A Distributed Algorithm for Multiple-Metric Link State QoS Routing Problem

Hakim Badis[1,2] and Khaldoun Al Agha[1,2]
[1]LRI Laboratory, University of Paris XI, Orsay, France
[2]INRIA Laboratory, Rocquencourt, France
{badis,alagha}@lri.fr

Abstract: Multimedia applications often require guaranteed quality of service (QoS) and resource reservation, which has raised a number of challenging technical issues for routing. Nevertheless, the QoS routing algorithm must be simple because a costly procedure does not scale with the size of the network. In this paper, an efficient QoS routing method based on Lagrange relaxation is presented, which provides a polynomial heuristic solution to the Delay and Bandwidth Constrained Least Hop path problem (DBCLH) and Delay, Bandwidth and Loss probability Constrained Least Hop path problem (DBLCLH). An algorithm analysis and numerical evaluation are presented. It is shown that our polynomial algorithm is close to the optimal solution computed by an exponential algorithm.

I. INTRODUCTION

Routing protocols usually characterize the network with a single metric such as hop-count or delay, and use shortest path algorithms for the path computation. It is becoming increasingly clear that such routing protocols are inadequate for multimedia applications, such as video conferencing, which often require guaranteed QoS. For a network to support QoS requirements, routing must explicit information on resources available in the network so that applications can make proper resource reservation. To achieve that, it is necessary for routing to have a more complex model of the network, taking into account important network parameters such as bandwidth, delay and loss probability.

The complexity of the algorithm is a function of the number of metrics treated and of their composition rules. Wang and Crowcroft [1] proved that the problem of finding a path subject to constraints on n additive and m multiplicative metrics is *NP-complete* if $n+m \geq 2$. Bandwidth can be taken into account by pruning links. Therefore, algorithms of acceptable complexity can only handle bandwidth and one additional metric. In the literature, most of the algorithms do not try to solve this complex problem, instead they define simpler problems. In this paper, we present an algorithm called Lagrange Relaxation-based Hop-count (LRH) that provides a polynomial solution for three-metrics formulated as DBCLH and four-metrics formulated as DBLCLH. Lagrange relaxation is a common technique for calculating lower bounds, and finding good solutions. First Held and Karp raised with this technique for the Traveling Salesman Problem in [2], [3].

In link-state routing protocols, packets are forwarded hop-by-hop at each node. Each node maintains a routing table with next hops for all destinations, and this table is usually computed periodically in response to routing updates. When a packet is received, hop-by-hop routing requires only a table lookup to find the next hop and send the packet to it. Link state routing can use fully distributed computation algorithms [4], which has lower memory requirements for the routers. In this paper we present a distributed algorithm for multiple-metric link state QoS routing.

The remain of this paper is organized as follows. In section II, we present the DBCLH problem and the proposed method which provides a polynomial heuristic solution. Section III describes the DBLCLH problem and the single mixed concept. In section IV, we show simulation results of our algorithms by comparing them to an exponential algorithm. We propose in V to adapt the LR based Hop (LRH) algorithm for Ad hoc networks. We conclude in VI.

II. THE DBCLH PROBLEM

It is clear that two additive and one concave metrics problem is *NP-Complete*. In this case, we take hop-count and delay as additive metrics (we can take any two additive metrics) and bandwidth as concave metric. We consider the Delay and Bandwidth Constrained Least Hop path problem (DBCLH). Given two constants, the minimum bandwidth $\Delta_{bandwidth}$ and the maximum delay Δ_{delay}. The Delay and Bandwidth Constrained Least Hop path problem (DBCLH) is to find a path p from i to r minimal for a hop-count, satisfying $del(p) \leqslant \Delta_{delay}$ and $\mathrm{Bw}(p) \geqslant \Delta_{bandwidth}$. The formal description is:

$$\min\{\mathrm{hop}(p) : p \in P(s,t) \text{ and } \mathrm{del}(p) \leqslant \Delta_{delay} \text{ and } \mathrm{Bw}(p) \geqslant \Delta_{bandwidth}\} \qquad (1)$$

Where $P(s,t)$ is the set of paths from the source node s to the destination node t, and hop(p) is the hop-count. Our heuristic is based on the Lagrange Relaxation algorithm.

We must decide the precedence among the metrics in order to define the best path. The delay has two basic components: queuing delay and propagation delay. The queuing delay is more dynamic and traffic-sensitive, thus bandwidth is often more critical for most multimedia applications. If there is no sufficient bandwidth, queuing delay and probably the loss rate will be very high. So, we define the precedence as bandwidth and then the propagation delay. Our strategy is to find a path that satisfies the bandwidth constraint, and when there is more than one, we choose the one with shortest hop and under the delay constraint. We can transform the problem (1) to

$$\min\{\mathrm{hop}(p) : p \in P'(s,t) \text{ and } \mathrm{del}(p) \leqslant \Delta_{delay}\} \qquad (2)$$

Where $P'(s,t)$ is the set of paths satisfying $\mathrm{Bw}(p) \geqslant \Delta_{bandwidth}, \forall p \in P'(s,t)$.

Theorem 1: *A path has a width no less of $\Delta_{bandwidth}$, if and only if each link in the path has a bandwidth no less than $\Delta_{bandwidth}$.*

Proof: if each link in p has a bandwidth no less than $\Delta_{bandwidth}$, it is obvious that $\text{width}(p) \geq \Delta_{bandwidth}$. Suppose that $\text{width}(p) \geq \Delta_{bandwidth}$ but there is a link (i, j) with Bw_{ij} less than $\Delta_{bandwidth}$. We then have $\text{width}(p) = \text{Bw}_{ij} < \Delta_{bandwidth}$, which contradicts the assumption that $\text{width}(p) \geq \Delta_{bandwidth}$.

Theorem 1 implies that any links with a bandwidth less than $\Delta_{bandwidth}$ are not parts of the path we want. Hence to find paths satisfying $\text{Bw}(p) \geqslant \Delta_{bandwidth}$, we eliminate any links with a bandwidth less than $\Delta_{bandwidth}$ so that any paths in the resulting graph satisfy $\text{width}(p) \geq \Delta_{bandwidth}$.

Lagrange relaxation is the common technique for calculating lower bounds, and finding good solutions for this problem. We neglect the constraining conditions (this is the relaxation), and build them into the objective function. The solutions feasible to the original problem can certainly suit the relaxation conditions as well, so we can get a lower bound of the original problem. If the path found is not feasible to the constraining conditions, we increase the dominance of it in the modified hop function, enforcing the solution to approach to the optimal solution. Moreover, decrease the difference between the obtained lower bound and the optimum of the original problem as well. The presented **Lagrange Relaxation based additive Metric** (LRM) algorithm consists of the heuristic that minimizes (we take the hop as additive metric) $\text{hop}_\lambda := \text{hop} + \lambda \times \text{del}$ modified objective function. For a given (fixed) λ, we can easily calculate the minimal path (p_λ). If $\lambda = 0$ and $\text{del}(p_\lambda) \leq \Delta_{del}$ we found an optimal solution for the original problem as well. If $\text{del}(p_\lambda) > \Delta_{del}$ we must increase λ, to increase the dominance of the delay in the modified objective function. So, we increase λ while the optimal solution of hop_λ suits the delay requirements. With the help of Lagrange relaxation we have an algorithm which can find the optimal λ for a given source destination pair. Moreover it gives an estimation for the optimal solution, although we have no guarantee of finding the optimal solution, we always get a bound for the solution, which tells that at most how far is it from the optimal solution. To find the value of λ that gives the best result, we use the following Claim.

Claim 1: Let $L(\lambda) := \min\{\text{hop}_\lambda(p) : p \in P'(s,t)\} - \lambda \times \Delta_{del}$. Then $L(\lambda)$ is a lower bound to problem (2) for any $\lambda \geq 0$.

Proof: let $p*$ denote the optimal solution of (2). Then

$$\begin{aligned} L(\lambda) &:= \min\{\text{hop}_\lambda(p) : p \in P'(s,t)\} - \lambda \times \Delta_{del} \\ &\leq \text{hop}_\lambda(p*) - \lambda \times \Delta_{del} \\ &= \text{hop}(p*) + \lambda(\text{del}(p*) - \Delta_{del}) \\ &\leq \text{hop}(p*). \end{aligned}$$

Proves the claim.

To obtain the best lower bound we need to maximize the function $L(\lambda)$, that is we are looking for the value $L^* := max_{\lambda \geq 0} L(\lambda)$, and the maximizing λ^*. Now, some properties of the function $L(\lambda)$ are given. The simple proofs are left to the reader.

Claim 2: L is a concave piecewise linear function, namely the minimum of the linear functions $\text{hop}(p) + \lambda \times (\text{del}(p) - \Delta_{del})$ for all $p \in P'(s,t)$.

Claim 3: For any $\lambda \geq 0$ and hop_λ-minimal path p_λ, $\text{del}(p_\lambda)$ is a supgradient of L in the point λ.

Claim 4: Whenever $\lambda < \lambda^*$, then $\text{del}(p_\lambda) \geq \Delta_{del}$ and if $\lambda > \lambda^*$, then $\text{del}(p_\lambda) \leq \Delta_{del}$ for each hop_λ-minimal path p_λ.

Claim 5: A value that maximizes the function $L(\lambda)$ if and only if there are paths p_{hop} and p_{del} which are hop_{λ^*}-minimal and for which $\text{del}(p_{hop}) \geq \Delta_{del}$ and $\text{del}(p_{\text{del}}) \leq \Delta_{\text{del}}$. (p_{hop} and p_{del} can be the same, in this case $\text{del}(p_{\text{del}}) = \text{del}(p_{\text{hop}}) = \Delta_{\text{del}}$).

Claim 6: let $0 \leq \lambda_1 < \lambda_2$, and $p_{\lambda_1}, p_{\lambda_2} \in P'(s,t)$ λ_1-minimal and λ_2-minimal paths. Then $\text{hop}(p_{\lambda_1}) \leq \text{hop}(p_{\lambda_2})$ and $\text{del}(p_{\lambda_1}) \geq \text{del}(p_{\lambda_2})$.

These two latter Claims together give that the λ^* maximizing the function $L(\lambda)$ gives the best modified cost function, that is λ^* is the smallest value for which there exists a hop_{λ^*}-minimal path p_{del} which satisfies the metric2 constraint. The LRM algorithm is as follows:

Step 1: Set $\text{del}_{ij} = \infty$ and $\text{hop}_{ij} = \infty$, if $\text{Bw}_{ij} < \Delta_{bandwidth}$

Step 2: $p_{\text{hop}} := \text{Dijkstra}(s, t, \text{hop})$;

Step 3: If $\text{hop}(p_{\text{hop}}) = \infty$ then return "there is no solution"; the algorithm is complete.

Step 4: If $\text{del}(p_{\text{hop}}) \leq \Delta_{\text{del}}$ then return p_{hop}; the algorithm is complete.

Step 5: $p_{\text{del}} := \text{Dijkstra}(s, t, \text{del})$;

Step 6: If $\text{hop}(p_{\text{del}}) > \Delta_{\text{del}}$ then return "there is no solution"; the algorithm is complete.

Step 7: $\lambda := \frac{\text{hop}(p_{\text{hop}}) - \text{del}(p_{\text{del}})}{\text{del}(p_{\text{del}}) - \text{del}(p_{\text{hop}})}$;

Setp 8: $p_{\text{Update}} := \text{Dijkstra}(s, t, \text{hop}_\lambda)$;

Step 9: If $\text{hop}_\lambda(p_{\text{Update}}) = \text{hop}_\lambda(p_{\text{hop}})$ then return p_{hop}. Otherwise,
If $\text{del}(p_{\text{Update}}) \leq \Delta_{\text{del}}$ then $p_{\text{del}} := p_{\text{Update}}$. Otherwise, $p_{\text{hop}} := p_{\text{Update}}$;

Step 10: Go to step 7;

Step 2 calculates shortest path on hop_λ modified hop function with Dijkstra algorithm by setting λ at 0. $\text{Dijkstra}(s, t, \text{hop})$ return a hop-minimal path between the nodes s and t. In step 3, if there is no path from s to t that can fulfill the bandwidth requirement, then $\text{hop}(p_{\text{hop}}) = \infty$. In step 4, if the path found meets the delay requirement Δ_{del}, this is the optimal path, and the algorithm stops. Otherwise, step 5 calculates the shortest path on the delay. If the obtained path suits the delay requirements (Δ_{del}), a proper solution exists. Otherwise, there is no suitable path from s to t that can fulfill the delay requirement, so the algorithm stops. In step 7, by using the claim 5, we can calculate λ because p_{hop} and p_{del} are hop_λ-minimal. In this case, the only possible λ is $\lambda = \frac{\text{hop}(p_{\text{hop}}) - \text{hop}(p_{\text{del}})}{\text{del}(p_{\text{del}}) - \text{del}(p_{\text{hop}})}$. After in step 8, we find a

hop_λ-minimal path p_{Update}. If $\text{hop}_\lambda(p_{\text{Update}}) = \text{hop}_\lambda(p_{\text{hop}})$ then p_{hop} and p_{del} are hop_λ-minimal, so the optimal path is p_{del}. Otherwise, we replace either p_{hop} or p_{del} with p_{Update} according the delay constaints.

Theorem 2: *The final path constructed by LRM for a given source s and destination t does not contain any loops.*

Proof: By contradiction. Let s-a-b-t the optimal path found by LRM the optimal λ (Figure 1).

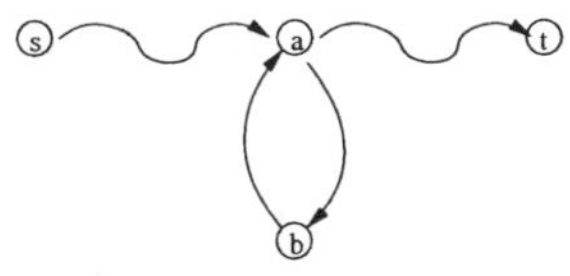

Fig. 1. A loop involving node a and node b

s-a-b-t is the shortest path in term of met_λ i.e.,

$$\forall p \in P(s,t)\ \text{hop}_\lambda(p) > \text{hop}_\lambda(s\text{-}a\text{-}b\text{-}t) \qquad (3)$$

We have $\text{hop}(s\text{-}a\text{-}t) \leq \text{hop}(s\text{-}a\text{-}b\text{-}t)$ and $\text{del}(s\text{-}a\text{-}t) \leq \text{del}(s\text{-}a\text{-}b\text{-}t)$. So, $\text{hop}(s\text{-}a\text{-}t) + \lambda \times (\text{del}(s\text{-}a\text{-}t) - \Delta_{\text{del}}) \leq \text{hop}(s\text{-}a\text{-}b\text{-}t) + \lambda \times (\text{del}(s\text{-}a\text{-}b\text{-}t) - \Delta_{\text{del}})$ (4)

Equation (3) and (4) contradict each other. This completes the proof.

Theorem 3: *The LRM algorithm terminates after $O(m\log^3 m)$ iteration, so the running time of the algorithm is $O(m^2\log^4 m)$* (Dijkstra algorithm runs on $O(m\log n)$ time and $n << m$).

Because there are only finite number of different path, the algorithm finds the optimal λ in a finite number of steps.

III. The DBLCLH problem

There is no routing algorithm in literature that treat four-metrics. We consider the Delay, Bandwidth and Loss probability Constrained Least Hop path problem (DBLCLH). Here, we use the single mixed concept. This single mixed metric combines delay and loss probability, but uses the absolute value of the logarithmic transmission-success probability function ($slog$) instead of the loss probability to avoid complex composition rules. $slog(p) = |log(1 - loss(p))|$. Furthermore, we assume that routes cannot have more than 90% of loss probability, that is, $0 < slog(p) < 1$. As ms unit is appropriate [5] to represent delay, we assume that delay is integer. As a consequence, we have a simple single-metric representation of delay and loss, where the integer part is delay and the decimal part represents loss. Each arc (i,j) in the path is assigned the value $\text{sm}_{ij} = \text{slog}_{ij} + \text{del}_{ij}$. When the arc (i,j) is inexistent, then $\text{sm}_{ij} = \infty$. Let $\text{sm}(p) = \text{sm}_{ij} + \text{sm}_{jk} + ... + \text{sm}_{qr}$. Given three constants, The minimum bandwidth $\Delta_{bandwidth}$, The maximum delay Δ_{delay} and the maximum logarithmic transmission-success probability $\Delta_{success}$, where the maximum single mixed metric is $\Delta_{ms} = \Delta_{delay} + \Delta_{succes}$. The Delay, Bandwidth and Loss probability Constrained Least Hop path problem (DBLCLH) is to find a path p from i to r that is minimal for a hop-count, satisfying $\text{sm}(p) \leqslant \Delta_{sm}$ and $\text{Bw}(p) \geqslant \Delta_{bandwidth}$. The DBLCLH's formal description is: $\min\{\text{hop}(p) : p \in P(s,t)$ and $\text{sm}(p) \leqslant \Delta_{sm}$ and $\text{Bw}(p) \geqslant \Delta_{bandwidth}\}$. The DBLCLH problem is *NP-complete*. By replacing del by sm, the LRM algorithm for the DBLCLH problem is similar that LRM algorithm for the DBCLH described before.

Theorem 4: *The final path constructed by LRM algorithm for the DBLCLH problem does not contain any loops.*

Proof: see the proof of the theorem 2.

IV. Simulation Results

We used simulation for our evaluation of the performance of the LRM method in case of two, three and for metrics. We built 3 random networks with 100 nodes (Figure 2).

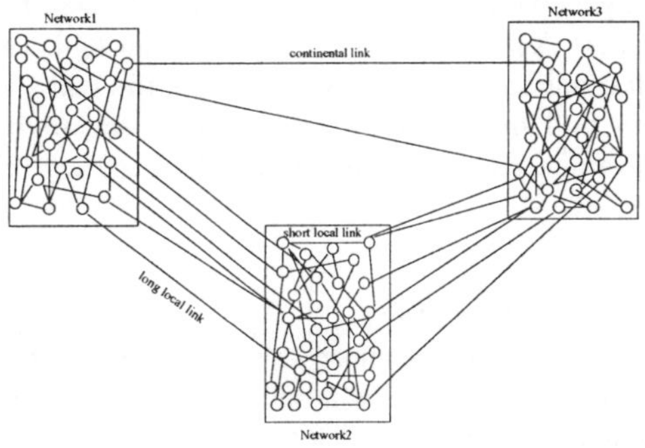

Fig. 2. Topology model

A random generator was used to create links interconnecting the nodes [6]. The output of this random generator is always a connected network in which each node's degree is at least 2. We adjust the parameters of the random generator carefully to obtain realistic network topologies with an average node degree of 4, which is close to the average node degree of current internetworks. For the LRM method, we consider the hop metric as the cost (LRM method is valid for any two additive metrics). The cost value on links varies from 1 to 30 based on uniform distribution. The propagation delay on links is selected from three ranges to resemble delay characteristics of nationwide network e.g. in the US. The first range (1-8 ms) represents shortest local links; the second range (8-12 ms) represents longer local links, while the third range (20-30 ms) represents continental links. On each network, 20% of nodes have longer local links and 2% have continental links.

A. Average Cost

Figure 3 shows the average cost of the paths found by the LRM algorithm compared with the Constrained Bellman-Ford (CBF) [7] algorithm that gives the optimal solution between a source and a set of destination nodes, but the running time is exponential.

Based on uniform distribution, the Cost Value on Links (CVL) varies from 1 to 20, 1 to 25, and 1 to 30. The LRM algorithm found almost the same paths as CBF. For $\Delta_{delay} < 15$, it is impossible to find short paths for all source destination

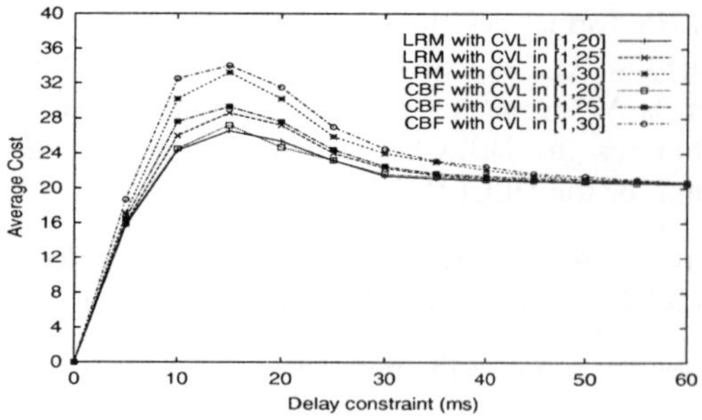

Fig. 3. Average cost

pairs with a small Δ_{delay}, thus the cost of them are also small. As the delay bound increases the algorithm find more and more (longer) paths, therefore the average of cost of the paths increases. At $\Delta_{delay} = 15$, the algorithm can find path for all source destination pairs. For $\Delta_{delay} > 15$, the LRM algorithm is able to find the paths with lower costs and higher delay. So, the average cost of the paths will decrease with the delay bound.

B. Average Number of Steps

We use the average number steps to measure the running time of the algorithm. The number of steps represents the events when the algorithm changes the contents of the heap. The curve of the Figure 7 is obtained by varying the CVL from 1 to 20.

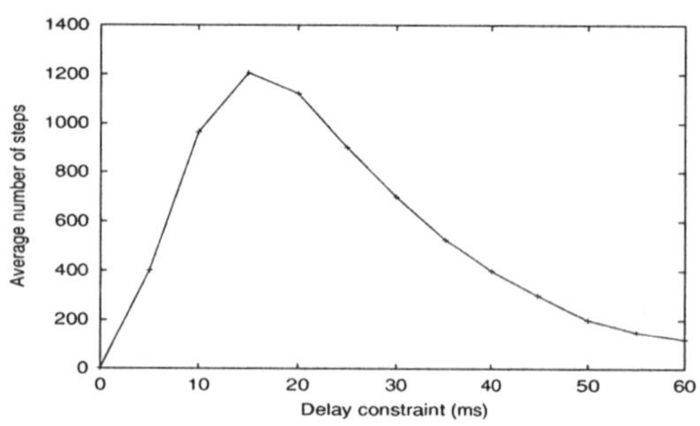

Fig. 4. Average number of step

For $\Delta_{delay} < 15$, as the LRM algorithm finds more and more paths, the number of Dijkstra executions increases. After $\Delta_{delay} > 15$, the LRM algorithm can find the optimal solution after the execution of the first Dijkstra because the minimum cost paths computed can more and more satisfy the delay bounds.

V. Future work: DBCLH problem in QOLSR protocol

Most routing protocols for Mobile Ad hoc NETworks (MANETs) [8], such as OLSR [9], are designed without explicitly considering QoS of the routes they generate. The number of hops is the most common criterion adopted by such proposed routing protocols. It is necessary to take into account important network parameters such as bandwidth and delay. To achieve that, we have developed the QOLSR protocol (OLSR based QoS) [10] that propagates MPR's delay and bandwidth information to each node in the network. Now, we propose to adapt the LRM algorithm for link-state protocols.

OLSR protocol is an optimization over the classical link-state protocol, tailored for Mobile Ad hoc NETworks (MANETs). The key concept used in the protocol is that of multipoint relays (MPRs). The idea of MPR is to minimize the overhead of flooding messages in the network by reducing duplicate retransmission in the same region. In route calculation, The MPRs are used to form the route from a given node to any destination in the network.

Let $G = (V, E)$ be the network with $|V|$ nodes and $|E|$ arcs and $p = (i, j, k, ..., q, r)$ a directed path. When the arc (i, j) is inexistent or j is not a MPR of i (due to the OLSR routing mechanism), then $\mathrm{del}_{ij} = \infty$ ($\mathrm{sm}_{ij} = \infty$), $\mathrm{hop}_{ij} = \infty$ and $\mathrm{Bw}_{ij} = 0$ for the DBCLH problem (DBLCLH problem). The application of LRM algorithm will be only over the MPR nodes.

VI. Conclusions

QoS routing provides better QoS guarantees to applications and improves the network resource utilization. Nevertheless, routing algorithms have to be carefully designed or the complexity will compromise their implementation. This paper presents a new algorithm called LRM for the DBCLH and DBLCLH problems. There are no routing algorithms in literature that treat four-metrics. Our routing algorithm can treat two-metrics (two additive metrics), three-metrics (two additive metrics and one concave metric) and four-metrics (two additive metrics, one concave metric and loss probability or transmission-success probability) and provides a polynomial heuristic solution. It is shown that our polynomial algorithm is close to the optimal solution computed by an exponential algorithm.

References

[1] Z. Wang and J. Crowcroft, "Quality of service routing for supporting multimedia applications," *IEEE Journal on Selected Areas in Communications*, vol. Vol.14, no. 7, pp. 1228-1234, September 1996.

[2] M. Held and R. Karp, "Quality of service routing for supporting multimedia applications," *Operation Research 18*, vol. pp. 1138-1162, 1970.

[3] R. Karp and M. Held , "The traveling salesman problem and minimum spanning trees, Part II," *Mathematical Programming 6*, vol. pp. 62-88, 1971.

[4] J. Garcia-Luna-Aceves, "A Unified Approach to Loop-Free Routing Using Distance Vectors or Link States," *In Proc. Sigcomm'89, Texas, USA*, September 1989.

[5] A. Fei, G. Pei, R. Liu and L. Zhang, "Mesurements on delay and hop-count of the Internet," *In IEEE GLOBECOM'98 - Internet Mini-Conference*, 1998.

[6] H. Salama, "Multicast Routing for Real-time Communication on High-Speed Networks," *PhD thesis, North Carolina State University, Department of Electrical and Computer Engineering*, 1996.

[7] R. Widyono, "The Design and Evaluation of Routing Algorithms for Real-time Channels ," *Technical Report TR-94-024, University of California at Berkeley*, June 1994.

[8] "http://www.ietf.org/html.charters/manet-charter.html."

[9] T. Clausen and P. Jacquet, "Optimized Link State Routing Protocol," *In IETF Internet Draft, draft-ietf-manet-olsr-11.txt*, July 2003.

[10] H. Badis, A. Munaretto, K. Al Agha and Guy pujolle, "QoS for Ad hoc Networking Based on Multiple-Metric: Bandwidth and Delay," *MWCN'03, Singapore*, October 2003.

ZONEPOW: A NOVEL APPROACH TO POWER CONTROL IN AD-HOC NETWORKS

SHACHINDRA SHARMA

Department of Computer Science, University of Illinois at Urbana-Champaign

R. S. SREENIVAS

Department of General Engineering, University of Illinois at Urbana-Champaign, 1308 West Main Street, Urbana, IL, 61801, USA

This paper presents an attempt to address the problem of power control at the network layer by proposing a novel protocol called ZONEPOW for Ad Hoc Mobile Networks. The aim of ZONEPOW is primarily to increase the overall traffic carrying capacity of the network. The solution to the power control problem is based on (i) the search for a common power level limited to discrete zones that proactively guarantees connectivity within the zone, and (ii) an appropriate, reactive forwarding procedure for any out of zone traffic. Being a hybrid power control protocol ZONEPOW is suited for ad-hoc networks and it is particularly attractive in networks where communication between hosts is mostly localized and communication across zones is comparatively rare.

1. Introduction

Space and Battery life are two critical resources in ad hoc and sensor networks and hence power control becomes important because it results in: (a) increased spatial reuse and reduced contention for floor acquisition at the data link (MAC) layer, (b) increased community throughput or the bit hauling capacity of the network, (c) less battery consumption. The significant works done to address the issue at the network layer are COMPOW [7], PARO [10], CLUSTERPOW [9] etc. Schemes using topology control include **Rudolpho and Meng** [16], **Bahl, Wang** et al. [11]. Other approaches in [12], [13] take the residual battery capacity into account. Solutions at the MAC layer include PCMA by **Bhargavan, Monks** et al. [14]. **Vaidya** et al. [15].

Our work draws inspiration from the common power level communication approach followed in COMPOW. The advantages of common power level have already been presented in the paper [7]. The need for this work arises from the following infeasibilities in COMPOW. As stated earlier COMPOW employs DSDV [2] network wide whose control overhead offsets the spatial reuse advantage. COMPOW also has the daemons running at different power level granularities, which deteriorates the advantage of having common optimal power level Pathological topologies in sparsely dense networks lead to choosing of much high power levels in COMPOW.

Claim: Power level search is not a cumulative but a discrete approach.

According to ZONEPOW the network is not seen in its holistic pictures but as a cumulative splicing of nodes lying in segregated zones, which forms the network topology. The search for a common power level will hence be limited to these discrete zones and the networks proactive route maintenance traffic will also be pruned to the size of different zones which will be decided by the number of hops at a given power level.

Zone Routing Protocol ZRP [3] methodology strikes us as an alternative to pure proactive routing mechanism in COMPOW. It provides for stable and up to date in-zone routes and reduces the network wide control traffic due to the on demand out of zone routes.

2. Protocol Description

2.1 Phase I Network Discovery Initiation

By the end of this phase we are guaranteed to have an undisputable network discovery initiator node termed as SOURCE who has the onus of commencing the network discovery. For this purpose any existing leader selection mechanism, which guarantees a real time successful leader discovery, can be employed (Suggestions [17], [18]). The protocol needs the discovery of the leader or SOURCE only at the time of network discovery initiation. The leader has no special role to play for the rest of the lifecycle of ZONEPOW run.

2.2 *Phase II Route Discovery*

This phase is constitutes the discovery of the network routes on the basis of zone radius and optimal zone power level and ensures: (a) the division of network into zones, (b) the assignment of the status to the nodes, (c) the discovery of common and optimal power levels specific to a zone, (d) network connectivity, (e) knowledge of routes at the nodes.

The SOURCE, initiates network discovery by beaconing with the network discovery HELLO packets at the ***minimum power*** granularity and waits for a response from any hosts, who might have been able to hear the packet. A predefined threshold number of nodes are required to acknowledge this request for the SOURCE to decide that the power level is appropriate for zone discovery and declare it as the zonal power level. The SOURCE and discovered nodes will then update their routing tables and zone power level will be followed for all subsequent discoveries in the zone. The discovered nodes set a flag, which tells them that they do not have to respond to any subsequent route discovery requests at greater or equal power and they then proceed to search for the nodes in their single hop vicinity at the zone power level. This process of the discovery of first zone will terminate in two circumstances, either when the host, which had sent the HELLO packet at the zone power level, finds no new neighbor and there is a corresponding time out or when the discovered node lies at the predefined maximum number of hop count (similar to ZRP) in other words the zone radius. This ensures the determination of the first zone boundaries and the corresponding nodes where the search terminated are termed as the zone's PERIPHERIES. They form the interface of the sub-network to the external world and any out of zone traffic in future will be sent to them. Any subsequent intra-zone communication will abide to the zonal common power level. In the other scenario when the threshold number of nodes does not respond to the SOURCE's discovery packet the SOURCE has to **hike up its power granularity** on the request time out. This means if the previous HELLO packet was sent at P_{min} it will now be sent at the next available higher power level i.e. P_{next}. In case the threshold number of nodes responds, the zone discovery proceeds as described otherwise the SOURCE repeatedly hikes up the power granularity. Threshold at maximum power blast can automatically be reset to one to avoid branching.

After the discovery of the first zone the designated PERIPHERIES are required to discover their part of the network. In a sense they now have to play the role of the SOURCE and have to discover their own zones. The PERIPHERIES follow the same process of route discovery as the SOURCE the only difference being that here that any node, which has already been discovered by another PERIPHERY or SOURCE, will stop responding to any future discovery requests at equivalent or higher power. This is a recursive process and the algorithm terminates in case none of the nodes discovered in the previous iteration is able to discover any new node even when they resorted to the maximum power level. It can be seen that here our perspective digresses from ZRP with respect to peripheries and overlapping zones since overlap and power control are two obtuse aspects with respect to each other.

2.3 *Phase III Route Maintenance*

This phase commences once the network has been discovered, the zones have been formed and routing information has been updated. It ensures the maintenance of up to date low delay intra-zone routes by virtue of the underlying proactive protocol (DSDV with respect to this paper) that requires all the nodes that are a part of the zone to beacon periodically sending their routing information so that the freshest routes are maintained. Besides this the phase takes care of the mobility. We do not expect a large scale movement at small interval of time; so within safe practical limits we assume the network to be pseudo static as is the underlying assumption in most of the routing protocols. The three possible movements are as follows:

2.3.1 MEMBER movement handling

The knowledge of movement of the node comes from the timeout of the periodic update from the node (reference DSDV) and the parent zone adjusts its routing information by sequence number based update information exchange. Now the moving node when it stops can land into situations: a) It hears a periodic beacon from some zone. It will the respond at the beacon's power level and will try to get incorporated in the zone it hears from as a member greedily accepting the invitation from lower power zone. b) In case it moves farther away from a zone than is the zone radius of the nearest zone it will beacon saying "Hello I am new" waiting for response and increasing its power level on timeout. In case of an ACK from host/hosts this node will choose the zone whose zone power is equal or higher than the power level of its request ensure minimum zone distortions.

2.3.2 PERIPHERY movement handling

If the network branch does not terminate at the moving periphery its movement has a larger impact on the network and hence there has to be additional signaling to have a replacement. The hosts within a single hop from that peripheral node ideally should be the candidate to take over as the new periphery. To deal with the movement following scenarios need to be addressed: (a) the zone power in the two bordered zones is different one being higher than the other; (b) the zones have the same power. To address this, the nodes promiscuously hear the beacons broadcasted by the different zone neighbors checking if these hosts are at one hop from the periphery. The nodes realizing that the periphery has moved out of the zone check if they have cached information and whether the other zone is at the higher or lower power. In case the other zone's zone power is more than this zone the node send the beacon at the higher power equivalent to the other zone's power and declare that it is the new periphery. After one node declares that it is the periphery the search terminates. In case the other zone is at the lower power then the nodes wait because it is mostly likely that someone from the other side will contact them due to information cached at those nodes and there is a timer maintained beyond which this zones nodes will themselves send the beacon at the lower power indicating to the other nodes that they want to take over as the periphery. The first node to do so will be chosen as the new periphery. In case both the zones are at the same power level then the first potential periphery to claim that it is the new periphery is assigned the role. In case there is no neighbor of the other zone that has been promiscuously heard the nodes wait for sometime and initiate afresh discovery for the other zone nodes that have the moving periphery as their one-hop neighbor. Nodes in the single hop neighborhood of the moving periphery communicate and inform each other once they have discovered other zone and declare themselves as peripheries.

2.3.3 SOURCE movement handling

If the moving SOURCE is also a periphery to the zone the movement is treated like periphery movement other wise it is considered as member movement.

2.4 Issue Resolution

There is a potential for collision in case the bordering zones have different zone power since there is a possibility of potential disturbers not being silenced by virtual carrier sensing (RTS/CTS did not reach the node). To deal with this the periphery always sends it control messages at the highest power level amongst the bordered zones and data messages at the corresponding zone power level alleviating the possibility of collisions. Moreover other nodes track the number of collisions and promiscuously cache bordering zone's power levels hiking their control message power beyond a threshold. For increasing the efficiency we propose the rule of greedy change of allegiance according to which whenever a host in the network hears a lower power blast from a different zone's periphery it responds back changing its allegiance to the "better claimer". A tradeoff to obtain more even and correct zone distributions and to reduce collisions we suggest bringing in a priority for lower power levels presently incorporated by a delay, which favors the searches at lower power levels. Another issue is in case of biased directional route discovery where there is a possibility of network getting branched in pathological topologies. To deal with this the peripheries and the source is required to maintain information about nodes that can be reached at higher than zone power levels and make checks if all corresponding nodes have been found after allowing appropriate time for network search.

3. Simulation and Analysis

The test-bed consists of the homegrown simulator for running COMPOW and ZONEPOW and the network consists of varying number of nodes being spread randomly in different topologies over a square grid of the area of 550 * 550 square meters. The number of nodes considered for the simulation runs were 15, 25, 35, 50, 65, 90, 100, 120, 140, 175, 200 and 225. For each topology 14 runs are made and the power level is averaged over these. There are two fold savings for ZONEPOW in comparison to COMPOW in terms of power savings and control traffic overhead. Our efficiency metric for power savings is given by:

Average Power-Level in ZONEPOW / Common Power-Level in COMPOW.

The proactive traffic in COMPOW in comparison to ZONEPOW increases with the square of number of nodes because the size of the routing packet is O(N) and every node is required to beacon the control information every periodic time T. The efficiency is measured by:

Total number control traffic packets ZONEPOW / Total number of control traffic packets COMPOW

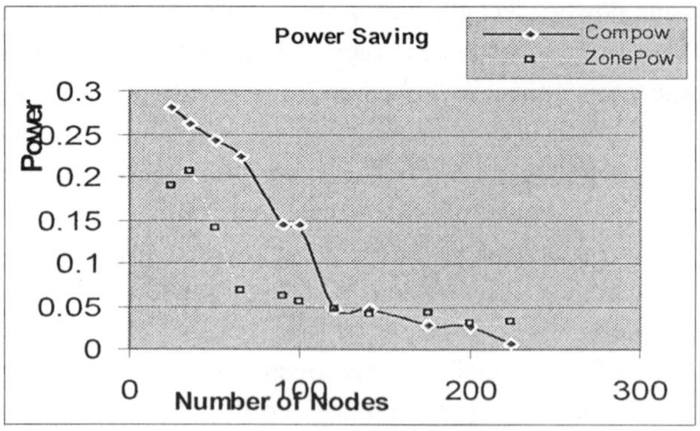

Figure 1 Power saving comparison ZONEPOW and COMPOW

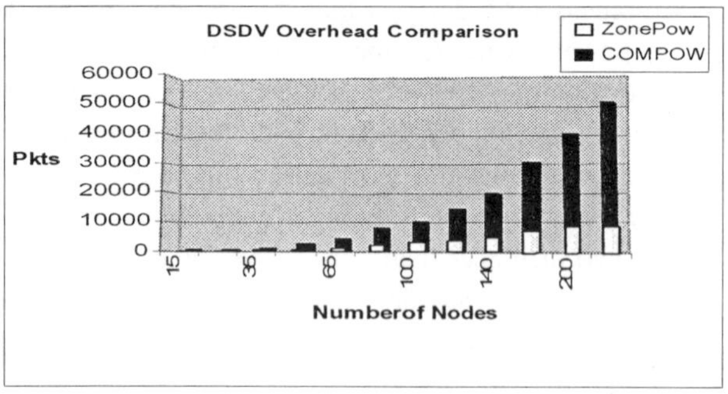

Figure2 Comparison of control traffic overhead in ZONEPOW and COMPOW

As can be seen in the Figure 1 power savings are considerable for ZONEPOW for sparse topologies. Till about 90 nodes for this topology ZONEPOW performs significantly above COMPOW on an average because of the fact that random sparse topologies often result in one or a bunch of nodes being sprayed unevenly from the rest of the network leading COMPOW to choose the higher power level. This is reduced as the number of nodes increases, where COMPOW and ZONEPOW performance tends to converge to a common optimal power level. In fact although theoretically it shouldn't happen towards very high node densities COMPOW even gets better at times because ZONEPOW is subject to issues related to synchronization and experimental errors.

4. Conclusion

ZONEPOW has been inspired by COMPOW, which attempted to come up with a solution by a common power level of transmission for a network but lost the scalability due to high proactive traffic and proclivity of the protocol to choose higher power level in case of networks with uneven density. ZONEPOW for lower node densities leads to considerable savings in terms of space and higher densities saves on overhead of control traffic in comparison to COMPOW. Moreover ZONEPOW follows a hybrid approach whose advantages are already known for Mobile Ad Hoc networks. The special status to PERIPHERIES imparts to ZONEPOW scalability and adaptability tending towards the hierarchical protocols, which are specially suited for the modern day internet architecture. There is an associated overhead but the advantages of the power control are prominent enough to justify the approach for the solution to power control problem.

References

1. C. E. Perkins, E. M. Royer, and S. Das, "Ad hoc on demand distance vector (AODV) routing." draft-ietf-manet-aodv-07.txt, 2000.
2. Highly Dynamic Destination-Sequenced Distance-Vector Routing (DSDV) for Mobile Computers ACM SIGCOMM, vol.24, no.4, October 1994.
3. The Zone Routing Protocol (ZRP) for Ad Hoc Networks Haas, Z.J., Pearlman,M.R., Samar, P., IETF Internet Draft, draft-ietf-manet-zone-zrp-04.txt,July 2002.
4. Intrazone Routing Protocol (IARP) Haas, Z.J., Pearlman, M.R. and Samar, P., IETF Internet Draft, draft-ietf-manet-iarp-02.txt, July 2002.
5. Interzone Routing Protocol (IERP) Haas, Z.J., Pearlman, M.R. and Samar, P., IETF Internet Draft, draft-ietf-manet-ierp-02.txt, July 2002.
6. Determining the Optimal Configuration for the Zone Routing Protocol Haas, Z.J., Pearlman, M.R., IEEE JSAC, August, 1999.
7. S. Narayanaswamy, V. Kawadia, R. S. Sreenivas, and P. R. Kumar. Power Control in Ad-Hoc Networks: Theory, Architecture, Algorithm and Implementation of the COMPOW protocol. In European Wireless 2002.

8. Piyush Gupta and P.R Kumar, "Critical power for asymptotic connectivity in wireless network," in Stochastic Analysis, Control, Optimization and Applications Eds. 1998, pp. 547–566, Birkhauser, Boston.
9. Vikas Kawadia and P. R. Kumar. In IEEE INFOCOM, 2003. Power Control and Clustering in Ad hoc Networks.
10. J Gomez, A. T. Cambell, M. Naghshineh, C. Bisdikian, T.J. Watson Power-Aware Routing Optimization Protocol (Paro) Internet Draft, draft-gomez-paro-manet-00.txt, work in progress, June 2001.
11. R. Wattenhofer, L. Li, P. Bahl, and Y.-M. Wang.Distributed Topology Control for Power Efficient Operation in Multihop Wireless Ad Hoc Networks in INFOCOM 2001.
12. C.-K. Toh. Maximum Battery Life Routing to Support Ubiquitous Mobile Computing in Wireless Ad Hoc Networks. IEEE Communications Magazine, 39(6):138-147, June 2001.
13. D Kim, J.J. Garcia-Luna-Aceves, K Obraczka, J Cano and P Manzoni. Performance Analysis of Power Aware Route Selection in Mobile Ad Hoc Networks.
14. J. P. Monks, V. Bharghavan, and W. mei W. Hwu. A Power Controlled Multiple Access Protocol for Wireless Packet Networks. In INFOCOM 2001.
15. E.-S. Jung and N. H. Vaidya. An Energy EfficientMAC Protocol for Wireless LANs. In INFOCOM 2002, June 2002.
16. V Rudolpho and T Meng. "Minimum energy mobile wireless networks", IEEE Journal on Selected Areas in Communications, vol. 17, pp. 1333-44, 1999.
17. G. Singh. Efficient leader election using sense of direction. Distributed Computing 10 (1997).
18. N. Malpani, J. L. Welch, and N. H. Vaidya. Leader Election Algorithms for Mobile Ad Hoc Networks. In Proc. 4th Intl. Workshop on Discrete Algorithms and Methods for Mobile Computing and Communications, Boston, MA, 2000.

A LOAD AWARE, LOCATION AIDED, AD-HOC ON-DEMAND ROUTING ALGORITHM

Tahmid Quazi and Stephen McDonald
School of Electrical, Electronic and Computer Engineering
University of Natal
King George V Avenue
Durban, South Africa, 4041
Email: quazit@nu.ac.za, mcdonalds@nu.ac.za
Tel: +27(0)31 2602756
Fax: +27(0)31 2602713

Abstract: A mobile ad-hoc network is an autonomous network of mobile routers that are connected via wireless links. These networks have dynamic topologies and have links that are bandwidth and energy constrained. There are many proposed schemes that solve the problem of routing packets in a dynamic ad-hoc network environment. Recent comparison work via simulations has shown that the on-demand routing protocols show better results than most proposed table driven algorithms. In this paper certain disadvantages of the on-demand protocols are presented and it is shown that if they are overcome the performance of the protocols will be enhanced. The proposed protocol uses location information of the hosts to limit the region in which a route discovery is attempted. This is an improvement on the highly redundant flooding technique used by most on demand routing algorithms. In addition to the hop count metric used by on demand schemes, the protocol uses route load as a metric to determine routes to destinations. Simulation results show that the proposed changes enhance the performance of an on-demand scheme thereby making it more scalable.

Key Words: ad hoc, on demand, load, location.

1. INTRODUCTION

A mobile ad-hoc network (MANET) is a network of mobile computers that are connected via wireless links. The nodes have limited transmission range, so each node may have to act as a router, making the network a multi-hop network. Due to the mobility of the nodes the network topology may change dynamically. In addition, the nodes have limited resources such as power and bandwidth. All these issues have to be taken into consideration in the design of protocols for MANETs. A routing protocol has to be fast enough to manage the constantly changing network topology, yet it must do so while remaining conservative in its use of resources available to the host. There are many applications for a MANET. Some of the commercial uses for MANETs are search and rescue communication systems. The military applications such as the mechanized battle-field and sensor networks are of particular interest of the authors of this paper [7], [17]. An infrastructureless network of wireless devices that allows voice and data communication for situational awareness in a battle-field environment would be invaluable.

The design of routing protocols has received much attention in the recent past. Many protocols have been proposed and they can be categorized into two classes. There are the pro-active or table driven schemes [1], [2], [3] and the reactive or on-demand routing protocols [4], [5]. Recent comparisons of routing protocols via simulations [6], [7] have quantified the advantages of on-demand routing protocols over table-driven methods.

The good performance of on-demand protocols in terms of low routing overhead and high delivery ratio makes them attractive for ad-hoc networks. The on-demand protocols typically have a two phase operation namely route discovery and route maintenance. Most of these schemes use flooding in the route discovery phase of their operation. There is redundancy created by using this technique since the route request is sent to nodes far from the vicinity of the source and destination nodes. The broadcasting nature of flooding also causes collisions and contentions in the transmission of packets. Making this route search algorithm more efficient would reduce routing overhead of the on-demand routing protocols. A further characteristic that is common in most on-demand algorithms is that the hop count is used as the routing decision metric. The authors of the AODV on-demand routing protocol have stated [8] that to make such protocols more efficient and scalable the load in the network has be to be taken into account. The focus of this work is to study the effect of adding query localization to the route discovery process of an on-demand routing protocol as well as investigating how load checking and balancing, affects the performance of the on-demand methodology.

The layout of the rest of the paper is as follows. Section 2 gives a brief description of AODV's on-demand scheme as well as mentioning the areas that possibly require improvement. The algorithm of the proposed protocol can be found in section 3. The simulation setup and results are in section 4 and 5. The paper then concludes with future work mentioned in section 6.

2. OVERVIEW OF AODV'S ON-DEMAND SCHEME

An on-demand scheme only creates a route to the destination once a packet requesting a route is generated. These protocols generally have lower routing overhead when compared to table driven schemes since the only effort is in maintaining routes that are actively used. When a request for an unknown destination is generated the source node initiates a route discovery by sending a route request packet. This packet is broadcast to the host's immediate neighbouring nodes. Each intermediate node that receives the packet creates a reverse route entry in its routing table for the source node and the last hop node taken by the route request. This reverse route is later used to forward route reply packets that arrive from the destination node. The intermediate node then broadcasts the route request packet to its neighbours. Once the route request arrives at the destination, a route reply packet is unicast by the destination node using its reverse route to the source. All intermediate nodes are then responsible for passing the route reply back to the source node. An intermediate node replies to route request on behalf of the destination if it has an active route to the node. Thus, during the route discovery phase, the protocol uses the flooding technique to ensure that the route request arrives at the destination.

Flooding is a robust method for getting the route request packet to every possible node in the connected component network. However, in a large and highly mobile network, considerable routing overhead is incurred by this technique. This reduces the advantage that on-demand algorithms have over table driven ones. Thus if the flooding can be made more efficient the routing overhead would be lowered. This would result in other advantages such as reducing network congestion as well as reducing the effect of the broadcast storm problem [9] (since route requests are broadcast packets). Query localization in ad-hoc networks has been examined in [10], [11] and [12]. The application in mind for the on-demand protocol being designed is for soldiers communicating in an automated battlefield. In this application situational awareness is of such importance that it is foreseen that each node will be equipped with a GPS module that will provide location information. If an ad-hoc router (which could be any node in the network) has prior knowledge of the location information of the destination nodes, it could use this information for the localization process. The method proposed in this work is based on ideas presented in [13]. The aim of the query localization is to bring the route request packet physically closer to the destination with each hop and prevent it from traversing to unnecessary parts of the network. The details of the location aided flooding implementation can be found in section 3.

Many of the on-demand ad-hoc protocols that have attained popularity use the hop count as the decision criterion for a better route. This method of using the least hop count has a drawback in that there exists the possibility of creating congestion at certain nodes in the network. This adversely affects data throughput as packets are dropped from overloaded protocol queues. The routing delay performance is also affected, as packets remain queued at the highly loaded nodes for a considerable amount of time. Recent simulation studies have shown that protocols using the least hop count as a metric suffer from poor performance when the traffic density of the network increases [8]. The work in [14] and [15] emphasizes the notion that doing load checking for routing traffic in a MANET has significant advantages over shortest path methods. Load checking not only prevents congestion at certain nodes in the network but also attempts to balance the network traffic amongst the nodes in the network. If the load is balanced there will be fewer critical nodes in the network, and possibility of power depletion at such nodes due to the burden of high load.

What is proposed in this work is the integration of a load metric to the hop count metric used in on-demand protocols. The metric, along with the hop count, is taken into account when deciding on the best routes between source and destination.

3. DESCRIPTION OF PROPOSED PROTOCOL

The protocol proposed in this work is called, Tactical AODV (TAODV) and is a modification of the Ad-hoc On-demand Distance Vector (AODV) routing protocol [4]. It is similar to AODV in that the routing information for each route to a destination is maintained in a distributed fashion in the routing tables of the nodes in the network. The protocol only creates routes to destination nodes when such routes are requested by the generation of data packets for the destination. Routes are only maintained as long as they are being actively used. If a source node does not have a route to the destination, it initiates a route discovery. The data packets are transmitted once a route is found. If during the transmission of data the route is broken, it is repaired before further transmission can continue.

The route localization used herein is an optimization of the flooding technique used by on-demand algorithms. If available, the location information of the destination node is used to determine if an intermediate node should rebroadcast a route request packet. It will only rebroadcast the packet if it is deemed to be closer to the destination than the node from which it received the route request packet. This method aims to prevent route request packets from traversing to unnecessary sections of the network i.e. nodes that are not in vicinity of the path between the source and destination pair. Preventing route requests from reaching such areas will result in a reduced protocol routing overhead.

The dissemination of a node's location data occurs in an on-demand manner. There is no periodic transmission of the location data and thus other nodes in the network will only know of a nodes whereabouts if they have communicated with it, or acted as a router for any of its routes. The location information of a source and destination node is piggy-backed with each route request and route reply packet respectively.

The route localization is implemented as follows. When the route request is generated for the destination node, the source node inspects its location cache to see if it has a location entry for the destination. This is likely if it has either communicated with the destination previously or acted as a router for it. If the location entry is found, the positional information of the destination (its *x* and *y* coordinates), are used to calculate the distance to it using equation 1.

$$D_{sd} = \sqrt{(\Delta x)^2 + (\Delta y)^2} + \lambda \quad (1)$$

D_{sd} is the distance from the source to the destination node while Δx and Δy represents the difference in the x and y location respectively. Here λ is a factor that takes into account the approximation of the distance measure and is given by equation 2.

$$\lambda = v \times (t_c - t_d) \quad (2)$$

Where t_c is the current time and t_d is the timestamp of the location information of the destination in the source node's cache. v is the specified maximum speed that a node can move. The measure given by equation 2 is the worst case scenario in terms of the distance to the destination node. It is a measure that is similar to the *expected zone* concept proposed in Location Aided Routing (LAR) protocol [13].

The distance calculated by the source node (D_{sd} in Figure 1) is included in the route request packet broadcast to its neighbours. The timestamp of the location information (t_{sd}) that was used to calculate D_{sd} is also one of the fields in the packet. When a node between the source and destination receives the packet, its first action is to query its location cache for an entry for the destination node. If an entry is found, the timestamp of the entry (t_{id}) is compared with the location information timestamp in the route request packet (t_{sd}).

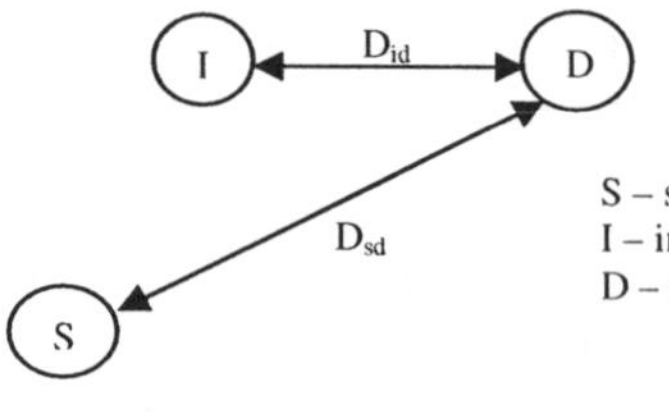

Figure 1 Re-broadcast decision

If t_{id} is newer than t_{sd}, it implies that the intermediate node's location information for the destination is more current. The intermediate node then calculates the distance from the source to the destination (D_{sd}) using its information for the destination and the location information of the source from the route request packet. The distance field in route request packet is updated accordingly. If t_{id} is not newer, the distance field remains unchanged. The intermediate node then calculates its distance to the destination node and compares this value with the source to destination distance. If D_{sd} is found to be larger than D_{id} (the intermediate node is closer to the destination than the node from which the route request arrived), the intermediate node broadcasts the route request packet. The distance metric to the destination is equated to D_{id}, narrowing the localization region with each hop. If the condition is not met the packet is dropped. During the localized route request process, if any node has no entry for the destination in it's location cache, it does not execute the localization algorithm. The node instead broadcasts the route request as it is done in a blind flood. If the location aided route request fails to secure a route to the destination after a certain number of attempts subsequent route requests are done using the blind flood method.

The load checking operation of the protocol is done at each of the intermediate nodes that processes a route request. The length of the protocol queue at each node is taken as a measure of the load at a node. When an intermediate node receives a route request, the first decision that is made at the node is whether it will get involved in the route. The node inspects its protocol queue and if it is near capacity the intermediate rejects the route request by dropping the packet. This will prevent already congested nodes from further overload. After the initial load check, if the intermediate node decides to act as a router for the source, it first creates a reverse route in its route table entry for the source node. This entry will be used to unicast reply packets back to the source node. The intermediate node then executes the load checking algorithm. The method used in this work is similar to one of the route selection procedures in the Dynamic Load Aware Routing (DLAR) protocol [14]. There is a *load* variable in the route request packet (*q_thresh_count*) that counts the number of nodes between the source and destination that have their protocol queues loaded with packets above a certain threshold τ. The variable is initialized to zero by the source node when it generates the packet. Prior to broadcasting the route request packet, an intermediate node inspects the length of its protocol queue. If the queue length is above the threshold value τ it increments the *load* variable in the packet. The variable keeps its previous hop value if the queue length is below the threshold value.

The load checking process is executed with each hop that the route request packet takes on route to the destination node. The routing protocol proposed here specifies that the route request travels to the destination node and no intermediate node is permitted to reply to the route request. Minimum hop count algorithms such as AODV, DSR specify that, if intermediate nodes have a cached route to the destination, they should send a route reply on behalf of the destination node. However, there are certain disadvantages to this methodology. Intermediate nodes replying to route requests generate a flood of route reply packets, which causes significant routing overhead. The reply packets are unicast transmissions which add to the congestion at each intermediate node in the routing path. The least hop count, shortest path method also results in certain routes in the network overlapping which creates congestion at certain nodes, creating bottlenecks.

The heuristic used to select routes is a combination of the load information and the hop count. There are essentially three conditions which determine if a route is better than another.

Case 1: A route is considered better if the *load* variable of a newly arrived route is lower than that of the previous route.

Case 2: The *load* variable is equal between the routes being compared. In such a situation the route with the lower hop count is deemed a better route.

Case 3: If the newly arrived route has a higher hop count than the previous route, it is only recognized as a route with a better metric, if the *load* variable is lower. In this case, although a longer route in terms of hop count is chosen, it is a route that is less loaded. Through experimentation it is noticed that including this case improves the performance of the protocol.

In most on-demand protocols (such as AODV) duplicate route request packets at all nodes are silently discarded. However since this is also done at the destination node, there is no mechanism to accept further, possibly better routes. In TAODV the destination node accepts more than one route request from a source. When the destination node first receives a route request it replies by generating a route reply packet. Further replies are only generated if the route request packets come along routes that have better routing metrics. The route reply packet is unicast to the source node via the reverse route that was set up by initial process of the route discovery. The arrival of the route reply packet prompts the source node to transmit the data packets it has in its protocol queue, using the newly discovered route. If a route reply with a better route arrives subsequent to this, the source node accordingly changes the transmission route to the destination.

4. SIMULATION SETUP

The simulation environment used in this work is the widely used NS-2 network simulator [16]. The code for the TAODV protocol is a modified version of AODV code that is within the NS-2 environment. However necessary changes had to be made to derive the location information of a node in the simulation space to aid with the query localization process.

The objective of the simulations is to measure the ability of the protocols to react to changing topologies while maintaining a high data transfer between source and destination nodes. Thus each protocol is subjected to a variety of load conditions in terms of the mobility of the nodes and the traffic pattern in the network. A network consisting of 50 nodes in a 1400m X 450m space is used for the simulations. The transmission range of a node was set to 250m which is a reasonable distance for a wireless card using the 802.11b standard. The mobility of the nodes followed the "random waypoint" model generated by the group at CMU [6]. Simulations ran for a 1000 seconds with varying pause times. The maximum speed chosen for the mobility model was 8m/s. The data traffic sources chosen for the simulation were constant bit rate (CBR) sources. In the 50 node network there were 20 sources and 30 connections.

5. SIMULATION RESULTS

PACKET DELIVERY RADIO

The packet delivery ratio is defined as the percentage of packets generated by source nodes that are received by the destination nodes. The packet delivery ratio result presented in Figure 2, shows that the performance of TAODV is comparable and better than that of AODV. The proposed algorithm performs better under high mobility conditions. Since congestion is avoided, the improvement in the delivery ratio is due to fewer packets being dropped from overloaded protocol queues.

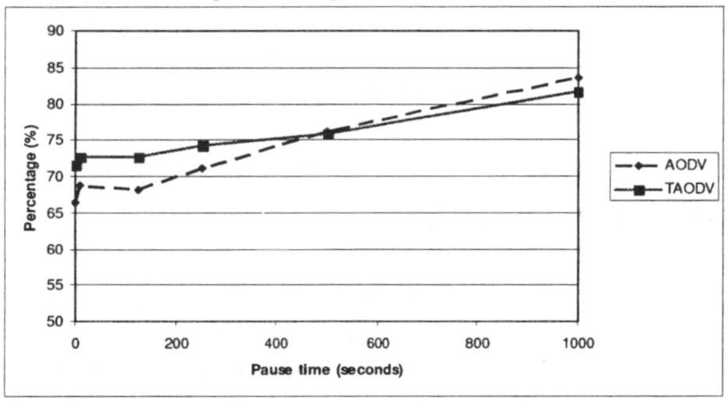

Figure 2 Packet Delivery Ratio

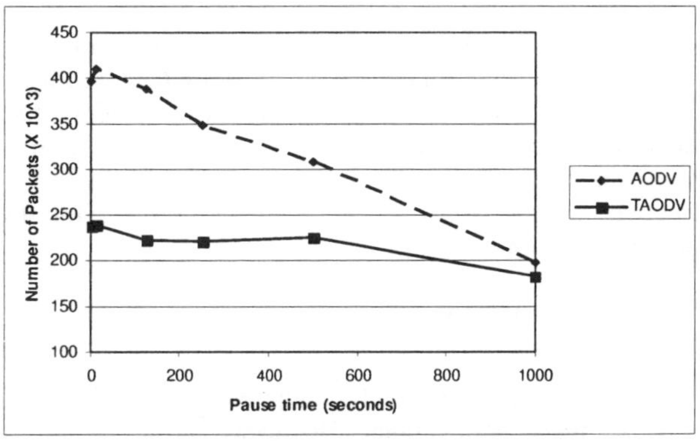

Figure 3 Routing overhead in number of packets

ROUTING OVERHEAD

The routing overhead incurred by a protocol is used as a measure of a protocols internal efficiency. The overhead is presented as the number of routing data packets transmitted during a simulation. The advantage shown by the proposed protocol is in terms of the routing overhead. As can be seen in Figure 3, TAODV incurs significantly less routing overhead, notably at high mobility conditions. The reduction in the number of packets is due to the route localization action of the algorithm. Since route requests are prevented from reaching unnecessary portions of the network the total number of routing packets is reduced.

AVERAGE END TO END DELAY

The average end-to-end packet delivery delay results (Figure 4) shows that TAODV delivers data packets with less delay than AODV under high mobility conditions. The difference is greater than 100ms at low pause times. This result is expected as the load checking algorithm aids in preventing delays due to packets being queued at congested nodes in the network. The load balancing results in reduced latencies in the delivery of data packets.

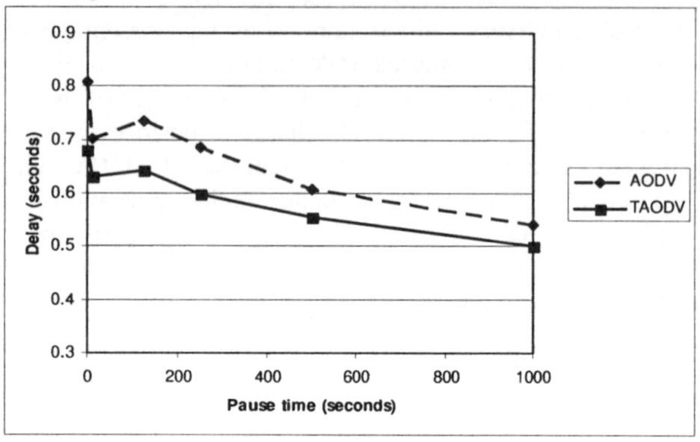

Figure 4 End to End packet delivery delay

6. FUTURE WORK

The aim of this work was to show that doing a localized route request flood provides a more efficient route discovery process. However, TAODV currently only spreads the location information in an on-demand basis i.e. nodes only know the location of other nodes if they either communicated with them or acted as routers for them. In future other methods of disseminating host's location and speed information to nodes more efficiently will be investigated. The load metric used in this algorithm only considers queue length while other effects of load such as queue length of neighbours and local signal to noise ratio would be a more effective measure of load, and thereby improve the load checking algorithm

7. CONCLUSION

With readily available mobile computers, MANETs will be used in many applications. However, these applications demand that the routing protocol for the mobile network be adaptive, distributed and scalable. Most of the on-demand routing schemes proposed in literature solve the routing problem in a more efficient manner than table driven ones. This is due to their lower routing overhead which conserves the resources available to a host in an ad-hoc network. What has been proposed with this work is that the routing overhead can be even further reduced if a more intelligent flooding technique is used instead of the blind flooding method. The results also show that performing load balancing increases the performance of on-demand algorithms. With lower routing overhead and lower routing delays, the proposed on-demand protocol is more scalable than other schemes.

REFERENCES

[1] C. E Perkins and P Bhagwat. Highly Dynamic Destination-Sequenced Distance-Vector Routing (DSDV) for Mobile Computers. Comp. Commun. Rev. Oct. 1994.pp 234-44

[2] Ching-Chuan Chiang and Mario Gerla. Routing and Multicast in Multihop, Mobile Wireless Networks. In Proceedings of ICUPC '97

[3] Tsu-Wei Chen and Mario Gerla. Global State Routing: A New Routing Scheme for Ad-hoc Wireless Networks", IEEE ICC '98, June 1998.

[4] C. E. Perkins and E. M. Royer, "Ad-hoc On-Demand Distance Vector Routing", Proceedings of the Second IEEE Workshop on Mobile Computer Systems and Applications, February 1999, pp. 90-100.

[5] D. B. Johnson and D. A. Maltz, "Dynamic Source Routing in Ad Hoc Wireless Networks", Mobile Computing, T. Imielinski and H. Korth, Kluwer, 1996, pp. 153-181

[6] J. Broch, D. A. Maltz, D. B. Johnson, Y-C. Hu and J. Jetcheva, "A performance comparison of multi-hop wireless ad hoc network routing protocols", Proceedings of ACM/IEEE MOBICOM '98, October 1998, pp. 85–97.

[7] P. Johansson, T.Larsson, N.Hedman. "Scenario-based Performance Analysis of Routing Protocols for Mobile Ad-hoc Networks", Proceedings of the IEEE/ACM MOBICOM 99, August 1999.

[8] Samir R. Das, Charles E. Perkins and Elizabeth M. Royer, "Performance comparison of two on-demand routing protocols for ad hoc networks", Proceedings of IEEE INFOCOM'2000, March 2000.

[9] Y.-C. Tseng, S.-Y. Ni, Y.-S. Chen, J.-P. Sheu "The Broadcast Storm Problem in a Mobile Ad Hoc Network", Wireless Networks 8, 153-167, 2002.

[10] R. Castaneda, S. R. Das and M. K. Marina "Query Localization Techniques for On-Demand Routing Protocols in Ad Hoc Networks" , Wireless Networks 8, 137-151, 2002.

[11] Skold, M. Using location information to reduce the route search in tactical radio networks, Military Communications Conference, 2001. MILCOM 2001. Communications for Network-Centric Operations: Creating the Information Force. IEEE , Volume: 1 , 2001 367 -371 vol.1

[12] Tripathi, S.M,Ghosh, R.K. An implementational study of certain heuristics for the performance enhancements of AODV, Parallel, Distributed and Network-based Processing, 2002. Proceedings. 10th Euromicro Workshop on 2002 395 -402

[13] Y. B. Ko and N. H. Vaidya, "Location-Aided Routing (LAR) in Mobile Ad-hoc Networks", Proceedings of ACM/IEEE MOBICOM '98, October 1998.

[14] Lee, S.-J., Gerla, M, Dynamic load-aware routing in ad hoc networks. Communications, 2001. ICC 2001. IEEE International Conference on , Volume: 10 , 2001

[15] Y. Yi , M. Gerla and T. J. Kwon, The Selective Intermediate Nodes Scheme for Ad Hoc On-Demand Routing Protocols, To appear in Proceedings of IEEE ICC 2002

[16] Kevin Fall and Kannan Varadhan (Eds.), "*ns* notes and documentation", 2002, available from http://www-mash.cs.berkeley.edu/ns/.

[17] J. Jubin, J.Tornow, "The DARPA Packet Radio Network Protocols", Proceedings of the IEEE, Jan 1987

GATEWAY-BASED MULTICAST PROTOCOL FOR MOBILE AD HOC NETWORKS

Jiawei Xie, Sukumar Nandi, and Anil K. Gupta
School of Computer Engineering, Nanyang Technological University, Singapore-639798
Email: xie_jiawei@pmail.ntu.edu.sg, assukumar, asgupta@ntu.edu.sg

Abstract: This paper presents a new multicast protocol, called Gateway-Based Multicast Protocol (GBMP). Receivers are grouped into local groups, in which a group-shared tree, whose root is called a Gateway Node, connects receivers and intermediate nodes. The gateway nodes and multicast sources make up a mesh. A hop-limited Receiver-Initiated Attach Procedure is introduced for multicast receivers to cope with link breakages. A new technique, Anonymous Receive Mode, is proposed to improve transmission efficiency. Simulation results show that our protocol outperforms ODMRP.

Key Words: Anonymous Receive Mode, Mesh, Multicast.

1. INTRODUCTION

A Mobile Ad hoc NETwork (MANET) consists of mobile wireless nodes, and the network topology changes constantly. Many multicast protocols for the MANETs have been proposed. ODMRP (On-Demand Multicast Routing Protocol) [1] is a simple and effective mesh-based multicast protocol. Mobility-based Hybrid Multicast Routing (MHMR) [2] and Adaptive Dynamic Backbone Protocol (ADB) [3] construct a hierarchical topology with mesh amongst cluster heads and backbone nodes respectively while maintaining tree topology in local areas. In ADMR (Adaptive Demand-Driven Multicast Routing) [4], multicast data packet receipt is not restricted to a specific branch. In this paper, we introduce a new MANET multicast protocol, Gateway-Based Multicast Protocol. The rest of the paper is organized as follows. Section 2 introduces the Gateway-Based Multicast Protocol algorithm. Section 3 evaluates its performance. Section 4 concludes this paper.

2. GATEWAY-BASED MULTICAST PROTOCOL (GBMP)

The GBMP utilizes a hierarchical topology, as adopted in MHMR and ADB, for the delivery of data packets from multicast sources to multicast receivers. The following terminologies are used in this paper:

- **Local group**: is a set of multicast receivers and Forwarding Group (FG) [5] nodes, which are close to each other (e.g. within 3 hops), for a long enough period of time.
- **Multicast group leader**: is the multicast source node that has the lowest ID among all the multicast sources in a multicast group.
- **Gateway node**: is a multicast receiver node that serves as the root of the local group-shared tree.

2.1 Data structures

- **Multicast Source Table**: stores all the multicast source IDs for every multicast group a node belongs to.
- **Route Table**: stores the routing information.
- **FG node status Table**: stores the status of whether a node is a FG node or not for a certain multicast group.
- **Responsibility Table (Rsp Table)**: stores all the downstream nodes' IDs that a node is responsible for their receipt of data packets.

2.2 Multicast topology creation

GBMP uses a hierarchical topology. The lower level, called a local group, consists of multicast receivers and intermediate nodes. These intermediate nodes have the same functionality as FG nodes, thus, in the following text, we name these nodes also as FG nodes. The topology used in a local group is a group-shared tree whose root is a multicast receiver node. This root node is also called a Gateway Node (GN). A mesh is established amongst GNs of different local groups and the sources.

When a multicast source wants to send data packets to the multicast group, it broadcasts a Global Join Query (GJQ) periodically. The format of GJQ is:

Global Join Query (Message Type, Multicast Group ID, Global TTL, Global Hop Count, Local TTL, Sequence Number, Gateway Node ID).

In the GJQ, the field Local TTL is initialized with a pre-defined value INVALID_TTL and the Global TTL is set to a large enough value that enables the GJQ to be forwarded to all the nodes in the network. The Local TTL is used by the receivers to setup local groups. When a multicast receiver receives a GJQ originated by the multicast group leader, if the value of this field is equal to INVALID_TTL, the receiver is eligible to become a GN. It sets the Local TTL to MAX_LOCAL_GROUP_RANGE, which is a limit for the range of a local group (the default value is 3), puts its own ID into the field Gateway Node ID showing that it has become a GN, and broadcasts this message (if the GJQ comes from a source other than the multicast group leader, it is simply forwarded without the processing of constructing the local groups). After that, the GN sends a Global Join Reply (GJR) back to the source. The format of GJR is as follows:

Global Join Reply (Multicast Group ID, Source ID, Next Hop ID)

Receiving a GJR, a node checks if its own ID is equal to the Next Hop ID. If this is the case, this node becomes an FG node. It then chooses its own next hop address to forward this GJR until it reaches the multicast source.

If in the GJQ initiated by the multicast group leader, a multicast receiver finds that the value is less than or equal to MAX_LOCAL_GROUP_RANGE, it knows that it is within the range of a local group. The multicast receiver stores the ID of the GN and the path towards it in the route table, decreases both the Local TTL and the Global TTL by one and forwards it until the Global TTL reduces to zero. If the value of the Local TTL is equal to zero, which means this GJQ has reached the border of a local group, and therefore, before forwarding this GJQ, the Local TTL is set to INVALID_TTL again.

Multicast receivers in a local group should send a Local Join Reply (LJR) to its GN in order to set up the local group-shared tree. The process of constructing a local group is illustrated in Figure 1.

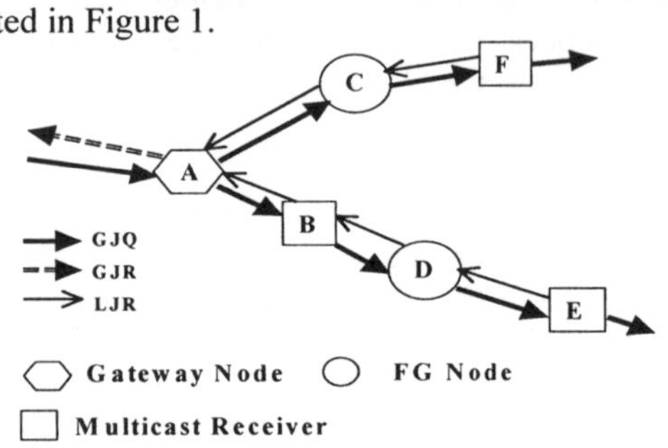

Figure 1. Local group construction

In between two successive GJQs from the multicast group leader, the GN sends out the Local Join Query (LJQ) to maintain the local group topology. The nodes belonging to this local group are required to send back LJR to the GN.

Note that the construction of the local groups is only triggered by the GJQs coming from the multicast group leader, and hence, the GNs correspond only to the multicast group leader. On the other hand, the GNs are in charge of sending the GJR to all the multicast sources. In this way, a topology consisting of a global mesh and the local trees is constructed.

When a non-multicast receiver node receives a GJQ, if the Local TTL has the value of INVALID_TTL, it re-broadcasts this message without any modification to it; otherwise, it performs the same actions as a receiver, which is within the range of a local group, does.

By using the above mechanism, nodes, which tend to move together or are close to each other, can be grouped.

2.3 Repair procedure

To tackle the mobility, a repair mechanism called Receiver-Initiated Attach Procedure (RIAP) is implemented. This requires that while receiving data packets, the multicast receiver stores the sequence number of the data packets and the time at which they are received. However, during a session, if there is no data packet coming from the application layer, the multicast sources should send "keep-alive" packets [4]. These keep-alive packets are also sequenced in accordance with the data packets. For simplicity, we define the keep-alive packet to be a special kind of data packet. In the following text, the terminology "data packets" includes both the normal data packets and keep-alive packets.

Using the information, a multicast receiver node is able to estimate the interval between two successive data packets. Based on the estimate, it then sets up a timer to monitor the receipt of the next data packet. When this timer expires and no new data packet has been received, a counter "NumPktCscLost" (number of packet consecutively lost) is incremented by one. When this counter reaches an upper threshold (set to PKT_LOSS_MIN_THRES, which is equal to 3, initially), the RIAP is initiated. Before initiating the RIAP, the multicast receiver first deletes the information pertaining to its GN. Then it broadcasts an Attach Request (AReq) with limited TTL (a typical value is 2 hops).

FG nodes as well as multicast group members are allowed to respond to AReq provided that the values of their respective counters "NumPktCscLost" are less than 3. However, these nodes have different priorities to respond to the AReq. Higher priority means lesser backoff time before sending an Attach Acknowledgement (AAck): the FG nodes have a higher priority than the non-FG multicast receiver nodes; the FG nodes that are not in Anonymous Receive Mode (ARM) have a higher priority than those in ARM (ARM is explained in the next subsection). Before sending the AAck, a node waits for a random delay according to its priority. During the wait, if it does not receive any AAck meant for the node that initiated the AReq, the node sends out the AAck. The node sending the AAck is called Attach Point Node (APN).

The receiver that initiated the AReq responds to the first AAck it receives by sending an Attach Notification (ANtf) towards the APN. The next node towards the APN, after receiving the ANtf, becomes an FG node for this multicast group, and then forwards the ANtf to the APN. When the APN receives the ANtf, if it is in ARM, it leaves this mode.

After broadcasting an AReq, if no AAck is received, the initiating node increases the threshold for counting consecutive packet loss by PKT_LOSS_DELTA (a typical value is 3), and then restarts the NumPktCscLost counter. This procedure goes on until the threshold reaches a given maximum value PKT_LOSS_MAX_THRES. After this, an AReq is sent only after PKT_LOSS_MAX_THRES number of consecutive packet loss is detected, which is repeated until the multicast receiver receives a data packet again.

2.4 Anonymous Receive Mode (ARM)

RIAP may introduce extra nodes into the FG, causing more redundancy in packet delivery. We introduce the technique of Anonymous Receive Mode (ARM) to solve this problem. To implement this, in each data packet (normal data packet and keep-alive packet) header, we add three fields: last node ID, previous node ID and a Boolean variable bInARM. The "last node ID" contains the ID of the node that sends the current data packet. The "previous node ID" contains the ID of the node from which the *last node* (specified by the last node ID) receives this data packet. If bInARM is set to FALSE, it means that the node that sends this packet is not in

ARM; otherwise, it is in ARM and this packet is used as a Heartbeat. The field bInARM is initially set to FALSE.

The ARM utilizes the Passive Acknowledgement (PsvAck) technique. When an FG node receives a non-duplicate data packet, it moves the content of the data header field "last node ID" to the field "previous node ID", puts its own ID into the "last node ID" and then broadcasts this data packet. When an FG node receives a duplicate data packet and if in the "previous node ID" it finds its own ID, this implies that this FG node is responsible for the data receipt at one of its neighboring (downstream) nodes, then it stores the "last node ID" in its Rsp Table, and discards the packet. Otherwise, this duplicate data packet is simply discarded. A multicast receiver node that does not have any downstream node broadcasts a message Data Receive Acknowledgement (DRA) containing its upstream node's ID to provide an explicit acknowledgement for data packets received. After receiving the DRA, the upstream node would be responsible for the receiver.

Whenever an FG node, say A, forwards a data packet, it waits for a given duration for PsvAck from the nodes that it is responsible for, say B. If in the PsvAck from B, node A finds, in the "previous node ID", another node's ID, say C, that means the node B is now starting to receive data packets from node C, then node A deletes the entry for node B from its Rsp Table. If node A has forwarded MAX_TIMES_NO_PSV_ACK (the default value is 3) number of data packets, and it has not received any PsvAck from node B, A deletes the entry for node B from its Rsp Table. After every deletion, node A checks if its Rsp table is empty. If so, it enters Anonymous Receive Mode (i.e. it needs not forward the data packets frequently), and broadcasts a message "Enter ARM" within one hop. When node A's upstream node, which is responsible for node A's data receipt, receives the "Enter ARM", it lengthens the time out duration for receiving the PsvAck from node A for the following data packets. The FG node, A, which is in the ARM, will forward data packets from its upstream node less often, for example, one data packet every 2 seconds, which is commonly known as heartbeat. In each heartbeat (data packet or keep-alive packet), the bInARM is set to TRUE.

An FG node in ARM, after sending a heartbeat, checks if there are some new nodes in the neighborhood receiving data packets from it by listening to the PsvAck or DRAs. If so, this FG node leaves the ARM silently. It then forwards every following data packet it receives, setting the field bInARM in the data header to FALSE. When its upstream node receives the PsvAck whose data header field bInARM is set to FALSE, the time out duration for waiting for the PsvAck is shortened.

2.5 Data delivery

As mesh is created amongst GNs and sources, a moderate redundancy is there to guarantee data delivery. When data packets reach the GNs, the GNs are in charge of forwarding them within the local groups. We choose the same strategy as described in the ADMR for data delivery within the multicast tree, i.e, we do not restrict that the packet delivery must be on the specific branches. A node may receive data packets from any branch. This loosening of the data receipt improves reliability and simplifies the protocol implementation.

2.6 Membership management

Nodes are free to enter or leave the multicast group at any time. A node may enter the multicast group by broadcasting an Attach Request message or by responding to a GJQ by sending a GJR or LJR. A multicast receiver node may leave the multicast group silently. A multicast source node is also allowed to leave the multicast group at any time. If it wants to do so, it must flood a message "Source Leave" to the whole network. Every node should have a source table storing IDs of all the multicast source nodes in a multicast group. When the multicast group leader leaves, the multicast source, whose ID is the least larger than the former leader, becomes the multicast group leader.

3. PERFORMANCE EVALUATION

We evaluate the performance using GloMoSim [6]. In our simulation model, there are 50 mobile nodes in an area of 1000m x 1000m. From these nodes, one source and 19 receivers are randomly chosen. The subgroup range is limited to 3 hops. The heartbeat interval applied in ARM is 2 seconds. The mobility model is random-way-point. A node moves at a speed distributed uniformly between 1 m/sec to some maximum speed, which varies from one simulation to another. The pause time is 5 seconds. The radio transmission range is 250 meters. The propagation path-loss model is free space. IEEE 802.11 is used as the MAC layer protocol. Constant Bit Rate model is used at the application layer, and data packets, which are 512 bytes long, are sent with an interval of 250 milliseconds. Each simulation runs for 1000 seconds. All multicast members retain their membership throughout the simulation. The multicast source starts transmitting CBR traffic at the 5th second of the simulation time and stops transmitting at 995th second. For ODMRP, we use the default values used in GloMoSim library: 3 seconds for Join Query interval and 4 seconds for the lifetime of a FG node. In our simulations of GBMP, we use 60 seconds for the interval of GJQ, and 9 seconds for that of LJQ.

3.1 Metrics

The following metrics are used to compare the performance of ODMRP and GBMP:

- **Packet delivery ratio (PDR)**
- **Number of data packets transmitted per data packet delivered**: indicates the multicast efficiency of the protocol
- **Normalized packet overhead**: the total number of all the data and control packets transmitted divided by all the data packets delivered to the multicast receivers

- **Consecutive packet loss**: the distribution of the number of packets lost consecutively by all the receivers

3.2 Simulation results

Figure 2 demonstrates the packet delivery ratio (PDR). It shows that even at high speeds, GBMP can achieve higher packet delivery ratio. This is because, in GBMP, a receiver is free to attach to other FG nodes or receivers, which do not suffer from large consecutive packet loss. This increases the likelihood that the receivers can receive packets soon again.

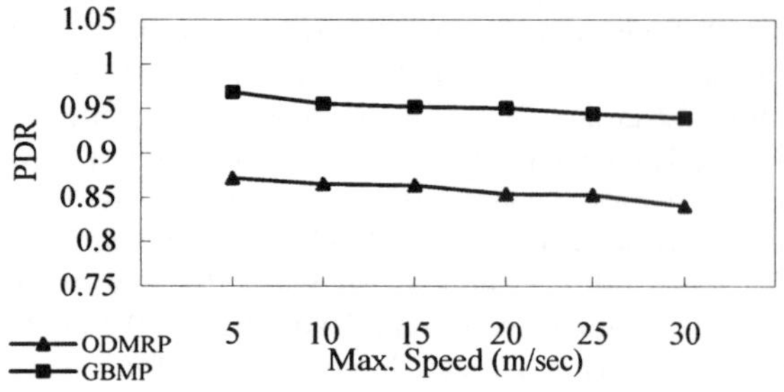

Figure 2. Packet Delivery Ratio

Figure 3 shows the transmission efficiency of the data packets. GBMP is transmitting less data packets to achieve higher packet delivery ratio because we use the ARM technique, by which the unnecessary data multicasting is reduced considerably.

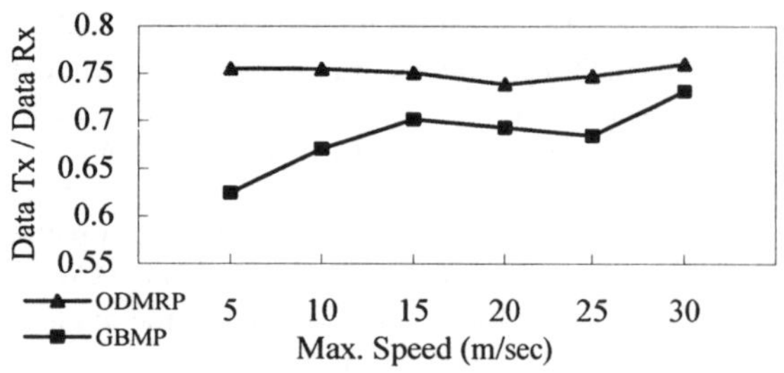

Figure 3. Data Packet TX per Data Packet Delivered

Normalized Packet Overhead demonstrated in Figure 4 shows that the cost of achieving higher packet delivery ratio in GBMP is quite low.

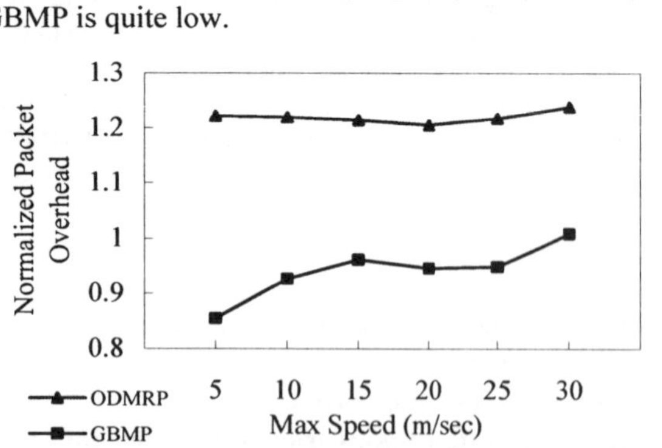

Figure 4. Normalized Packet Overhead

In some real-time applications, the occurrence of consecutive packet loss is a critical issue. As shown in Figure 5, when the maximum speed is 20 m/sec, three or less data packets are consecutively lost in most of the cases in GBMP due to hop-limited Receiver-Initiated Attach Procedure adopted; while in ODMRP, the number of consecutive packet loss is significant up to 12 packets.

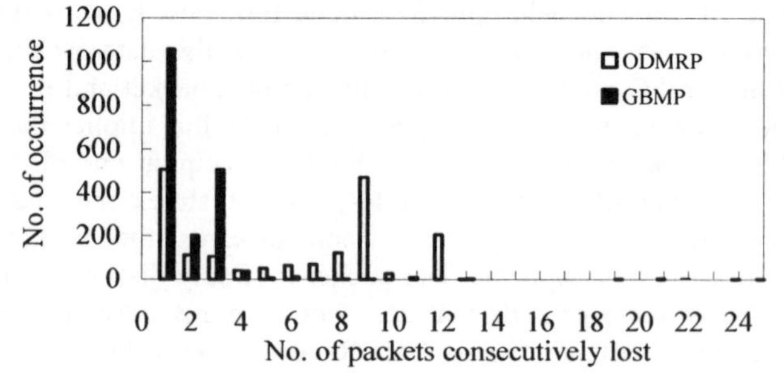

Figure 5. Occurrence of Consecutive Packet Loss (20m/s)

4. CONCLUSIONS

We have proposed an efficient multicast routing protocol, called Gateway-Based Multicast Protocol. A hierarchical topology is constructed for the data delivery. A local repair mechanism, Receiver-Initiated Attach Procedure, is used to ensure high data packet delivery ratio. Also, Anonymous Receive Mode is used to improve transmission efficiency. GBMP has low overheads and achieves high packet delivery ratio even in high mobility scenarios.

REFERENCES

[1] Y. Yi, S.-J. Lee, W. Su and M. Gerla, "On-Demand Multicast Routing Protocol (ODMRP) for Ad Hoc Networks", Internet-Draft, "draft-ietf-manet-odmrp-04.txt", Nov. 2002.

[2] B. An, S. Papavassiliou, "A Mobility-Based Hybrid Multicast Routing in Mobile Ad-hoc Wireless Networks", Proc. IEEE Military Communications Conference (MILCOM2001), pp. 316-320, Oct. 2001.

[3] C. Jaikaeo and C. –C, Shen, "Adaptive Backbone-Based Multicast for Ad hoc Networks", IEEE International Conference on Communications (ICC), New York City, April 28-May 2, 2002.

[4] J. G. Jetcheva and D. B. Johnson, "Adaptive Demand-Driven Multicast Routing in Multi-Hop Wireless Ad Hoc Networks", In Proceedings of 2001 ACM International Symposium on Mobile ad hoc networking & computing 2001, Long Beach, CA, USA

[5] C.-C. Chiang, M. Gerla and L. Zhang, "Forwarding Group Multicast Protocol (FGMP) for Multihop, Mobile Wireless Networks", ACM Baltzer Journal of Cluster Computing: Special Issue on Mobile Computing, page 187-196, 1998.

[6] Global Mobile Information Systems Simulation Library, UCLA, http://pcl.cs.ucla.edu/projects/glomosim

DIP3S: A DISTRIBUTED ROUTING ALGORITHM FOR POWER-CONSERVING BROADCASTING IN WIRELESS AD HOC NETWORKS

VIKAS P. VERMA and AMIT A. CHANDAK and HUNG Q. NGO

Computer Science and Engineering Department, 201 Bell Hall, State University of New York at Buffalo, Amherst, NY 14260, USA.

Given a set of ad hoc network nodes and a distinguished source node, the minimum-energy broadcasting problem is to assign appropriate power levels to all nodes in the network so that all nodes are reachable from the source, and that the total power usage is as small as possible. In the centralized version of the problem, one finds such power assignment given the entire network topology. In the distributed version, a power assignment is found by exchanging information between neighboring nodes.

In this paper, we proposed a new distributed algorithm called the *distributed incremental power with potential power saving* (**DIP3S**) to construct a power-effective broadcast tree. Simulations show that the new algorithm not only works better than known distributed algorithms, but also better than the centralized version of the broadcast incremental power (BIP) algorithm, and the newly proposed algorithm based on relative neighborhood graph (**RBOP**).

1 Introduction

Energy conservation in ad hoc networks is of paramount importance. In many typical ad hoc network applications such as mobile conferencing, emergency services, and battle field communications, multicasting and broadcasting are the most natural communication primitives. Multicasting is a fundamental generalization of both unicasting and broadcasting. On the other hand, many unicast and broadcast protocols, whether or not they are power-aware, rely on broadcasting to communicate routing data, to discover routes, etc. Consequently, the problem of devising energy-efficient broadcast algorithms and protocols are of tremendous interest in ad hoc wireless networking. This problem has received special attention from various researchers in recent years [2,3,4,5,11,6].

The power-aware broadcasting problem we are facing is assumed to be *source initiated,* Here, we assume omnidirectional antennas are used so that given a power assignment at a node, say u, all nodes within a certain radius of u are reachable from u in a single hop.

As in all previous works, we assume that v is reachable from u if the power assigned at u is at least d_{uv}^{α}, where d_{uv} is the Euclidean distance between u and v, and $\alpha \in [2,5]$ is the propagation loss exponent.

We would like to reduce the total power in the power assignment vector. This problem can be casted in terms of building a broadcast tree rooted at the source [2,3,4]. The trees are *directed rooted trees*, also called *arborescences*, where there is a directed edge from u to v if and only if the power assigned at node u is large enough to reach v in one transmission hop.

Although the literature on this problem is quite large, there is little done on the more desirable distributed solutions to this problem. This paper proposes a simple and effective distributed protocol to construct an energy efficient broadcast tree.

The rest of this paper is organized as follows. Section 2 reviews previous work. Section 3 gives the main idea of the protocol and section 4 gives simulation results which show that the algorithm works better than most known algorithms.

2 Related works

Wieselthier et al. [4] studied several heuristics and studied their performances by simulations. The algorithms under their study include: MST (minimum spanning tree), BIP (Broadcast Incremental Power), and SPT (Shortest Path Tree). Wan et al. [2] and Clementi et al. [8] gave upper and lower bounds on the performance ratios of several of these heuristics. Cagalj et al. [11] came up with an algorithm called EWMA (Embedded Wireless Multicast Advantage) which tries to modify an MST to form a better power assignment.

All the algorithms mentioned above are centralized. Recently, Julien Cartigny et al. [6] presented the related neighborhood graph (RNG) broadcast oriented protocol (RBOP), which is distributed.

Generalized versions of the problem have also been considered in [7,9,5].

3 DIP3S main ideas

The algorithm we propose in this paper is called the *distributed incremental power with potential power saving* (**DIP3S**) algorithm. DIP3S iteratively builds a broadcast tree starting from the tree consisting of only the root node.

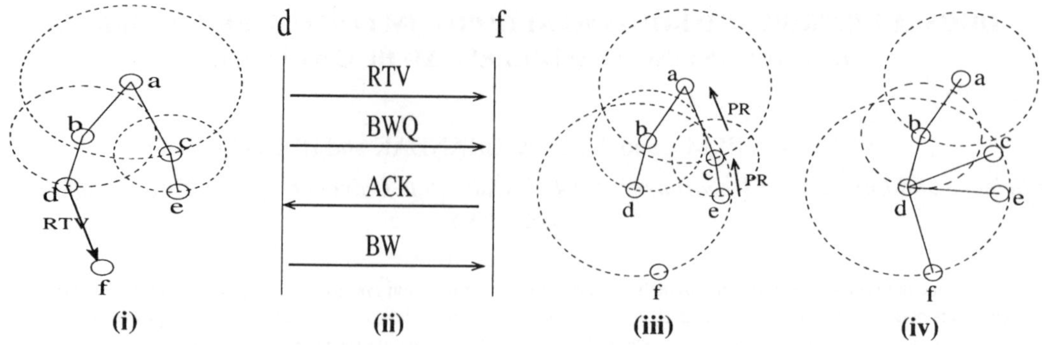

Figure 1. Message communication in a typical node expansion

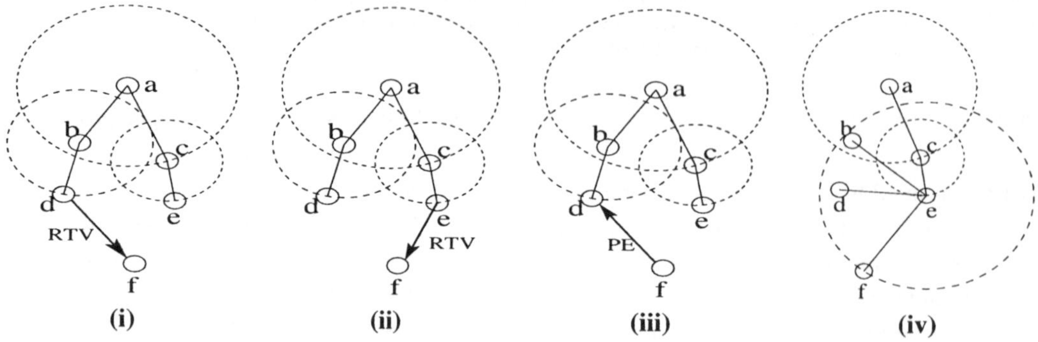

Figure 2. Preemption process in DIP3S

Each node maintains a list of power values required to cover any of its neighbor.

At each "iteration", every node independently computes an object function value (OFV), which is the amount of incremental power required to cover its nearest uncovered neighbor. (This step of the algorithm is similar to that of BIP.) Nodes in the current tree set a timer proportional to OFV^2. At the end of the timer, a node "expands" its power to cover at least one uncovered node. The newly covered node becomes the child of the covering node and repeats the process. Several mechanisms are implemented so that nodes get updated information on their newly covered neighbors, and re-compute their OFVs.

The messages required to complete a covering process are illustrated in Figure 1 and Figure 2. Figure 1(i) shows the state of a sample broadcast tree at a particular iteration. Node a is the source node. Nodes b and c are current children of a, according to the current power assignment. Similarly, d and e are b's and c's child, respectively. As soon as d gets covered by b it calculates its OFV to cover node f. Node d sends a *report timer value* (RTV) message to f, and sets a timer proportional to OFV^2. The RTV packet contains the value of the OFV. If any of a, b, c, e is capable of reaching f, they would do the same.

The message exchanges following the RTV message are shown in Figure 1(ii). After the timer expires d sends a *become white query* (BWQ) message to f. (Here, a node is *white* if it is in the current tree, and *black* otherwise.) Node f then decides to send an *acknowledgement* (ACK) or a *negative-acknowledgement* (NACK) to d. An ACK is sent to a node with lowest OFV, basically telling that node: "please cover me." A NACK is sent to all other nodes.

In the current example f sends an ACK. Upon the ACK reception node d sends a *become white* (BW) message to f, and increases its power range to cover f. It also covers nodes c and e within this range and hence makes them his children too. Nodes c and e, after overhearing this, switch their parent to d and send a *reduce power* (RP) to their previous parents. Node a reduces its power to just enough to cover b and node c reduces its power to zero. The final broadcast tree is shown in Figure 1(iv).

Although a timer of OFV^2 seems to be large, in reality we never have to wait for OFV^2 until a new node is covered. This is accomplished via the concept of *preemption*, as illustrated in Figure 2. Node f receives an

RTV message from e following the RTV message from d. It then compares the OFV values in both the RTV messages and selects node e to cover itself. It then sends a *preemptive* (PE) message to node d following which node d recalculates its OFV. Figure 2(iv) shows the final broadcast tree.

4 Simulation results

We evaluate our idea with extensive simulations. Four algorithms DIP3S, BIP, EWMA and RBOP are to be compared. A fifth algorithm, called IP3S, which is a generalized global version of DIP3S is also implemented. In IP3S, the OFVs are computed in a way which allows a few new nodes to be covered at the same time, and the potential power reduction is also taken in to account when computing OFVs.

The simulation are done using **YacSim** [10], an event driven parallel simulator for static nodes. The number of nodes varies from 10 to 100 with an increment of 10 at each step. All these nodes are uniformly distributed on a $10 * 10$ square. For each network instance, three different values of the propagation loss exponent α were tested: 2 and 4.

The maximum power level for nodes, p_{max}, is set to be enough to cover 25%, 50%, 75%, and 100% of the region. Here, to cover $x\%$ of the region means p_{max} can only be used to reach the nodes within a distance of $x\%$ of the maximum possible distance of any two nodes. This is the length of the diagonal of the 10×10 square, which is $10\sqrt{2}$. The source node is selected randomly among all the nodes.

For each simulation configuration - a combination of network size n, propagation loss exponent α, and node's power level p_{max} - we generate 100 random instances and run the above four algorithms on them.

The performance metric we adopt is the total power of the broadcast tree. To compare among different algorithms, we use the idea of *normalized power* [4].

We present the results for the cases when $\alpha = 2, 4$ and maximum power levels 75%, 50%, 25% as shown in Figures 3, 4, 5. The graphs follow similar trend when the max power level is set to 100%.

The simulations consistently show that, for all values of α, for varying p_{max} and network sizes, DIP3S works better than centralized BIP, RBOP, and almost as good as centralized EWMA. Moreover, (centralized) IP3S works better than all known centralized algorithms

References

1. C. E. Perkins, *Ad Hoc Networking*. New Jersey, USA: Pearson Education, Dec 2000.
2. P.-J. Wan, G. Calinescu, X.-Y. Li, and O. Frieder, "Minimum-energy broadcast routing in static ad hoc wireless networks," in *Proceedings of the Twentieth Annual Joint Conference of the IEEE Computer and Communications Societies (INFOCOM)*, vol. 2, pp. 1162–1171, IEEE, 2001.
3. J. E. Wieselthier, G. D. Nguyen, and A. Ephremides, "Algorithms for energy-efficient multicasting in static ad hoc wireless networks," *Mobile Networks and Applications*, vol. 6, no. 3, pp. 251–263, 2001.
4. J. E. Wieselthier, G. D. Nguyen, and A. Ephremides, "On the construction of energy-efficient broadcast and multicast trees in wireless networks," in *Proceedings of the Nineteenth Annual Joint Conference of the IEEE Computer and Communications Societies (INFOCOM)*, vol. 2, pp. 585–594, IEEE, 2000.
5. W. Liang, "Constructing minimum-energy broadcast trees in wireless ad hoc networks," in *Proceedings of the third ACM international symposium on Mobile ad hoc networking & computing (MOBIHOC)*, (Lausanne, Switzerland), pp. 112–122, ACM Press, 2002.
6. J. Cartigny, D. Simplot, and I. Stojmenovic, "Localized minimum-energy broadcasting in ad-hoc networks," in *Proceedings of the Twenty-Second Annual Joint Conference of the IEEE Computer and Communications Societies (INFOCOM)*, IEEE, 2003.
7. I. Caragiannis, C. Kaklamanis, and P. Kanellopoulos, "New results for energy-efficient broadcasting in wireless networks," in *Proceedings of the 13th International Symposium 2002 on Algorithms and Computation (ISAAC '02)*, pp. 332–343, Springer Verlag, Lecture Notes in Computer Science, Nov 2002.
8. A. E. F. Clementi, P. Crescenzi, P. Penna, G. Rossi, and P. Vocca, "On the complexity of computing minimum energy consumption broadcast subgraphs," in *STACS 2001 (Dresden)*, vol. 2010 of *Lecture Notes in Comput. Sci.*, pp. 121–131, Berlin: Springer, 2001.
9. S. Guha and S. Khuller, "Improved methods for approximating node weighted Steiner trees and connected dominating sets," *Inform. and Comput.*, vol. 150, no. 1, pp. 57–74, 1999.

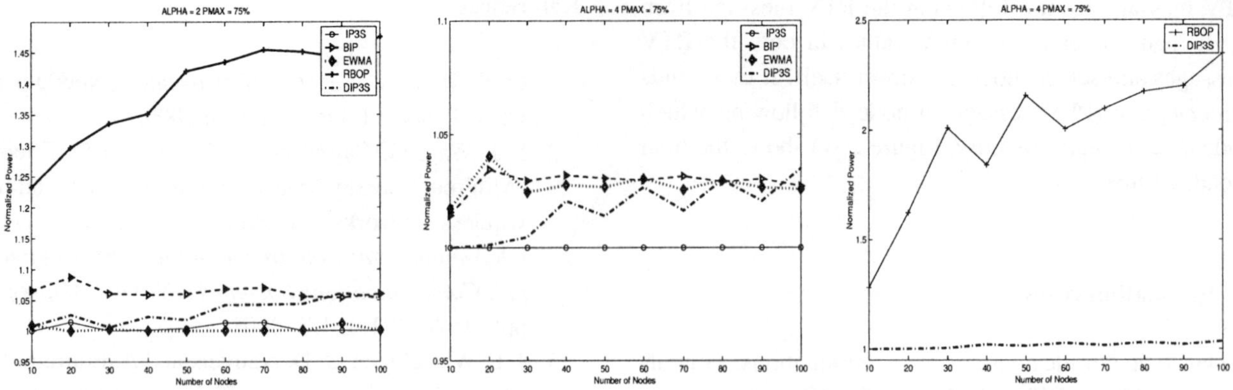

Figure 3. p_{max} is enough to cover 75% of the region

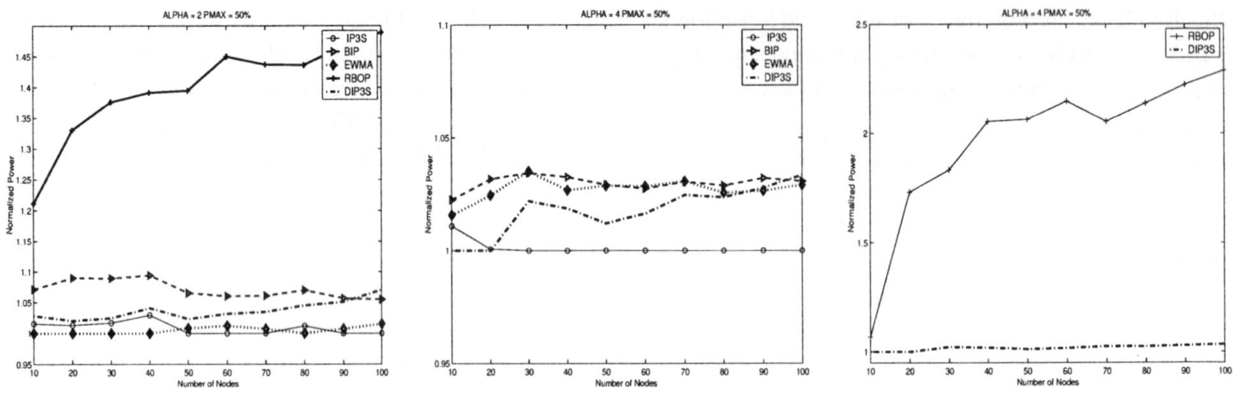

Figure 4. p_{max} is enough to cover 50% of the region

10. A. M. Law and W. D. Kelton, "Simulation Modeling and Analysis," McGraw Hill, 1991.
11. M. Cagalj, J.-P. Hubaux, and C. Enz, "Minimum-energy broadcast in all-wireless networks: Np-completeness and distribution issues," in *Proceedings of the 8th annual international conference on Mobile computing and networking (MOBICOM)*, (Atlanta, Georgia, USA), pp. 172–182, ACM Press, 2002.

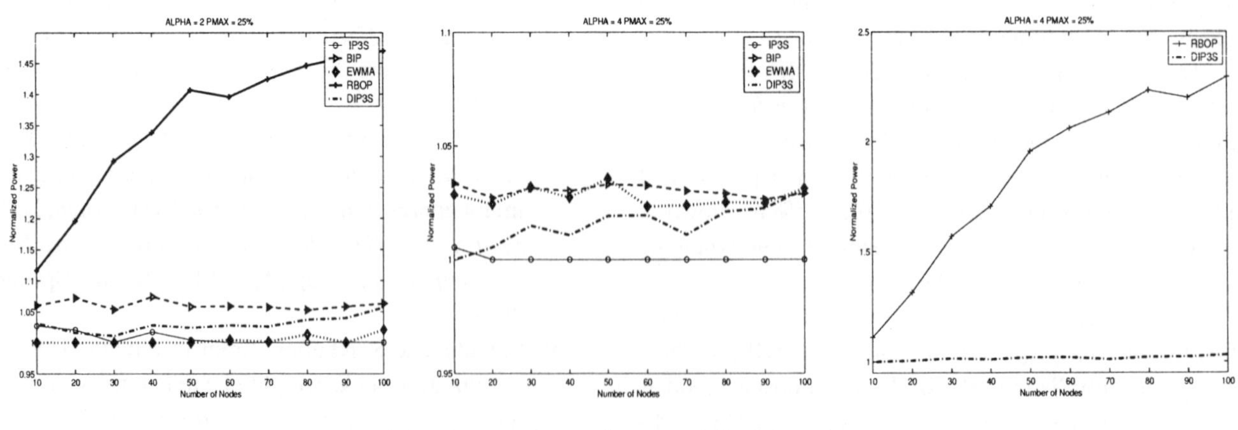

Figure 5. p_{max} is enough to cover 25% of the region

SESSION 5A:

SMALL AREA NETWORKS — PICONET

A Novel Traffic and Energy Aware Bridge Management Policy for Bluetooth Scatternets[†]

Ranganath Duggirala, Roy L. Ashok, Dharma P. Agrawal
OBR Center for Distributed and Mobile Computing, ECECS
University of Cincinnati – Cincinnati, Ohio 45221-0030, USA
Email: {duggirrt, royai, dpa}@ececs.uc.edu

Abstract– **Bluetooth is a short-range, packet radio system designed to connect various low power, resource constrained devices. Devices form autonomous ad hoc networks of up to 8 active devices, referred to as piconets. Piconets may be inter-connected by one or more bridge devices to form a scatternet. Bridges have an added responsibility of sharing their active time between the piconets that they are connected to. This could lead to a disproportionate energy usage pattern at the bridge nodes, as compared to other nodes in the piconet. When a bridge node dies, it could create a partition in the scatternet, leading to packet loss and delays. In this paper, we consider two related issues. 1. Enhance the lifetime of the scatternet by simply improving the lifetime of the bridge. 2. Extend the lifetime of the bridge by making a fair distribution of the role of being a bridge among different devices in the piconet. To achieve these goals, we propose a policy for managing bridge devices based on both prevalent traffic conditions and the energy available at devices using a weighted credit scheme.**

Keywords– Bluetooth, Bridge, Energy Efficiency, Networking, Management, Scatternet.

I. INTRODUCTION

Bluetooth [1] is a simple, low-cost, low-powered wireless packet radio system for short-range point-to-point and point-to-multipoint voice (Synchronous Connection Oriented or **SCO** links) and data (Asynchronous Connectionless or **ACL** Links) transfer. It is based on a master-slave polling scheme wherein a single master periodically polls all its slaves. A master can have at the most seven active slaves – forming an ad hoc network, referred to as a *piconet*. The master always polls the slave in an even slot and the slave *must* respond in the following odd slot. Large Ad hoc networks may be formed by inter-linking these individual piconets to form a *scatternet*.

Initially, Bluetooth enabled connectivity among disparate peripheral devices like cameras, scanners, printers, PDAs, etc. Today the focus has shifted to emerging areas like sensor networks [4, 5], where the low cost and low power nature of Bluetooth has become attractive. A sensor network might potentially consist of several piconets interconnected to form one big scatternet. Bluetooth was not really developed to handle networks of such a large scale and as such there are several issues that have cropped up, namely interference, scheduling, load distribution and energy efficiency. In this paper, we look at a way to extend the lifetime of a scatternet, by improving the energy utilization and the distribution of traffic at certain crucial connecting nodes.

The core Bluetooth specification clearly defines the setting up and operation of a piconet but only mentions the concept of a scatternet, and as such, issues concerning a scatternet are being actively researched. A scatternet is formed by the interconnection of two or more piconets, which share at least one device. This device has the responsibility of forwarding data packets from one piconet to the other. We henceforth refer to this device as a *bridge*. These nodes act like switches relaying data from one piconet to another. Consequently, they expend proportionally more energy than other devices and are likely to die out faster. This leads to problems like partitions in the scatternet when this node fails. It is therefore important to use these bridges in a manner that reduces the energy expended, and thus increase the scatternet lifetime.

There are two major issues. The first is the energy available at the designated bridge nodes. The second is the traffic flowing towards these bridges. For example, a single piconet (say P_1) is connected to two other piconets P_2 and P_3, via bridges B_2 and B_3 respectively. Suppose B_3 is lower in energy than B_2 and the traffic towards piconet P_3 is also very light, while the traffic towards B_2-P_2 is very high. If we had considered only energy, we would have assigned an additional bridge to the B_3-P_3 link, in order to prevent it from breaking down completely. However, considering the traffic conditions, it is more prudent to assign an extra bridge to the B_2-P_2 link. In essence, this is what our bridge management policy tackles – how to assign an extra bridge to an inter-piconet connection based on the prevailing traffic and energy conditions. All masters in the scatternet execute the policy. It co-relates the traffic towards a piconet, the energy of the bridge nodes and the number of bridges assigned to a particular inter-piconet connection, using a weighted credit scheme. It constantly monitors all the bridges and dynamically reassigns them to prevent over-usage of any single device.

The rest of the paper is organized as follows. Section II describes the Bluetooth technology in depth and the motivation for the paper. Related work is described in section III. Our protocol *T*raffic and *E*nergy *A*ware *Bri*dge *M*anagement (TEABriM) is discussed Section IV, followed by extensive simulation in Section V and conclusion in VI.

II. BLUETOOTH OVERVIEW AND MOTIVATION

The core protocol specifications and architectures are defined and published by the Bluetooth SIG [1]. In this section, a brief overview of Bluetooth is presented. Bluetooth operates in the ISM band between 2.402 GHz and 2.483 GHz in the USA and most European countries. It serves as the baseline for the IEEE 802.15.1 [3]. A total of 79 RF channels of 1

† This work has been supported by the Ohio Board of Regents Doctoral Enhancement Funds and the National Science Foundation under grant CCR-0113361

MHz width are defined, where the raw data rate is 1 Mbit/s. A Time Division Duplex (TDD) technique divides the channel into 625μs slots and, with a 1Mbit/s symbol rate; a slot can carry up to 625 bits. Transmission occurs in packets, where DHx (without FEC) and DMx (with FEC) packets are usually employed for data traffic, while HVx packets for voice applications. Here, *x* represents the number of slots and can be equal to 1, 3 or 5.

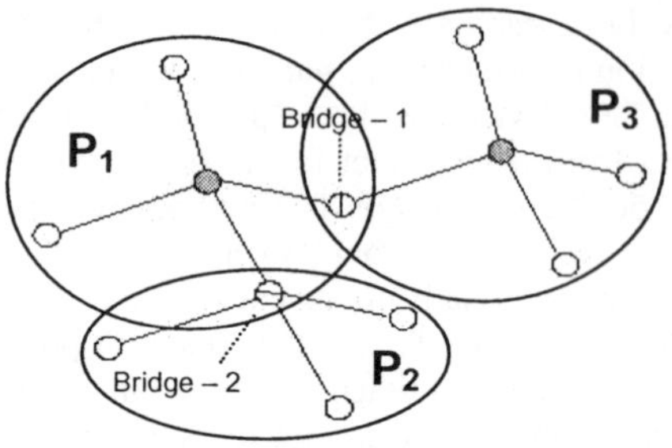

Figure 1: Example Scatternet Setup undertaken by TEABriM.

Multiple piconets are linked together to form a scatternet (Figure 1). Two piconets communicate with each other by time-sharing one or more *bridge* devices. These bridges may either be slaves in both piconets or a master in one and a slave in another; but cannot be master in both. Recent research [6] has shown that master-slave configurations result in reduced throughput. We therefore consider bridges that operate only in the slave-slave mode.

The bridge bears the responsibility of a switch, buffering incoming data packets, switching to another piconet and relaying the buffered packets to a new master. They achieve this by making use of any of the low power modes (HOLD, SNIFF or PARK). For example, assume that a given node, say node 1, is shared as a bridge between piconets 1 and 2. The way bridging works is, node 1 goes into say SNIFF mode for *t* slots in piconet 1, during which time it is active in piconet 2. This way, while active in one particular piconet, the bridge buffers data packets intended for the next piconet, and delivers them when it actually switches to the next piconet. This means that, they are always transmitting, receiving or switching between piconets. While this might work for low to even medium traffic conditions, it essentially drains the bridge energy and at high traffic loads, the bridge may be overwhelmed, suffering buffer overflows, packet drops and increased end-to-end delay. A bridge node that is drained of power will die, disrupting inter-piconet traffic (either temporarily or permanently) causing a heavy loss of packets. While this may be inevitable, we must ensure that the device (and hence the scatternet) is alive for as long as possible. Therefore, we believe that, in order to achieve energy efficiency in a Bluetooth scatternet, an effective policy for *Bridge Management* is needed [13]. The goal of the policy is to prolong the life of the scatternet by increasing the lifetime of the bridges, and to share the responsibility of being a bridge among capable devices. The list of capable devices is dependent on a number of factors like resource constraints, local security policies, etc. and may be formed jointly by scheduling mechanism and the bridge management policy. The difference between a scheduling algorithm and bridge management is that the former decides the time and piconet in which a bridge has to be available, while the latter decides which nodes to allocate (at the points in time specified by the scheduling algorithm) based on energy and user specific policies. A single piconet may be connected to more than one other piconet (Figure 1). Each inter-piconet connection may have different traffic loads, and the bridges used in these connections will certainly have different energy/resource constraints. In order to balance out the load, certain high traffic load or low energy links may warrant an extra bridge.

Most Scatternet scheduling algorithms aim at fair utilization of bandwidth and lower packet delays, assume only one bridge device between any two piconets. In our simulations (Figure 2) we have clearly shown that adding a second bridge is advantageous in that, it reduces packet drops and latencies, and drastically decreases the energy usage at any one device, simply because the load is shared. Therefore the Bridge Management Policy developed in this paper dynamically assigns a second bridge to inter-piconet connections that require them. It will also keep track of energy usage and remove critical devices, replacing them with more able slaves. The goal of the Bridge Management Policy is to evolve a comprehensive mechanism of assigning, removing and re-assigning devices to the different inter-piconet connections, by making an assessment of traffic and energy conditions at the bridge nodes. Of course, this policy works only when there are *some* bridge devices to manage. This means that each master will maintain an updated list of slaves that are capable of forming bridges with the various piconets that it may be connected to. This list is referred to as the Prospective Bridges Set (PBS). Each piconet in the scatternet will run this policy and to avoid control message overhead, the energy values of every device are piggybacked together with the existing Bluetooth ACK packet. Using this policy, we hope to introduce the concepts of fairness and shared responsibility among capable devices, thereby increasing the lifetime of the scatternet.

III. RELATED WORK

Some papers aim at improving the Bluetooth performance which has the desired side effect of reducing energy consumption. To the best of our knowledge there is no work dealing specifically with bridge management. Bluetooth has certain in-built energy savings mechanisms like HOLD, SNIFF and PARK [1], but these are pretty much useless for bridges, which actually use them as a way of being inactive in a particular piconet while being active in another. In [7], power is saved by scheduling the occurrence of the SNIFF slots and the length of each occurrence. [8] discusses power control by using cost metrics associated with routing and switching the roles of the master and slave. In our current work, we concentrate on sharing the bridge responsibility and

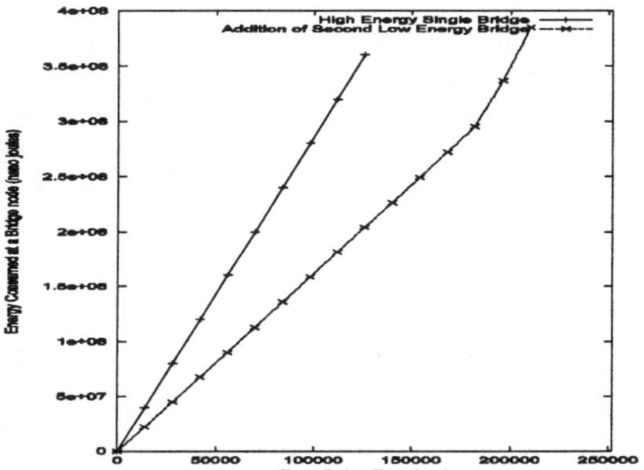

Figure 2: Performance comparison of one bridge and two bridge cases

responding to energy changes based on a node's capability, the relative power levels of other prospective bridge slaves and on traffic conditions. Our goal is to increase the scatternet lifetime, and we achieve this by extending the lifetime of the bridge. It is a flexible policy that can be used with any scheduling scheme [7, 9, 10] or routing mechanism [8]. It is unique in that the approach itself is based on energy management, directly dependent on relative energy levels of various nodes and specifically targeted at improving the lifetime of bridge nodes, as opposed to previous work which targeted scheduling or routing. The focus in our approaches is the bridges themselves.

IV. *TEABriM* : *T*raffic and *E*nergy *A*ware *Bri*dge *M*anagement Policy

TEABriM works on a weighted credit scheme that combines the traffic and the energy usage at an inter-piconet connection. Each inter-piconet connection has a variable credit value that is evaluated in every *transaction*. This transaction is defined as the period for which a node buffers data packets in one piconet, switches to the other, relays the buffered packets and then switches back to the piconet. TEABriM is executed at the master every time a transaction occurs. The occurrence of the transaction in time is left to the scatternet scheduling algorithm. Each master maintains the Prospective Bridge Set (PBS). It uses the slaves in this set to assign or remove bridges from an inter-piconet connection. For the purpose of illustrating the TEABriM algorithm, we will use the following scenario (Figure 1). There are three piconets, P_1, P_2, and P_3 where P_1 has a connection to both P_2 and P_3. These three piconets are part of a large scatternet, whose specific connectivity is not important. Initially, each of the connections P_1-P_2 and P_1-P_3, have only one bridge. Every device in the PBS fixes a Critical Energy Value (CEV), beyond which it cannot operate as a bridge and must be removed from the PBS. A Credit Threshold Value (CTV_{ij}) and Credit Accumulated Value (CAV_{ij}) is associated with each inter-piconet connection. In every transaction (defined above), the master first checks if any of the existing bridges have reached their CEV, and removes those and replaces them with members from the PBS (if there are any left). It then checks if the credits accumulated (CAV_{ij}) for that inter-piconet connection (P_i - P_j) have exceeded the threshold set for it (CTV_{ij}). If it has not, the CTV_{ij} then it implies that, that particular inter-piconet connection is able to handle its load with its current energy settings and does not need any help. If it does exceed the CTV_{ij}, then it will assign a bridge from the PBS to that connection and reset the value of the CAV_{ij} to zero. In the transactions following this, there is a very high probability that the bridge switching transactions will occur in a disjoint manner and that it may not always be possible to provide them with the extra bridge in a timely manner. For example, if the P_1-P_2 (having a single bridge) switch transaction finishes first, and the CAV_{12} has breached the CTV_{12} we would have to assign it a bridge. But the bridge to be assigned is with a transaction that has not yet finished. In such a case we cannot do anything but allow P_1-P_2 to continue operating on a single bridge for another transaction, the length of which is a dynamically adjusted shorter period to coincide with the switching back of P_1-P_3. This allows a smaller period of high load on P_1-P_2 even though it has already been determined that it needs an extra bridge to operate effectively. When P_1-P_3 finishes its transaction, the CAV_{13} is once again checked to see if it has exceeded the CTV_{13} in which case the link continues with two bridges. It may be argued that this procedure is not fair to P_1-P_2. Fairness here is not related to how often a particular inter-piconet connection is given a second bridge, but is a direct consequence of the prevailing traffic conditions. In this case the CTV_{12} has been breached despite the fact that it had two bridges, which shows that there is an extremely high load on that connection and it is only fair to continue a two bridge operational scenario. If however, the CAV_{12} has not exceeded the CTV_{12} then the bridge is removed from that connection and reserved on a priority basis for P_1-P_2. In a similar fashion this may be extended to cover more than two inter-piconet connections. The entire scheme depends on how the credits for the CAV_{ij} are calculated. This is briefly outlined below. The CAV has two components, *Traffic* and *Energy*. The method for calculating the individual units is specific to application scenarios. One recommended way of calculating this (used in our simulations) would be to develop a simple unit of *traffic* and *energy* and increment those fields each time the unit conditions are satisfied. For example, the *traffic unit* could be a single packet in the case of a single bridge inter-piconet connection and two packets in the case of a double bridge connection. This means that in the single bridge case, every time a packet arrives at the bridge the traffic credit is incremented by one, and in the double bridge case only when both bridges have received a packet each (hence two for that connection). This has an advantage in that it forces the *Traffic* credit component to depend on the number of bridges handling the traffic as well, giving a more realistic credit value. Each device can set the unit for the energy component since it will know what resources are at its disposal. For example, each energy unit could be 1% of the remaining energy. So in every transaction, whenever the energy consumed is equal to 1% of the remaining energy, the *energy* component is incremented by 1. In the dual bridge case, the *energy* part simply is the sum of the energy parts of the individual bridges. At the end of the transaction, the accumulated energy and traffic credits are added to form the net Credit Accumulated Value (CAV_{ij}) for that inter-piconet connection in a particular transaction. It should be noted that the units of energy or traffic may also be related to the software configurations and applications that the bridge devices are running. By keeping the credit value local to a transaction, varying traffic loads are easily represented.

V. PERFORMANCE ANALYSIS

Performance of our scheme has been evaluated using a Bluetooth simulator developed in Java using the *simJava* discrete event simulation package [11]. The baseband, link manager and a simple round robin polling scheme has been implemented. Mobility i.e. nodes joining and leaving the piconet during the simulation has not been modeled. In [12] it has been shown that the ratio of reception to transmission is usually 1:1.4 and the energy expended for transmitting 1 bit is 115 nJ. We have reused these results in our simulations. The results presented reflect control conditions where the bridges are simply either receiving or transmitting data and are not involved in other tasks that drains energy like printing, scanning, taking pictures, etc.

Figure 2 gives an accurate comparison of the way the energy is depleted when a single high energy bridge connects two piconets and in the presence of a second very low energy bridge to share the load. It actually outperforms by transmitting about 66% more packets, with about 43% reduction in energy usage.

The performance of TEABriM has been tested from the point of view of a piconet (P-1) in Figure 1. A Constant Bit Rate (CBR) traffic of 57.6 Kbps between piconets 1 and 2 and 138.2 Kbps between piconets 1 and 3 have been used to evaluate the scheme. Figure 3 shows the energy consumption patterns of TEABriM. The graph shows curves for four different slaves whose total energy progressively goes down from high to very low. The energy is "spread out" with the higher power devices doing a proportionately greater amount

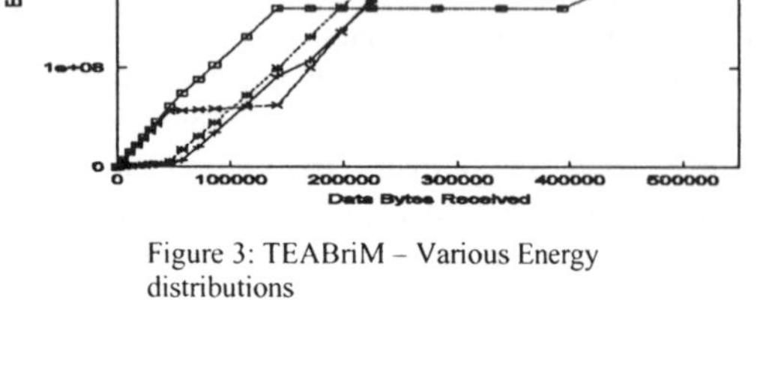

Figure 3: TEABriM – Various Energy distributions

of work by taking on the role of a bridge more often. As a result, the lifetime of the bridges and hence of the scatternet is also substantially increased. As we can see, the number of packets handled by the bridges for the same total energy has increased by almost 90%. Intuitively, when this is applied on a global scale in a big scatternet, inter-piconet connections are also extended by some bridge or the other and paths are somehow guaranteed.

VI. CONCLUSION AND FUTURE WORK

Bridges are an important part of the scatternet as they provide connectivity between the piconets and therefore end up with a greater amount of work than the other slaves. Saving energy at the bridge nodes will increase the lifetime of these networks preventing partitions in the scatternet. In this paper we have specifically looked at the concept of bridge management. Our policy co-relates the traffic towards a piconet, the energy of the bridge nodes and the number of bridges assigned to a particular inter-piconet connection.

Since nothing has been assumed about the scheduling algorithm, TEABriM is flexible enough to be used with any of them, thereby adapting to various scenarios. We have shown that our scheme is extremely efficient and effective in reducing the energy consumed at the bridge nodes and increasing the overall throughput. Most importantly, they do not require any modifications to the core Bluetooth protocol specifications. Future work would include intelligent bridge management by looking at the scatternet as a whole or in parts while assigning bridges. Even though the scheduling algorithm is independent of bridge management, it may be possible to extract performance improvements if the two can be combined in a meaningful way. We are currently looking into these issues.

REFERENCES

[1] Bluetooth SIG, http://www.bluetooth.com.

[2] D. Agrawal and Q-A. Zeng, "Introduction to Wireless and Mobile Systems," Brooks/Cole Publishing, 436 pages, ISBN 0534-40851-6, 2003.

[3] C. Bisdikian, "An Overview of the Bluetooth Wireless Technology", *IEEE Communications Magazine*, December 2001, pp. 86-94.

[4] O. Kasten and M. Langheinrich, "First Experience with Bluetooth in the smart-its Distributed Sensor Network," In *Proc. of the Workshop in Ubiquitous Computing and Communications (PACT)*, October 2001.

[5] F. Siegemund and M. Rohs, "Rendezvous Layer Protocols for Bluetooth-enabled Smart Devices," in *Proc. of International Conference on Architecture of Computing Systems*, April 2002.

[6] G. Zussman and A. Segall, "Capacity Assignment in Bluetooth Scatternets – Optimal and Heuristic Algorithms," ACM/Kluwer Mobile Networks and Applications (MONET), Special Issue on Advances in Research of WPAN and Bluetooth Enabled Networks, 2003, In press.

[7] Ting-Yu Lin and Yu-Chee Tseng, "An Adaptive Sniff Scheduling Scheme for Power Saving in Bluetooth," *IEEE Wireless Communications*, December 2002.

[8] Balakrishna J. Prabhu and A. Chockalingam, "A Routing Protocol and Energy Efficient Techniques in Bluetooth Scatternets," *IEEE ICC* 2002.

[9] S. Baatz, M. Frank, C. Kuhl, P. Martini and C. Scholtz, "Bluetooth Scatternets:An enhanced Adaptive Scheduling Scheme," *IEEE Infocom* '02.

[10] N. Johansson, F. Alriksson and U. Jonsson, "JUMP-mode a Dynamic window-based Scheduling Framework for Bluetooth Scatternets," *ACM MobiHoc* 2001.

[11] Simjava, http://www.dcs.ed.ac.uk/home/simjava/doc/index-all.html

[12] Mark Stemm and Randy H Katz, "Measuring and reducing energy consumption in handheld devices," *IEICE Transactions on Communications*, Vol E8-B, No.8, Aug. 1999.

[13] Ranganath Duggirala, Roy L. Ashok, Dharma P. Agrawal "Energy Efficient Bridge Management Policies for Inter-Piconet Communication in Bluetooth Scatternets", Proceedings of IEEE Vehicular Technology Conference, October 6-9, 2003, Orlando, Florida, USA, in press.

Implementing a Self-organizing Wireless Sensor Network: Experiences and Challenges

Joo Ghee Lim, Kim Loon Chee, Hee Boon Leow, Yew Kwan Chong, PK Sivaprasad and SV Rao
Institute for Infocomm Research, Singapore Science Park II, Singapore
Fax: +65 67795441
{limjg, cheekl, leowhb, chongyk, pksiva, raosv}@i2r.a-star.edu.sg

Abstract: Wireless sensor networks have specifically unique characteristics with regards to energy, memory, computational and complexity limitations. This paper shares the experiences gained in designing and implementing a wireless sensor network when we take these considerations into account. We also highlight some challenges faced in the implementation of the network. The hardware that is developed provides a realistic and flexible platform for future implementation and testing of relevant higher layer schemes.

Key Words: Sensor networks, design, implementation

1. INTRODUCTION

In recent years, there has been great interest in the development of small, embedded devices that organize and communicate among themselves in a distributed manner, forming wireless sensor networks. Applications include military surveillance, inventory control and environmental monitoring. Much of the research has been focused on investigating the performance of the communication schemes, e.g. routing, data dissemination and positioning, at the analytical modeling and simulation level.

On the physical implementation front, some testbeds have been set up to evaluate ad-hoc and sensor networks [1-4]. Apart from [4], most of these testbeds implement ad-hoc networking using notebooks and PDAs with IEEE 802.11b as the underlying layer. The rationale is to give the researchers freedom in testing out networking protocols over a stable Medium Access Control (MAC) and physical layer. However, these nodes may not accurately reflect the true sensor network environment (e.g. radio propagation, range and channel conditions). In addition, new MAC protocols could not be tested on these testbeds.

This paper presents the work done in designing and implementing a wireless sensor network. The merits of this paper are the discussions of the challenges met and experiences gained during the process of developing the different aspects of the sensor network. We will discuss these challenges and experiences with regards to the different layers of our implementation. Sections 2, 3 and 4 describe the physical, MAC and routing layer respectively. Section 5 provides an overview of the operating system that binds these layers together. We discuss the overall system implementation in Section 6 before concluding the paper.

2. PHYSICAL LAYER

A sensor node, in its most basic form, consists of a microcontroller connected to a RF transceiver. The microcontroller serves as the "brain" of the node. It contains the decision logic that processes the information the node receives, both from the external environment as well as other nodes. The transceiver is the "ear and mouth" of the node with which it is able to communicate with other nodes. Often, a sensory module (the "eye") is also present, which captures the information from the environment (e.g. image, temperature). Fig. 1 shows a block diagram of the hardware.

Power consumption and flexibility were key considerations when deciding the hardware. The need for low power consumption within the nodes is a well-known fact in sensor network research. Flexibility of hardware configuration is needed because, as a testbed, it would be desirable for the nodes to be easily configured to suit different schemes, e.g. communication range, sleeping modes, etc.

The Texas Instruments' MSP430F149 microcontroller [5] was chosen because of the following features: low power consumption, multiple power-saving modes and a 16-bit architecture. We used the Chipcon CC1000 [6] chipset for the transceiver because it has adjustable transmission power and channel frequency, and has RSSI sensing capability. These hardware features, besides being low powered, provide the flexibility needed to test the variety of communication protocols.

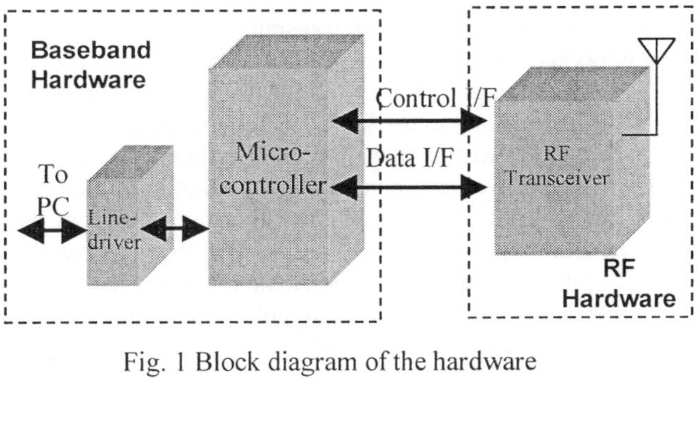

Fig. 1 Block diagram of the hardware

Besides the microcontroller and transceiver, a RS-232 line-driver was also included in each node to make it possible to connect the node to an external personal computer (PC) for testing and monitoring purposes. The MAX 3317 line-driver from Maxim [7] was selected due to its low power and operating voltage.

One of the challenges faced when we implemented the hardware was the variation in the nodes' performance due to deviations in the components' tolerance values. The deviations caused disparities among different nodes in terms of transmission range and receiver sensitivity. Careful calibration process was put in place where we fine-tuned the transmission power of each node to give a consistent performance among the nodes.

In addition, the printed circuit board (PCB) miniaturization also had an effect on the monopole antenna's omni-directional performance. Careful consideration during the PCB design phase helped to reduce this effect and we were able to achieve an antenna far-field response that approached omni-directional performance.

3. MAC LAYER

The MAC layer manages the communication channels to avoid collisions and errors. In order to simplify node operations and reduce power consumption, our MAC layer uses a common message structure, shared between the routing and application layers.

We use a TDMA-like mechanism to schedule transmission as this allows the node to switch off its radio transceiver during idle time slots, reducing the power consumption. However, we need to find a computationally inexpensive way of synchronizing the timings among the nodes. This is to eliminate errors caused by the drifting of times slots between nodes. We introduce a novel way of using *relative time slot referencing* within individual nodes that is able to provide synchronization (a requirement in most TDMA systems) but does away with the need to maintain a global timing.

Following is a brief discussion on how *relative time slot referencing* works – each node transmits messages at regular time intervals (frames). Each frame is further sub-divided into smaller intervals (slots). Within each frame, the node will send its message at the same slot and all other slots will take reference from this transmission slot (known as slot 0). Synchronization between a node and its neighbor is done by monitoring the slot (relative to slot 0) where messages are received from this neighbor. Any drift in this receive slot will be updated in the node's neighbor records. If this slot drifts too near to the transmission slot, or another occupied slot, a recovery action will be activated.

This relative time slot referencing mechanism provides a simple and distributed way of maintaining time synchronization among all the nodes within the network, thereby reducing the computational cost significantly.

Maintaining a record of the neighbors' characteristics (neighbor identity, relative transmission slot, etc) also provides us with a simple way of mobility detection at the software level. Self-mobility detection generally requires the presence of complex peripherals, e.g. video, motion sensors, or algorithms to deduce that the node has indeed moved. We implement a simple mobility detection mechanism by monitoring the change in the number of new and missing neighbor nodes over a time period. We deduce that a node has moved from its original position if the number of new and missing neighbors over a given time goes beyond a threshold value.

4. ROUTING LAYER

Above the MAC layer sits the routing layer. The function of the routing protocol is to discover a route from one node to another. Besides power and memory constraints, a notable characteristic of wireless sensor networks is that the identities and locations of destination nodes are usually known[1]. In our implementation, we have only a single destination node that serves as a host system, where all other nodes will forward their information. Our routing scheme makes use of this fact to simplify the routing operations.

In general, routing schemes in ad hoc-like networks can be classified into proactive and reactive routing. Briefly, proactive routing requires routes to be always available. Route maintenance is activated every time a link fails. In reactive routing, a node only request for a route to a destination when it has something to send. A key challenge in designing the routing scheme is balancing the tradeoff between responsiveness and computational cost of route maintenance.

Proactive routing is often frowned upon because it is deemed too costly to exchange routing information whenever a node moves or a communication channel fails. However, this may not always be the case in sensor networks. Unlike ad-hoc networks, sensor nodes in certain applications hardly move. Hence, route changes due to node movements are greatly reduced. Moreover, there only exist a finite number of destinations where all information will be sent. It makes sense to establish and maintain a route to each of these destinations. Thus, proactive routing is used. We further

[1] Most applications involve a large number of sensor nodes that forward information to a small, finite number of host systems that record and keep track of the information.

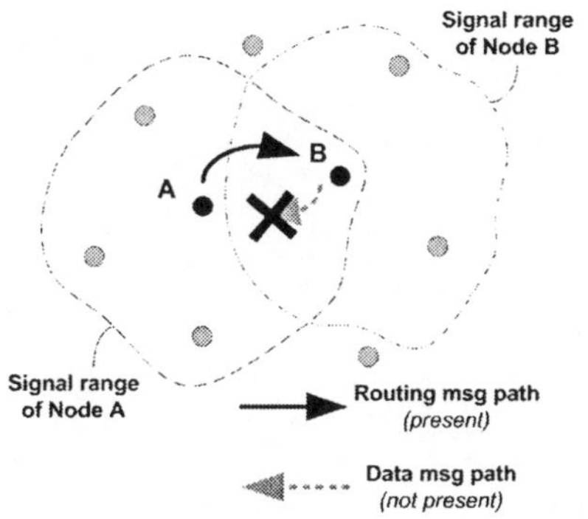

Fig. 2 Asymmetrical communication between nodes A and B

reduce the effects of channel failures by maintaining multiple routing paths to the destinations.

Another challenge we faced involves node behavior under RF boundary conditions. When a node lies near the limit of another node's RF range, the communication may become asymmetrical. This is due either to the weak, inconsistent signal strength, or the non-uniformity of the signal coverage. Routing information messages travel in the opposite direction to actual data messages that eventually use the route. Near the boundary of the node's signal range, the presence of a routing message path may not always mean that the data path (actual route) is available.

Figure 2 gives an illustration of such an example. When node A broadcasts a routing information message, node B receives it and sets up a path to the destination through node A. However, any data message sent by node B via A would be lost since node A is not within the signal range of B. One of the ways to reduce this effect is to provide for redundancy in data delivery by maintaining multiple paths and sending the same message over different paths, which will lead to additional overheads. Thus, a balance has to be struck between energy efficiency and reliability of information delivery.

5. OPERATING SYSTEM

Besides the hardware and communication layers, another important feature of embedded devices is the operating system (OS), which schedules tasks and manages the resources (e.g. memory, peripherals, registers) of the device. The aim is to design a simple OS that is able to manage all the specific requirements of the resource-limited sensor node. Besides the TinyOS [4], there are few OSs that meet our specific requirements. For flexibility and greater source-code control, we decided to develop a simple OS to support all the requirements of the nodes.

The OS we developed is based on an Event-Driven Scheduler (EDS). It provides a basic framework to integrate the different node resources without incurring high processing and memory overhead. In general, each module (MAC, routing, sensory, etc.) is realized as a task. Each task has its own associated event list that is managed as a linked-list by an *Event Manager*. This keeps the OS modular, giving us the flexibility to add in new modules (e.g. other sensory functions) in future. A *Memory Manager* manages a pool of memory buffers. If data or messages are needed to be passed along with any event, the Event Manager will request for the buffer from the pool. When the memory is no longer needed, it will be returned to the pool. This central management of memory prevents any unauthorized access of memory or accidental memory corruption.

A *Controller Resource Manager* provides a means of the microcontroller's functionality – the RF module interface, UART and a timer mechanism. Finally, a *Scheduler* allocates the CPU time to the different tasks to service their events in a round-robin manner. When no event is present, the Scheduler will invoke the microcontroller to enter low-power mode to conserve energy.

In the design of the EDS, we concluded that a suitable OS is important in the whole operation of a sensor node. Key features of such an OS include energy-efficiency, simple tasks management, flexibility, as well as fair and robust management of resources.

6. OVERALL SYSTEM IMPLEMENTATION

Finally, a sensor network is made up of a collection of nodes working together within a system that processes the information exchanged. The overall system will manage, process the information gathered by the nodes and display it in a clear manner to any user.

Our implementation of the embedded device is flexible enough to cater to many different system applications and test scenarios. With our design, we created 40 nodes each calibrated for a range of 5m and running on 2 AAA-sized alkaline batteries. Each node has an average power consumption of approximately 20mA. Fig. 3 shows a snapshot of one of these nodes.

Using these nodes, we set up a sensor network testbed with the specific task of implementing a distributed location identification system. In this system, nodes within the network region exchange information with one another to discover some location-related information. This information is forwarded (making use of the routing scheme) from each node to a single destination, which consists of a host system that processes, stores and displays this information in a relevant form. This allows any user to identify the location of the nodes within the network region at any give time. Following are the various physical components of the system:

Fig. 3 Snapshot of a sensor node

Fig. 4 Snapshot of a testbed implementation

- **Reference nodes** – These are implementations of the nodes with fixed and known locations.
- **Mobile nodes** – These are implementations of the nodes that are free to move and will predict its location within the network, based upon its distance (in terms of hop counts) from the reference nodes.
- **Reader node** – The reader is identical in physical implementation to all other nodes, except that it is connected to a host system via a RS-232 cable. The location information from all the mobile nodes is eventually forwarded to the reader for processing.
- **Host System** – The host system consists of a notebook that runs an application program that collects all the location information for processing. It also includes a graphical user interface to display the location of the nodes within the network region.

Fig. 4 shows a snapshot of our system implementation. Our platform provides a useful testbed for future testing of network protocols as well as system implementations.

CONCLUSION

This paper presents the experiences gained and challenges faced when designing and implementing a wireless sensor network. The key characteristics of the sensor network are simplicity and energy-efficiency. We believe that these considerations must be present in all levels of the design and implementation, from the node hardware to the network protocols and finally, the overall system.

Physical conditions influence greatly the design of a sensor network. One of the key challenges faced involves minimizing the effects of the physical environment on the RF pattern. Multpath effects, especially in an indoor environment will affect the performance of the sensor node. We found that the hardware design, e.g. the placement of the antenna with respect to the other components plays a part in changing the omni-directionality of the RF signal pattern.

The purpose of this paper is to give a broad overview of the work we have done in this area of sensor network. More details of the various aspect of the implementation will be the subject of future papers. In spite of the challenges, some of which are still being looked into, the implementation provides for us a realistic and flexible platform to test and implement future sensor network-specific communication schemes. It also gives us a first hand look at the conditions affecting this new area of network research.

REFERENCES

[1] D. Maltz, J. Broch, and D. Johnson, "Lessons from a full-scale multihop wireless ad hoc network testbed," IEEE Personal Communications, 8(1): pages 8-15, February 2001.

[2] H. Lundgren, D. Nielsen, J. Nordstrom, and E. Tschudin, "A large-scale testbed for reproducible ad hoc protocol evaluations," IEEE WCNC, 2002.

[3] S. Sanghani, T. Brown, S. Bhandare and S. Doshi, "EWANT: the emulated wireless ad hoc network testbed," IEEE WCNC, 2003.

[4] J. Hill and D. Culler, "A wireless embedded sensor architecture for system-level optimization," U.C. Berkeley Technical Report, 2001.

[5] Texas Instuments, *MSP430x1xx Family User's Guide (Rev. C)*, SLAU049C, 2003.

[6] Chipcon AS, *SmartRF® CC 1000 Preliminary Datasheet (rev. 2.1)*, 2002.

[7] Maxim Integrated Products, MAX3316-MAX3319 RS-232-Compatible Transceivers Datasheet, 2000.

A GREEDY ALGORITHM TO CONFIGURE AND RECONFIGURE BLUETOOTH SCATTERNETS

WANDERLEY RAVAGNANI JR., EDMUNDO R. M. MADEIRA, JACQUES WAINER

Institute of Computing
UNICAMP - University of Campinas
Campinas, SP 13083-970, Brazil
{wanderley.junior, edmundo, wainer}@ic.unicamp.br

This paper presents an heuristic greedy algorithm to configure and reconfigure Bluetooth scatternets, in order to satisfy the specific communication requirements of each pair of network devices. The paper also presents simulation results that show the efficiency of the algorithm.

1 Introduction

Bluetooth is an specification for wireless, short range, low-cost communication through radio links. Typically, in Bluetooth applications, multiple piconets, which can contain at most 8 devices, can overlap in a same area forming an ad-hoc system called *scatternet*, where the piconets that compose this system should not be frequency-synchronized [6]. Piconets in a scatternet can communicate sharing one (or more) node(s) with the other piconets. The shared node is denominated a *bridge*.

A Bluetooth channel has 1 Mbit/s throughput for communication among the devices in the same piconet. However, the throughput among piconets is not specified by the standard and it depends on the time that a node takes to move from a piconet to another and how long a node is active in each piconet. These values depend mainly on a scheduler policy among piconets [1,4,7].

The usual solution to the Bluetooth scatternet formation problem [3,5,2] is a distributed algorithm that arbitrarily assigns each device to a piconet and determines bridges to link different piconets. Such algorithm does not take into consideration the communication requirements among the different devices that will form the scatternet. Clearly, if the communication requirements are known, an arbitrary assignment of devices to piconets is not appropriate since two devices that have high communication need between them can be placed in different piconets, or in different piconets that have no direct connection (a bridge) between them. The algorithm proposed here is an heuristic, centralized algorithm that uses a greedy search with backtracking, that given a N nodes and the communication requirements (T_{ij}) of each pair of devices i, j, will construct a scatternet that support such communication requirement, or fail. The algorithm will not necessarily return an optimal configuration for the scatternet, and may fail even when there is a possible solution.

The scatternet reconfiguration problem is the situation in which a scatternet is configured but there is an increase in the throughput between at least two devices in such a way that the resulting throughput is not supported by the current scatternet configuration. The scatternet must be then reconfigured. We will show that even in the case of scatternet reconfiguration it is better to run the greedy algorithm from scratch than to use some from of local search from the current configuration to obtain the new configuration.

Due to the complexity of the problem, some restrictions were made to model Bluetooth scatternets:

- any two devices in the scatternet can communicate directly;
- the scatternet should be connected;
- a bridge will only connect two piconets so it does not become overloaded;
- the master of a piconet is never a bridge;
- since the issue of the throughput between two piconets is still an open research topic, we assumed that the maximum throughput supported by one bridge that connects any two piconets is 100 kbits/s. This figure is consistent with the work [4] which tested different scheduling policies for the bridges.

This paper is organized as follows. Section 2 describes the algorithm, and the Section 3 presents the results of the simulations using the algorithm in the configuration and reconfiguration of Bluetooth scatternets. Finally the Section 4 discusses the conclusion and future work.

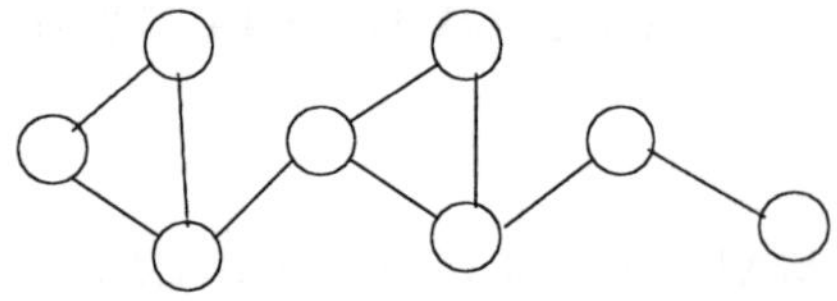

Figure 1. The triangle-edge topology

2 The Algorithm

The idea of the greedy algorithm is to place the devices that need to communicate the most among themselves inside of a same piconet. If it is not possible, these devices are placed in different piconets that have a common bridge in order to generate less traffic among piconets. The algorithm receives as input the number of devices (N) and the matrix of communication requirement (T) among them.

The algorithm will construct a scatternet in a topology we call Triangle-Edge. This topology will first construct a triangle of piconets, followed by an edge, followed by a triangle and so on (see figure 1). In such topology, all piconets have two or three bridges (with the exception of the last which may have only a single one). Such topology was chosen because it gave the best results when compared to some other fixed topologies such as ring, or binary tree.

The algorithm structure is discussed below. Space constraints do not allow us to show the pseudocode.

1. For each device i, a list containing the other N-1 devices is created, in decreasing order of throughputs between i and the other network devices. A pointer Pi will initially point to the first element of the list of the device i. When the device i is chosen to be included into the network, the device pointed by Pi will be its candidate to be the next element to enter the network.

2. Bridges are selected among the devices that have the least communication need with the other devices, so their own needs would not decrease its throughput as a bridge. Bridge candidates are stored in a bridge list and they will be inserted in the network at the end of the algorithm. We defined the bridges as being the devices that has a throughput need of less than 10 Kbits/s.

3. The first device (i) to be included in the network is the one that has highest throughput need with another device. The first piconet is created and device i is inserted there. Device i is removed from all the other lists. The pointer Pi associated with the device i will then point to the next candidate to be included into the network.

4. Then, the choice of the other devices to enter in the network is made as follows:

4.1 The device pointed by Pi, denoted by c_i, will be chosen if it satisfies the following two conditions:

a) including c_i in the current piconet (np) will result that the sum of the internal throughput of np is still less than 1 Mbit/s minus the extra-piconet throughput of np.

b) including c_i in np will result that the extra-piconet throughput of np is still less than its maximum extra-piconet throughput (100 Kbits/s times the number of bridges in np). However, this condition will only be accepted if the position of c_i in np (considering that a piconet is mathematically represented as a vector) is larger than the value Lim_{np} which we call the piconet np (**break condition**). If its position is less than this break value, this position is stored in $\mathrm{Lim\_prox}_{np}$ so that backtracking can be performed later. The break value Lim_{np}, as well as the need of its use will be explained below.

4.2 The chosen device d_c is the device that respected the condition *(item a)* and the break condition *(item b)*. If only the break condition was not satisfied, then d_c will be the device that needs to communicate more with the other devices already inserted in np, besides respecting the *item a*. When d_c is inserted in the piconet np, the queue F_{dc} is initialized, making P_{de} to point to the first element of F_{dc}, besides all the other queues are updated removing from them the chosen device (d_c);

4.3 If no candidate was chosen, or the number of devices already inserted in the piconet np is equal to ($8 - lim\_j$), or the break condition occurred, this piconet is closed and a new piconet is created. Besides this, if still remain devices to be inserted in the network, the number of bridges and the maximum number of no-bridge devices ($8 - lim\_j$) are calculated for the next piconet and the algorithm returns to the *item a*;

5. When the number of inserted devices is equal to N minus the number of chosen bridges, or when the chosen bridges become exhausted to form the network topology, the remaining devices (if they exist) are inserted in the end of the bridge queue (*Bridges*), then the bridges are inserted in the network and the

remaining devices are inserted in new piconets added in the end of the network, when the algorithm finishes.

The complexity of this greedy algorithm is $O(N^3)$. The break condition (Lim_{np}) mentioned above limits the number of devices inside each piconet p_i; without this condition one would insert in the piconet some devices that has very little communication need to the devices already present there. This may lead to the creation of a new piconet p_{i+1} whose devices will have a throughput requirement to the last element inserted in p_i which may exceed the extra-piconet throughput between $p-i$ and p_{i+1}. Thus by adjusting the break condition, a device that does not need to be placed in p_i will not be inserted there.

In order to find a suitable break value for each piconet, the procedure described below is executed. This procedure uses the following steps: initially the network is generated using the greedy algorithm with the break value (Lim_{np}) equals to 7 Kbits/s for all piconets. The greedy algorithm returns the last position where could occur a previous break to Lim_{np} (Lim_prox_{np}) for each piconet of the network. Then, the break value is reduced from Lim_{np} to Lim_prox_{np} and the greedy network is generated for each piconet that has its intra-piconet or extra-piconet throughput above its maximum limit. If after the algorithm analyzes all the piconets, if the communication requirements of the network are still not satisfied, the algorithm comes back the last analyzed piconet that may still be broken ($\text{Lim}_{np} \neq \text{Lim\_prox}_{np}$) and again the break value is reduced and the greedy network is generated for each piconet below it. This procedure finishes when the communication requirement of the network is totally satisfied or when no piconet can be broken anymore.

Using this procedure, the total complexity of the greedy algorithm becomes exponential. However, as it will be seen in the next section, the algorithm execution time will be limit in 1,5 seconds.

3 Results

The tests were performed in a Pentium II 300Mhz computer with 192 Mbytes of RAM, using Windows 98. The algorithms were implemented in C.

Configuration: In the configuration experiments we created a random scatternet generator, which can generate networks in two different topologies triangle-edge and random networks with at most three bridges for each piconet. The random network generator outputs just the communication requirements and the total number of devices. This is used as input to the configuration algorithm, which may find a solution for the requirement (with possibly a different topology), or may not find a solution, in which case we will say that the algorithm failed, since by construction there is a solution to the configuration problem.

The random network generator receives two parameters, the number of piconets and the topology. We organize the results according to these two parameters. 500 cases were generated for each situation.

The configuration algorithm was limited to execute for at most 1.5 seconds, after which it was considered that the algorithm did not find a solution, and thus failed. The results are reported in table 1. The second line of the table refer to generated networks that were in the triangle-edge topology. Thus that line in some way shows the error introduced by the "greedy" aspect of the algorithm: the topology used to configure the network is the same as the one used to generate it, and thus failures seems more related to the fact that the algorithm is heuristic. The third line refer to networks generated in the random but bounded topology, and thus the differences between the second and third lines are due mainly to a "topological gap" that is, to the fact that the greedy algorithm does not make a search on the space of possible topologies.

Reconfiguration: To simulate the reconfiguration problem, we used the generator to create a feasible network, randomly selected a pair of devices and increased their communication requirement so that it would either overload the intra-net or the internet throughput. But as opposed to the configuration problem in which the resulting communication requirement was by construction feasible (had a solution), in this case we do not know if the resulting communication requirement is feasible.

Thus, instead of measuring the satisfiability of the algorithm, we compared it to a local search algorithm, which will swap any two non bridge devices from different piconets, in order to improve the throughput. The algorithm will chose the swap that improves the throughput the best and from there

Table 1. Satisfaction results for the greedy algorithm

Num. Piconets	3	4	5	6	7	8	9	10	Total
Triangle-edge (%)	100,0	96,4	95,0	95,2	91,6	89,2	82,8	81,8	**91,5**
Random (%)	98,2	92,6	88,0	88,0	80,0	76,8	73,6	67,8	**83,13**

Table 2. Reconfiguration results

Num. Piconets	3	4	5	6	7	8	9	10	Total
Satisfied - Local Search(%)	8,0	12,4	17,2	8,8	13,6	9,8	13,8	10,4	**11,75**
Satisfied - Greedy (%)	36,0	40,4	38,6	35,2	38,8	37,2	35,8	32,2	**36,78**

continue the local search until a solution is achieved or until no swap improves the throughput. The local search algorithm was chosen because is makes use of the "old" network topology.

The results are in table 2. They show that for all sizes, to run the greedy algorithm from scratch is always better than to perform a local search using the "old" configuration as basis.

4 Conclusion

This paper presents a greedy algorithm to configure and reconfigure Bluetooth scatternets when information about the communication requirements among the devices is given. The algorithm is based on greedily placing devices that need to communicate the most between themselves in the same piconet, but with some form of backtracking that allows one to control when to stop adding to a piconet and start the formation of a new one. The algorithm constructs the scatternet in a fixed topology, which we called triangle-edge.

The results showed that the algorithm is fast (it was limited to run for 1.5 seconds in a not powerful architecture) so that it could conceivably run in one of the Bluetooth devices. The algorithm is also efficient: on average, for networks of up to 10 piconets, the algorithm did not find a solution in less than 20% of the cases. The algorithm is also on the average 3 times as better than a local search algorithm to solve the problem of reconfiguring a scatternet. Unfortunately we cannot compare this algorithm with others, both to configure and to reconfigure scatternets, since the examples in the literature do not cover the case used here in which the communication requirements taken into consideration in the configuration.

Future work will possible include some sort of "topology search" to lessen the gap shown in Table 1, and the inclusion of constraints that for example will forbid two devices from belonging to the same piconet, for example because they are too far away from each other.

References

1. P. Johanson, Y. Lee, M. Gerla, and M. Kazantzidis. Bluetooth - an enabler for personal area networking. In *IEEE Network, Special Issue in Personal Area Network*, September-October 2001.
2. Ching Law, Amar K. Mehta, and Kai-Yeung Siu. Perfomance of a new bluetooth scatternet formation protocol. In *ACM Symposium on Mobile ad Hoc Networking and Computing*, October 2001.
3. Lakshmi Ramachandran, Manika Kapoor, and Alok Aggarwal. Clustering algorithms for wireless ad hoc networks. In *SIG Mobile*, 2001.
4. Andrs Rcz, Gyrgy Mikls, Ferenc Kubinzky, and Andrs Valk. A pseudo random coordinated scheduling algorithm for bluetooth scatternets. In *ACM Symposium on Mobile ad Hoc Networking and Computing*, pages 193–203, October 2001.
5. Theodoros Salonidis, Pravin Bhagwat, Leandros Tassiulas, and Richard LaMaire. Proximity awareness and ad hoc establishment in bluetooth. In *Infocom*, 2001.
6. SIG. *Bluetooth Core Specification, version 1.1*, May 2002. http://www.bluetooth.com.
7. Godfrey Tan and John Guttag. A locally coordinated scatternet scheduling algorithm. In *IEEE Conference on Local Computer Networks (LCN)*, November 2002.

Potential Interference of Ultra-Wide Band Transmitters on HipeLAN/2

N.Simeonidis (nps4@kent.ac.uk), K. Haider
(kh15@kent.ac.uk) and H.S.Al-Raweshidy
(H.Al-Raweshidy@kent.ac.uk)

Communication Systems Division,
Department of Electronics, University of Kent,
Canterbury, Kent, UK, CT2 7NT
Tel: 00441227823396 Fax: 00441227456084

Abstract: A UWB communication Transmitter with PPM modulation scheme in the presence of AWGN was simulated in order to determine the interference effect to a traditional 16QAM, 64QAM, QPSK and BPSK narrowband system. The interference highly depended on both the spectrum of UWB signal and the center frequency of the narrowband system.

1. INTRODUCTION

1.1 Ultra-Wideband

A new uprising technology that focuses in wireless communication systems is UWB (Ultra-Wideband). It is a promising technology both in telecommunications and radar that is defined by its large fractional bandwidth. UWB systems are based on very short pulses of radio frequency energy. It offers great Multipath immunity and has no problem penetrating obstruction-rich environments. Therefore it can be used for a great variety of indoor wireless applications. UWB transmits information by modulating the position of very fast (sub-nano second, one trillionth (10^{-12}) pulses and can still maintain a high data rate. Because these transmitted pulses are so narrow, the spectrum is very broad, covering hundreds of MHz or even GHz. At 5GHz UWB has a fractional bandwidth of 25% (1.25GHz) of the centre frequency. A very important area that researchers are looking into is the potential interference that UWB can cause to other communication or non communication systems, i.e. Wireless LAN, GPS, Amateur Radio etc.

1.2 HyperLan2

High Performance Radio Local Area Network type 2 (HiperLAN/2) is an upcoming standard, which is being specified by the ETSI/BRAN project. It shall provide high-speed communications of up to 54 Mbps between mobile terminals and various broadband infrastructure networks, HiperLAN/2 system operates at the 5GHz region. HiperLAN/2 system is likely to be deployed in a wide range of environments, such as buildings, exhibition halls, airport, industrial buildings and outdoor deployment. By using various modulation schemes and puncturing of convolutional codes, the data rate can be varied. Seven physical layer modes are specified, of which the first six are compulsory, where as the last which uses 64 QAM is optional.

Mod	Modulation	Code Rate	Bit Rate	B/Sym
1	BPSK	1/2	6Mbps	3
2	BPSK	3/4	9Mbps	4.5
3	QPSK	1/2	12Mbps	6
4	QPSK	3/4	18Mbps	9
5	16QAM	9/16	27Mbps	13.5
6	16QAM	3/4	36Mbps	18
7	64QAM	3/4	54Mbps	27

Figure -1- HiperLan2 Modulation Schemes

Both of these technologies Ultra-Wideband and HiperLAN/2 could be deployed in same environment, for example in an office, an airport, onboard an aircraft, or even in the street. Therefore and since both technologies could be working in the same frequency range (5GHz) interference caused by Ultra-Wideband system to HiperLAN/2 would make the bit error rate (BER) increase and this will have a direct effect on the number of users that a single AP (Access Point) would be able to serve, In the worst case the wireless devise will loose connection to the AP.

UWB and HiperLAN/2 physical models are modelled in Matlab simulation package. The hiperLAN/2 model will consist of three blocks which are the transmitter, channel model and the receiver end according to the ETSI standard specification. UWB model will consist of only the transmitter part as this will act as additional interference. Simulation will be run for the HiperLAN/2 system having only the AWGN channel acting as noise. The value of BER will be plotted and then compared to that when the UWB is added to the channel interference.

This paper will show the level of interference introduced by UWB on a HiperLAN/2 system by plotting graphs of BER against Eb/No for each HiperLAN/2 mode. From the measurements the magnitude of the interference cause by UWB will be visualized and the appropriate transmission rate would be determined in order to reduce the interference level and hence the effect upon the bandwidth per user would be minimized.

As the UWB communication systems are being studied, its performance depending on the monocycle's shape the modulation method became a primary concern. It is also important to see its interference to the traditional narrowband systems. In this report, we simulated the Eb/No performances of a UWB Transmitter on a HPL2 Transceiver.

1.3 Motivation

As the UWB communication systems are being studied, its performance depending on the monocycle's shape or the modulation method became of primary concern. It is also important to see its interference to the traditional narrowband systems. In this report, I simulated the BER performances of a UWB system with various conditions and compared them

2. ULTRA-WIDEBAND PROCEDURE

In the following section we derive the Ultra-wideband monocycle shape from the Gaussian equation.

1. UWB monocycle shapes

1) Gaussian

a) Equation

$$w(t)=\sqrt{Ep}*\frac{2.1245}{\sqrt{\tau}}*e^{-32(\frac{t-Tc}{\tau})^2} \quad (1)$$

Where,

Ep: Pulse Energy (J),

τ: Impulse width parameter (s)

Tc: Time shift (s)

In addition, the width of UWB signal's nonzero part was defined as Impulse width (s).

b) Pulse shape and Energy spectral density

Ep = 1 (J), τ = 0.6 (ns),

Width = 0.6 (ns)

Gaussian second derivative

a) Equation

$$w(t)=\sqrt{Ep}*\sqrt{\frac{8}{3\tau}}*\{1-4\pi(\frac{t-Tc}{\tau})^2\}*e^{-2\pi(\frac{t-Tc}{\tau})^2} \quad (2)$$

b) Pulse shape and Energy spectral density

Ep = 1 (J), τ = 0.2877 (ns), Width = 0.6(ns)

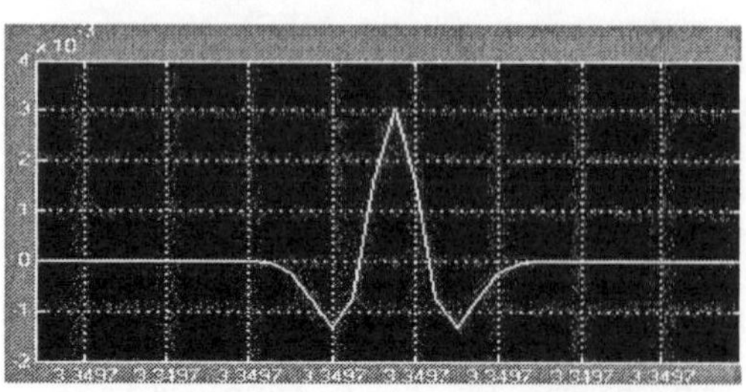

Figure -2- Ultra-Wideband Pulse

2.1 Performance Evaluation

This is what one of the final models (64QAM) looks likes in Simulink Matlab when we run the simulations to determine the potential interference that UWB causes on HiperLAN/2. Figure -3-

This model consists of a HiperLAN/2 Transmitter a AWGN channel and a Receiver. Just before the AWGN channel you can see the block named UWB. This contains the UWB transmitter and the rest of the necessary circuit to be able to mix the two signals together so the entire simulation can run with no errors.

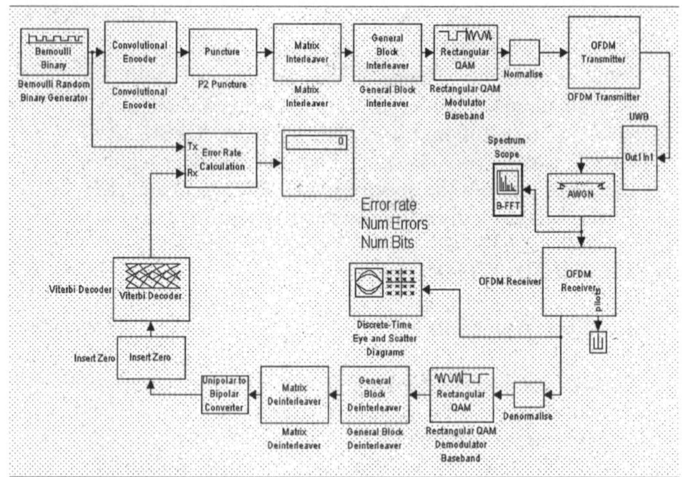

Figure -3- Complete Model in Simulink

Four models were simulated in total for comparison to show the interference caused by an Ultra-Wideband transmitter to a HyperLan2 Access point. In Figure 4 you can see the topology of HyperLan2 network that is being used by 5 Pc users. In the same area there is also an Ultra-Wideband transmitter that is the source of potential interference to the number of users that the HylerLan2 Access point could serve.

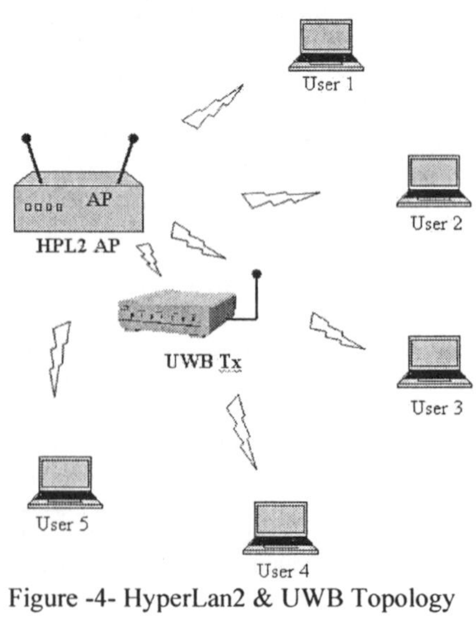

Figure -4- HyperLan2 & UWB Topology

3. SIMULATION RESULTS

The models used to simulate the potential interference from an Ultra-Wideband Transmitter on a HyperLan2 access point were:

1) A HyperLan2 Mode 6 transceiver using 16 QAM modulation scheme with a UWB Transmitter using PPM modulation with a AWGN power level of 0.05W (Figure 5)

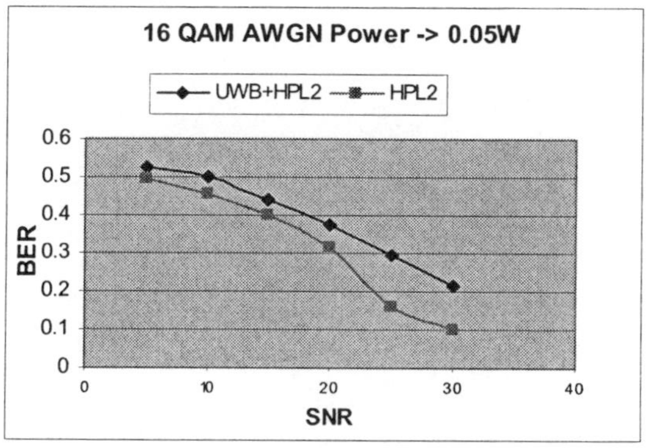

Figure -5- 16QAM SNR vs. BER

2) A HyperLan2 mode 7 transceiver using 64 QAM Modulation scheme with a UWB Transmitter using PPM modulation with a AWGN power level of 0.1W (Figure 6)

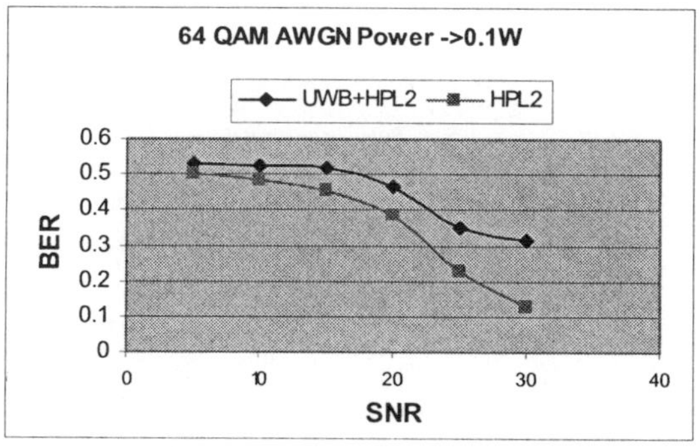

Figure -6- 64QAM SNR vs. BER

3) A HyperLan2 mode 4 transceiver using QPSK as a modulation scheme with a UWB Transmitter using PPM modulation with a AWGN power level of 0.5W (Figure 7)

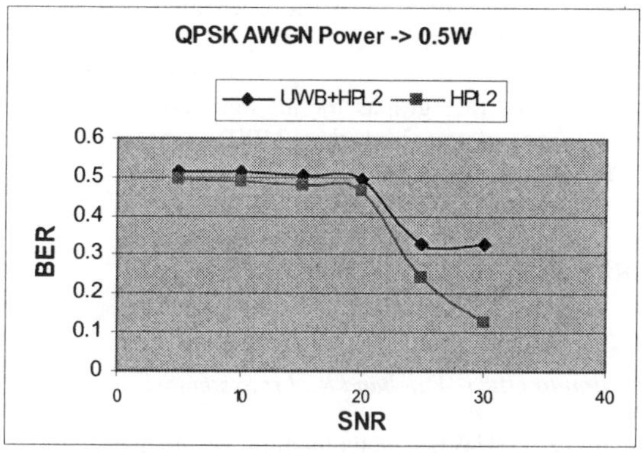

Figure -7- QPSK SNR vs. BER

4) Finally a HyperLan2 mode 1 transceiver using BPSK as a modulation scheme with a UWB Transmitter using PPM modulation with a AWGN power level of 0.75W (Figure 8)

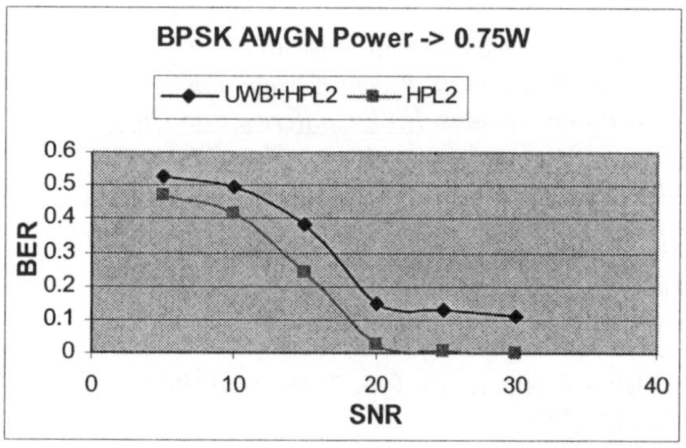

Figure -8- BPSK SNR vs. BER

4. CONCLUSION

In this paper, an overview was presented of the potential interference of an Ultra-Wideband Transmitter on a HyperLan2 Access Point, highlighting the difference between the four different modulations schemes that HyperLan2 uses to transmit its Data. Then an Ultra-Wideband signal was introduced to the system and several simulations were run using different levels of AWGN channel noise and various channel power levels, In order to determine the HyplerLan2 signal degradation. After plotting the Signal to noise ratio over the bit error rate graphs for all four modulation schemes it was determined that UWB causes the HyperLan2 signal to degrade.

REFERENCES

1. James D. Taylor, 1995 Chapter One, CRC Press, , *Introduction to Ultra-Wideband Radar Systems*

2. Youngjae Kim, UWB Communication Performance Simulations, Uwb Report1, Stanford University

3. Li Zhao, Alexander M. Haimovich, ECE Dept. New Jersey Institute of Technology, Newark *The Capascity of an UWB Multiple-Access Communications Systems.*

4. Ewan Frazer, Dave Harmer (Thales Research & Technology, Reading, Uk) *UWB Applications and Interference (UWB Colloquium 23rdJuly 2002)*

5. Robert Fontana, Aitan Ameti, Edward Richley, Lance Beard, Dennis Guy. *Recent Advances in Communications Systems. IEEE Conference on Ultra Wideband Systems and Technologies, May 2002 Baltimore, MD*

6. Robert J. Fontana, Multispectral Solutions, Inc Germantown, Maryland USA. *Recent Trends in Ultra Wideband Communications Systems.*

7. ETSI BRAN, HiperLAN Type 2, Physical (PHY) layer, TS 101 475, April 2000

8. ETSI BRAN, HiperLAN Type 2; Data Link Control (DLC) layer Part 2: Radio Link Control (RLC) Sublayer, TS 101 761-2, April 2000.

9. www.hiperlan2.com

10. www.ultrawidebandplanet.com

11. www.multispectral.com

SESSION 5B:
APPLICATIONS

Media Synchronization Quality of Packet Scheduling Algorithms for Wireless and Mobile Networks

Azzedine Boukerche[†*] and Harold Owens II[††]
[†]SITE, University of Ottawa [††]University of North Texas
E-mail: boukerche@site.uottawa.ca E-mail: hjo0004@cs.unt.edu

Abstract

Wireless multimedia synchronization is concerned with distributed multimedia packets such as video, audio, text and graphics being played-out onto the mobile clients via a base station(BS) that services the mobile client with the multimedia packets. In this paper, we study the media synchronization quality of packet scheduling algorithm for wireless and mobile networks using hard hand-off scheme. We present a set of simulation experiments we have carried out to study their effect in terms of buffer usage, overlow and underflow at the mobile host while mobile hosts request and play-out of multimedia packets.

1 Introduction

Wireless communication technological advancements have created a new paradigm known as mobile distributed multimedia systems. In a mobile distributed multimedia system, diverse packets can be simultaneously manipulated. Packets such as text, images, audio and video can be played out on a mobile client via a base station. Some multimedia packets are time dependent upon each other[3]. Typically, during play-out on the mobile client, media components must be kept in temporal order. This ordering process is known as media synchronization. There are two kinds of timing aspects for constrained elements: 1) intra-media continuity is subject to a real-time-constraint in handling media packets and 2) inter-media continuity is subject to temporal correlation during playback of media packets[3].

To ensure quality of service of multimedia services and to minimize end-to-end multimedia packet delay for the mobile client, our multimedia synchronization scheme (Mosynch) considers both intra-stream synchronization and inter-stream synchronization [1]. Mosych assume hard-hand off scheme. In the future, we will study the investigate both Hard and soft Hand scheme [2].

This paper considers transmission of live audio stream and the corresponding video stream from the distributed multimedia servers to base stations which service the mobile clients. In this paper, we wish to investigate the effect of packet scheduling algorithms at the base station on overall network synchronization. We investigate four packet scheduling algorithms: First-In-First-Out (FIFO), Priority-Queuing (PQ), and Round-Robin (RR) and Weighted Fair-Queuing (WFQ) queuing.

2 Media Synchronization Algorithm

The media synchronization algorithm (Mosych) [1] assumes that the network can provide sufficient resources to deliver multimedia packets to the mobile client. The multimedia data consists of many multimedia packets that may be progressively transmitted over the network. Mosync does not use a global synchronization clock. In the MoSync algorithm, the servers time stamp each multimedia packet, with current local time to allow the base station to calculate round trip delay, jitters and inter-arrival time. The multimedia packets carries the server's number and packet sequence number but these additions are negligible so multimedia synchronization is the primary focus of the this paper. The synchronization model consists of K scalable server nodes and L mobile clients. A base station can communicate with multiple mobile clients because the communication between the base station and mobile client is wireless.

Mosync uses three simple protocols. In case one, to avoid buffering of early arrival multimedia packets sub-stream the start-up protocol is used. This allows evaluation of the round trip delay for different sub-streams. In case two, end-system jitters and network jitters is counteracted. Both end-system and network jitters must be countered to avoid underflow or overflow at the mobile client. For multimedia packets to play-out smoothly the servers data transmission rate must match the available service rate at the mobile client. The overall goal when dealing with network and end-system jitters, is to keep the average delay, buffer underflow and buffer overflow at an acceptable minimal level. In case three, to solve the average delay changes, clock drift and server dropouts exponential smoothing forecasting is used[1]. When a mobile client receives a multimedia packet, the mobile client calculates the expected arrive time of the next multimedia packet and sends the calculation to the base station. The base station then forwards the information to the servers. In both case two and three the Mosync algorithm uses either the pessimistic or optimistic synchronization protocol when requesting and receiving multimedia packets from the servers. With MoSync's pessimistic protocol, the base station updates the servers on each

*Dr A. Boukerche was partially supportd by the CRC- Canada Research Program, and Texas ARP/ATP Advanced Research Program.

multimedia packet request made by the mobile client. With MoSync's optimistic protocol, the server sends the multimedia packets directly to the mobile client without continuous updates being made by the base station once the first multimedia packet from the server arrives to the mobile client. The Mosync algorithm operates at the network's application layer.

The MoSync algorithm uses three types of nodes: server, Quasi-receivers and receivers[1]. The algorithm contains K servers, N base stations and L mobile clients, where K servers are type servers, M base stations are type Quasi-receiver and the L mobile clients are type receivers. The mobile client synchronizes the multimedia packets, received from the multimedia server, by reporting buffer usage to the base station and updating the base station with multimedia packet arrival time differences. After a base station has information about the arrival time of each packet, it calculates the synchronization time for the next packet. A scheduler, in the server, manages the on-time transfer of the subframes, as a part of the frames to the mobile client. The mobile client requests multimedia service from the multimedia servers, through the base station. When a base station requests the first group of multimedia packets, it sends the synchronization point information to all servers[1]. The base station can receive messages from the mobile client and from neighboring base stations. The mobile client messages are for requesting multimedia packets from servers and done messages to let base stations know all packets have been received by the mobile client. The neighboring base station messages are used to inform the base station if its neighbor is on or off, to set-up hand-offs to neighboring base stations. It is the mobile client's job to calculate the latest arrival time and the differences between every multimedia packet's arrival time, the buffer usage and update its serving BS with the new data.

• **Hard Hand-off Scheme:** Hand-offs in a wireless network are classified as Soft Hand-offs and Hard Hand-offs. In this paper, we focus upon a hard hand-off where the connection between the mobile client and servicing BS, is broken before new services are established with the new cell's BS. The mobile client can only be connected to one BS at any time, with a hard hand-off, because frequencies between base stations must be different to avoid interference between mobile client signals. Because a mobile client can only be connected to one BS at a time, during hand-off, delays between services are introduce into the network. Hard hand-off is mostly used in FDMA and TDMA[9].

Space limiation preclude us to include the results we have obtained for Soft Hand-off schemes [1].

3 Packet Scheduling Algorithms

In this paper, we wish to investigate how packet scheduling algorithms can affect media synchronization in a wireless ad mobile environment.

• **First-In-First-Out (FIFO):** In today's wired and wireless systems, First-In-First-Out (FIFO) is the most common scheduling algorithm. There is only one queue and all packets are treated equally. Packets are stored at a buffering location and processed on a first come first serve basis. Note that FIFO does not recognize priority or classes of traffic.

• **Priority Queuing (PQ):** Some multimedia requests, like on-demand-video, requires large bandwidth and guaranteed processing order of multimedia packets. In this special case, Priority Queuing(PQ) can be used to guarantee the bandwidth, memory and processing time for a particular high priority data flow. PQ sets a certain number of priority classes according to some parameter. Priority-based algorithms may help to reduce delays for high priority classes that are bandwidth sensitive, such as video, audio and interactive traffic.

• **Round Robin (RR):** Lately, there has been an increase in multimedia traffic such as audio and video. The heavy demand for multimedia traffic increases QOS needs to enhance the multimedia communication. Packet scheduling in the networks play an important part in the QOS of each client in the network. These multimedia services are bursty and bandwidth hungry. To ensure fairness between processing of packets in the network, we choose to investigate the effect of Round Robin(RR) packet scheduling on our media synchronization scheme.

• **Weighted Fair Queuing (WFQ):** Multiple traffic flows can be identical in priority and their network resource usage. With the multimedia traffic of today, assigning a priority or guaranteeing play-out time of multimedia packets is not enough to ensure high QOS for a client. The Weighted Fair Queuing(WFQ) algorithm combines the PQ and RR into a dynamic fair queuing that divides the available bandwidth among traffic queues base on weight or priority. In today's networks, video packets are given a higher priority than audio packets, audio packets are given a higher priority than graphic packets and graphic packets are given a higher priority than text during play-out of multimedia packets. WFQ ensures that each packet is treated fairly according to its weight. Packets of the same priority is stored in the same buffer. Multimedia packets of the same priority are played-out in a round robin order. Hence, satisfactory response time is a result of WFQ fairness.

4 Simulation Experiments

We developed a discrete-event model to simulate a cellular wireless multimedia system, on a wireless and wired network. Our assumption is that the communication between servers and base stations are wired and connections between mobile client and base stations are wireless. In our model, there are 300 channels available, for 60 cells, with system load equally distributed over all cells. Each mobile client has a buffer size at least 3 times that of the largest multimedia packets and at most six time as large as the multimedia packets received from the server. In this simulation, we employ the hard hand-off in AMPS/GSM[7].

We performed several tests using many jitters and resynchronization. We evaluate MoSync's performance using a nonuniform jitters for all the 4 scheduling schemes, FIFO, PQ, WFQ and RR. In the nonuniform case, we assume exponential distribution of network delay jitters. This distribution was used in earlier studies of the MoSync algorithm[1]. For our experiments, there are multimedia media delay ranges as follows: 1) 0-200 msec 2) 0-400 msec 3) 0-600 msec. The ranges are exponentially distributed, with mean values of 20, 40 and 60, respectively. The upper bound delay is set in the nonuniform distribution to prevent failure caused by large delays in the stimulation. The mean communication session is set to 20 multimedia packets and the mean buffer size of a mobile client is set to three times the size of the multimedia packets[1]. Our results will show that each range of delay jitters cause the network to behave differently.

Let us now turn to our results.

Figure 1 illustrates the buffer usage for nonuniform case, where *Buffer Usage* measures the synchronization behavior for each mobile client. As we can see, at 20 ms nonuniform delay jitters, all packet-scheduling algorithms synchronize equally, requiring the same amount of buffer space to play-out smoothly. With small amounts of nonuniform network jitters, the network can easily counter these delays by pushing delay to the upper most layers where multimedia packets are buffered by the BS and played-out onto the mobile client smoothly. At 40 ms nonuniform, PQ and FIFO requires at least six times less multimedia packets than WFQ and RR packet scheduling. Both RR and WFQ use a fair-scheduling strategy to play-out packets at the BS. The fair-scheduling, used by RR and WFQ increases turn-around time of multimedia packets. In a hard hand-off, packets need to be processed quickly to avoid retransmission due to hard hand-off, that causes packets to be dropped during the mobile client's transfer between BS. With PQ, we process the packets with the smallest play-out time first, ensuring better average turn-around time of multimedia packets than RR, WFQ and FIFO. With FIFO, packets received at a burst are usually transmitted at a similar burst. At 40 ms nonuniform, FIFO and PQ allow for smooth play-out of multimedia packets at the mobile client. At 60 ms nonuniform, PQ and FIFO require half or less number of multimedia packets to play-out smoothly at the mobile client. The four packet scheduling algorithms are all affected by the large delay introduced into the network, but PQ is affected the least because, on average, we process more multimedia packets at the BS than FIFO, WFQ and RR because we process the packets with the smallest play-out time first. PQ reduce delay for a high-priority class that is both bursty and low rate, relative to link bandwidth (video stream, interactive traffic, etc), without reducing the throughput of lower priority classes[6].

Figure 2 illustrate the overflow rate for the nonuniform case. The Overflow Rate measures the average number of multimedia units that overflow a mobile client buffer. At the network's application layer, multimedia packets are buffered and handle as continuous data streams to allow for play-out of multimedia packets without jitters. The object is not to overflow the mobile clients buffer, which would cause data in buffers to be rendered useless. It is much easier to control overflow then to control underflow. For example, we can introduce a delay great enough to eliminate all overflow at the mobile client, to the maximum 100 multimedia packets request as illustrated in Figure 2. This works in theory, but in application, this will under-utilize the network's resources and not allow for smooth play-out of multimedia packets at the mobile client. The MoSync algorithm, however, is very efficient in preventing overflow. The Mosync Algorithm performs both intra-stream and inter-stream synchronization to allow on time delivery of multimedia packets and prevention of early multimedia packet arrival to mobile client. The Mosync Algorithm's ability to adjust to network delay jitters and prevent mobile client's buffer overflow is illustrated in Figure 2.

Figure 3 illustrates the underflow rate for the nonuniform cases. The *Underflow Rate* measures the average deficiency in the clients buffer of the number of multimedia units currently needed for smooth play-out. In Figure 3, at 20 ms nonuniform delay jitters, all algorithms perform well and produce no underflow because the network shifts the small delays to the application layer where the multimedia packets are buffered at the BS and played-out on the mobile client, without jitters. In Figure 3, with nonuniform delay jitters of 40 msec distributions, PQ produces a great deal of underflow at higher number of requests, while FIFO, RR and WFQ remain almost constant. With PQ, we must process the multimedia packets with the smallest play-out time, which allows for faster turn-around time, but less throughput. At 60 msec distributed nonuniform delay jitters, PQ produces a lot of underflow at the mobile client, while FIFO,WFQ and RR remain almost constant because a larger number of mobile clients are able to play-out their multimedia packets. With PQ, with a particular data flow, only the mobile clients with higher priority requests are able to play their multimedia packets out on-time. At greater delays, the RR and WFQ scheduling produces little or no underflow, caused by the ability to spread out a burst of multimedia packet request among flows, fairly, allowing for more mobile clients to play-out the multimedia packets smoothly.

These results imply that RR, WFQ and FIFO algorithms provide better intra-stream and inter-stream synchronization.

5 Conclusions

In this paper we presented media synchronization Quality of packet scheduling algorithms for wireless and mobile networks. We analyze their behavior, based on the QoS of the mobile client in terms of buffer usage at the mobile client, as well as underflow and overflow buffer at the mobile client while mobile hosts request and play-out of multimedia packets.

We present a set of simulation experiments to evaluate their perforamnce and study the effect of packet scheduling on our wireless multimedia synchronization protocol usinh hard hand-off scheme. Our results indicate that that the play-out of multimedia packets, by the BS onto the mobile client, affects the overall network synchronization and can be used to enhance the QoS at the mobile client by allowing on-time delivery of multimedia packets to the mobile client.

We plan to extend this work, and study the effect of packets

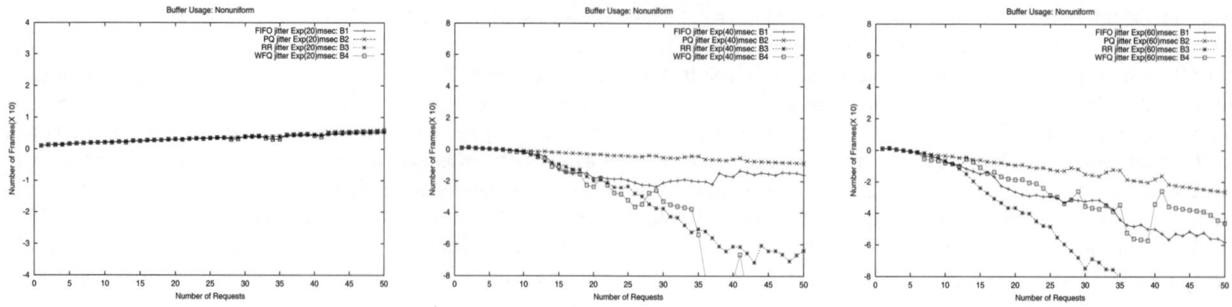

Figure 1: Buffer Usage for hard hand-off: Nonuniform Case

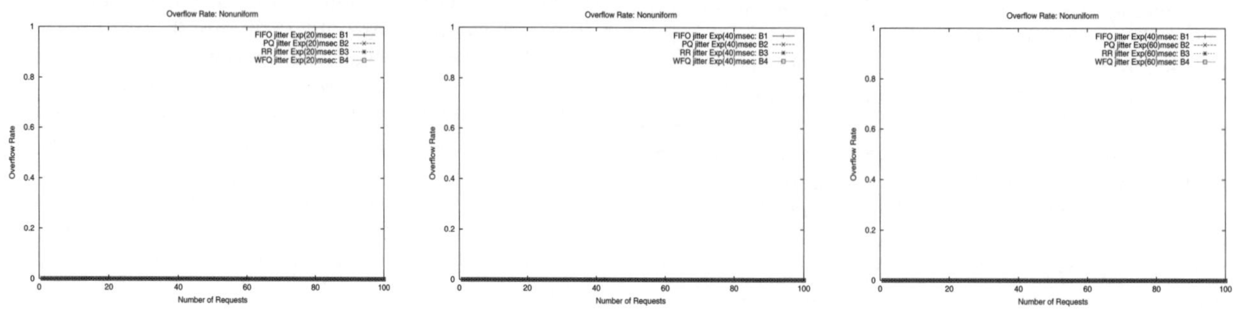

Figure 2: Overflow for hard hand-off: Nonuniform Case

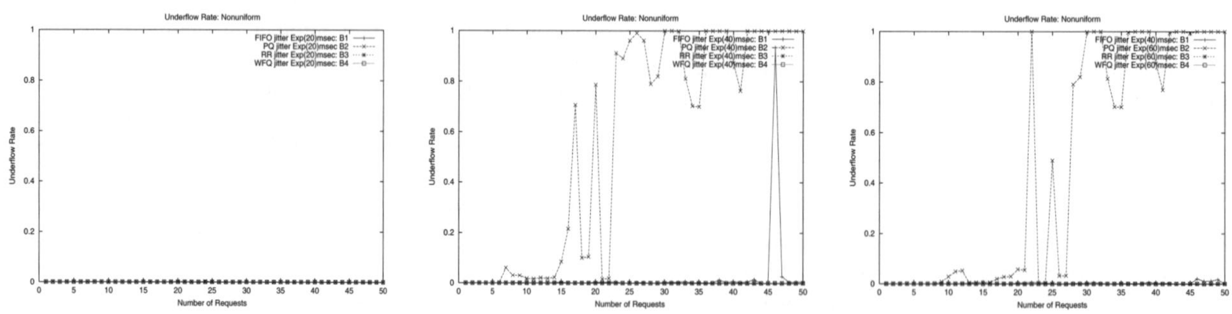

Figure 3: Underflow for hard hand-off: Nonuniform Case

scheduling strategies using soft hand-off schemes.

References

[1] A. Boukerche, Hong Sungbum and T. Jacob, An Efficient Synchronization Algorithm for Cellular Wireless and Mobile Multimedia Systems, *9th IEEE/ACM Int'l MASCOTS Symposium*, pp. 89-96, 2001.

[2] A. Boukerche, "Wireless Multimedia Synchronization Protocols", Technical Report, In Progress.

[3] N.U. Qazi, M Woo, and T.D.C. Little, "A Temporal Reference Framework for Multimedia Synchronization", IEE J. Selected Areas in Comm. Vol. 14, No. 1, 1996

[4] L. Lamont, L. Li, R. Brimont, and N.D. Georganas, "Synchronization of multimedia Data for a Multimedia News-on-Demand Application", IEEE J Selected Areas in Comm., vol. 14, no. 1, pp. 264-278, 1996.

[5] T. Znati, R. Simon, and B. Field, "A network-Based Scheme for Synchronization of Multimedia Streams", Technical Report TR95-02, Dept. of Compt. Sc, Univ. of Pittsburg.

[6] Sally Floyed, Barbara Denny, Van Jacobson, Scott Shenker and Lixia Zhang "Packet scheduling Research", DARTnet II Meeting, March 1995.

[7] U. Black, "Second Generation Mobile and Wireless Networks", New Jersey" Prentice Hall, 1999.

[8] Daniel Moss'enand Ha Ly "User to Network QoS Parameter Transformation in Networked Multimedia Systems" Workshop on Resource Allocation Problems in Multimedia Systems(with RTSS'96), Dec 1996.

[9] Sanchin Abhyankar "Handoff in Mobile Wireless Network" Nov, 2000

PROACTIVE ENERGY-AWARE VIDEO STREAMING TO MOBILE HANDHELD DEVICES

SHIVAJIT MOHAPATRA & NALINI VENKATASUBRAMANIAN
Dept. of Information & Computer Science,
University of California, Irvine, CA 92697-3425, USA
E-mail: mopy,nalini@ics.uci.edu

In this paper, we present a adaptive middleware solution for power-aware video streaming to mobile handheld devices. Using knowledge of dynamic device parameters(e.g residual energy), changing network noise levels and device mobility patterns, the middleware dynamically determines video streaming properties(e.g video burst sizes) to maximize the user QoS(reduce packet loss) while optimizing power consumption of the device. Specifically, we propose a **proactive** video transcoding and streaming technique that uses global system knowledge to effect power optimization, by transitioning the wireless NIC to sleep mode. Our performance results indicate that such a proactive scheme can reduce the energy consumption of network cards by as much as 70%, even in the presence of network noise and provide graceful video adaptations for better user experience.

1 Introduction

Improving the service lifetimes of low-power mobile devices through effective power management strategies can facilitate optimization of user experience for streaming video on to handheld devices. To achieve this, a system should be able to dynamically adapt to global system changes, such that the entire duration of a requested video is streamed to the user at the highest possible quality, while meeting the power constraints of the user's low-power device. We achieve such an optimal balance between power and performance, by introducing a notion of "*Utility Factor* U_F" for a system, and optimizing the U_F for the system. This approach precludes the system from aggressively optimizing for power at the expense of performance and vice-versa; thereby providing an optimized operating point for the system at all times.

We assume the system model depicted in Fig. 1. The system entities include a multimedia server, a proxy server that utilizes a directory service, a mobility manager service and a realtime video transcoder, the wireless access point and users with mobile low-power wireless devices. The dotted arrows represent the noise at the access point. The multimedia servers stream videos to clients and all communication between the handheld device and the servers are routed through the proxy server, which can transcode the video stream in realtime. An adaptive middleware executes on both the handheld device(lightweight) and the proxy, and performs two important functions. On the device, it obtains residual energy information and feeds it back to the proxy, updates its mobility information on the directory service and relates the video stream parameters and network related control information to underlying layers(OS). On the proxy, it performs a energy-aware video stream adaptation through realtime transcoding. It also regulates the video transmission over the network based on the noise level and the residual power at the device. The proxy utilizes the mobility manager to predictively determine future network noise levels. The proxy implements an adaptation control module that performs the proactive adaptations and admission control.

2 Energy-Aware Video Streaming

In this section, we first identify discrete video transcoding levels for handheld computers. This significantly reduces the realtime video transcoding costs as the transcoding parameters no longer have to be identified in real time. We then briefly describe a *reactive* proxy-based adaptation scheme and suggest a *proactive* scheme that significantly improves the U_F for the system. Finally we present our characterization of the network traffic regulation for energy-aware video streaming. The "*Utility Factor* U_F" is a measure of "user satisfaction" and we specify it as follows: given the residual energy E_{res} on a handheld device, a threshold video quality level (Q_A : $Q_{MAX} \geq Q_A \geq Q_{MIN}$) acceptable to the user, and the time of the video playback T, the U_F of the system is non-negative, if the system can stream the

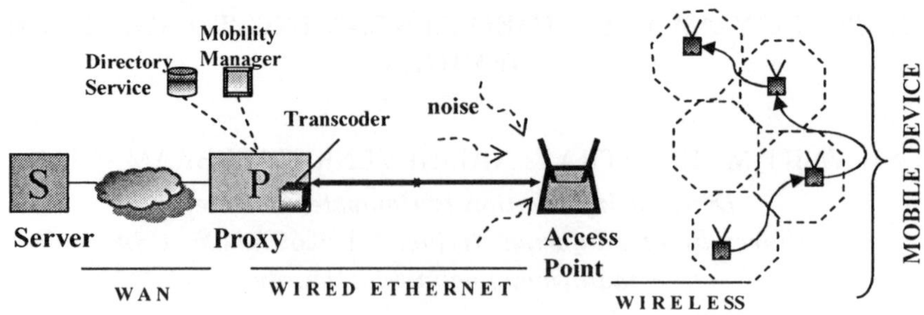

Figure 1. **System Architecture**

highest possible quality of video to the user such that the time, quality and the power constraints are satisfied; otherwise U_F is negative. Let P_{VID} denote the average power consumption rate of the video playback at the handheld and Q_{PLAY} be the quality of video streamed to the user by the system. Using the above notation, we define U_F as follows:

$$U_F = \begin{cases} Q_{PLAY} - Q_{MIN} & \mathit{IFF}\ P_{VID} * T < E_{RES} \\ & \mathit{and}\ Q_{PLAY} \geq Q_A \\ -1 & \mathit{Otherwise} \end{cases}$$

2.1 Energy-Sensitive Video Transcoding

We conducted extensive experiments to determine video transcoding levels that have a significant bearing on energy. To validate our assessments, we also conducted an extensive survey to subjectively assess [3] the human perception of video quality on handhelds(iPAQ). We briefly present some of our findings: (i) *It is hard to programmatically identify video quality parameters(a combination of bit rate, frame rate and video resolution) that produced a user perceptible change in video quality and/or a noticeable shift in power consumption in handhelds.* (ii) *For all the video streams on handheld devices, it was enough to use just three standard intermediate formats(e.g SIF(320x240), Half SIF(340x160) and Quarter SIF(160x120)) for frame resolution values. Other resolutions did not produce a perceptible quality change or power uptake compared to the nearest SIF encoded video with similar bit and frame rates.* Based on these conclusions, we identified eight dynamic video stream transformation parameters (Table 1) for our proxy-based realtime transcoding and use the profiled average power consumption values to perform our adaptations.

2.2 Reactive vs. Proactive Adaptation

In traditional proxy-based *reactive* adaptation schemes [1], the proxy reacts to dynamically changing system conditions by adapting video streams (by either lowering or improving stream quality) for improved performance. In this approach, a change in system state is first detected (possibly due to dropped packets, increased noise/congestion levels or low-power at the device), at a potential loss of performance (video jitter). A subsequent adaptation is then performed to react to the global system change, by dynamically exploring an optimal search space for an optimized operating point. To improve performance in such dynamic environments, we propose a *proactive* adaptation scheme in which the proxy "proactively" predicts future system conditions and determines an optimized operating point in advance. Specifically, the middleware at the proxy exploits its cognizance of the optimized network transmission search space, the profiled energy consumption levels of transcoded video streams, the residual power at the client device and the predicted dynamic network noise levels within each cell, to "proactively" adapt the video stream to maximize the end user experience. For example, the proxy middleware can predict the noise/congestion level of a cell just before a user moves into the cell, and determine how to adapt the stream as the user enters the cell. The "mobility manager" can be used to track users in the target cell. Two factors significantly influence the performance of the schemes: dynamic noise levels within each cell and the velocity of the device. We therefore compare the schemes using these criteria. To represent the dynamic noise levels we characterize the noise induced by each mobile device as the following distributions: *constant, random, poisson, uniform.*

Quality	Transformation Parameters	Avg. Power (Windows CE)	Avg. Power (Linux)
(Q8) Like Original	SIF, 30fps, 650Kbps	4.42 W	6.07 W
(Q7) Excellent	SIF, 25fps, 450Kbps	4.37 W	5.99 W
(Q6) Very Good	SIF, 25fps, 350Kbps	4.31 W	5.86 W
(Q5) Good	HSIF, 24fps, 350Kbps	4.24 W	5.81 W
(Q4) Fair	HSIF, 24fps, 200Kbps	4.15 W	5.73 W
(Q3) Poor	HSIF, 24fps, 150Kbps	4.06 W	5.63 W
(Q2) Bad	QSIF, 20fps, 150Kbps	3.95 W	5.5 W
(Q1) Terrible	QSIF, 20fps,100kbps	3.88 W	5.38 W

Table 1. **Energy-Aware Transformations for Compaq Ipaq 3650 with bright backlight. Cisco 350 Series Aironet WNIC card**

2.3 Energy-Aware Video Stream Regulation

In this section, we characterize a proxy-based traffic regulation mechanism to reduce energy consumption of the wireless network interface. We exploit the fact that wireless network interface(WNIC) cards consume significantly less energy in the *sleep mode* (0.184W) than in the *idle/recv mode*(1.34W/1.435W). In our scheme, the middleware at the proxy, buffers the transcoded video and transmits I seconds of video in a single burst along with the time τ=I for the next transmission as control information. The device then uses this control information to switch the interface to the active/idle mode at time $\tau + \gamma \times D_{EtoE}$, where γ is an estimate between zero and one and D_{EtoE} is the end-to-end network delay with no noise.

Let B be the average video transmission bit-rate, F the video frame rate, S_N the packet size used by the underlying network protocol. Let there be N users in the system. We model each "*other*" user as a Poisson process that injects packets into the network with the packet inter-arrival times following an exponential distribution with rate λ, with a density function of $f(t) = \lambda.e^{-\lambda t}$. Therefore, the number of packets introduced into the network by each user in an interval 't' has an expected value $E(t) = \lambda.t$. Assume that the AP employs a simple round robin service for transmitting packets and let L_{AP} be the buffer length(in number of packets) available at the AP for downstream traffic. Let BW_s, BW_d be the bandwidth available to the wireless device for transmitting and receiving data; T_{AP}, P_d be the queueing transmission delay per packet of the AP and propagation delay of the wireless network respectively. Using the above characterization, the number of packets per frame $\alpha = \frac{B}{8 \times S_N \times F}$ and the total number of packets transmitted in one burst $P_b = \alpha \times F \times I$.

To simplify the analysis, we further assume that there are no loss of packets at the AP due to weak signal strengths or collisions. However, packets do get dropped if the AP buffer capacity (L_{AP}) is less than the number of arriving packets. A queueing theory analysis is used to predict packet loss rates at the AP. If T_b is the time taken to transmit the burst by the proxy, then the total expected number of packets received at the AP in that interval is $\sigma = P_b + \sum_{k=1}^{k=N-1} E_k(T_b)$, where E_k is the expected value of noise from by user k. The worst case expected transmission delay experienced by the last packet in the burst is $D = \sigma \times T_{AP}$. Using the above approach of bursty transmissions, the total "*sleep*" time(δ) of the network interface can be calculated as $\delta = \tau - (D + \gamma \times D_{EtoE})$, neglecting the propagation delay of the final packet. As significant power gains are only achieved when the network interface is in the sleep mode, the total power savings are $P_{saved} = \delta * (P_{IDLE} - P_{SLEEP})$. Observe that large values of I can result in packet losses at the access point and/or buffer overflows at the device. The above analysis can be used by an adaptive middleware to calculate an optimal I(burst length) for any given video stream and noise level. Other approaches for power management of NIC cards have been studied in [2,1].

3 Experimental Results

We simulated the overall system(Fig. 1) with the adaptation mechanisms and energy-aware streaming for measuring the *Utility factor(U_F)* for the system.

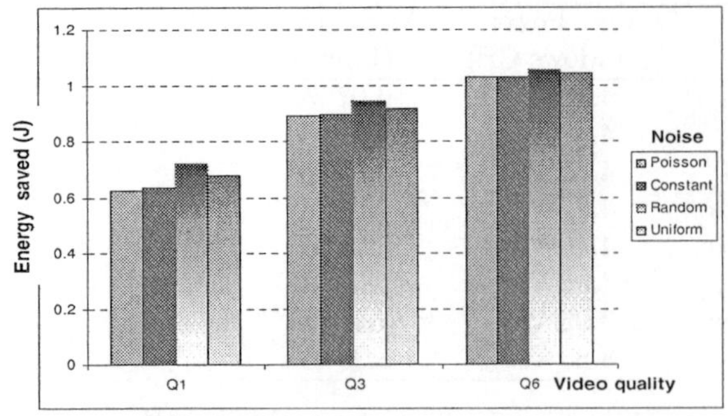

Figure 2. Impact of Noise Patterns

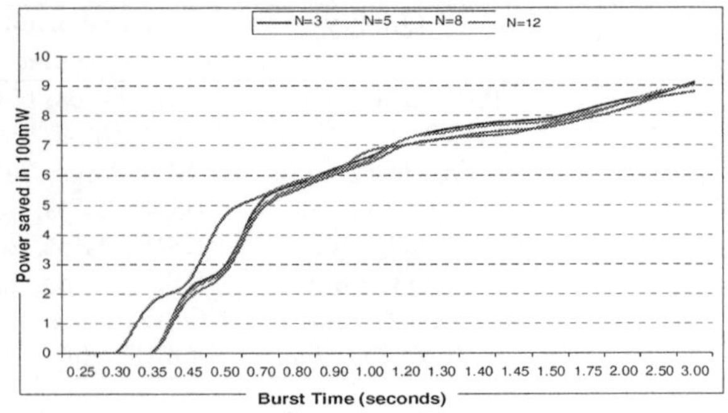

Figure 3. Power Saved vs Burst Time

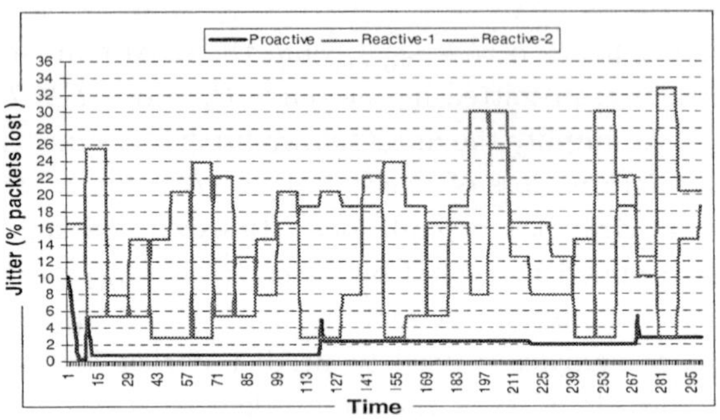

Figure 4. Proactive vs. Reactive

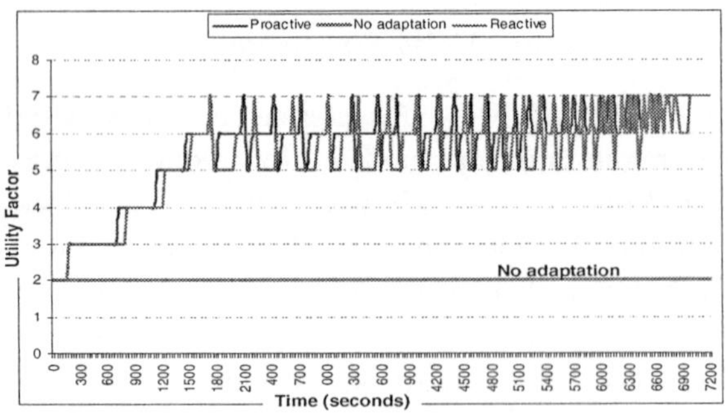

Figure 5. U_F vs. Adaptation scheme

Our simulations were based on measurements made for a Compaq iPAQ 3650 and a 206Mhz Intel StrongArm processor. The iPAQ used a Cisco 350 Series Aironet 11Mbps wireless PCMCIA NIC for communication. In our simulator, a constant end-to-end network delay of 400ms was used; the wired and wireless ethernet bandwidths were set to 10Mbps and 8Mbps(effective B/W, 802.11b is capable of higher throughput) and γ was set to a 0.85. The transmission delay of the wireless access point was also fixed at 400μs per packet. First, for each video quality(Q1-Q8), we varied the video burst time(100ms to 20sec) and the network noise levels(users), the network packet size (200bytes to 700 bytes) and measured the power savings. An incremental mobility model [5] was implemented. Fig. 2 shows the impact of *noise patterns* on the maximum energy saved for different video quality levels. Lower quality streams consume less power, but for a given video quality the distribution used to model the noise has little impact. Fig. 3 shows the variation of power savings with video burst sizes at different noise levels. Using these values a search space is generated for ascertaining optimized video burst times for a particular video quality and noise level. Fig. 4 shows that the proactive adaptation results in a much smoother video when compared to reactive adaptation. Finally, we see in Fig. 5, that using a proactive adaptation scheme a significant improvement in the U_F of the overall system can be achieved.

References

1. Prashant Shenoy and Peter Radkov, "Proxy-Assisted Power-Friendly Streaming to Mobile Devices", in MMCN 2003.
2. Surendar Chandra, "Wireless Network Interface Energy Consumption Implications of Popular Streaming Formats", in MMCN 2002.
3. ITU-R Recommendation BT-500.7, "Methodology for the subjective assessment of the quality of television pictures", ITU Geneva Switzerland.
4. S.Mohapatra, R.Cornea et.al., "Integrated Power Management for Video Streaming to Mobile Handheld Devices", in ACM MM 2003.
5. Z. Haas, "A new routing protocol for the reconfigurable wireless networks", in IEEE Personnal Comm. 1997.

DEVELOPING PDA FOR LOW-BITRATE LOW-DELAY VIDEO DELIVERY

K.-W. Roger Cheuk S.-H. Gary Chan
K.-W. Alice Mong C.-M. Martin Lee S.-S. Shirley Sy
Department of Computer Science
Hong Kong University of Science and Technology
Clear Water Bay, Kowloon
Hong Kong
Email: {rcheuk, gchan}@cs.ust.hk

Abstract— **Pervasive wireless multimedia applications often require Personal Digital Assistants (PDAs) for processing and playback. The capability of PDAs, however, are generally much lower than desktop PCs. When these devices are used to play back video delivered over a network from a desktop server, their buffers can easily overflow, seriously degrading the video quality. In this paper, we report our implementation of some special stream processing techniques to deal with the capability mismatch between a PC and PDAs for low-delay live video streaming. These techniques are, the Selective Packet Drop (SPD) algorithm, the Game API (GAPI) optimization and the speed adaptation algorithm. All of them can be easily implemented. We show that our system provides much better video quality than systems without our techniques.**

I. Introduction

With their continuous reduction in size and weight, Personal Digital Assistants (PDAs) such as Pocket PCs and Palms have grown in popularity over the past years. Nowadays, PDAs offer multimedia capabilities, network functionalities, and a suite of programming kits for application development.

We have developed a low-delay real-time video delivery system for pervasive multimedia applications such as surveillance, interactive distance learning, video conferencing and so on. It is based on packet-based video, and users roaming in a wireless LAN can ubiquitously access and display video streams with their PDAs at any time. In the system, a desktop PC captures the video, compresses it into the H.263+ format and finally encrypts the compressed video with a key (we have used the RC4 symmetric key encryption algorithm due to its low processing requirement) before streaming through a wireless LAN using UDP to PDAs for display, as illustrated in Fig. 1. Users are able to view the compressed video with a Pocket PC (we have used Pocket PC instead of Palm due to its higher processing and multimedia capabilities).

Streaming video over the Internet and wireless medium presents many challenges [1], [2]. Specifically, PDAs has the following limitations:

• Limited buffering capability — PDAs do not have as large storage as desktops. As a result, how its limited buffer is managed impacts greatly on the system performance. As a matter of fact, in low-delay real-time video, buffering should not be large. When a buffer becomes full, packets have to be dropped or discarded intelligently in order to guarantee high video quality.

• Relatively low processor speed — Despite the continuous improvement in processor speed, PDAs still lag far behind desktops or laptops. This limitation becomes apparent in processing-intensive low-delay video applications.

The relatively low processing speed of PDAs often causes serious degradation in video quality delivered to users, due to the so-called "Speed Mismatch Problem": although video encoding is generally more computationally-intensive than video decoding, the video bandwidth can easily overwhelm the mobile devices, if the encoding rate is not properly chosen or controlled. When this happens, the client's buffer tends to overflow, leading to bursty packet drops. To make the matter worse, many video compression standards such as H.263+ use inter-coded P frames to reduce temporal redundancy in order to conserve bandwidth. As the packets corresponding to these P-frames are dropped, errors propagate to the subsequent frames, leading to poor video quality. We show in Fig. 2 what happens when the encoding frame rate is much higher than the decoding frame rate if the client employs a simple drop-tail buffering policy. The server is a PIII PC with 128MB RAM and the client is an iPaq 3600 Pocket PC. As the server "outpaces" the PDA, high packet drop rate and error propagation result. Clearly, the video quality is unacceptable. This example shows that addressing the mismatch problem is important to offer quality video.

This work was supported, in part, by the Sino Software Research Institute at the HKUST (SSRI01/02.EG21) and the Competitive Earmarked Research Grant from the Hong Kong Research Grant Council (HKUST 6014/01E and HKUST 6199/02E).

Correspondence to: Dr. S.-H. Chan

Fig. 1. Our wireless video delivery system.

Fig. 2. Video quality without applying any stream processing techniques.

To address this problem, we have implemented a number of special processing techniques. Our techniques are general enough to work with many error recovery schemes (such as [3], [4], [5] and references therein) and many network-adaptive rate control schemes [6], [7]. Regarding video, we have used a GOP structure of IPPPP...IPPPP..., with the GOP size chosen so as to tradeoff startup delay, error propagation and bandwidth requirement (we have used a GOP size of 10 in this paper). The techniques we have implemented in the system are:

• Client-side implementation:

– Selective Packet Drop (SPD) algorithm: SPD drops packets to retain only the most important few frames (i.e., the leading P-frames) in the buffer upon the arrival of an I-frame. This is done in order to prevent buffer overflow and to synchronize the video delivered between a server and the clients.

– Game API optimization: In order to improve the decoding rate of PDA, we have used the Game API library for fast video frame display. We demonstrate that this technique can markedly increase the decoding rate as compared to simply using Win32 system calls. The processing gap between a desktop PC and a Pocket PC is narrowed due to reduced CPU usage in rendering activities.

• Feedback-based implementation: Note that even with SPD and GAPI, the video may appear "jerky" if a server outpaces a receiver by a large margin (mainly due to the fact that whenever an I-frame arrives, a burst of P-frames may be dropped.). In order to address this problem, the server can match the clients' processing capabilities based on feedback from clients. Clearly,

due to the feedback implosion issue, this solution is suitable for unicast and small-scale multicast applications (for large-scale multicast applications, a means of controlling the feedback rate has been suggested in [8]).
Adapting the encoding rate to the decoding rate has been discussed in [9], [10], [11]. Most of these algorithms require a receiver periodically and continuously feeding back its buffer state back to the server. This increases the network bandwidth and may cause the server to adapt too frequently, which is visually disagreeable. We implement a new algorithm in our system in which a client only updates the server whenever it is necessary. The speed matching algorithm is termed *Speed Adaptation Algorithm* (SA).
In SA, clients measure the decoding time it takes to decode its frames. Each client measures the decoding time of the first frame and reports a slightly increased decoding time as the target inter-frame time to the server (via a feedback TCP channel). Hence after, each client compute at each frame a new target inter-frame time based on the measured frame decoding time. Whenever there is a significant change in the target inter-frame time as compared to the last feedback value, the client updates the server with the new value. Based on this feedback, the server adjusts its encoding frame rate so as to match the speed of the client(s). Obviously, such a solution is effective when the processing capability of the PDAs does not vary much during a round-trip, which is usually the case.

By applying these techniques, we have successfully built a low-delay live video surveillance system with a Pocket PC over a wireless LAN. Our results show that excellent video quality can be achieved on a Pocket PC.

This paper is organized as follows. We first describe in detail these techniques in Sect. II. In Sect. III, we present our experimental environment and measurement results. We conclude in Sect. IV.

II. System Implementation

In this section, we describe in more detail the client-side implementations followed by the feedback-based implementation.

A. Client-side solutions

We have implemented the following two techniques in our video delivery system at the client-side, namely the Selective Packet Drop (SPD) algorithm and the Game API (GAPI) optimization.

A.1 Selective Packet Drop (SPD) algorithm

Note that the importance of video frames in a GOP sequence IPPPP... decreases from the first I frame to the last frame in that GOP. This is because each P frame in the GOP uses the previous frame as the reference frame. The Selective Packet Drop (SPD) algorithm aims to drop the trailing P-frames to prevent a buffer from overflowing while maintaining low video delay.

In SPD, packets are allowed to accumulate in the buffer as long as there is buffer space. To keep the delay low and to take into consideration of network delay jitter and fluctuations in the time required to encode/decode a frame, the n most important video frames (i.e., the leading frames at the head of the buffer queue) are retained in the buffer whenever an I-frame arrives. Clearly, higher network jitter and encoding/decoding variability require higher n, which also means a higher delay.

We use a dedicated thread to run the SPD. The module implemented is shown in Fig. 3. It starts by initializing the queue data structure and the enqueue flag. It then enters into a loop to handle all received packets by the SPD algorithm. When the first packet of an I-frame is received, all except the first n frames in the queue are dropped. The last block of the pseudo code enqueues the currently received packet if the enqueue flag is set. The purpose of the enqueue flag is to make sure that once a frame is dropped, it waits until the next I-frame arrives before accepting packets again as the subsequent P frames would be useless. If the queue cannot accommodate all packets of the frame at the tail of the queue, the function call to `RemoveCurrRecvFrame` is used to remove that frame's packets buffered in the queue.

```
SPDTHREAD(m, n)
Q ← Empty Queue of size m
EnqueueFlag ← 1
while 1
do READPACKET(P)
    if P is the first packet of an I-frame
      then RETAIN(Q, n)
               EnqueueFlag ← 1
    if EnqueueFlag = 1
      then if not FULL(Q)
               then ENQUEUE(Q, P)
               else REMOVECURRRECVFRAME(Q)
                    EnqueueFlag ← 0
```

Fig. 3. The Selective Packet Drop (SPD) algorithm.

TABLE I
Some Game API display-related functions.

Function	Description
GXOpenDisplay	Initialize the Game API library
GXCloseDisplay	Release the resources used by Game API
GXBeginDraw	Prepares the Game API before display
GXEndDraw	Release the resources allocated with GXBeginDraw after display is done
GXGetDisplayProperties	Obtain the hardware display capabilities usable by the Game API library

A.2 Game API Optimization

Microsoft has released a new set of API, the Game API (GAPI), specifically targeted for use in high-performance, real-time games on Pocket PC devices. It has a number of useful functions for fast video display. We have achieved a substantial increase in the decoding frame rate (nearly 40% in our experiments) when we change our display code from the Win32 graphics API to this set of GAPI. This improvement comes from the fact that these functions allow the display memory to be accessed directly. With GAPI, the decoding rate is significantly increased. Table I shows the list of the display-related functions available in the Game API.

B. Feedback-based solution

B.1 System Description

We show in Fig. 4 the system diagram of our server-side video-streaming subsystem with the feedback and speed adaptation architecture. It consists of the following components: the Synchronization Unit, the H.263+ encoder, the Network Module, the Token Bucket and the virtual Clock. The Synchronization Unit controls the rate at which a frame is captured by waiting for at least one token in the Token Bucket. When the unit obtains a picture and passes it to the H.263+ encoder, a token is removed from the token bucket. Tokens are inserted to the token bucket by a clock ticking at a rate specified by the most recent feedback data from the receiver. The clock is informed of the most updated token insertion interval by the network module. When the token bucket is full, no more tokens can be inserted. The network module buffers the compressed stream from the encoder and sends it to the user(s) via UDP. It may also poll the client(s) for feedback.

Regarding the client-side video-streaming subsystem (Fig. 5),

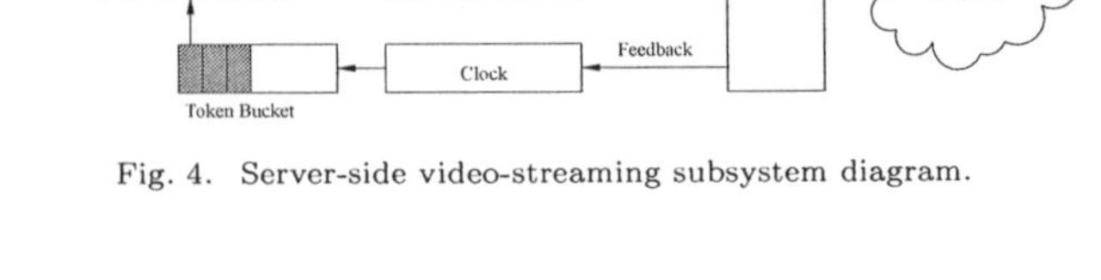

Fig. 4. Server-side video-streaming subsystem diagram.

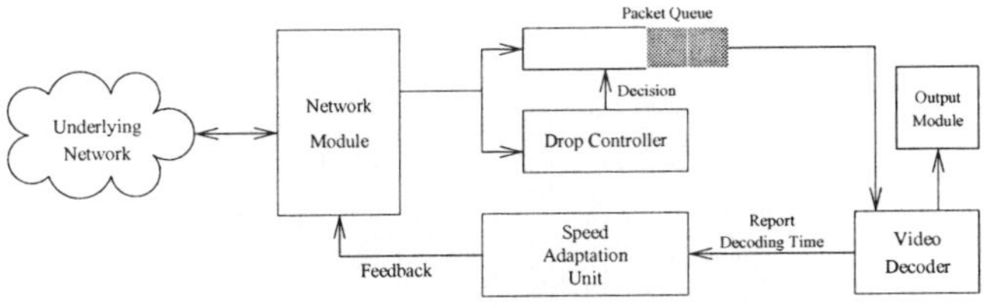

Fig. 5. Client-side video-streaming subsystem diagram.

it has a Packet Queue to buffer all the video packets from the network. The drop controller analyzes all the incoming video packets and decides when to drop packets. It notifies the packet queue of the dropping decision it needs to perform. Independently, the H.263+ video decoder reads packets from the packet queue and decodes them to output to the output module. It also measures the time to decode a frame and sends the data to the Speed Adaptation Unit. Using the data, the Speed Adaptation Unit decides whether to feed back to the server based on a formula given below (in Sect. II-B.2).

B.2 Speed Adaptation

Note that the achievable frame-rate at a client may fluctuate with time (due to background jobs and variations in frame decoding time). In order for the server not to outpace the client most of the time, the interval between successive encoded frames ("inter-frame" time) should not be set simply equal to the average time to decode a frame at the client, but slightly longer to account for the statistical fluctuation of the frame-rate at the client. In our system, each client hence reports to the server a requested inter-frame time using a mechanism similar to the round-trip time estimation as used in TCP, i.e., it first estimates the mean time to decode a frame and the mean deviation of it, and then requests an inter-frame time for the encoding process using these data.

The mean time to decode a frame is estimated as follows. For every frame i, $i \geq 1$, the decoder measures

- $T_t^{(i)}$ — the total time interval between the rendering of frame $i-1$ and frame i;
- $T_r^{(i)}$ — Obviously, $T_t^{(i)}$ consists of two parts: the busy time for the CPU to process the packets of frame i, and the idle time for the CPU to wait for the the constituent packets of frame i to arrive (mainly due to the blocking call used in UDP, `recvfrom`). Denote the total idle time spent in the function call `recvfrom` by $T_r^{(i)}$.

Obviously, the busy time a CPU takes to process the frame i is the limit the PDA can process frame i (the maximum decoding capability at that time), which is given by, $T^{(i)} = T_t^{(i)} - T_r^{(i)}$. $T^{(i)}$ is hence the inter-frame time for frame i which the Pocket PC is capable of processing[1]. The average time $\overline{T^{(i)}}$ to decode and display a frame can therefore be estimated as $\overline{T^{(i)}} = \alpha\overline{T^{(i-1)}} + (1-\alpha)T^{(i)}$, where $0 < \alpha < 1$. Clearly, a high α means that the adaptation "remembers" history and is resistant to changes or fluctuation; on the other hand, a low α means that the system is more susceptible to statistical fluctuation. We estimate the mean deviation of $\overline{T^{(i)}}$, $D_T^{(i)}$, as $D_T^{(i)} = \gamma D_T^{(i-1)} + (1-\gamma)|T^{(i)} - \overline{T^{(i)}}|$, where $0 < \gamma < 1$.

Each value $\overline{T^{(i)}} + \beta D_T^{(i)}$ is a target inter-frame time for feedback, where β is a positive constant. Choosing too large a β may unnecessarily reduce the frame rate and too low a β will increase the chance that the decoder drops packets due to statistical fluctuation (from our experiments, $\beta = 0.8$ performs quite well).

To reduce the amount of feedback, the target inter-frame time is sent selectively, i.e., when the target inter-frame time is significantly different from the previously feed back value. In our implementation, it is feed back when the difference between the current value of $\overline{T^{(i)}} + \beta D_T^{(i)}$ and the previously transmitted feedback value is larger by more than a certain fraction, f. As compared to the Continuous Media Player in which the encoder speed is adjusted by a mechanism similar to Additive Increase and Multiplicative Decrease (AIMD), our speed adaptation algorithm achieves a smoother rate fluctuation and lower feedback overhead [9].

In the server, it adapts the decoding speed by using a token bucket. The time between successive token insertion is equal to the most recent feedback value. The encoder has to obtain a token from the token bucket before encoding a frame; otherwise it is stalled for a token. In this way, the encoding rate is limited to the rate as suggested by the recent feedback data, but is unrestricted if the server's encoding frame rate is less than the achievable decoding frame rate at clients.

[1]We assume that the decoding thread is of highest priority possible, in which the execution time of other preempting threads/processes can be ignored. Other threads/processes are allowed to run when the decoding thread is waiting for a packet.

III. Experimental Results and Comparisons

We have implemented the aforementioned techniques to evaluate the effectiveness of our solutions. In this section, we first present our experimental environment followed by some illustrative measurements and results.

A. Experimental Environment

We have used the Foreman as a representative video sequence in our experiment (the use of other sequences show a similar trend and hence is not discussed). The sequence consists of 400 frames and is in the QCIF format. The frames are encoded in the H.263+ format before delivered over the network. The server-side video delivery program runs on a Pentium III 550MHz PC with 128 MB memory. The server is connected to a 100Mbps LAN. The mobile access point for offering wireless network connections is directly connected to the same LAN. The client-side program runs on an iPaq 3600 Pocket PC. Besides the wireless LAN card, no other additional hardware is installed to the Pocket PC.

Regarding the H.263+ settings, we have used a Quantization Parameter (QP) of 13, a search window size of 31 and a GOP size of 10. Error concealment is not used so as to eliminate its effect for fair comparisons. We use the following values in our experiment: $\alpha = 0.875$, $\beta = 0.8$, $\gamma = 0.75$, $T_a^{(0)} = 100$ms, $D_T^{(0)} = 3$ms, $n = 0$ (low-delay video), and $f = 10\%$.

The encoded video stream is transmitted packet-by-packet to the clients. Each encoded frame is divided into blocks of 1,024 bytes of UDP packets. The buffer at the decoder side is a FIFO queue accommodating up to 32 packets.

B. Illustrative Measurement Results

The evaluation is done by comparing the picture quality in terms of PSNR with and without our solutions. Note that since some packets may be dropped, we compute PSNR only for the displayed complete frames. We also compare the encoding frame rate with and without Speed Adaptation.

Note that there is one more issue that needs to be considered. As packet loss can be due to reasons other than speed mismatch between the server and a client (i.e., channel errors and congestion losses), it is necessary to isolate their effect from our experiments in order to draw meaningful conclusions. For all experiments, we independently maintain a log for packets transmitted and a log for packets received by the client application. By comparing the logs at the client-side and the server-side for each experiment, it is confirmed that the wireless environment has negligible error and there is no congestion losses throughout any of the experiments. We conclude that packet loss is due to the inability to process packets fast enough before buffer overflow.

B.1 Client-based solutions

In Fig. 6 we show the decoded video quality when none of our solutions is used (i.e., the NO OPTION setting). The gaps in the sequence means that there are dropped frames due to buffer overflow and propagation errors. The video quality decreases quite sharply once packets start to be dropped (at around frame 50). This is when the decoder buffer becomes full (as the encoder out-runs the decoder). With the buffer often being full, we also observe a rather high delay (in seconds) in the system.

The "landsliding" performance in video quality can be markedly improved when our SPD algorithm is used. This is

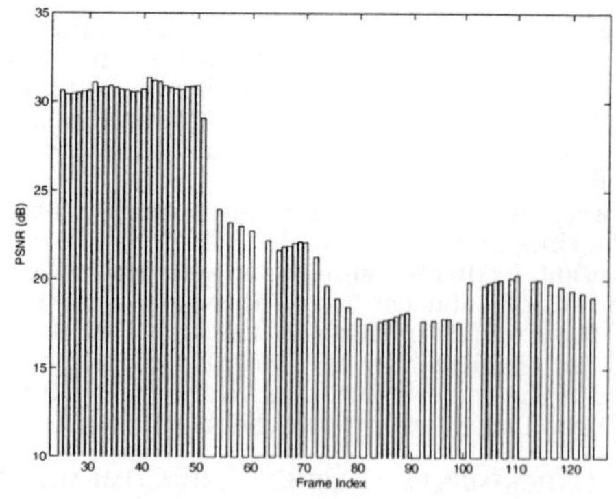

Fig. 6. PSNR for the decoded frames with the NO OPTION setting.

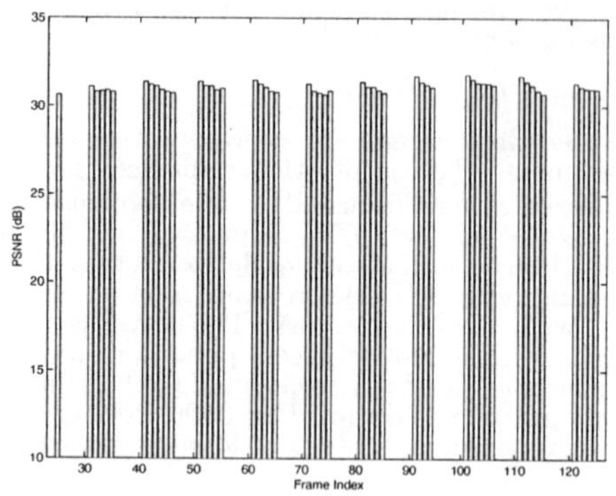

Fig. 7. PSNR of decoded frames with SPD.

shown in Fig. 7. Note that the PSNR is maintained at a high level. Since the arrival of an I-frame triggers the intelligent packet drop algorithm, some video frames are not displayed, as evident from the discontinuities in the figure. Basically, in each GOP, around half of the frames are dropped. Though PSNR and delay are greatly improved, the burst of frame loss leads to some "jerkiness" in the decoded sequence.

This missing frame problem can be solved by applying the GAPI optimization, as shown in Table II. In this table, the results for both cases are obtained (both cases also with the SPD algorithm turned on). By applying GAPI optimizations, the decoding frame rate increases substantially and hence reduce significantly the fraction of frame loss. Note that the combined method, i.e. the SPD algorithm and GAPI optimizations, requires implementation at the client-side only, making it suitable for multicasting video to a large number of PDAs.

B.2 Feedback-based solution

We finally present the performance of a feedback-based system with SPD and GAPI implemented at the client side PDA and the Speed Adaptation algorithm implemented at the server side. We show in Fig. 8 the corresponding PSNR as measured in the PDA. Clearly, the video quality has been greatly improved with much less dropped frames. The figure shows that our speed adaptation algorithm is effective in matching the speeds between the encoder and decoder. The end-to-end delay is also low, due to the synchronizing nature of our Selective Packet Drop algorithm.

TABLE II
OVERALL FRAME RATE STATISTICS FOR THE 400 FRAMES FOREMAN SEQUENCE

Options	Encoding frame rate (frame/s)	Decoding frame rate (frame/s)	Fraction of frame loss
SPD alone	34.59	17.41	48.25%
SPD with GAPI optimization	34.52	29.32	14.5%

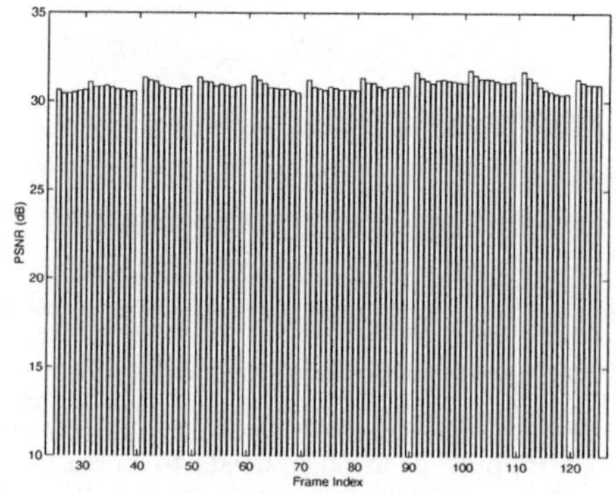

Fig. 8. PSNR of the decoded frames in a feedback-based system, with speed adaptation, SPD and GAPI optimization.

IV. CONCLUSIONS

In this paper, we have described our experience in implementing a low-delay video streaming system. Pocket PCs are generally much slower than desktop PCs, and hence the encoder may easily overwhelm the decoder, leading to packet drop and poor video quality (the so-called "Speed Mismatch Problem"). Stream processing techniques are needed to bridge the processing gap between the encoder and the decoder.

This speed mismatch problem can be addressed by our Selective Packet Drop (SPD) algorithm, the Game API optimization and the speed adaptation (SA) algorithm. The SPD algorithm drops the least important frames whenever an I-frame arrives to prevent buffer overflow while keeping the delay low. The Game API optimization renders video with a higher speed, hence narrowing the processing gap. The speed adaptation (SA) algorithm uses clients' feedback to match the encoding rate to the decoding capability of the clients.

Our experiments show that video quality is substantially improved with our techniques. The display frame rate with the Game API can be increased significantly. This shows that our techniques are useful and implementable for pervasive multimedia applications such as surveillance and distance-learning.

REFERENCES

[1] D. Wu, Y. T. Hou, W. Zhu, Y.-Q. Zhang, and J. M. Peha, "Streaming video over the Internet: approaches and directions," *IEEE Transactions on Circuits and Systems for Video Technology*, vol. 11, pp. 282–300, March 2001.

[2] D. Wu, Y. T. Hou, and Y.-Q. Zhang, "Scalable video coding and transport over broadband wireless networks," *Proceedings of the IEEE*, vol. 89, pp. 6–20, January 2001.

[3] T.-W. Lee, S.-H. Chan, Q. Zhang, W.-W. Zhu, and Y.-Q. Zhang, "Allocation of layer bandwidth and FEC for video multicast over wired and wireless networks," *IEEE Transactions on Circuits and Systems for Video Technology*, vol. 12, pp. 1059–1070, Dec 2002.

[4] A. Majumda, D. Sachs, I. Kozintsev, K. Ramchandran, and M. Yeung, "Multicast and unicast real-time video streaming over wireless LANs," *IEEE Transactions on Circuits and Systems for Video Technology*, vol. 12, pp. 524 –534, June 2002.

[5] B. Girod and N. Farber, "Feedback-based error control for mobile video transmission," *Proceedings of the IEEE*, vol. 87, pp. 1707–1723, October 1999.

[6] Q. Zhang, W. Zhu, and Y.-Q. Zhang, "Network-adaptive rate control with TCP-friendly protocol for multiple video objects," in *Proceedings of IEEE International Conference on Multimedia and Expo*, vol. 2, pp. 1055–1058, July/August 2000.

[7] J. Cabrera, A. Ortega, and J.I.Ronda, "Stochastic rate-control of video coders for wireless channels," *IEEE Transactions on Circuits and Systems for Video Technology*, vol. 12, pp. 496 –510, June 2002.

[8] J. C. Bolot, T. Turletti, and I. Wakeman, "Scalable feedback control for multicast viodeo distribution in the Internet," in *Proceedings of ACM SIGCOMM 94*, vol. 24, pp. 58–67, October 1994.

[9] L. A. Rowe and B. C. Smith, "A continuous media player," in *Proceedings of Network and Operating System Support for Digital Audio and Video*, pp. 376–386, 1992.

[10] A. Goel, M. H. Shor, J. Walpole, D. C. Steere, and C. Pu, "Using feedback control for a network and CPU resource management application," in *Proceedings of the 2001 American Control Conference (ACC)*, vol. 4, pp. 2974–2980, June 2001.

[11] S. Cen, C. Pu, R. Staehli, C. Cowan, and J. Walpole, "A distributed real-time MPEG video audio player," in *Proceedings of the 5th International Workshop on Network and Operating System Support for Digital Audio and Video*, pp. 151–162, 1995.

FEATURE EXTRACTION IN WIRELESS PERSONAL AND LOCAL AREA NETWORKS

RENE MAYRHOFER, HARALD RADI and ALOIS FERSCHA

Institut für Praktische Informatik, Johannes Kepler University Linz, Altenberger Str. 69, 4040 Linz, Austria
Email: {mayrhofer,radi,ferscha}@soft.uni-linz.ac.at

Context awareness is currently being investigated for applications in different application areas of mobile computing. The integration of Bluetooth and Wireless LAN technologies into a vast of mobile devices — ranging from smartphones and PDAs to portable computers – has made user context sensing based on those technologies a feasible and promising approach. In this paper, we study which of the Bluetooth and Wireless LAN technology features (like radio-signal strength, device address management, etc.) can be exploited to derive user context, and develop a procedure how low level sensor data can be brought to application level context information. We introduce a method to automatically classify heterogeneous sensor data features with supervised or un-supervised classification methods. By defining two operations, a distance metric and an adaptation operator, any feature can be used as input for the classifier and can thus contribute to context detection.

1 Introduction

Common to most of the visions for next-generation computing are the paradigms of mobile computing and context awareness;[7] additionally, they all agree that user interfaces should become less obtrusive and "smarter" with regards to adapting to the user. Our approach is to add context awareness to applications, allowing them to adapt to the situation (*context*) at hand. One of the more recent definitions of context by Dey[1] is *any information that can be used to characterize the situation of an entity, where an entity can be a person, place, or a physical or computational object*, which we adopt in this paper. The goal is to derive high level context information from low level sensor data by following a three-step approach: data acquisition, feature extraction and classification. Starting from low level sensor data based on Bluetooth and Wireless LAN (*WLAN*), appropriate features (e.g. SNR, MAC addresses in proximity, access points in range, ESSID) are extracted and classified to yield high level user context information (e.g. busy, traveling, in a meeting, in the office, at lunch, at home, telephoning, etc.).

However, these sensors do not yield single, numerical values but data with a more complex structure. Some important information that can be extracted from these sensors is categorical and non-atomic, e.g. the list of MAC addresses which are currently in communication range. Nonetheless, it could provide useful information for determining the current user context via an automatic classification of all available sensor signals. If non-atomic values should be used as input to a classification algorithm which can only work with numerical inputs, they need to be coded. The standard procedure for dealing with categorical data is to code each possible sensor value as binary input to the classification algorithm. This has been applied successfully to categorical data with a bounded set of values (e.g. department), but is problematic when the set of possible values is not known in advance or too large to be coded with binary inputs (e.g. WLAN MAC addresses would need 2^{48} input dimensions to cover all possible values). Therefore, coding categorical data as numerical inputs does not seem to be the best solution and a better method should be found.

2 Related Work

In this paper, we concentrate on feature extraction from wireless network interfaces — an overview of our research on context awareness and prediction of user context has been presented elsewhere.[5,2] Although feature extraction and classification are well-known fields of research, most publications only cover numerical, continuous features. In the field of context awareness of mobile devices, Van Laerhoven described a system with custom sensors (e.g. accelerometers, light sensors), using minimum, maximum, mean and variance as features.[4] Because the restriction to numerical features severely limits the choice of sensors, we introduce a model for utilizing heterogeneous features (e.g. list of MAC addresses and WLAN ESSID) in a common classification step. For a Kohonen SOM, a method has been described to cope with heterogeneous input values.[6] However, the general case of context detection with heterogeneous features does not seem to have been covered before and is the main focus of our work.

3 Concept

To derive knowledge about the device/user context from raw sensor data, we follow the following steps:

1. Sensor data acquisition: Sensors provide data streams (time series) of measurements. Usually some physical values like RF signals are the base for measurements, but more abstract sensors like the currently active application can also be utilized. Besides wireless network interfaces, other common sensors available on typical mobile devices include a microphone, a brightness sensor (for automatic control of display brightness) and information about being connected to the docking station. Additional sensors like GPS, GSM, compass, accelerometer, tilt, temperature or pressure sensors can easily be added by connecting them via wire or Bluetooth.

2. Feature Extraction: From raw sensor data, information can be extracted by domain-specific methods, yielding multiple condensed data values, which are called *features* F with samples $f \in F$. During feature extraction, the available data is transformed, allowing it to be interpreted better. Usually, simple statistical parameters like the mean $\bar{x}$, standard deviation σ or higher moments are used as features for time series. For wireless networks, special features like the signal strength, the current WLAN ESSID or the list of access points in range are more appropriate.

3. Classification: Several features extracted in the previous step together compose a *feature vector* $F_1 \times F_2 \times ... \times F_n$, which defines points $\langle f_1, ..., f_n \rangle \in F_1 \times F_2 \times ... \times F_n$ in the multidimensional *feature space*. The goal of classification is to find common patterns in the feature space, which are called *classes* or *clusters*. Because a feature vector should possibly be assigned to multiple classes with certain *degrees of membership* (the "probability" that the feature vector belongs to a class), we utilize soft classification / soft clustering approaches. These approaches can be defined as mapping a feature vector of n different features to degrees of membership $u \in [0; 1]$ for m different classes: $F_1 \times F_2 \times ... \times F_n \rightarrow [0; 1]^m$

4 Feature Extraction

This paper concentrates on the second step and its interface to the third. Based on the specific feature extractor, a feature's class is, within this paper, defined as one of the following:

- nominal (categorical, qualitative): set of values on which no order relation has been defined. A special case are binary features with $F = \{0, 1\}$.
- ordinal (rank): set of values with a defined order relation.
- numerical (quantitative): ordered set of values with defined $+$ and $\cdot$ operations (an algebraic field). We can further distinguish according to the density of values in the set between discrete ($F \subseteq \mathbb{Z}$) and continuous ($F \subseteq \mathbb{R}$) features.
- interval: intervals instead of single values, e.g. $F \subseteq \mathrm{Pot}\,\mathbb{R}$.

When deriving context from wireless network interface data, we can identify a number of features from different categories, which seem to be relevant for mobile devices:

- Bluetooth: list of MAC addresses in range (nominal), number of MAC addresses in range (numerical/discrete)
- WLAN: list of access point MAC addresses in range (nominal), number of access point MAC addresses in range (numerical/discrete), current ESSID (nominal), associated to access point (binary), access point MAC address associated to (nominal), signal strength (numerical/continuous), transmission rate (numerical/discrete)
- GSM: list of GSM cells in range (nominal), number of GSM cells in range (numerical/discrete), current provider (nominal), current GSM cell (nominal), signal strength (numerical/continuous)

This list is not exhaustive, but should provide a meaningful view of the wireless networks context around the mobile device.

The feature vector $\langle f_1, ..., f_n \rangle \in F_1 \times F_2 \times ... \times F_n$ formed by an arbitrary combination of these features is highly heterogeneous, making it necessary to cope with the different types and semantics of the feature space dimensions in the classification step. From our comparison of different classification methods, we concluded that only two operations are necessary on an abstract feature F (this matches the results of other research[6]). The first is a similarity measure or *distance metric*, which has to be defined on every feature, i.e. on every dimension of the feature space.

$$d : F \times F \to [0;1]$$
$$\delta = d(f_1, f_2)$$

defines the distance between two samples of the feature F, normalized to $[0;1]$; the normalization is not necessary, but eases the classification step. A general distance metric has to fulfill non-negativity, identity, commutativity (symmetry) and the triangle inequality. We would like to note that the range of values in F can change (increase) during runtime, thus the distance of two given samples can also change due to the normalization. Although this is no problem with the classification methods we are investigating, others might need modifications. Additionally, the second operator adapts/modifies a point (in one dimension) and is necessary for supervised and un-supervised learning.

$$\alpha : F \times F \times [0;1] \to F$$
$$f' = \alpha(f, g, a)$$

modifies the sample $f \in F$ towards $g \in F$ by a learning/adaptation factor $a \in [0;1]$. With these two operators, supervised and un-supervised classifiers like the Kohonen Self-Organizing Map (*SOM*) or K-Means clustering can be applied to any feature which defines them.

In the following, we will give example definitions of both operators for a selection of the features listed above.

1. signal strength: The Bluetooth, WLAN or GSM signal strength is a numerical, continuous variable; therefore, we can apply the L1 (Manhattan) metric:

   $$d(f_1, f_2) := \frac{|f_1 - f_2|}{F_{max} - F_{min}}$$

   and

   $$\alpha(f, g, a) := f + (g - f) \cdot a$$

   for samples $f_1, f_2, f, g \in F$ with maximum and minimum values F_{max} and F_{min}.

2. associated to access point: This is a binary variable with values $f \in \{0, 1\}$; the distance operator can thus be simplified to the equality relation:

   $$d(f_1, f_2) := \begin{cases} 1 \; if \; f_1 = f_2 \\ 0 \; if \; f_1 \neq f_2 \end{cases}$$

   A simple adaptation operator could be defined as

   $$\alpha(f, g, a) := \begin{cases} f \; if \; a \leq 0.5 \\ g \; if \; a > 0.5 \end{cases}$$

   which only sets it to one or the other value, depending on the current value of the adaptation factor a. A better, although more complicated variant is to implement the operator with state so that it internally sums up a and only modifies the feature value after a reaches a certain threshold.

3. number of MAC addresses in range: The number of Bluetooth peers or WLAN access points in range is a numerical, discrete variable; we define

   $$d(f_1, f_2) := \frac{|f_1 - f_2|}{F_{max}}$$

   and

   $$\alpha(f, g, a) := \begin{cases} \lceil f + (g - f) \cdot a \rceil \; if \; f \geq g \\ \lfloor f + (g - f) \cdot a \rfloor \; if \; f < g \end{cases}$$

   for a number of already detected, different MAC addresses F_{max}.

4. list of MAC addresses in range: The list of Bluetooth peers or WLAN access points in range is a nominal, non-atomic variable. Due to the non-atomicity, there are various ways for coding; but each value $f \in F$ can be seen as a set of addresses. For a list of addresses, we apply the Hamming distance (the number of different addresses)

   $$d(f_1, f_2) := |(f_1 - f_2) \cup (f_2 - f_1)|$$

   when f_1 and f_2 are sets. The adaptation operator can be arbitrarily complex; our current operator changes, according to the adaptation factor a, a randomly selected fraction of the different addresses in f and g to the addresses in g by adding and/or removing addresses in f. For performance reasons, we chose to code the list of MAC addresses as a bit vector, where already seen addresses correspond to bit positions.

5 Classification

Using these definitions, arbitrary classification methods can be used to classify feature vectors and thus derive context from wireless sensor data. Because of the resource limitations of mobile devices, it is not possible to record all sensor data; thus, batch algorithms which iterate over the whole data set at once can not be applied. On-line algorithms incorporate each input sample as soon as it arrives from the sensors and are therefore suited better. Another issue for context recognition with mobile devices is that algorithm parameters must either be constant or self-adaptive during run-time; a continuously decreasing "learning rate" as it is used in many algorithms like the SOM during its learning phase prevents an algorithm from running constantly without interruptions. For embedded systems, a distinction between learning and recognition phases does not seem to be appropriate.

Currently, we use the Growing Neural Gas[3] (*GNG*) clustering algorithm for our experiments because it offers a number of advantages over other methods, notably an unlimited number of clusters, un-supervised classification and clusters with arbitrary shapes. A short comparison of suitable clustering algorithms shows that GNG seems to be a good choice for mobile devices.[5]

All of the features listed above contribute to the classification and can therefore increase the quality of the context detection. Every feature has an unique view of the environment, yielding additional information that others can not provide. The combination of the features not only provides spatial context (in terms of qualitative localization), but also other aspects like the number of people standing nearby (if they carry Bluetooth-capable devices, which is becoming common). This can be used to determine non-spatial context, e.g. to detect a meeting situation.

6 Conclusion

We have considered to exploit Bluetooth and WLAN technologies as sensors for deriving user context. Current generations of mobile devices like smartphones or PDAs are equipped with these technologies, making them commonly available. In our approach, low level sensor data is transformed to high level context information in three steps: data acquisition, feature extraction and classification. Because some extracted features are not numerical, but categorical or even non-atomic, standard approaches to code them for the classification step do not seem appropriate. Experimental data has been collected using Bluetooth and WLAN as spatial proximity sensors over 10 days and was analyzed with K-Means and SOM classification algorithms using standard input coding. The preliminary results of our experiments suggest the use of a different input coding, e.g. the approach presented in this paper. An interface for using our feature extraction code including the different distance metric and adaptation operations in Matlab is currently being created and will allow direct, quantitative comparisons with standard algorithms. For future work, we will concentrate on the labeling of context classes by the user and on prediction of future user context, which will allow the development of proactive applications.

References

1. A. Dey, G. D. Abowd, and D. Salber. A context-based infrastructure for smart environments, 1999.
2. Alois Ferscha. Contextware: Bridging physical and virtual worlds. In *Ada-Europe*, volume 2361 of *Lecture Notes in Computer Science*, pages 51–64. Springer, 2002.
3. B. Fritzke. A growing neural gas network learns topologies. In G. Tesauro, D. S. Touretzky, and T. K. Leen, editors, *Advances in Neural Information Processing Systems 7*, pages 625–632. MIT Press, Cambridge MA, 1995.
4. K. Van Laerhoven. Combining the Self-Organizing Map and K-Means clustering for on-line classification of sensor data. In *ICANN*, pages 464–469, 2001.
5. Rene Mayrhofer, Harald Radi, and Alois Ferscha. Recognizing and predicting context by learning from user behavior. In *Proceedings of The International Conference On Advances in Mobile Multimedia (MoMM2003)*. Austrian Computer Society (OCG), September 2003. *(To appear)*.
6. S. Negri and L. Belanche. Heterogeneous kohonen networks. In José Mira and Alberto Prieto, editors, *Proceedings of the 6th International Work-Conference on Artificial and Natural Neural Networks, IWANN 2001*, Lecture Notes in Computer Science, pages 243–252. Springer, 2001.
7. A. Schmidt and M. Beigl. There is more to context than location: Environment sensing technologies for adaptive mobile user interfaces. In *Workshop on Interactive Applications of Mobile Computing IMC'98*, 1998.

SESSION 6A:
MOBILITY

POLICY BASED MOBILE IP HANDOFF DECISION (POLIMAND) USING GENERIC LINK LAYER INFORMATION*

STEFAN AUST, DANIEL PROETEL, NIKOLAUS A. FIKOURAS, CARMELITA GÖRG

Department of Communication Networks (ComNets)
Otto-Hahn-Allee, NW1, 28359 Bremen, Germany
Phone: +49 421 218 8264, Fax: +49 421 218 3601
E-mail: {aust|daf|niko|cg}@comnets.uni-bremen.de

CORNEL PAMPU

Siemens AG
Information & Communication Mobile Networks
Siemensdamm 62, 13623 Berlin, Germany
Phone :+49 30 386 20265, Fax : +49 30 386 31110
E-mail : Cornel.Pampu@siemens.com

There is currently a large variety of heterogeneous wireless access networks and cellular systems available, capable of providing data services. However, such systems require that users remain connected to a same network or terminate all communications when changing the point of attachment during handoff to another network or access technique. This paper describes link layer parameters which can be used to control Mobile IP (MIP) handoffs for seamless mobility. Based on a predefined Generic Link Layer (GLL) Mobile IP handoffs can be forced and can be used by different MIP systems, such as Hierarchical Mobile IP (HMIP) and Fast Mobile IP (FMIP). This paper describes a first implementation of a policy based Mobile IP handoff decision (POLIMAND) which uses link layer parameters for MIP handoff decisions.

1. Introduction

The approach of Mobile IP (MIP) handoffs that can be controlled by a Generic Link Layer concept, which is described in this paper, is part of IPonAir project funded by BMB+F (German government). To provide seamless mobility between heterogeneous access networks, the IPonAir system architecture is based on Mobile IP [1], and its Hierarchical Mobile IP extension [2].

However, the decision of Mobile IP handoffs between heterogeneous subnets will have to be more complex. The handoff decision should be based on application-specified policies and user requirements as well as periodic measurements of the quality of the underlying network connectivity. In the following, a concept for Mobile IP handoffs based on a GLL approach which supports the network access by different WLAN cards in a single Mobile Node is presented.

2. Description of Mobile IP Handoffs using Generic Link Layer Information

The concept of a GLL is motivated by the requirements of wireless communication with a seamless and lossless cooperation between different access networks and supports three additional functions [3]:

- Cooperation with mobility management functions and protocols
- Context transfer at handover
- Lossless reconfiguration

Moreover, the GLL defines three different interfaces:

- The physical layer interface
- The network layer interface
- The control interface

The interface of the GLL are defined and the interaction of the Generic Link Layer with mobility protocols, e.g. Mobile IP, as well as the handover performance have to be analysed.

* This work was done within the framework of the 'IPonAir' project which is partly funded by the German Ministry of Education and Research (BMB+F). See also http://www.iponair.de.

3. Available Network Parameters for a Generic Link Layer Definition

In the following a list of important and required network parameters [4] is presented to reflect the theoretical overview and related work for policy based Mobile IP handoff decision. Figure 1 contains an overview where the Generic Link Layer is located and shows the interaction with lower layers, e.g. WLAN networks, GSM, GPRS, 3G and beyond.

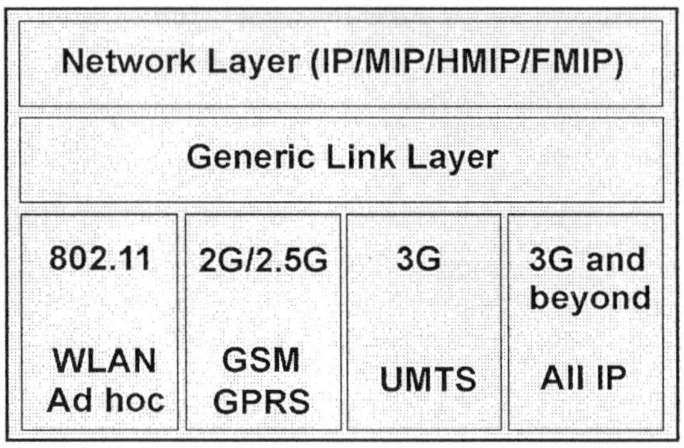

Figure 1: The Generic Link Layer (GLL)

Table 1 contains an overview of different link layer related information for different bearer technologies, IEEE802.11, GSM, W-CDMA and HIPERLAN/2. This overview is an abstract of the collection of network parameters for different bearer systems and outlines that there are parameters available for networks which can be used to define a GLL, e.g. signal strength [4].

Table 1: Available link layer information in different bearer access networks [4]

Link Layer Trigger	IEEE 802.11b	GSM	W-CDMA	HIPERLAN/2
Signal strength downlink	yes	yes	yes	yes
Signal strength uplink	no	yes	yes	yes
S/N downlink	yes	no	yes	yes
S/N uplink	no	no	yes	yes
BER downlink	no	yes	no	no
BER uplink	no	yes	no	no
FER downlink	yes	not specified	yes	yes
FER uplink	no	not specified	yes	yes

Table 2 contains a general list of parameters considered in wireless network systems. It defines different groups of parameters, namely the Link Information, the Environment Information, the Neighbourhood Information and the Link Layer Management Information.

Table 2: List of available network parameters

Group	Attributes	Parameters/Characteristics
Link Information		
	Status	
		Connected - associated with an access point in a subnetwork
		Pending - attempting to associate itself with an access point in a sub network
		Disconnected - not associated with any access point in any sub network
		Idle - powered down
	Quality	
		General Quality - a single percent value representing signal quality
		Signal Strength - in dB
		Noise/Silence Level
		Discarded Packets - number of packets lost due to poor link quality
		Retransmissions - number of packets retransmitted due to poor link quality
		Packets unable to decrypt
	Security	
		Encryption type
		Security Associations presence or not of security association with access point
Environment Information		
	Network	
		Frequency/Channel
		Protocol Type - IEEE 802.11/Bluetooth/etc.
		Power Consumption
		Cost - Economical cost of using this link
		Access Point and MAC - MAC address and/or unique ID
		Network Identity
		Bandwidth
		Throughput - when link is not in use
	QoS	
		Round Trip Time - RTT to communication peer and represents the distance
		Intserv/Diffserv
Neighbourhood Information		
	List of adjacent access points and routers	
Link Layer Management Information		
	Signal Level Threshold	

The Link Information is a collection of parameters that describes the status and quality as well as the loss characteristics of a single link. The Environment Information is a collection of parameters that provide information on neighbouring candidate access points and routers. The Neighbourhood Information which contains mainly a list of all access points and respective routers in the vicinity of the Mobile Node.

4. Policy based Mobile IP Handoff Decision (POLIMAND)

Mobile IP handoffs are mainly based on the reception of agent advertisements which are being frequently sent by the Mobility Agents. To accelerate the handoff process it is required to start a MIP registration process through a combination of MIP signalling and link layer hits.

Policy based Mobile IP handoff decision (POLIMAND) can be defined as the possibility to establish a Mobile IP handoff which is based on a function that consists of several input values of the link layer. Therefore, POLIMAND is more efficient using link layer information from a defined Generic Link Layer due to the fact that information from GLL are useful for POLIMAND within different Mobile IP network infrastructures and Mobile Nodes (e.g. cell phones, PDAs, Notebooks).

It is possible to define a policy for different decisions for Mobile IP handoffs. This is required for different available access networks, e.g. WLAN, GPRS, and UMTS. The policy defines when the current access networks is no longer available or it does not fulfil the requirements or predefined quality parameters. In this case the policy is the source for the decision which access network has to be used by the Mobile Node.

5. Implementation Details of POLIMAND

In the following a first implementation of POLIMAND is described. In this approach the S/N ratio is the policy for a Mobile IP handoff decision. The S/N ratio is the most important value in WLAN systems that can be measured by the Mobile Node to receive information about the radio conditions within the wireless network system.

Therefore, it has to be defined a threshold when the S/N ratio is not adequate. A Mobile IP handoff to a mobile network will be forced if the S/N ratio falls below a predefined threshold. Moreover, a hysteresis has to be defined to avoid ping-pong effects which can occur if physical radio conditions for the old and the new access networks are about equivalent.

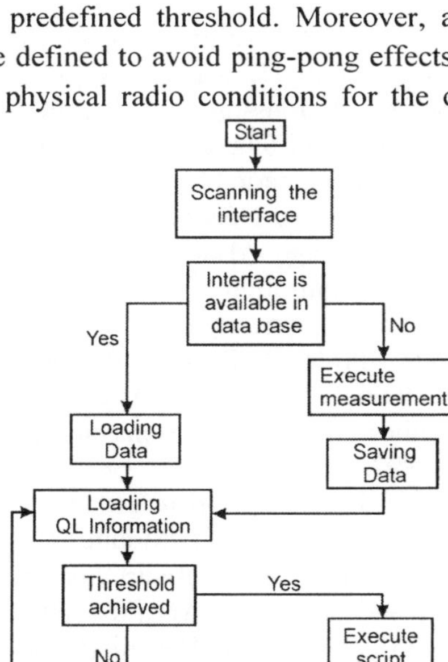

Figure 2: Flow chart of the algorithm for WLAN PCMCIA card measurements

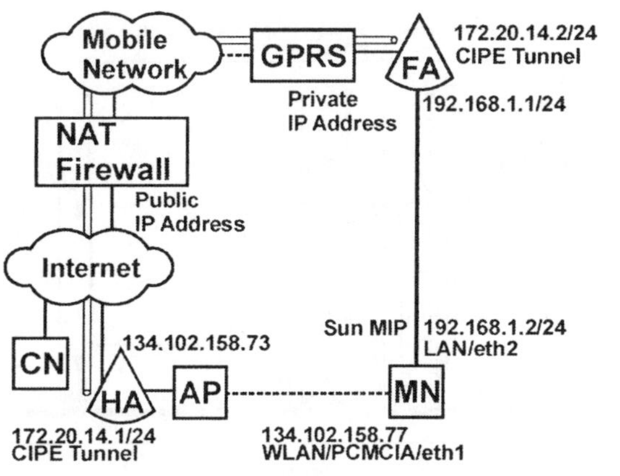

Figure 3: System architecture (Mobile IP testbed)

To support a wide spectrum of different WLAN PCMCIA cards a concept has to be developed to measure dynamically the different WLAN cards and to fix the hysteresis for each WLAN card to consider the conditions of each WLAN card. Figure 2 presents the algorithm which has been defined to measure different WLAN PCMCIA cards used by a single Mobile Node.

Especially for Linux operating systems, the Quality Link is the most important parameter that has to detect as the source for Mobile IP handoffs. This parameter is provided by the Wireless Extensions and can be used as it is described in [5].

Figure 3 presents the system overview which has been realized to investigate the algorithm that is described above. The complete testbed for Mobile IP handoffs between GPRS and WLAN that has been used for this measurement is described in [6].

6. Results

To show the handoff scenario for Mobile IP without layer 2 hints measurements have been done and the results are depicted in figure 4. The measurement results present the handoff process that has been measured while the Mobile Node moves (between 0,2m/s and 0,5m/s) in WLAN coverage and leaves the coverage area after 75 seconds. While the Mobile Node leaves the WLAN coverage the handoff process is not reliable and high packet loss between 25s and 75s can be observed.

Furthermore, figure 4 presents high packet loss that can be observed when the Mobile Node moves back into the WLAN coverage after 115s. Between 115s and

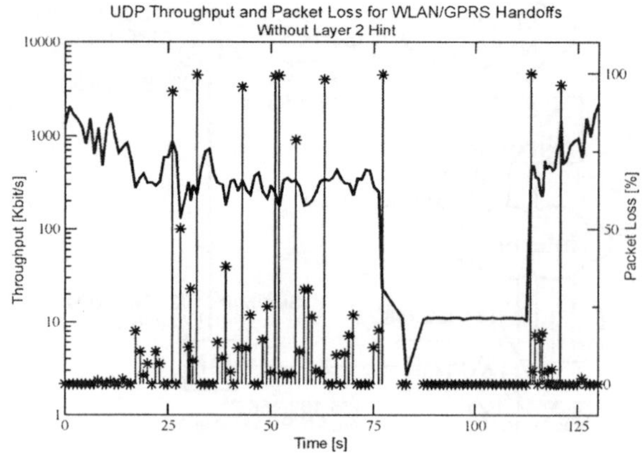

Figure 4: UDP throughput measurement including packet loss without layer 2 hint

120s high packet loss during the handoff from GPRS to WLAN can be detected.

The analysis of POLIMAND is depicted in figure 5. The Mobile Node leaves the WLAN coverage area and a handoff occurs when the WLAN signal strength reaches the low watermark of the threshold (50s). In this case POLIMAND forces a Mobile IP handoff that leads to a more reliable handoff and low packet loss that can be observed in figure 5.

The comparison of figure 4 and figure 5 presents the performance of POLIMAND. It increases the performance of vertical Mobile IP handoffs up to sevenfold when 100% packet loss is considered. Moreover, it can be shown that the handoff process is more reliable due of a well defined hysteresis that can be observed in figure 5.

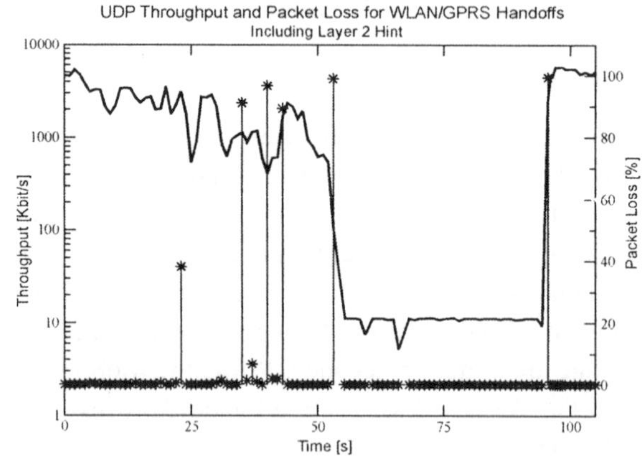

Figure 5: UDP throughput measurement including packet loss with layer 2 hint

7. Conclusions

This paper presents a list of link layer parameters which can be used for MIP handoffs and can be more efficient using Generic Link Layer (GLL) information in order to avoid handoff latency and packet loss. An implementation of policy based Mobile IP handoff decision (POLIMAND) has been presented which uses a hint from the link layer that provides reliable vertical MIP handoffs between GPRS and WLAN. Furthermore, POLIMAND implements an algorithm that can be used for different WLAN cards in a Mobile Node. It has been shown that POLIMAND outperforms Mobile IP handoffs without link layer hint.

The first implementation of POLIMAND uses signal quality to force handoffs between WLAN and GPRS networks. Further enhancements of POLIMAND which consider more network parameters and user requirements are subject to further research.

Acknowledgments

The authors would like to thank J. Sachs from Ericsson (EED), Germany for valuable comments within the 'IPonAir' project.

References

1. C. E. Perkins. IP Mobility Support. RFC 3344, Internet Engineering Task Force, August 2002.
2. E. Gustaffson, A. Jonsson, C. E. Perkins, draft-ietf-mobileip-reg-tunnel-07.txt, work in progress.
3. J. Sachs, "A Generic Link Layer for Future Generation Wireless Networking", ICC, May 11-15, 2003, Anchorage, Alaska, USA.
4. A. Festag, "Optimisation of Handover Performance by Link Layer Triggers in IP-Based Networks: Parameters, Protocol Extensions and APIs for Implementation", Version 1.0, August 2002.
5. J. Tourrilhes, "Wireless Extensions for Linux", January 1997.
6. S. Aust, D. Proetel, A. Könsgen, C. Pampu, C. Görg, "Design Issues of Mobile IP Handoffs between General Packet Radio Service (GPRS) Networks and Wireless LAN (WLAN) Systems", WPMC 2002, Honolulu, Hawaii (USA), October 2002.

MINIMISING HANDOVER OVERHEAD IN HOST BASED ROUTING IP WIRELESS ACCESS NETWORKS

M. A. O. Barry and H. S. Al-Raweshidy
Department of Electronics, University of Kent,
Canterbury, Kent, CT2 7NT,
United Kingdom.
E-mail: mb43@kent.ac.uk, H.Al-Raweshidy@kent.ac.uk

Abstract: In order to reduce handover delay and packet loss in a Host Based Routing (HBR) domain, the flow of packets to the mobile host (MH) are diverted at the cross-over node (CoN)[1-3]. This is done using path update packets that create the new path from the CoN to the new location of the MH. An increase in depth of the network due to an expanding topology will further aggravate these overheads. Our proposal is an architectural extension to the *path update* process adding multi-homed bi-casting base stations (MHOBS) at certain depths, and point, in the topology where the overheads are most sever [4]. Simulation results show that handover overheads due to path updates to the CoN are eliminated using MHOBS.

Key Words: HBR, Micro Mobility, Mobile IP, Cellular IP, MHOBS.

1. INTRODUCTION

One of the challenges faced by the future envisaged All-IP wireless access networks is the smooth handover of MHs running delay sensitive applications. This is especially true of real-time applications such as streaming and interactive media applications. Handovers are required to be executed with minimal associated latency and packet loss. Most handovers are local by nature, and are confined mostly within the same domain. Therefore the mobility management of the handover can be limited to a local or *micro mobility* environment as opposed to the global or *macro mobility* environment. Therefore, dividing the mobility management problem into two layers of; *macro* mobility for global mobility and *micro* mobility for regional and local mobility means that the scope of the handover problem can be narrowed down.

Thus, within the micro mobility scope deploying the network over as wide an area as possible is desirable, as this keeps the mobile user in the same network for a longer length of time. The problem arises when the micro mobility domain increases in size. That is to say; As the depth of the network topology increases, so would the round trip time to the CoN. This round trip time is the time during which packets are lost. Therefore minimisation of this time will result in less packets being lost. Our proposal is to add MHOBS at certain points in the wireless access network where the path update delay becomes severe. There are two situations where this may arise. The first is where the path update delay to the cross-over node (CoN) increases with an increase in the depth of the topology. The second situation is when the new path, between new base station (BS) and the CoN, is longer than the old path, between the CoN and the old BS. We call this the asymmetrical paths delays scenario.

In the next section IP micro mobility is reviewed citing some examples of the main protocols in the field. In the subsequent sections, our proposal is discussed followed by some results of simulations carried out.

2. IP LAYER MICRO MOBILITY

The most developed proposed standard by the Internet Engineering Task Force (IETF) for IP layer mobility is a protocol called Mobile IP (MIP) [5,6]. Yet MIP causes significant delays to traffic destined to MHs in a Micro mobility environment. This is due to MIP's requirement that a MH perform a registration after every change in its network point of attachment. In addition MIP incurs high latency on traffic due to its triangular routing to a potentially distant Home Agent [6]. These shortcomings in the Micro mobility environment make Mobile IP inherently less suited to frequent changes in the network point of attachment of MHs within a domain. And have, at least in part, motivated the development of so-called micro mobility protocols. A number of micro mobility protocols have been proposed. These protocols rely on the notion that frequent mobility by its nature is a local or regional phenomenon. Wireless access networks predominantly have a local or regional scope and therefore mobility management in all-IP wireless access networks will be appropriately managed by a micro mobility protocol [2]. Hence MIP is more suited to Macro mobility. The relationship between the macro and micro mobility with in the IP mobility paradigm is illustrated in Figure 1.

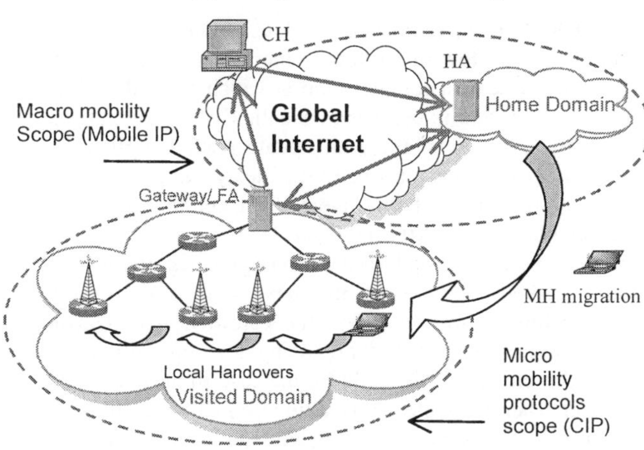

Figure 1- Macro and Micro Mobility scopes.

There are two main classes of micro mobility protocols that are disparate based on how traffic forwarding is managed within the domain. One class of the protocols uses host specific routing, also known as Host-Based Routing (HBR), to manage traffic forwarding in micro mobility domains [1]. Examples include Cellular IP (CIP) [2,7], and Handoff Aware Wireless Access Internet Infrastructure (HAWAII) [3]. The other class uses Tunnelling as in base MIP. Examples are Hierarchical MIP (HMIP) [8], and TeleMIP/Intra-Domain Mobility management Protocol (IDMP) [9]. Although our delay minimization can be applied to both of the two classes of protocols, as it is an architectural optimisation, an HBR protocol based on CIP was used in this investigation. This is because of the inherent advantage in HBR schemes manifest in the integration of routing and location management, [1,2] . This is in contrast to the two-tire approaches of tunnelling protocols that are derivatives of MIP. In a domain using a HBR protocol, soft-state routing functionality is used coupled with that of a cross-over node (CoN) in the mobility management of problems of location updates and traffic redirection, as a result of frequent MH handovers. A CoN is the node that lies on the closest intersection between the old and new paths to the MH. While Soft-state routes are those of MH specific routing information that include MH address. These soft state routes are created dynamically on a hop by hop basis with uplink control packets, called route or path updates, in an explicit manner. They are subsequently explicitly refreshed by the path updates as well as implicitly refreshed by uplink data packets from the MH. If the soft-sate cache entry is not refreshed by a certain route-update time they expire and are erased from the routing state in the respective node. In this manner HBR protocols enable the access network to combine routing with location management for active MHs, while at the same time the soft-state aspect makes the usually unacceptable scaling disadvantage of host-specific routes in the global Internet, an advantage in the local scope of the micro mobility domain [9].

3. PATH UPDATE MANAGEMENT

In a HBR domain, to reduce handover delay and resultant packet loss, the flow of packets to the MH is diverted at the CoN. This is nearest routing node at which the old and new paths intersect, and in the worst case the CoN is the gateway. Taking the average of the path update packet (uplink) and data packet (downlink); the handover delay can be defined as the round trip time of this average packet to traverse the topology between the MH and CoN.

This increase in the depth of the topology is significant since control packets that traverse the network form the new BS will take longer time to get to the gateway. All the while packets already transmitted are being lost through the old path at the old BS. This could be the case in an inter-subnet handover when the Gateway is acting as a CoN, as illustrated in Figure 2. The arrows pointing up, shown in Figures 2-6, represent uplink path update packet that the MH transmits after every handover, and the arrows pointing down represent downlink traffic flow to the MH.

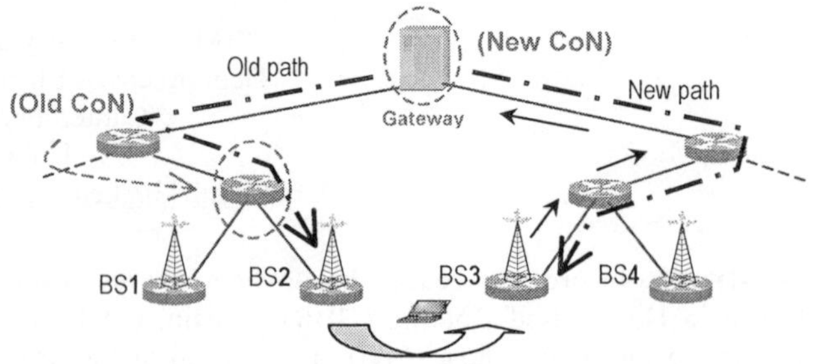

Figure 2- Inter-subnet handover in the access network.

During handover, the packet loss will be proportional to the to the round trip time to the gateway, and the downlink packet rate. Our proposal is to modify the topology by adding Multi-homed BSs (MHOBS) at these and other depths to counter act the effects of relatively distant CoNs on the handover process. MHOBS addition will ensure the elimination of packet loss, during transmission of the location update packet to the gateway. By eliminating the effect of the path update delay on the handover process altogether at that depth in the hierarchy. So the handover associated with the maximum path update delay, as shown in Figure 2, is replaced with 2 handovers with the minimum delay, as illustrated in Figure 3.

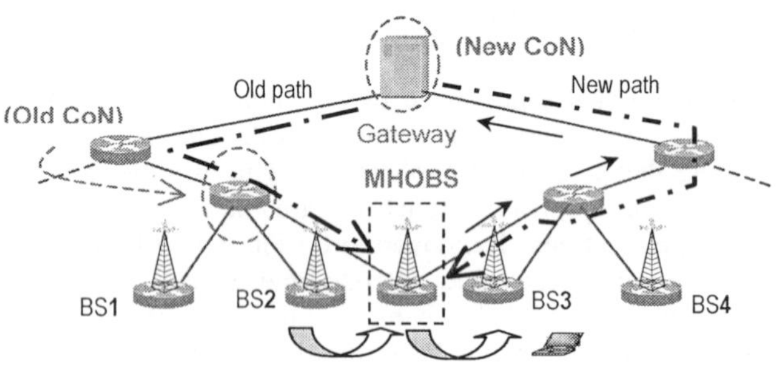

Figure 3- Inter-subnet handover using MHOBS.

3.1 Packet Loss Reduction due to Asymmetrical Path Delays to CoN

CIP semi-soft handover scheme has an option to bi-cast path updates from a MH to two BSs to facilitate fast handover [2,7]. We propose taking the bi-casting functionality up one level from the MH, to the BS regardless of the handover scheme whether hard or semi-soft. Since this is an architectural modification it can be implemented in any micro mobility protocol employing a tree based topology and not only limited to CIP. Unlike in CIP where bi-casting is implemented at the MH at every handover, in our proposal bi-casting is implemented at specifically placed MHOBSs to minimize route update times in due to any increase in the depth of the topology, as well as due to asymmetry of the lengths of the old and new paths to the CoN. Using a

MHOBS the time required to update paths to the new location of the MH is reduced to the minimum possible. Where there would usually be two BSs the added MHOBS is used to eliminate effect of the handover process when the gateway is the CoN, as well as provide a single point of convergence of packet to get to the MH from both old path and new path.

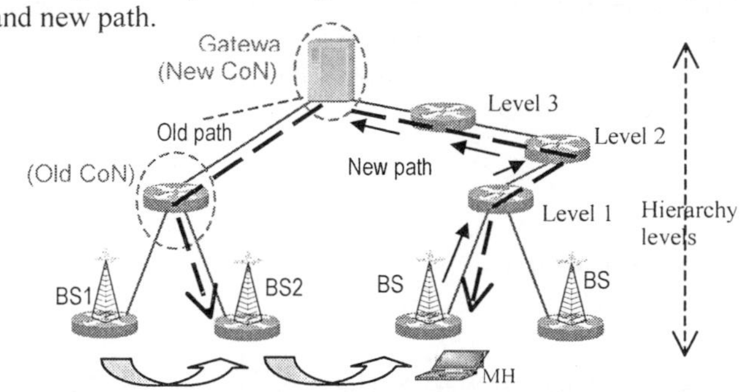

Figure 4- Asymmetry between old and new path augments the packet loss during path updates.

In this manner the MH can receive all the packets that are destined to it from the old path that would have other wise be lost as shown in Figure 4. Therefore the packets loss that would have been incurred while the new path is configured to the CoN are not lost due to the MHOBS as shown in Figure 5.

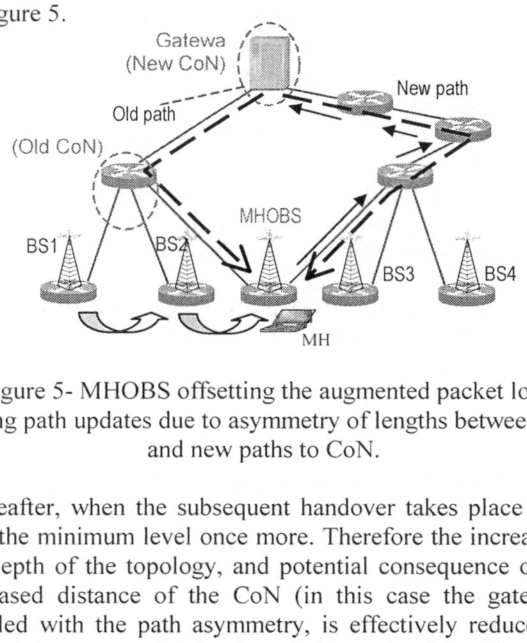

Figure 5- MHOBS offsetting the augmented packet loss during path updates due to asymmetry of lengths between old and new paths to CoN.

Thereafter, when the subsequent handover takes place CoN is at the minimum level once more. Therefore the increase in the depth of the topology, and potential consequence of the increased distance of the CoN (in this case the gateway) coupled with the path asymmetry, is effectively reduced to the minimum overhead of a CoN at level-1. Thus the handover is reduced from a single handover at a level-3 CoN (the Gateway) shown in Figure 4, to two handovers with CoNs a level-1 CoN shown in Figure 5.

The asymmetry between old and new path that would have led to an augmentation of the packet loss as shown in Figure 4, is offset by the MHOBS shown in Figure 5. The path update message traverse the longer path to the CoN , illustrated in the figure as separate arrows pointing upwards, while the old path remains active, all the while data packets that would have been lost are recovered form the old rout to the MHOBS.

4. PERFORMANCE EVALUATION

4.1 Simulation Environment

The model used for the simulations assumes cells with perfect overlapping coverage, zero transmission errors and zero propagation delay. Lower layer 2 handover is assumed to be smooth and instantaneous. Other parameters used to facilitate the simulations are bandwidth of wired and wireless links, latency of wired and wireless links, and average size of path update and data packets. Traffic was constant bit rate UDP to simulate voice application with a data rate of 25kbits/s. The model was built and simulated using OPNET Modeler [10]. Two different scenario sets are presented next.

4.1.1To-and-fro scenario simulations. Two models were simulated for comparison. One to show the performance of a topology with MHOBS, and the other topology without MHOBS. The topology with MHOBS is shown in Figure 2, while the one without is shown in Figure 3. The two models were identical in terms of network parameters. A session of 80s was simulated with 12 handovers every 5s, starting from 15s to 75s. In the simulation without MHOBS the handovers were to-and-fro between BS2 and BS3. While in the topology with MHOBS the handovers were from BS1 to MHOBS, BS2 and back the same way to BS1. The to-and-fro handover scenario method was used so as to bring out the performance improvements of the MHOBS.

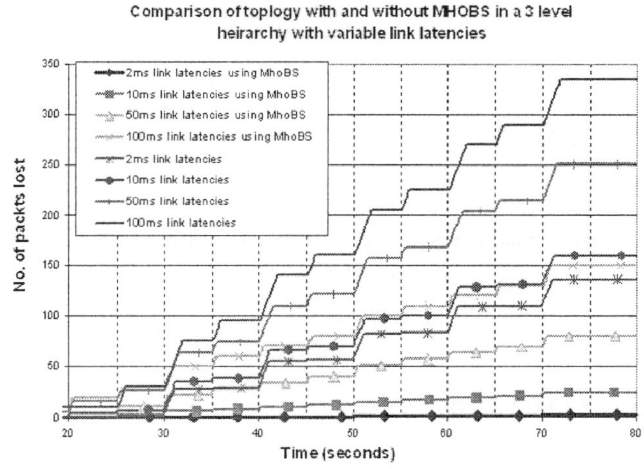

Figure 6- Packet losses for 8 simulations after 12 to-and-fro handovers.

As shown in Figure 6, the number of packets lost in total after the eight simulations (run with varying values for the link latency for each simulation). It can be seen that the MHOBS enabled topology resulted in loss of fewer packets in almost all the cases as compared to the topology without MHOBS. On average more than 100 additional packets were lost in simulations without MHOBS. The worst case packet loss for MHOBS was 150 packets compare with almost 350 without MHOBS. Indeed, the worst case packet loss for MHOBS was only slightly more than the best packet loss for the one without MHOBS. For the link latencies of 2ms about 140 packets were lost compared to the MHOBS packet loss of only 3 packets for the same link latencies.

4.1.2 Realistic simulation scenario. Another set of simulations were run for a more realist movement scenario. This time the movement was not restricted, and the handover were between all the BS in the respective topologies. These were similar to the ones shown in Figures 3 and 4 respectively, except that for the MHOBS scenario, there were 3 MHOBS and 4 BS. From Figure 7, it can be seen that the MHOBS in these more realistic simulations still perform better than the topologies without MHOBS. For example after 4 handovers of a 40s session, the 100ms link latency MHOBS dropped 155 packets compared to 175 packets in the simulation without MHOBS. While the improvement in reduction in packet lost is not as dramatic in these sets of simulations as compared to the to and fro sets, it still is a significant improvement for a more realistic scenario.

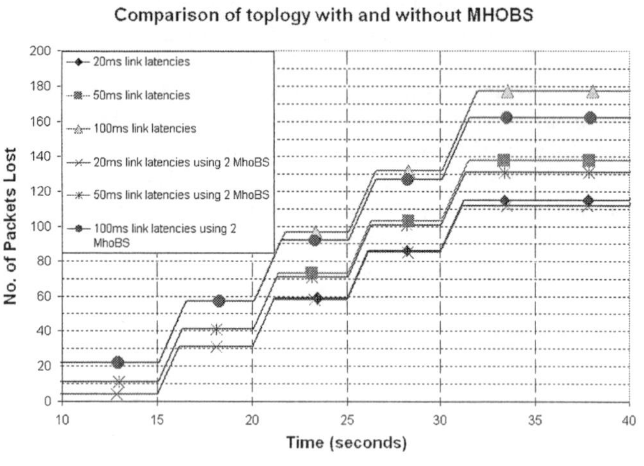

Figure 7- Packet losses for 6 simulations after 4 handovers in the realistic scenarios

5. CONCLUSION

In this paper, an overview was presented of the IP mobility paradigm, highlighting the difference between the macro and micro mobility approaches, with emphasis on the HBR approach. Then a novel path delay minimisation was introduced. Two possible application scenarios were discussed for use in All-IP wireless access networks. The first scenario is where the path update delay to the cross-over node (CoN) increases with an increase in the depth of the topology. The second situation is when the new path, between new base station (BS) and the CoN, is longer than the old path to the old BS. This latter scenario we called the asymmetrical paths delays scenario.

Thereafter simulation results presented showed that the proposed MHOBS-based approach resulted in less packet loss during handovers. This was more pronounced in the to-and-fro scenario when compared to other configurations. The simulations clearly show the extent of the improved performance.

REFERENCES

[1] K. D. Wong, H. Wei, A. Dutta, K. Young, "Performance of IP Micro-Mobility Management Schemes using Host Based Routing", Proceedings 4th International Symposium on Wireless Personal Multimedia Communications (WPMC'01) , 2001.

[2] A. T. Campbell, J. Gomez, S. Kim, Z. Tur´anyi, A. G. Valk´o, C-Y. Wan, "Internet micromobility", Journal of High Speed Networks, IOS Press, (11) pp177-198, 2002.

[3] R. Ramjee, T. La Porta, S. Thuel, K. Varadhan, L. Salgarelli, "IP Micro-mobility support using HAWAII", IETF draft, 2001. (work in progress) http://www.ietf.org/internet-drafts/draft-ietf-mobileip-hawaii-01.txt

[4] M. A. O. Barry and S. H. Al-Raweshidy, "Improving Hard Handover in HBR IP Micro Mobility Protocols Employing Tree Based Topologies", Proceedings of The Fourth International Conference on 3G Mobile Communication Technologies, pp 200-205, 2003.

[5] http://www.ietf.org/

[6] C. Perkins, 2002, IP Mobility Support for IPv4 , IETF RFC 3220, http://www.ietf.org/rfc/rfc3220.txt

[7] A. T. Campbell, J. Gomez, S. Kim, A. G. Valk´o, C-Y. Wan, and Z. Tur´anyi, "Design, Implementation, and Evaluation of Cellular IP". IEEE Personal Communications, vol. 7(4):pp42–49,2000.

[8] H. Soliman, C. Castelluccia, K. El-Malki, L. Bellier, "Hierarchical MIPv6 mobility management (HMIPv6)", Internet Draft , 2001. (work in progress) http://www.ietf.org/internet-drafts/draft-ietfmobileip-hmipv6-04.txt

[9] S. Das et al, "TeleMIP: Telecommunication-Enhanced Mobile IP Architecture for Fast Intradomain Mobility". IEEE Personal Communications, vol. 7(4):pp50–58,2000.

[10] http://www.opnet.com

Achieving Seamless IP Handover and Dynamic Neighborhood Discovery in IEEE 802.11 Networks

Paul Tan[1], QunYing Xie[2], Hoang M. Nguyen[1,2], Jaya Shankar[1], Winston Seah[1]
Email: {tanpaul, stuxieqy, hoangnm, jshankar, winston}@i2r.a-star.edu.sg

[1]Networking Department, Communications and Devices Division,
Institute for Infocomm Research (Member of A*STAR)
21, Heng Mui Keng Terrace. Singapore 119613

[2]Department of Electrical & Computer Engineering,
Faculty of Engineering, National University of Singapore
10, Kent Ridge Crescent. Singapore 119260

Abstract: To achieve seamless layer 3 handover, mobile nodes must be able to perform movement detection and neighborhood candidate access routers discovery quickly and efficiently. In this paper, we show how to extend the IEEE 802.11 management frames to carry extensible application-specific Information Elements, which allow access points to advertise the capabilities information of its associated network/provider and also to improve movement detection. We show that mobile nodes can dynamically discover neighboring candidate access routers or networks more quickly and efficiently, and also to obtain valuable capabilities information so as to select the 'best' target access router to initiate seamless handover. Lastly, we illustrate through NS-2 simulations, the performance evaluation for our proposed scheme, which attempt to achieve faster movement detection and neighboring Access Router/Access Point discovery/selection.

....

Key Words: seamless handover, mobile IP, 802.11, movement detection, neighborhood discovery

1 INTRODUCTION

The ability to stay connected on the Internet while on the move is becoming an essential requirement in the Mobile Internet space. At the same time, public wireless LANs using IEEE 802.11 technology are gaining popularity for providing high-speed wireless Internet services at various hotspots such as airports, hotels. The proliferation of high-speed wireless Internet access will also invite the possibilities for real-time multimedia applications.

In order to support mobility/portability across the Internet, IETF has proposed Mobile IPv6 [1] to provide continuous Internet connectivity while the mobile node (MN) changes its point of attachment to the Internet.

To achieve seamless mobility, handover latency needs to be reduced. We propose [9] to extend the IEEE 802.11 management frames to carry extensible application-specific Information Elements (IE), which allow access points (APs) to advertise information of its associated network/provider. The recommendation also improves in the movement detection (MD) mechanism by avoiding unnecessary Layer 3 (L3) messaging over the wireless link. More importantly, we believe that MN can dynamically discover neighboring candidate access router (CAR) or networks more quickly and efficiently, and also to obtain valuable capabilities information so as to select the 'best' target access router (TAR) to initiate seamless handover.

The remainder of this paper is structured as follows. In Section 2, we describe briefly the problems and some related works. In Section 3, we present our proposal to achieve fast movement detection and neighborhood discovery. Section 4 evaluates our proposed scheme to achieve faster MD using NS-2 Simulation. Finally, Section 5 concludes with a discussion of future studies.

2 RELATED WORKS

Mobile IPv6 [1] was originally designed as a simple and scalable wide-area mobility solution. During the handover, the MN is unable to send/receive IPv6 packets both due to its Layer 2/3 (L2/3) handover operations. This handover latency is unacceptable to real-time or delay sensitive traffic/applications.

In [1], the MD algorithm relies on the periodic Router Advertisements (RA) to enable the MN to determine their current location. MD, which is achieved through the use of Neighbor Discovery [2] requires routers to send unsolicited RA at a minimum interval of 3 seconds [2]. Upon receipt of a Router Solicitation (RS) message, the router must delay its response (i.e. RA) by a random time. The protocol also requires transmission delay of the initial solicitation. Such delay is simply not desirable in achieving seamless handover.

There are several related works/protocols [3-6] that attempts to minimize the handover latency. In [5], the authors propose a new low-latency handover method, where AP is used in a wireless LAN environment and a dedicated MAC bridge is jointly used to alleviate packet loss without altering the Mobile IP specifications. One of the advantages is that no modification is required on both Mobile IP and IEEE 802.11 specifications. It does not require any L2 triggers for MD. However, there are some scalability issues since it requires a MAC bridge to connect to all the subnets inside a domain. In Fast-Handovers for Mobile IPv6 [6], the objective is to reduce the time necessary for that a MN to sends or receives IPv6 packets. To achieve this objective, the protocol requires MN to configure in advance a new care-of-address (CoA) before it actually moves into the new sub-network. However, it assumes that the current AR must be capable of mapping this new L2 identifier into the IP address of the new AR.

3 PROPOSAL

3.1 INSTANTANEOUS MOVEMENT DETECTION

In this section, we describe how to achieve instantaneous MD by configuring the Extended Service Set Identifier (ESSID) of an infrastructure 802.11 network to include the L3 Identifier/Information of its associated AR or network. We recommend that APs to include the subnet prefix information for their associated AR (Subnet-Router-Anycast-Address) into the ESSID. However, to improve the IEEE 802.11 inter-cell handover performance, several APs belonging to different administrative domains (different subnets) may choose to share similar ESSID. Therefore, as MN moves across these cells, it only needs to initiate re-association instead of the plain association process, which can take significantly longer. In this case, instead of embedding network prefix information into the ESSID, which is specific to each subnet, AP can include Network-Prefix-Information object using the Application-Specific Information Element into the beacon frames. The table below shows an example of a list of beacon information advertised by the neighboring APs received by the MN on a specific wireless interface with an identifier, mniid.

Subnet Prefix	ESSID	AP-MAC	RSSI	Care-of-Address	Status
Prefix1	Essid1	MAC1	X dbm	Prefix1:mniid	-
Prefix2	Essid2	MAC2	X dbm	Prefix2:mniid	ACTIVE
Prefix3	Essid3	MAC3	X dbm	Prefix3:mniid	-
Prefix4	Essid4	MAC4	X dbm	Prefix4:mniid	-

Table 1: Beacon Lists

3.2 DYNAMIC CAR DISCOVERY

MN also constantly listens for any beacon frame advertised by any neighboring APs. When MN discovers a new AP/Advertisement, it stores the advertised capabilities information, which is encapsulated inside the Network-Capabilities-Information Object into its Neighborhood CAR list.

MN can store these discovered information ordered by either the signal strength or other parameters (e.g. pricing). To achieve/assist fast handover operation, information such as L2/L3 address mapping of the AR can be obtained from the Access-Router-Information Object.

```
      MN   AP1   AP2   AR1   AR2   HA    CN
      |     |     |     |     |     |     |
      |*****|     |     |<------data------|
      |     |<---data---|     |     |     |
      |<----|     |     |     |     |     |
      |     |     |     |     |     |     |
  (1) |***********|     |     |     |     |
  (2) |====>|     |     |     |     |     |
      |     |==RtSolPr=>|     |     |     |
      |     |     |     |=HI=>|     |     |
      |     |     |     |<HACK|     |     |
      |     |<===HACK===|     |     |     |
      |<====|     |     |     |     |     |
      |====>|     |     |     |     |     |
      |     |====F-BU==>|     |     |     |
      |     |<==F-BACK==|     |     |     |
      |<====|     |     |<------data------|
      |     |     |  forward Pkts   |     |
  (3) |     |     |     |---->|     |     |
      |     |     |<----------|     |     |
  (4) |~~~~~~~~~~~|     |     |     |     |
      |<----------|     |     |     |     |
      |     |     |     |     |     |     |

**** L2 Beacon
~~~~ L2 Association
---- Data
==== Fast-Handoff Messages
```

Figure 1: Fast Handover Sequence

Subnet Prefix	ESSID	Cost	Fast-HD	QoS	Security
Prefix1	Essid1	0 unit	N	N	N
Prefix2	Essid2	1 unit	N	N	Y
Prefix3	Essid3	2 units	N	Y	Y
Prefix4	Essid4	3 units	Y	Y	Y

Table 2: Neighborhood CAR list

Figure 1 shows a sequence of fast-handover operations that allows a MN to handover to a newly discovered subnet. During the Neighborhood Discovery Phase (1), MN discovers neighboring APs or networks and creates a list of CARs with their advertised information (Table 2). Optionally, during this phase, MN can configure in advance new CoA for each discovered APs/Subnets. In the Panic Phase (2), if the current associated AP's signal strength drops or if the MN discovers AR2, which offers 'better' service/pricing, MN can immediately perform a handover to this new AP/AR. We assumed here that AP2 is associated with a different subnet; MN will know that it has moved to a new network using the cached information. Assuming MN is still attached to AP1, it quickly initiates fast-handover (2) using the cached information of AP2 (AR2-L2-ID, AR2-L3-ID). At (3), packets are forwarded from the AR1 to the AR2. At (4), MN has established its L2 association with AP2.

3.3 EXTENSIBLE APPLICATION-SPECIFIC IE

In this paper, we propose how to extend the current IEEE 802.11 management frames to carry extensible application-specific IE. For example, the IEEE 802.11 beacon frame with the newly proposed IE is shown in the Figure 2.

The Extensible Application-specific IE allows future applications/protocols to encapsulate their information into the IEEE 802.11 frame easily. We believe that valuable information can then be easily included in the IEEE 802.11 management frames and discovered by roaming MNs. The above application-specific IE consists of the following fields: Element ID, Length, and Extensible-Application-Specific Information.

```
 0                   1                   2                   3
 0 1 2 3 4 5 6 7 8 9 0 1 2 3 4 5 6 7 8 9 0 1 2 3 4 5 6 7 8 9 0 1
+-+-+-+-+-+-+-+-+-+-+-+-+-+-+-+-+-+-+-+-+-+-+-+-+-+-+-+-+-+-+-+-+
|         Frame Control         |           Duration            |
+-+-+-+-+-+-+-+-+-+-+-+-+-+-+-+-+-+-+-+-+-+-+-+-+-+-+-+-+-+-+-+-+
|                      Destination Address                      |
+-+-+-+-+-+-+-+-+-+-+-+-+-+-+-+-+-+-+-+-+-+-+-+-+-+-+-+-+-+-+-+-+
|                               |        Source Address         |
+-+-+-+-+-+-+-+-+-+-+-+-+-+-+-+-+-+-+-+-+-+-+-+-+-+-+-+-+-+-+-+-+
|                        Source Address                         |
+-+-+-+-+-+-+-+-+-+-+-+-+-+-+-+-+-+-+-+-+-+-+-+-+-+-+-+-+-+-+-+-+
|                      Basic Service Set ID                     |
+-+-+-+-+-+-+-+-+-+-+-+-+-+-+-+-+-+-+-+-+-+-+-+-+-+-+-+-+-+-+-+-+
|                               |       Sequence Control        |
+-+-+-+-+-+-+-+-+-+-+-+-+-+-+-+-+-+-+-+-+-+-+-+-+-+-+-+-+-+-+-+-+
|    Capability Information     |        Beacon Interval        |
+-+-+-+-+-+-+-+-+-+-+-+-+-+-+-+-+-+-+-+-+-+-+-+-+-+-+-+-+-+-+-+-+
|           Timestamp           |      SSID (2-34 octets)       |
+-+-+-+-+-+-+-+-+-+-+-+-+-+-+-+-+-+-+-+-+-+-+-+-+-+-+-+-+-+-+-+-+
|                 Supported Rates (3-7 octets)                  |
+-+-+-+-+-+-+-+-+-+-+-+-+-+-+-+-+-+-+-+-+-+-+-+-+-+-+-+-+-+-+-+-+
.                                                               .
|            Extensible-Application-Specific IE                 |
.                                                               .
+-+-+-+-+-+-+-+-+-+-+-+-+-+-+-+-+-+-+-+-+-+-+-+-+-+-+-+-+-+-+-+-+
|                     Frame Check Sequence                      |
+-+-+-+-+-+-+-+-+-+-+-+-+-+-+-+-+-+-+-+-+-+-+-+-+-+-+-+-+-+-+-+-+
```

Figure 2. 802.11 Beacon Frame with new Extensible-Application-Specific IE

```
 0                   1                   2                   3
 0 1 2 3 4 5 6 7 8 9 0 1 2 3 4 5 6 7 8 9 0 1 2 3 4 5 6 7 8 9 0 1
+-+-+-+-+-+-+-+-+-+-+-+-+-+-+-+-+-+-+-+-+-+-+-+-+-+-+-+-+-+-+-+-+
|   Element ID  |    Length     |            ..                 |
+-+-+-+-+-+-+-+-+-+-+-+-+-+-+-+-+                               +
|                                                               |
+               Application-Specific Object                     +
|                                                               |
+-+-+-+-+-+-+-+-+-+-+-+-+-+-+-+-+-+-+-+-+-+-+-+-+-+-+-+-+-+-+-+-+
```

Figure 3. Extensible-Application-Specific IE

As defined in [8], the element ID field is 1 octet long and specifies the type of IE. The length field is 1 octet long and specifies the length (number of octets) of the application-specific Information field. Lastly, the application-specific object field carries the specific application object using the following structure.

```
 0                   1                   2                   3
 0 1 2 3 4 5 6 7 8 9 0 1 2 3 4 5 6 7 8 9 0 1 2 3 4 5 6 7 8 9 0 1
+-+-+-+-+-+-+-+-+-+-+-+-+-+-+-+-+-+-+-+-+-+-+-+-+-+-+-+-+-+-+-+-+
|    App ID     |    Length     |            ..                 |
+-+-+-+-+-+-+-+-+-+-+-+-+-+-+-+-+                               +
|                                                               |
+             Application-Specific Information                  +
|                                                               |
+-+-+-+-+-+-+-+-+-+-+-+-+-+-+-+-+-+-+-+-+-+-+-+-+-+-+-+-+-+-+-+-+
```

Figure 4. Generic Application-Specific Object Structure

We have proposed several application-specific objects [9] (Network-Prefix, AR-Info and Network-Capability) to achieve some of our intended objectives.

4 PERFORMANCE ANALYSIS

4.1 SIMULATION MODEL

We use NS-2 version 2.1b9 as our simulation tool to evaluate the performance between plain-Mobile IPv6 and our proposed enhanced-Mobile IPv6. The simulation topology is shown in Figure 5.

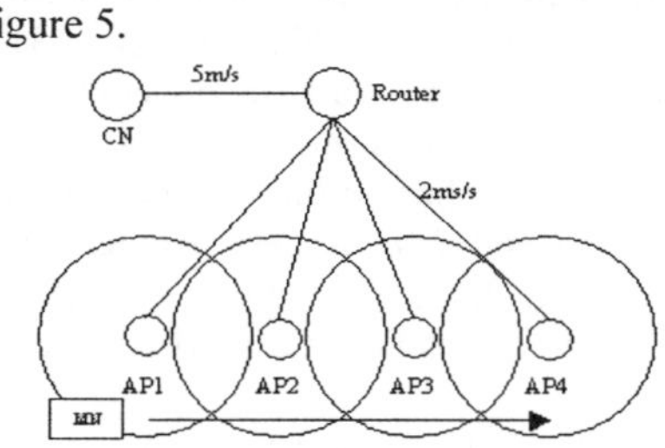

Figure 5: Simulation Network Topology

Parameter	Value
CBR Packet Size	100 bytes
CBR Interval	5ms
Link Delay (CN-Router)	5ms
Link Delay (Router-APs)	2ms
AP Separation Distance	200m
AP Overlapping Area	10m
Simulation Duration	600s
MN Speed	5m/s

Table 3: Simulation Parameters

In this simulation, we investigate the average handover latency and packet loss probability for both schemes as a function of both the RA and L2 Beacon Interval. The NS-2 MAC implementation was modified to implement our proposed scheme. As the MN moves into an overlapping area of two APs, the MN will hear an extended-beacon from the new AP.

In our simulation, the reception of the new AP's beacon indicates movement into a possibly new subnet and also triggers the start of the registration process (pre-registration). Clearly, the extended-beacon frame can assist the MN to achieve faster MD.

4.2 ANALYSIS

Figure 6 shows the average handover latency experienced by MN when it moves across several APs at a speed of 5m/sec. In the plain Mobile IP scheme, the performance of the handover depends on how frequent or quickly a MN can detect its movement into a new subnet.

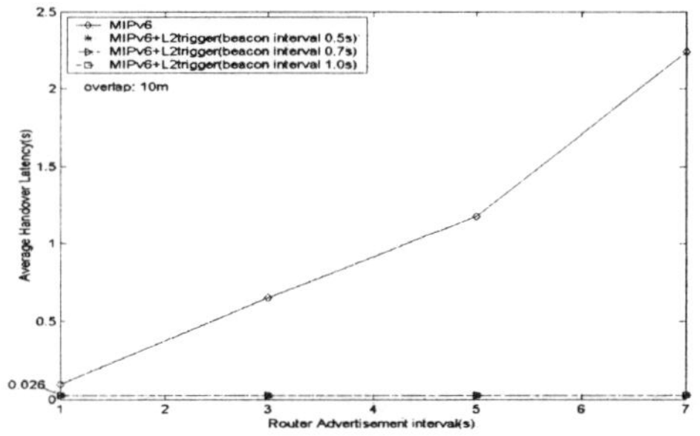

Figure 6: Average Handover Latency vs RA Interval

This is based on the normal MD schemes using RA/RS message exchanges. To reduce the latency, the RA must be transmitted at a much higher rate. On the other hand, our proposed MD scheme does not depend on the RA interval. Since network-prefix information is encapsulated into the IEEE802.11 beacon frames, MNs can now quickly detect its movement using this extended-beacon frame, which is currently transmitted periodically.

Figure 7 shows the results of packet loss probability for various RA and Beacon intervals. Similarly, the loss rate for the plain Mobile IPv6 depends on the rate in which RAs are sent.

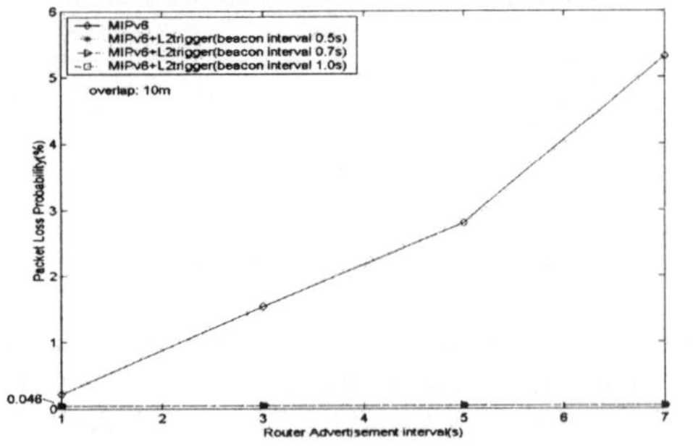

Figure 7: Packet Loss Probability vs RA Interval

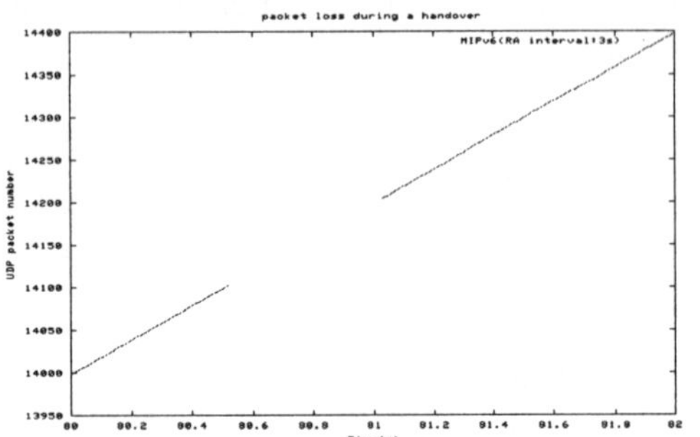

Figure 8: UDP Received Sequence Number (plain-MIP)

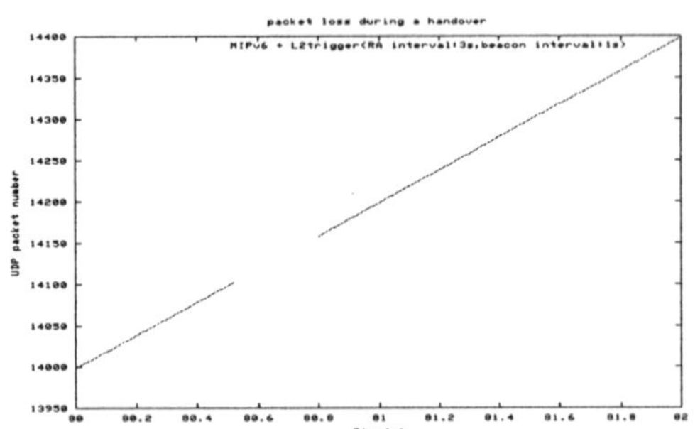

Figure 9: UDP Received Sequence Number (Proposed)

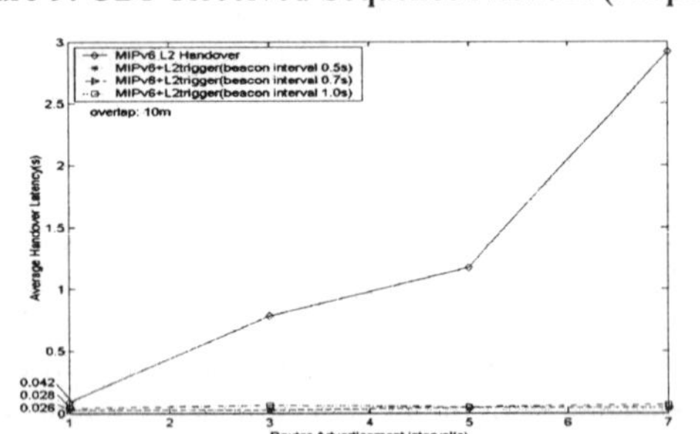

Figure 10: Average HO Latency (Random Speed)

In Figure 8 and 9, we can see some improvement for the UDP communication between the MN and CN. However, the scheme can only reduce the latency for MD. To improve the registration delay further, intra-domain mobility protocol (e.g. Hierarchical Mobile IP) can be used. In Figure 10, the MN is configured to move across the APs at a random speed pattern. Similarly, the improvement is almost identical to the earlier simulation results. The simulation results have shown that there is an improvement for both the handover latency and the packet loss probability as the single MN moves across overlapping several APs at a constant speed of 5m/s. The ability to advertise L3 network information in L2 messages, allows MN to quickly detect its movement into a new subnet. We can leverage on the existing IEEE802.11 infrastructure to aid MNs in faster movement detection. However, it would be desirable if the sending rate of the extended-beacon frames can be configured in such a way that it adapts itself to the current environment. For example, in a highly mobile environment, the beacon rate should be increased accordingly. This helps to reduce the overheads over the wireless link.

5 CONCLUSION

In this paper, we proposed a scheme to achieve seamless layer 3 handover through faster movement detection and dynamic neighborhood discovery. We have successfully modified the NS-2 simulation code to embed network-prefix information and other proposed information elements (e.g. cost/AR) into the beacon frames. We also showed through simulations, the performance comparison between plain Mobile IP and our proposed scheme. The results indicate some positive improvement in terms of the handover latency and the packet loss probability. However, it would be beneficial to analyze the performance/impact when more mobile nodes moving at a higher speed are introduced. In our future work, we would also like to include fast-handover [6] mechanisms (e.g. bi-casting/tunneling) into our proposed scheme.

REFERENCES

[1] D. Johnson, C. Perkins and J.Arkko, Mobility Support in IPv6, draft-ietf-mobileip-ipv6-18.txt, May 2002.

[2] T.Narten, E. Nordmark and W.Simpson, Neighbor Discovery for IPv6, RFC 2461, Dec 1998.

[3] R. Ramjee, et al. "HAWAII – A Domain-Based Approach for Supporting Mobility in Wide-Area Wireless Networks", IEEE ICNP 1999, pp. 283- 292.

[4] A. G. Valko, A. T. Campbell, J. Gomez, Cellular IP, Internet Draft, draft-valko-cellularip-00.txt, Nov 1998.

[5] Hidetoshi Yokota , Akira Idoue , Toru Hasegawa , Toshihiko Kato, "Link layer assisted mobile IP fast handoff method over wireless LAN networks", ACM Mobicom, Sept 02, Georgia, USA, pp. 131 – 139.

[6] Dommety, G., et. al., Fast Handovers for Mobile Ipv6, draft-ietf-mobileip-fast-mipv6-05.txt.

[7] Trossen, D., Krisharmurthi, G. Chaskar, H., Kempf, J, Issues in Candidate Access Router Discovery for seamless IP-level handoffs, draft-ietf-seamoby-cardiscovery-issues-04.txt, work in progress, Oct 2002.

[8] IEEE, "Part II: Wireless LAN Medium Access Control (MAC) and Physical Layer (PHY) Specifications", 1999.

[9] Paul Tan, Recommendations for achieving seamless IPv6 handover in IEEE 802.11 networks, draft-paultan-seamless-ipv6-handover-802-00.txt,Feb 2003.

NEW MECHANISM FOR MOBILE IPV6 FAST HANDOVER IN WIRELESS LAN

P. S. KIM and Y. K. KIM

Mobile Platform Lab., Digital Media R&D Center,
Samsung Electronics Co., Ltd., Suwon, 442-742, Korea
Tel: 82-31-200-4635, Email: kimps@samsung.com

M. E. LEE

School of Information Technology,
Korea Univ. of Tech. & Edu., Chonan, 330-708, Korea

In this paper, a new fast handover mechanism in Mobile IPv6 is proposed for IEEE 802.11 wireless networks with several access routers. To implement the proposed mechanism, the beacon message used in L2 is defined newly by adding a specific subfield to the capability information field. In addition, Router Table Request/Reply messages used in L3 are defined newly. The proposed mechanism provides the faster acquisition of new access router's network information than the existing one, which might be advantageous for the low handover latency. In addition, the proposed mechanism might reduce amount of traffic in comparison with the existing one, which might be remarkable when there are many mobile nodes that are now connecting to current access router. Moreover, in the proposed mechanism, a mobile node can know whether it changes access router or access point, which can remove redundant traffic of the existing one.

1 Introduction

In recent, a fast handover mechanism [1] has been proposed for Mobile IPv6 [2,3,4,5] as a way to reduce a L3 handover latency.

In this paper, a new fast mechanism is proposed for Mobile IPv6. Actually, this mechanism is developed for IEEE 802.11 wireless LAN (WLAN) environment in [6] where there are several access routers (ARs) connected by access points (APs). For example, buildings such as offices, banks and hospitals can be considered. These buildings have several floors where there is one AR for each floors. Mobile hosts such as laptop computers and handheld PCs used by resident people within the building are considered as mobile nodes (MNs) in Mobile IPv6. Usually, MNs would not go out the building, but would move within the building.

To implement the proposed mechanism, the beacon message used in L2 is defined newly by adding a specific subfield to the existing reserved field. Using this message, a MN can know whether it changes AR or AP. In addition, Router Table Request/Reply messages used in L3 are defined newly. Using these messages, the MN can acquire network information about all ARs in ESS, such as ARs' IP addresses, prefix lengths, and identities, which is performed once only at the booting time and isn't performed in real-time communication. In real-time communication, this network information allows the MN to formulate a new care-of address and to be able to communicate immediately when the MN changes its point of attachment from the current AR (CAR) to the new AR (NAR). That is, the proposed mechanism can omit a Router Solicitation for a Proxy message and a Proxy Router Advertisement message used in the existing one [1], in real-time communication. This means that the proposed mechanism provides the faster acquisition of NAR's network information than the existing one, which might be remarkable for low handover latency. In addition, the omission of above two messages might reduce amount of traffic when there are many MNs that are now connecting to the CAR. Moreover, in the proposed mechanism, the MN can know whether it changes AR or AP while the existing one cannot. Thus, when the MN changes AP, the proposed mechanism can remove redundant traffic of the existing one.

2 IEEE 802.11 WLAN Environment with Several Access Routers

This paper considers an IEEE 802.11 WLAN environment where there are serveral ARs. All APs connected to ARs and MNs are considered to belong to the same Extended Service Set (ESS), and thus have same ESS identity (ESSID). For example, buildings such as offices, banks and hospitals can be considered. These buildings have several floors where there is one AR for each floors. Mobile hosts such as laptop computers and handheld PCs used by resident people within the building are considered as MNs. Usually, these MNs would not go out the building, but would move within the building. As shown in [1], all ARs in this WLAN share each other's network information, such as ARs' IP addresses, prefix lengths, and identities, and thus may have information table that will be called the Router Table as shown in Fig. 1. Note that the AP can know its AR's identity (ARID) by a system administrator's presetting.

A change of AP on same subnet does not require a change of AR because AR is L2 reachable from the MN connected to any APs. This case is outside the scope of L3 handover mechanisms. However, a change of AP between different subnets requires a change of AR, because the MN would be attaching to a different subnet. In this case, a L3 handover mechanism would need to be invoked. Therefore, it should be required for the MN to know whether it changes AP or AR.

	ARID	IP Address	Prefix Len
AR1	0x01	3ffe:2e01:2a:4::1	64
AR2	0x02	3ffe:2e01:2a:4::2	64
AR3	0x03	3ffe:2e01:2a:4::3	64

Figure 1. Router Table shared by ARs

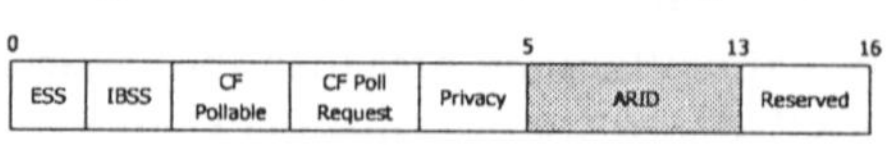

Figure 2. Modified Capability Information field

3 Main Works

3.1 Beacon Message Format

In order to implement the proposed mechanism, the beacon message used in L2 is defined newly. The beacon message contains information such as timestamp, beacon interval, capability information, SSID, etc [6]. In this paper, the capability information field is modified by adding the specific subfield "ARID" to the reserved field as shown in Figure 2. Since this field uses the existing reserved field, other fields are not affected. Using this beacon message, the MN can know whether it changes AR or AP, which will be explained later.

3.2 Router Table Request/Reply Message Formats

In order to implement the proposed mechanism, Router Table Request/Reply messages used in L3 are defined newly. These messages are made from the Internet Control Message Protocol for IPv6 (ICMPv6) in [7]. Then, using these messages, the MN performs the Router Table Request/Reply at its booting time.

Fig. 3 shows the Router Table Request message. This message is used by the MN to request network information about all ARs from one of them. The description of each fields are as follows:

- ICMPv6 Type : 160 (or any available value)
- Code : 0
- Identifier : MUST be set by the sender so that replies can be matched to this request.
- Reserved : MUST be set to zero by the sender and ignored by the receiver.

In the IP header of this message, the source address is the MN's address and the destination address is the address of the all ARs multicast address.

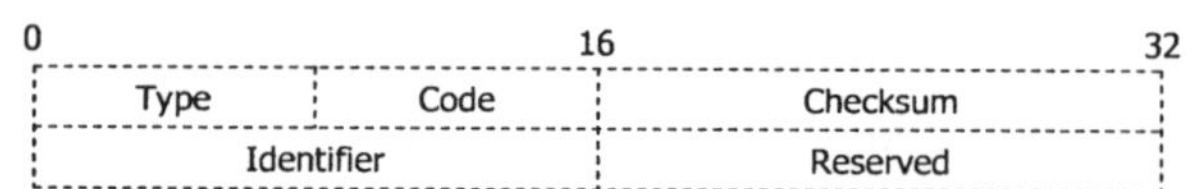

Figure 3. Router Table Request message format

Fig. 4 shows the Router Table Reply message. This message is used by the AR to reply network information about all ARs in ESS to the MN. The description of each fields are as follows:

- ICMPv6 Type : 161 (or any available value)
- Code : 0
- Checksum : The ICMP checksum.
- Identifier : Copied from Router Table Request.
- Number of ARs : The number of ARs' addresses.
- ARID# : ARs' identities.
- Prefix Length : The prefix length of AR's address.

In the IP header of this message, the source address is the AR's address and the destination address is the MN's address.

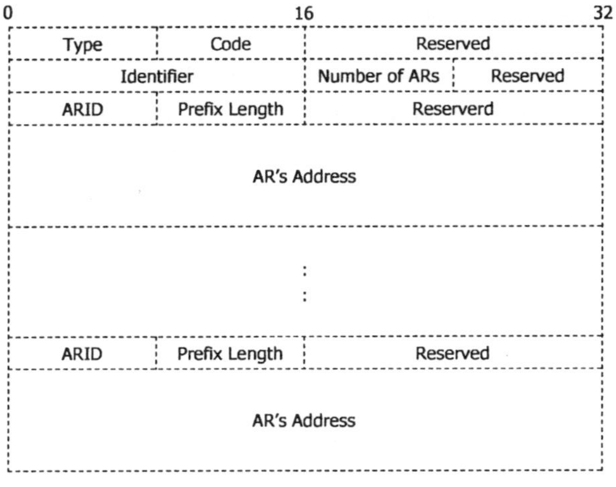

Figure 4. Router Table Reply message format

3.3 Operation Procedure

The operation procedure of the proposed mechanism is described and the comparison with the existing one [1] is shown. According to the specification of IEEE 802.11 WLAN [6], it is required that the MN sets the ESSID before operation. At the same time, in this paper, it is required that the MN sets a flag additionally to determine whether it conforms to the proposed mechanism. That is, if its ESS supports the proposed mechanism, the MN sets the flag and conforms the proposed mechanism. Otherwise, the MN will not set the flag and can conform the existing one in [1].

When the MN is booting, it sends a Router Table Request message using all routers multicast address to current subnet in oder to acquire network information about all ARs in ESS, ARs' IP addresses, prefix lengths, and identities. In response to Router Table Request message, the current AR (CAR) on the current subnet sends a Router Table Reply message using the Router Table shown in Fig. 1. Then, the MN receives this reply message and caches the Router Table in this reply message. Note that this Router Table Reqeust/Reply is performed once only at the booting time and thus isn't performed in real-time communication.

Assume that the MN communicates with the corresponding node (CN). In real-time communication, when the MN moves, the "trigger" may arrive from specific L2 events that might determine the need for handover. In this paper, this trigger itself is not specified in detail. Since this trigger is based on the beacon message given by AP, the MN can know the ARID of its AP from the trigger. Then, the MN checks this ARID using the Router Table cached previously, in oder to determine whether the MN changes AP or AR. If the MN changes AP, it continues real-time communication via the CAR. Otherwise, the MN finds the new AR (NAR) corresponding to ARID from the Router Table, and formulates a new care-of address (NCoA) using the prefix of the NAR's address.

On the other hand, in the existing mechanism [1], when the trigger containing the AP's identity arrives from specific L2 events, the MN sends a Router Solicitation for a Proxy (RtSolPr) message to the CAR in order to acquire network information of the NAR in real-time communication. This message contains the identity of its prospective AP. To response to RtSolPr message, the CAR finds the network information about the NAR corresponding to AP's identity from the Router Table, and sends a Proxy Router Advertisement (PrRtAdv) message with the network information about the NAR to the MN. Note that these RtSolPr and PrRtAdv are performed even if the MN changes AP without change of AR. Then, the MN receives this reply message and determines whether the MN changes AP or AR. If the MN changes AP, it continues real-time communication via the CAR. Otherwise, the MN formulates a NCoA using the prefix of the NAR's address.

After then, in both the proposed mechanism and the existing one, the MN associates its current CoA (CCoA) with NAR's IP address for forwarding purposes using a Fast Binding Update (FBU). Then, the CAR sends a Fast Binding Acknowledgment (FBACK) message to the MN and the NAR. The FBACK message confirms whether NCoA could be used, only after which the MN must use NCoA on the new subnet. At this time, the CAR sends a Handover Initiate (HI) message to NAR, after looking up the IP address corresponding to ARID supplied by the MN using its Router Table. This FBU/FBACK and HI/HACK allows that packets arriving at the CAR can be tunneled to the NAR. When the MN has confirmed its NCoA and then completes the Binding Update procedure with its CN, the MN continues real-time communication with the CN using NCoA as its source IP address via the NAR.

The Fig. 5 and 6 show the operation procedure for the proposed mechanism and the existing one.

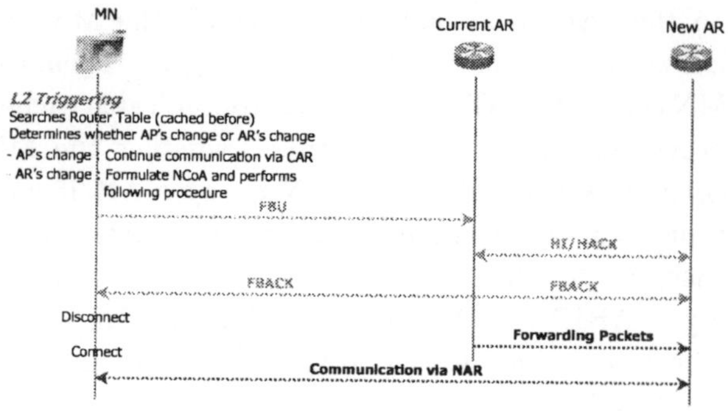

Figure 5. Operation procedure for proposed mechanism

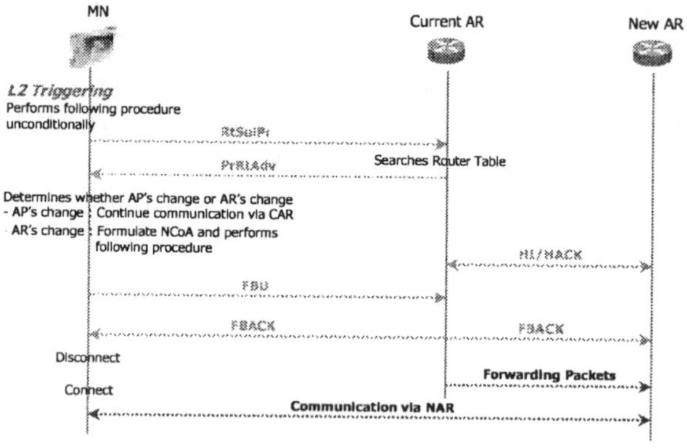

Figure 6. Operation procedure for existing mechanism

3.4 Advantages over Existing Mechanism

The proposed mechanism provides several advantages over the existing one in [1]. As shown in Fig. 5 and 6, the proposed mechanism can omit RtSolPr/PrRtAdv messages used in the existing one, in real-time communication. Thus, the proposed mechanism provides the faster acquisition of NAR's network information than the existing one, which might be advantageous for the low handover latency. In addition, the omission of above two messages might reduce amount of traffic in comparison with the existing one, which might be remarkable when there are many mobile nodes that are now connecting to the CAR. Moreover, in the proposed mechanism, the MN can know whether it changes AR or AP, while the existing one cannot. Thus, when the MN changes AP, the proposed mechanism can remove redundant traffic of the existing one.

However, when there are many ARs on the IEEE 802.11 WLAN, the size of the Router Table may increase. Then, the more memory is required for the MN to cache the Router Table, which may be a weak point of the proposed mechanism.

4 Conclusions

This paper has proposed the new Mobile IPv6 Fast Handover mechanism in IEEE 802.11 WLAN with several access routers. To implement the proposed mechanism, the beacon message used in L2 is defined newly and Router Table Request/Reply messages used in L3 are defined newly. The proposed mechanism provides several advantages. The proposed mechanism provides the faster acquisition of NAR's network information than the existing one, which might be advantageous for the low handover latency. In addition, the proposed mechanism might reduce amount of traffic in comparison with the existing one, which might be remarkable when there are many MNs that are now connecting to the CAR. Moreover, in the proposed mechanism, the MN can know whether it changes AR or AP, which can remove redundant traffic of the existing one.

References

1. R. Koodli, *Fast Handovers for Mobile IPv6.* IETF Draft:draft-ietf-mobileip-fast-mipv6.txt, July 2003.
2. D. B. Johnson, C. E. Perkins, and J. Arkko, *Mobility Support in IPv6.* IETF Draft:draft-ietf-mobileip-ipv6-24.txt, July 2003.
3. X. P. Costa and H. Hartenstein, "A simulation study on the performance of mobile IPv6 in a wlan-based cellular network," *Computer Networks*, vol. 40, pp. 191–204, 2002.
4. H. C. Chao, Y. M. Chu, and M. T. Lin, "The implication of the next-generation wireless network design: cellular mobile IPv6," *IEEE Transactions on Consumer Electronics*, vol. 46, no. 3, pp. 656–663, 2002.
5. I. Okajima, N. Umeda, and Y. Yamao, "Architecture and mobile IPv6 extensions supporting mobile networks in mobile communications," in *IEEE Vehicular Technology Conference*, pp. 2533 -2537, 2001.
6. ISO/ICE, *Wireless LAN Medium Access Control (MAC) and Physical Layer (PHY) Specifications.* ANSI/IEEE Std 802.11, 1999.
7. A. Conta and S. Deering, *Internet Control message Protocol (ICMP) for the Internet Protocol Version 6 (IPv6) Specification.* IETF RFC 2463, December 1998.

A Mobility-Based Adjustment of Binding Update Lifetime for Signaling Cost Reduction in Hierarchical Mobile IPv6*

Myung-Kyu Yi and Chong-Sun Hwang
Dept. of Computer Science & Engineering, Korea University
E-mail: {kainos,hwang}@disys.korea.ac.kr
Tel: +82-2-924-0547, Fax: +82-2-953-0771

Abstract

This paper focusses on adapting the lifetime value of binding update for mobile nodes in a hierarchical Mobile IPv6. In our proposal, mobile node's lifetime can be adjusted from minimum value to maximum value according to mobile nodes's mobility pattern. The simulation result presented in this paper shows that our proposal can reduce the signaling cost for binding update by dynamic allocation of binding update lifetime.

Keywords: mobile IP, mobile computing, location management, mobility, MIPv6

1 Introduction

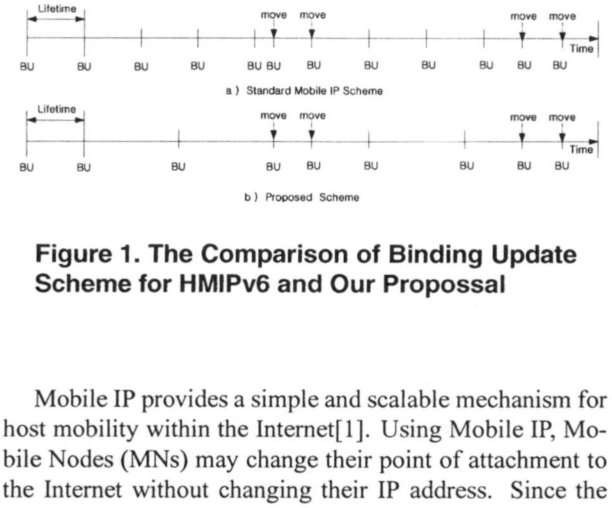

Figure 1. The Comparison of Binding Update Scheme for HMIPv6 and Our Propossal

Mobile IP provides a simple and scalable mechanism for host mobility within the Internet[1]. Using Mobile IP, Mobile Nodes (MNs) may change their point of attachment to the Internet without changing their IP address. Since the frequent emission of Binding Requests (BRs) and Binding Updates (BUs) would result in a relative high signaling overhead, determining the lifetime value of binding cache is the most important issue in Mobile IPv6 (MIPv6)[1]. This paper focusses on adapting the lifetime value of BU for MNs in a Hierarchical MIPv6 (HMIPv6)[2]. In our proposal, MN's lifetime can be adjusted from minimum value to maximum value according to MN's mobility pattern. Fig. 1 shows the comparison of BU scheme for MIPv6 and our proposed scheme. While Fig. 1 (a) shows an IETF standard Mobile IP[2] scheme, Fig. 1 (b) shows the our proposed scheme. In this figure, the line represents the MN's elapsed time and the interval of the line means the BU lifetime of the MN. Also, the vertical arrow indicates that MN changes a point of attachment. As shown in Fig. 1 (a) and (b), an MN sets an its BU lifetime to initial value whenever an MN changes a point of attachment. Notice that an MN sends a BU to the Home Agent (HA) or Correspondent Node (CN) with static BU lifetime in the case of IETF Mobile IP as shown in Fig. 1 (a). However, when an MN does not change its point of attachment, our proposal support dynamic allocation of BU lifetime by increasing the lifetime of BU as shown in Fig. 1 (b). Thus, our proposal can reduce the number of BU and BR message more than the number of BU and BR message presented in [2] by dynamic allocation of BU lifetime.

*This work was supported by grant No. R01-2002-000-00235-0 from the Basic Research Program of the Korea Science & Engineering Foundation

2 Dynamic Allocation of Binding Update Lifetime

2.1 System Model

In this paper, we define a *domains* that are designated site as the highest level of our mobile network. It can be an ISP network, a campus network, a company network, a set of LANs, or even a single LAN. As most of the existing HMIPv6 schemes, we use a Mobility Anchor Point (MAP) in order to deploy two level of hierarchies. A MAP is used by the MN as a local Home Agent (HA) and maintains binding per MNs currently visiting the site. Our system model is similar to that in IETF hierarchical scheme[2].

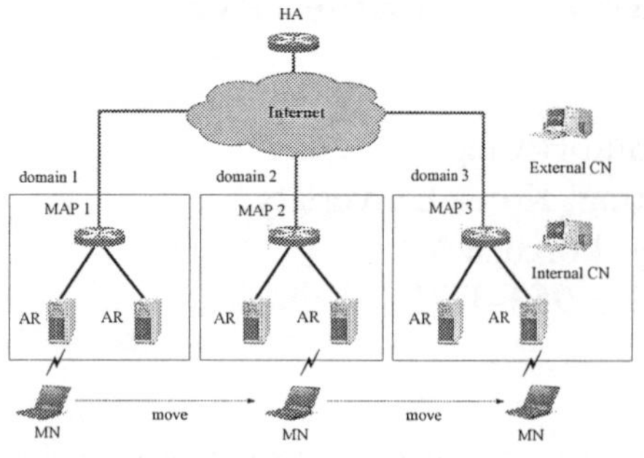

Figure 2. The Proposed Architecture for Dynamic Allocation of Lifetime

2.2 Data Structure

To reduce signaling cost of BU using the dynamic allocation of lifetime, we define data structures for our proposal as follows. This data is stored at each MN.

- LT_L: It is a lifetime value of local BU for the MAP or internal CN.
- LT_R: It is a lifetime value of regional BU for the HA or external CN.
- LC: The number of local BU messages during the MN stays within a subnet.
- RC: The number of regional BU messages when an MN stays within a domain.
- LT_L^{min}: It is the minimum lifetime value of local BU for the MAP or internal CN.
- LT_L^{max}: It is the maximum lifetime value of local BU for the MAP or internal CN. This value can be defined in terms of security considerations.
- LT_R^{min}: It is the minimum lifetime value of regional BU for the HA or external CN.
- LT_R^{max}: It is the maximum lifetime value of regional BU for the HA or external CN. This value can be defined in terms of security considerations.
- ω_L: It is the weight for lifetime value of local BU.
- ω_R: It is the weight for lifetime value of regional BU.
- ω_L^H: It is the previous weight for local BU lifetime.
- ω_R^H: It is the previous weight for regional BU lifetime.

To reduce BU signaling cost, we introduce two lifetimes managing the two COAs. One is the local lifetime LT_L for LCoA, while the others is the domain lifetime LT_R for RCoA. To reduce the BU bandwidth, the local BU that has been sent to a MAP and internal CNs contains only one lifetime value, LT_L, while the regional BU that has been sent to the HA and external CNs contains the two lifetimes, LT_L and LT_R. Also, each MN maintains a profile that characterizes the behavior of MN as shown in Fig. 3. Each profile contains MN's historical information such as past visiting subnet or domain identifier, weight, and the average number of BU messages. Therefore, an MN determines weight depending on the profile when it visits the subnet or domain again. A more detailed algorithm for deriving weights will be described in the next section.

ID	Average Residence Time / Movement	Past Rate of Increase(ω_L^H/ω_R^H)	Average Number of BU	Prefix
Domain 1	1 (hour)	3	12000	FEC0:0:0:1::/64
Subnet 1	15 (min)	1/3	6000	FEC0:0:0:0:3D80::/61
Domain 3	3 (hour)	5	46000	FEC0:0:0:3:/64
Subnet 11	140 (min)	4	4600	FEC0:0:0:0:3DA0::/61
Domain 4	0.5 (hour)	1	16400	FEC0:0:0:2:/64
Subnet 6	30 (min)	1/2	27000	FEC0:0:0:0:3DF0::/61

Figure 3. User Profile Based On the Mobility Pattern

2.3 Dynamic Control of Weight

When an MN enters a new subnet or domain, it sets ω_L and ω_R to initial value. As time goes by, ω_L and ω_R can be adjusted by comparing the current number of BU per unit time with average number of BU messages. The historically estimated weight is calculated with the equation described as shown in Fig. 4. The ω_L^H and ω_R^H represent the historically and previously estimated weight that can be obtained MN's profile, respectively, and the ω_L and ω_R represent weight which is ordinarily estimated weight. Moreover, the α is variable within the range from 0 to 1. The parameter α controls the relative weight of initial value and past history. In detailed, the α is the weight to increase or decrease lifetime of BU according to MN's mobility patterns. If $\alpha = 1$, then $\omega_L = \omega_L \pm 1$. Consequently, an MN may adjust higher or lower lifetime of BU more than initial value. Otherwise, if $\alpha = 0$ then, $\omega_L = \omega_L^H$, and only the most recent weight that obtained by profile. The initial ω_L can be defined as a constant or as an overall average wight depending on the mobility pattern. The ω_R can be calculated as same manner of ω_L. Whenever an MN moves out of a domain or subnet, it records a value of ω_L or ω_R in its profiles.

```
Dynamic_Adjustment_of_Weight( ) {

    IF (current number of local BU > past number of local BU )
        ωL = α × (ωL + 1) + (1 − α) × ωL^H;
    IF (current number of local BU < past number of local BU )
        ωL = α × (ωL − 1) + (1 − α) × ωL^H;
    IF (current number of regional BU > past number of regional BU )
        ωR = α × (ωR + 1) + (1 − α) × ωL^H;
    IF (current number of regional BU < past number of regional BU )
        ωR = α × (ωR − 1) + (1 − α) × ωR^H ;
}
```

Figure 4. Dynamic Adjustment of Weight

2.4 Local Binding Update Procedure

Algorithm 1 : Local Binding Update

```
Procedure LocalBindingUpdate( ) {
 Before an MN send a binding update {
  IF(LC == 1) THEN
     LT_L = LT_L^min;
  ELSE {
     LT_L = LT_L^min × LC × ωL;
     IF(LT_L > LT_L^max) THEN
       LT_L = LT_L^max;
  }
  LC = LC + 1;
  An MN send a local binding update with an LT_L;
 }
}
```

When an MN moves into a new MAP domain, it needs to configure two CoAs: an RCoA and an LCoA. And then, an MN sets the LT_L and the LT_R to LT_L^{min} and LT_R^{min}. And then, the MN must perform local BU with MAP and internal CN and regional BU with HA and external CN. In our proposal, whenever an MN moves within a subnet, the MN sets a value of LC to 1. Similarly, whenever an MN moves out of a domain, the MN sets the value of LC and RC to 1. As a result, the value of LC and RC can be initialized according to micro-mobility or macro-mobility. Before expiration of a binding cache entry for an MN, the internal CN can initiate a refresh of the binding by sending a BR towards the MN in HMIPv6[2]. Receiving this BR the MN may reply with a local BU. In our proposal, whenever an MN sends a local BU to the MAP or internal CN, it must perform *LocalBindingUpdate( )* procedure. To illustrate more detail, if the value of LC is equal to 1, the MN sets the value of LT_L to LT_L^{min}. Otherwise, the MN calculates new lifetime for local BU as follows.

$$LT_L = LT_L^{min} \times LC \times \omega_L \tag{1}$$

After computation of new LT_L, the MN compares LT_L with LT_L^{max}. If the LT_L is more than the LT_L^{max}, the MN sets the value of LT_L to LT_L^{max}. Finally, the MN sends a local BU to the MAP and internal CN with an LCoA and a new LT_L and increases the value of LC by 1. Whenever an MN moves within a domain, the MN sets the value of LC to 1 and sets the LT_L to the LT_L^{min}.

2.5 Regional Binding Update Procedure

Algorithm 2 : Regional Binding Update

```
Procedure RegionalBindingUpdate( ) {
 Before an MN send a binding update {
  IF(RC == 1) THEN
     LT_R = LT_R^min;
  ELSE {
     LT_R = LT_R^min × RC × ωR;
     IF(LT_R > LT_R^max) THEN
       LT_R = LT_R^max;
  }
  RC = RC + 1;
  An MN send a regional binding update with an LT_R;
 }
}
```

After registering with the MAP, the MN must register its new RCoA with its HA by sending a BU that specifies the binding (RCoA, Home Address) as in HMIPv6[2]. As similar to local BU, an MN must send a regional BU to the HA or external CN if it moves out of the domain or its lifetime LT_R is expired. Before expiration of a binding cache entry for an MN, the external CN can initiate a refresh of the binding by sending a BR towards the MN in HMIPv6. Receiving this BR the MN may reply with a regional BU. In our proposal, whenever an MN sends a regional BU to the HA or external CN, it must perform *RegionalBindingUpdate( )* procedure. To illustrate more detail, if the value of RC is equal to 1, the MN sets the value of LT_R to LT_R^{min}. Otherwise, the MN calculates new lifetime for regional BU as follows.

$$LT_R = LT_R^{min} \times RC \times \omega_R \tag{2}$$

After computation of new LT_R, the MN compares LT_R with LT_R^{max}. If the LT_R is more than the LT_R^{max}, the MN sets the value of LT_R to LT_R^{max}. Finally, the MN sends a regional BU to the HA and external CN with an RCoA and a new LT_R and increases the value of RC by 1. Whenever an MN moves out of a domain, the MN sets the value of RC and LC to 1 and sets the LT_L and LT_R to LT_L^{min} and LT_R^{min}.

restart it, without other connected applications noticing it. All inter-component connections are transparently re-established, and neither one of the applications loses any state information. We have found this feature useful during the testing, debugging, and upgrading of distributed applications.

A logical connection between PROCs is maintained even if the underlying physical connection changes. PROCs can move between hosts while maintaining persistent connectivity to other PROCs. The structure of the application does not change and the flow of data in the system is not interrupted. Using move notification information, the engines transparently re-establish inter-PROC connections over the physical connections between the hosts of the communicating PROCs. The seamless connectivity between DACIA components offers a great benefit to mobile users, who can move applications from one device to another without having to manually re-establish all connections to other parties. It can also provide transparency of the location of PROCs and users, if so desired.

Through *application parking* [5] (Figure 4), component mobility and persistent connectivity can be used to support the off-line operation of interactive applications. Initially developed in the context of groupware applications, application parking is suitable to any interactive distributed application. Using DACIA, a parked application can continue to maintain state and to participate, on a limited basis, to collaborations on the user's behalf, while the user is disconnected or is not active. When the user reconnects, eventually from a different place, she can take over the control from the parked application.

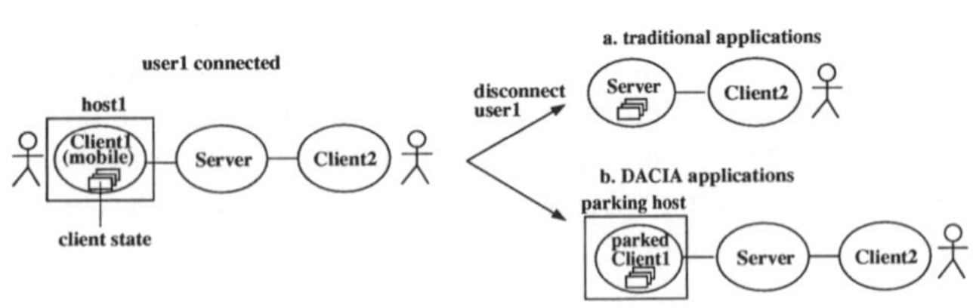

Figure 4. Application parking. (a.) In traditional groupware applications, when a user disconnects, its state has to be saved on a server. If the user later connects to a different host, the state is transferred between servers and between the new server and client. (b.) Using DACIA, the user can park her client to a fixed, connected host. While the user is disconnected, a parked client can continue to maintain state and its connections, and it can interact with collaborative partners.

Conclusions

Our work addresses the challenges of using reconfigurable component-based applications to support the specific needs of mobile and wireless environments. Solutions exist for maintaining connectivity between mobile hosts, that attach to different networks and change their IP addresses. However, maintaining logical connectivity between mobile applications has not been thoroughly investigated.

In this paper, we propose a solution for implementing component mobility so as to ensure reliable message delivery among components in the presence of mobility. Our design also provides support for persistent component connectivity when TCP/IP connectivity incurs transient failures or the IP address of the device changes. To ensure that messages are reliably and orderly delivered both during and after a component relocation, component moves are synchronized with message delivery. We discuss the applicability of component mobility and persistent connectivity to supporting off-line operation of interactive applications.

We are currently extending DACIA to address the security concerns of both mobile components and the hosts where they execute.

Acknowledgments

This work has been supported in part by the National Science Foundation under Grant No. ITR 0082851.

References

[1] G. D. Abowd. Software Engineering Issues for Ubiquitous Computing. In *Proceedings of the 21st International Conference on Software Engineering (ICSE '99)*, Los Angeles, CA, May 1999.

[2] A. Fox, S. D. Gribble, E. A. Brewer, and E. Amir. Adapting to Network and Client Variability via On-Demand Dynamic Distillation. In *Proceedings of the Seventh International ACM Conference on Architectural Support for Programming Languages and Operating Systems (ASPLOS '96)*, Cambridge, MA, Oct. 1996.

[3] X. Fu, W. Shi, A. Akkerman, and V. Karamcheti. CANS: Composable, Adaptive Network Services Infrastructure. In *Proceedings of the 3rd USENIX Symposium on Internet Technologies and Systems (USITS)*, San Francisco, CA, March 2001.

[4] T. D. Hodes, R. H. Katz, E. Servan-Schreiber, and L. Rowe. Composable Ad-hoc Mobile Services for Universal Interaction. In *Proceedings of the 3rd ACM/IEEE MobiCom*, Budapest, Hungary, Sep 1997.

[5] R. Litiu and A. Prakash. Developing Adaptive Groupware Applications Using a Mobile Component Framework. In *Proceedings of the 2000 ACM Conference on Computer-Supported Cooperative Work, (CSCW 2000)*, Philadelphia, PA, Dec. 2000.

[6] B. D. Noble, M. Satyanarayanan, D. Narayanan, J. E. Tilton, J. Flinn, and K. R. Walker. Agile Application-Aware Adaptation for Mobility. In *Proceedings of the 16th ACM Symposium on Operating Systems Principles (SOSP '97)*, Saint-Malo, France, Oct. 1997.

[7] C. Perkins. Mobile IP. *IEEE Communications*, 35(5):84–99, 1997.

[8] J. P. Sousa and D. Garlan. Aura: an Architectural Framework for User Mobility in Ubiquitous Computing Environments. In *Proceedings of the 3rd Working IEEE/IFIP Conference on Software Architecture*, Montreal, CA, August 2002.

[9] M. Weiser. Hot Topics: Ubiquitous Computing. *IEEE Computer*, Oct. 1993.

[10] M. Yarvis, P. Reiher, and G. J. Popek. Conductor: A Framework for Distributed Adaptation. In *Proceedings of the Seventh Workshop on Hot Topics in Operating Systems (HotOS '99)*, March 1999.

SESSION 6B:
ARCHITECTURE

Service Management Architecture for Providing AAA and Differentiated Service in Mobile IPv6

Shoubin Dong[1], Luliang Jiang[2], Liqun Tang[1] , Jian Ma[2], Ling Zhang[1]

[1] Network Research Center, South China University of Technology
510640 Guangzhou, China
[2] Nokia Research Center (China)
100013 Beijing, China

Abstract: Recent advances in the IPv6 for mobile network had led to extensive research for AAA and QoS in mobile environment. In this paper, we propose an extended AAA architecture that support for the authentication, authorization and QoS negotiation of the Mobile Nodes based on Internet protocols and standards for Mobile IPv6 environment. We also describe the implementation of a Bandwidth Broker for providing the Diffserv in MIPv6. Finally, we propose a framework of service management for MIPv6 to provide the integrated services of AAA and Diffserv. A prototype system has been constructed to demonstrate the design.

Key Words: Mobile IPv6, Service Management, AAA, Diffserv

1. INTRODUCTION

Recent advances in the IPv6 for mobile network had led to extensive research for Authorization, Authentication and Accounting (AAA) and Quality of Service (QoS) [1-3]. The key problem is how to optimize the allocation of network resources in mobile environment to meet the demands of QoS at different levels while maintaining the connection efficiently. When a mobile node visits a foreign network and requests services based on Service Level Agreements (SLAs), the Internet Service Providers (ISPs) must care about how to charge the mobile user. In addition, SLAs have to be renegotiated between home/foreign links and their ISPs. Thus, procedures of authentication, authorization and accounting must be provided.

While service management has become the key technique of the next generation network, we found that it is significant to develop a service management system for customers to apply for service and for operators to configure the routers and servers to meet the users' demands. It has been widely accepted that a good service manager would enables ISPs to effectively plan, provision, operate and bill for a wide range of value added services, while increasing revenues and reducing costs, which will consequently push the wide application of QoS on Mobile IPv6 network.

This paper presents the architecture of service management for providing the AAA and Differentiated Services to Mobile IPv6 users. To reach this goal, we propose a new AAA architecture based on the DHCP (Dynamic Host Configuration Protocol) [4], LDAP (Lightweight Directory Access Protocol), SLP (Service Location Protocol) [5] and Diameter [6] protocols in section 2. Section 3 discusses specific issues about Differentiated Service implementation - Bandwidth Broker by COPS (Common Open Policy Service) [7,8]. Section 4 proposes the implementation architecture of the service management system based on extended AAA architecture and bandwidth broker. We present our conclusion in Section 5.

2. EXTENDED AAA ARCHITECTURE

The basic AAA model for MIPv6 has been proposed in [6]. In the model, when a Mobile Node (MN) roams to a new visited domain, it will register to an AAA client and wait to be authenticated by the network service provider. The AAAv (AAA server in the visited domain) firstly checks whether the MN is a local domain user, if yes try to authenticate it, or transfers the identities from AAA client to AAAh (AAA server in the home domain of MN), the authenticate may be finished by AAAv or HA (Home Agent), which depends on the properties of identities of MN.

In MIPv6, the Mobile Node may use auto-configuration mechanisms to get a new IP address while visiting foreign domains. However, the MN can do anything at will if it acquires a valid IP address automatically. This has a tremendous impact on the network security and also causes problems in user accounting and billing.

Therefore, it is necessary to extend the basic AAA architecture for Mobile IPv6. T. Braun et al [1] suggested a Service Location Protocol (SLP) integrated AAA architecture. It seems that this model is benefit not only to prevent an MN from using networkresource illegally, but also to help integrating Differentiated Services (Diffserv) with mobile environments. However there are some drawbacks in this model, such as the application layer protocol SLP should be used before the DHCP assigns a valid IP to MN. Such communication schema may cause great change to the current protocol implementations of SLP and DHCP.

Here, we propose a new extended AAA model for Mobile IPv6 environment. The architecture is shown in Fig.1. The idea of this model is to provide an integrated AAA architecture based on current protocols and Internet standards, which makes this model more practical and feasible.

2.1 Network Access

The detail authentication process for MN's network access is operated in the following procedures. Each step is corresponding to the number marked in Figure 1.

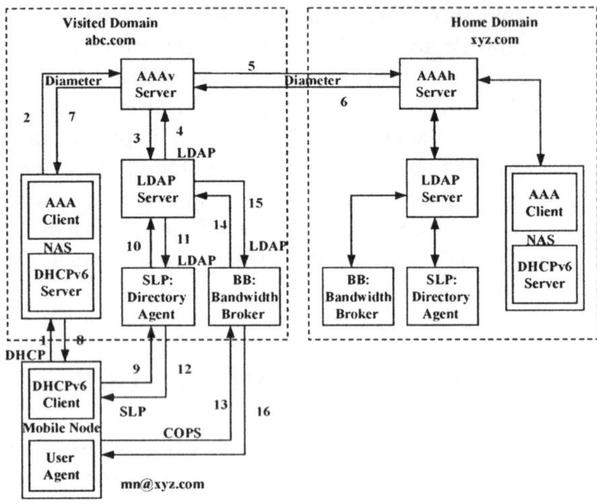

Figure 1: Extended AAA architecture

(1) When the MN visits the foreign domain (abc.com), it sends an attribute request to the DHCP server to obtain an IP address. Here suppose that stateless auto configuration in IPv6 routers has been disabled in advance and the DHCP server announce itself by broadcast.
(2) When the DHCP receives an attribute request, it will forward the authentication information to the AAAv server by Diameter.
(3) The AAAv server first checks its LDAP server to see whether the mobile user belongs to its own network.
(4) The LDAP Server respond with the information that if the user is a visitor.
(5) If the user is a visitor, the AAAv server contacts the external AAAh by Diameter to further verify the user identity.
(6) The home AAA server checks the validity of the user identity based on the confidential information and the application specifications (such as QoS) which may be indexed by username, password etc. Then, it gives a proper response to the AAAv server.
(7) When the foreign AAA server receives a positive reply from the home AAA server, it will not only store the SLA specification to establish a customer record in its LDAP database, but also inform the DHCP Server about the authentication result.
(8) If the user is valid, the DHCP will directly assign an IP to MN.

In our model, we use DHCP and Diameter integrated implementation schema. The message flows for authentication and authorization in the above model are illustrated in figure 2.

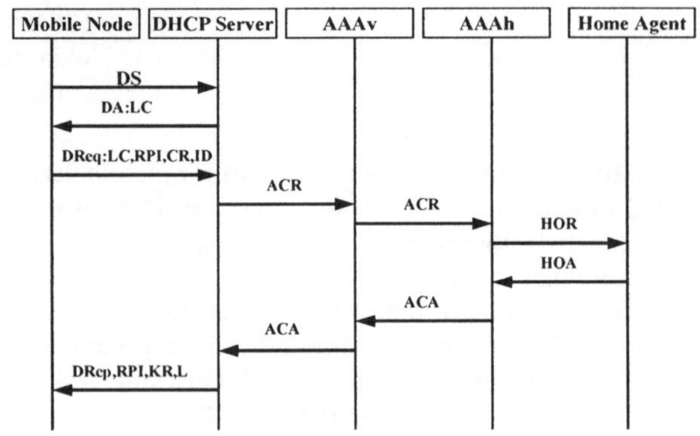

Figure 2: Message flow in AAA model

Where the commands codes are [9]:

- DS = DHCP Solicit
- DA = DHCP Advertise
- DReq = DHCP Request
- DRep = DHCP Reply
- LC = Local AAA Challenge
- RPI = Replay Protection Indicator used between client and AAAh
- CR = AAA Credential
- ID = Client Identifier
- KR = Key Reply
- L = Lifetime
- ACR = AAA Client Request (using an AAA protocol)
- ACA = AAA Client Answer (using an AAA protocol)
- HOR=Home-Agent-MIPv6-Request Command
- HOA=Home-Agent-MIPv6-Answer Command

2.2 QoS Negotiation

After the MN has successfully got access to the network, it might further desire to apply for QoS. First, it has to find an available Differentiated Services and what types of Differentiated Services are supported in the foreign domain, then to make a SLA with this domain. This procedure is called QoS negotiation.

To provide QoS, a Bandwidth Broker (BB) is introduced to act as the resource manager for each administrative domain. The basic operation of QoS negotiation in our model is described as followings:

(9) The MN sends an attribute message to the DA (Directory Agent in SLA) to request QoS.
(10) DA contacts local LDAP server for the authentication information of this user.

(11) The LDAP server checks that this user has been authenticated before.
(12) The DA will subsequently send an attribute reply to the MN, and gives the MN the IP address of the BB (Bandwidth Broker) server which supplying the Diffserv.
(13) Then, the MN communicates with the BB to request the available Diffserv.
(14) The BB may also need to contact local LDAP server to further verify the user as DA does.
(15) The LDAP server checks that this user has been authenticated before.
(16) Finally, the user is informed by the BB that the requested service is now available.

The QoS level may be modified because of various reasons, such as the charged price, the current network load and the available service type etc. If BB server in the visited domain could not provide the service the MN's acquired, it may need to re-negotiate with MN. The BB will give a list of supported Diffserv types in the request reply to the MN, and the MN chooses its favor Diffserv type among the various options. The subsequent procedure is the same as above.

3. BANDWIDTH BROKER

To assure the end-to-end QoS, a Bandwidth Broker (BB) is implemented to perform the policy control in Diffserv networks. The architecture of bandwidth broker is shown in figure 3. The policy control allows network operators to set up policy rules in policy repository. The Policy Decision Point (PDP) determines the QoS control based on the policy rules. The Policy Enforcement Point (PEP) within the network element controls the network element based on the determination of PDP, and the COPS protocol, is used for policy exchange between the PDP and the PEP [7,8].

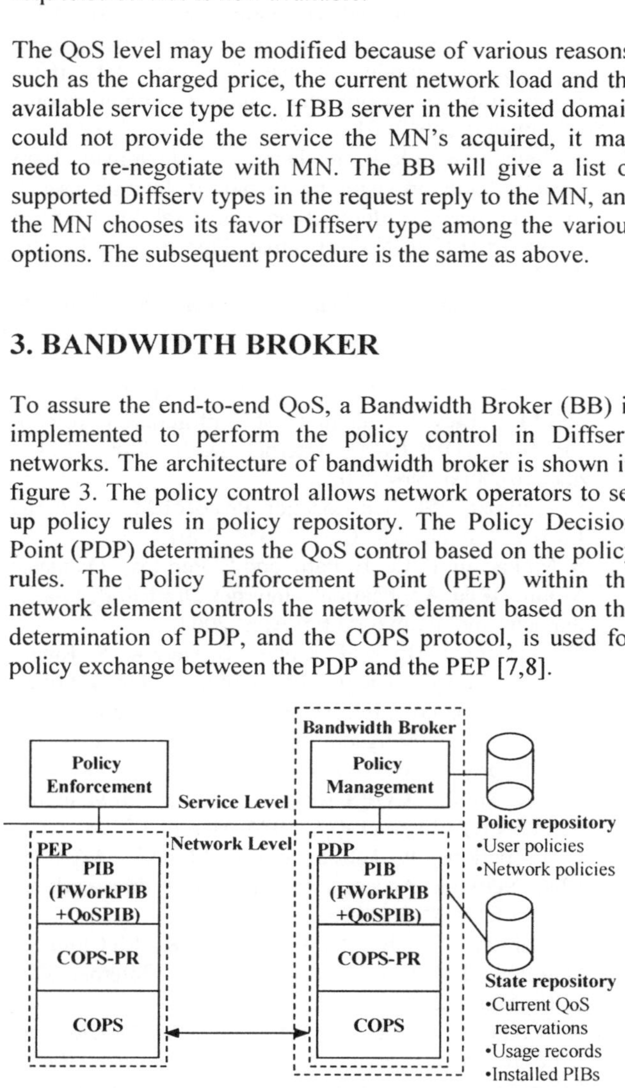

Figure 3: Architecture of controlled router and BB server

Firstly, the controlled router opens a COPS connection to the PDP server. The connection is maintained as long as both parties are up. In the COPS-PR model [8], after the connection is established, the PEP sends information about itself to the PDP in the form of a configuration request. It includes client specific information (e.g., descriptions of each of the PEP' s interfaces and capabilities of metering, queuing and scheduling), which is in the format of a named data structure called Policy Information Base (PIB)[10]. The configuration request is encapsulated into a COPS *request message*.

The BB receives this Request message and is notified of the configuration of the PEP. In the service level, the BB makes decision by downloading all provisioned policies that are currently relevant to that device from the policy repository. The policies data are used to install a PIB incarnation on the PEP, and they are in the format of a series of PIB tables defined in [10,11]. These PIB tables are encapsulated into COPS solicited *decision message* and sends to the PEP.

The decision is enforced in the PEP. We use TC (Traffic Control) [12] to implement the enforcement in our project. The install decision, which combines of a series of PIB tables, is translated into TC commands. They are executed accordingly. The result of execution, 'success' or 'failure', is send back to the BB by the COPS solicited *report message*.

After that, if user makes a new Resource-Allocation-Request (RAR) to the BB in the service level, the Policy Management (PM) first access the policy repository to ascertain whether the user is allowed to make such a request. If the request is valid, the policies required for this request are read from the repository and passed on to the PDP. The PDP checks whether existing bandwidth reservations allow for more reservations to be made, i.e. there is any bandwidth left to reserve in the network at the time. This kind of information is stored in a state repository database. If enough bandwidth is available, the PDP configures the PEPs, where the reservation traffic will pass, by an unsolicited install *decision message*. After the PEP returns a 'success' report message for this decision, BB updates the state repository database of the new reservations. Then, a positive Resource Allocation Answer (RAA) is sent back to the user to inform the RAR was successful. If either the user is not allowed to make such a RAR or there is no bandwidth available or the policy is not successfully install, the RAA will be negative. The RAR can be deleted afterward. This will turn in an unsolicited remove *decision message*.

When the connection is maintained, the PEP also uses unsolicited report message to provide periodic reporting of information for accounting and state monitoring purposes. Differ from the Report message mentioned above, the Report-Type of this Report message is "Accounting". A framework of this accounting PIB group is proposed in [13]. The provision class *frwkFeedbackUsageClasses* in this PIB group describes the packet counts, byte counts, last timestamp that usage. Thus the count and timestamp information of the forwarding packets can be monitored and recorded by the PEP and supplied to the PDP with the accounting type report message. The periodic reporting information of accounting can be used for billing or new decision purpose.

4. SERVICE MANAGEMENT SYSTEM

The final aim of our work is to establish integrated service management architecture for providing AAA, Differentiated Service and other services for Mobile IPv6. Based on the extended AAA architecture and bandwidth broker described above, we propose to build the Service Management System (SMS). The overall solution is shown in Fig. 4. There are 3 layers in this implementation architecture:

(1) Service Information Portal (SIP): to provide the portal to customers and system administrator;
(2) Service Level: to provide the key management functions;
(3) Network Level: to provide the supporting functions for service management.

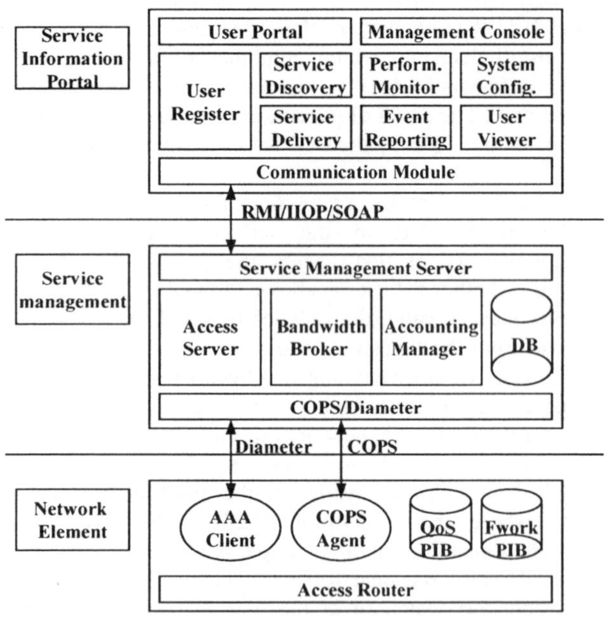

Figure 4: The implementation architecture of SMS

The service level is separated clearly from network level, thus to constitute a two-tier management hierarchy. In this two-tier model, the Bandwidth Broker (BB) acts as the resource manager and AAA server acts as Authorization, Authentication and Accounting server for each administrative domain. The COPS and Diameter protocols is used by service management system to control the network elements.

In order to well integrate the system components, we employ the advanced web service over J2EE (Java 2 Platform, Enterprise Edition) platform at SIP and service level. The protocols between these two layers may use RMI (Remote Method Invocation), IIOP (Internet Inter-ORB Protocol) or SOAP (Simple Object Access Protocol). The J2EE (http://java.sun.com/j2ee/) defines the service container architecture, and offers the advantages of a standard, extensible, component-based application model, which makes the SMS potential for real application.

5. CONCLUSION

This paper discussed the architecture and implementation of service management system for providing integrated service to Mobile IPv6 users. A prototype system based on J2EE distributed computing environment has been constructed, which works well and has demonstrated the flexibility and extensibility of the design.

REFERENCES

[1] T. Braun, L. Ru and G. Stattenberger, "An AAA architecture extension for providing differentiated services to mobile IP users", Proc. 6th IEEE Symposium on Computers and Communications, Hammamet, July 2001.

[2] S. Lepaja, R. Fleck, R. Donner, and Nguyen Nam Hoang, "QoS provisioning to mobile Internet users", 2nd European Conference on Universal Multiservice Networks (ECUMN02), Colmar, France, pp. 230-236, 2002.

[3] Z. Kan, D. Zhang, R. Zhang and J. Ma, "QoS in mobile IPv6", Proc. Int. Conf. on Info-tech and Info-net (ICII01), Volume 2, 2001.

[4] R. Droms, J. Bound, B. Volz, T. Lemon, C. Perkins, and M. Carney, "Dynamic Host Configuration Protocol for IPv6 (DHCPv6)", Internet Draft draft-ietf-dhc-dhcpv6-28.txt, Nov 2002.

[5] E. Guttman, "Service Location Protocol Modifications for IPv6", Internet RFC 3111, May 2001.

[6] S. M. Faccin, F. Le, B. Patil, and E. Perkins, "Diameter Mobile IPv6 Application", Internet Draft draft-le-aaa-diameter-mobileipv6-03.txt, April 2003.

[7] D. Durham, J. Boyle, R. Cohen, S. Herzog, R. Rajan, and A. Sastry, "The COPS (Common Open Policy Service) Protocol", Internet RFC 2748, January 2000.

[8] K. Chan, J. Seligson, D. Durham, S. Gai, K. McCloghrie, S. Herzog, F. Reichmeyer, R. Yavatkar, and A. Smith, "COPS Usage for Policy Provisioning (COPS-PR)", Internet RFC 3084, March 2001.

[9] C. E. Perkins, E. Tacsik, and T. Eklund, "AAA for IPv6 Network Access", Internet Draft draft-perkins-aaav6-06.txt, May 2003.

[10] R. Sahita, S. Hahn, K. Chan, and K. McCloghrie, "Framework Policy Information Base", Internet RFC 3318, March 2003.

[11] K. Chan, R. Sahita, S. Hahn, and K. McCloghrie, "Differentiated Services Quality of Service Policy Information Base", Internet RFC 3317, March 2003.

[12] Linux Advanced Routing & Traffic Control, http://lartc.org

[13] D. Rawlins, A. Kulkarni, K. H. Chan, M. Bokaemper, and D. Dutt, "Framework of COPS-PR Policy Information Base for Policy Usage Feedback", Internet Draft draft-ietf-rap-feedback-fr-pib-06.txt, April 2002.

PERSISTENT CONNECTIVITY AND RELIABLE MESSAGE DELIVERY IN COMPONENT-BASED MOBILE APPLICATIONS

RADU LITIU and ATUL PRAKASH

Department of Electrical Engineering and Computer Science
University of Michigan, Ann Arbor, MI 48109-2122, USA
Email: {radu,aprakash}@eecs.umich.edu

Abstract

Applications for mobile and wireless environments will need to support computing devices with a wide range of capabilities, varying network connectivity, and increasing mobility of users. We have designed a component-based framework for building adaptive distributed applications. Distributed applications are viewed as consisting of connected components that typically implement data streaming, processing, and filtering functions. Our framework provides mechanisms for run-time reconfiguration of applications to allow them to adapt to the changing operating environments. Components or whole applications can be moved to different hosts during execution, while maintaining seamless communication connectivity with other components. This paper presents techniques used to implement component mobility, ensure application consistency and reliable message delivery during component moves, as well as maintain persistent connectivity during transient network failures. We show potential benefits of maintaining seamless connectivity between mobile components.

Keywords: mobile users, mobile applications, dynamic application reconfiguration, persistent connectivity, application parking

1. Introduction

Mobile computing environments are characterized by a wide variety of computing devices, ranging from high-end machines to simple devices such as personal digital assistants (PDAs) or cell phones. Network connectivity can be highly dependent on the carrier's coverage, device capability in terms of power and hardware, and the location of the user. In this paper, we consider support for situations in which mobile users are running applications that are part of a distributed computing environment, being connected to services, data sources, and to groupware applications of other users. Unlike current environments, we consider new demands on the infrastructure where a user needs to move his applications, or their components, from one computing device to another (e.g., from a PDA to a desktop), while maintaining connectivity with other applications. It would be desirable to provide support to users so that they are able to move applications across devices while maintaining seamless communication connectivity with other applications (Figure 1).

Operating in a mobile environment also raises the problem of dealing with the inherent unreliability of mobile network connections and variations in connection quality. For many applications, it would be desirable that the underlying middleware layers mask transient network and communication failures.

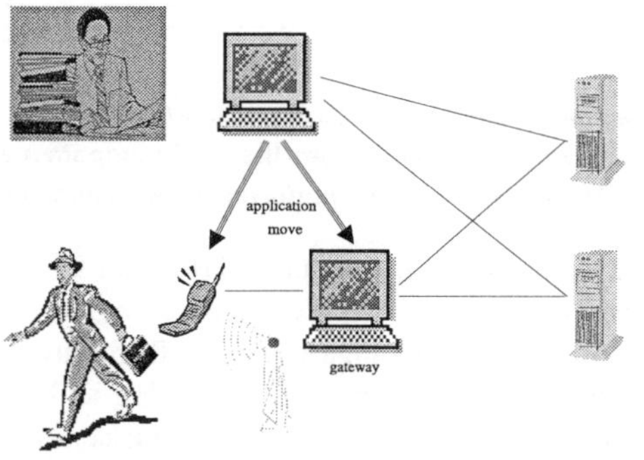

Figure 1. Applications need to support mobility of users, heterogeneity, and variations in available resources. They should be able to move from one computing device to another.

Mobile IP [7] addresses the problem of maintaining connectivity between moving hosts. As a computer moves from one wireless network to another, the computer's home IP address is maintained. When only some applications move across hosts, moving the IP address along with those applications would not be appropriate, since there might be other applications running on that machine that rely on the same mobile IP address.

To address the above issues, we have designed and implemented (in Java) a mobile component framework, called DACIA (Dynamic Adjustment of Component InterActions), for building and executing reconfigurable distributed applications. The flexible architecture proposed allows modular applications to dynamically change their structure and relocate components, while maintaining persistent connectivity with other components. Logical connections between components are maintained even when underlying TCP/IP connectivity is lost for short times (as it can happen in today's GPRS networks) or the device's IP address changes due to mobility.

Our work complements existing solutions proposed to address the variability and resource constraints encountered both in wired and mobile computing systems, such as Daedalus/BARWAN [4], on-demand dynamic distillation tech-

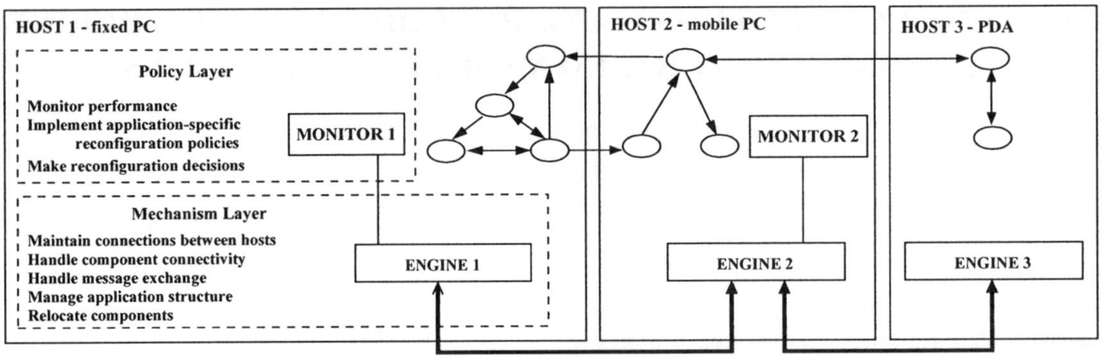

Figure 2. A DACIA distributed application is a directed graph of connected components (ovals represent components). An engine runs on every host. It manages the local components and the connections between components, both local and across different hosts. The monitor gathers performance data and implements application-specific relocation and reconfiguration policies.

niques [2], Odyssey [6], Conductor [10], and CANS [3]. It addresses some of the complex software engineering issues identified in supporting a high degree of flexibility in ubiquitous and pervasive computing [1, 8, 9]. We address part of that problem, so that distributed application components can be dynamically relocated without losing connectivity to other components. They can potentially be programmed to relocate according to a specified policy, such as moving with the location of the user, or to optimize the usage of currently available resources.

The rest of the paper presents an overview of the architecture of the system and outlines algorithms that implement component mobility and ensure application consistency and message integrity during component moves. We also show potential benefits of maintaining persistent connectivity between mobile components.

2. DACIA Architecture

DACIA is a framework for building adaptive distributed applications in a modular fashion, through the flexible composition of software modules implementing individual functions. A DACIA application can be seen as a directed graph of connected components. The links between components indicate the direction of the data flow within the application.

In DACIA, a component is a *PROC* (Processing and ROuting Component). A PROC can apply some transformations to one or multiple input data streams. It can synchronize input data streams; it can split the items in an input data stream and send them alternately to multiple destinations. A PROC is identified system-wide using a unique identifier.

The *engine* decouples an application and component-specific code and functionality from the general administrative tasks such as maintaining the list of PROCs and their connections, migrating PROCs, establishing and maintaining connections between hosts, and communicating between hosts. A DACIA distributed application (Figure 2) uses an engine on every host it runs on. We chose to use an engine per application per host, as opposed to sharing an engine running on a host between multiple applications, in order to minimize the cost of communication between PROCs and between PROCs and the engine.

PROCs communicate by exchanging *messages* through *ports* (Figure 3). Communication can be either *synchronous* or *asynchronous*. In the case of asynchronous communication, the messages received by a PROC are inserted into the PROC's *message queue*. Every PROC has a thread that handles the messages in the queue, usually in FIFO order. We designed and implemented the communication mechanisms in DACIA with the goal of minimizing the overhead of message exchange. DACIA provides a lightweight solution to local communication, by co-locating local PROCs within the same address space. Thus, message exchange translates into simple procedure calls.

The engine maps virtual connections between PROCs to either local or remote network connections, and handles data transfers accordingly. Multiple virtual remote connections between pairs of PROCs are multiplexed over a single network connection between two engines. The connectivity between remote PROCs is maintained as long as the corresponding engines are connected. Sharing physical connections reduces the cost of establishing network connections in a highly dynamic application, where PROCs often connect to each other or they are disconnected.

The engine of an adaptive application may work in conjunction with a *monitor*. While the engine provides the *mechanisms* used to build and reconfigure applications, the monitor implements the *policy* layer in a DACIA application. The engine is not aware about the semantics of the application. The monitor keeps track of the application performance, makes reconfiguration decisions, and instructs the engine accordingly. The engine is responsible for establishing and removing con-

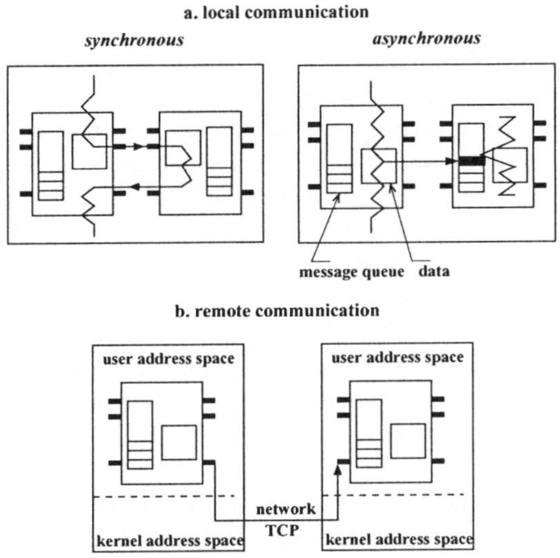

Figure 3. Inter-PROC communication.

nections between components and for moving components to other hosts.

An application can be dynamically reconfigured by reordering or relocating some components, or replacing a set of components with a different set of components, possibly connected in a different configuration. Components can be connected or disconnected. New components can be dynamically loaded, and existing components can be removed or relocated to different hosts. The performance of the application can be improved by changing the way different application components interact and their location of execution, thus taking advantage of the resources available system-wide. Our solution to runtime reconfiguration ensures that the application executes correctly both during and after the reconfiguration.

3. Component Mobility

One of the key features of our architecture is the ability to move components between hosts. Our work addresses the problem of capturing the state of a component and restoring it at the destination. To reduce the overheads of component movement, thus the amount of data that has to be moved, we do not transfer the execution state of a PROC (e.g., program counter, stack and registers content, thread state, etc). However, a moving PROC carries with it the state of its data members, the messages received and not handled yet, and the state of its connections. PROC migration happens at well-defined times with respect to the execution of the PROC. Before a PROC moves, if a message is currently being handled, the handling routine completes. When a PROC moves to another host, all messages left in its message queue move with the PROC.

A major challenge in implementing PROC mobility in DACIA has been to make the move transparent to communicating PROCs, allow other PROCs to send messages to a moving PROC, and ensure that these messages are delivered reliably to their destination. The FIFO order of message delivery and processing at the destination has to be preserved as components relocate to a new engine. Message exchange has to be synchronized with a PROC move.

The PROC is *locked* while it is moving, thus preventing other PROCs from sending messages to it. The message send operation blocks while the PROC is moving. The sender PROC (the corresponding engine) retrieves a *remoteProc* reference corresponding to the receiver, then sends the message to the new instance of the receiving PROC, at its new location. We avoid acquiring a lock corresponding to the receiving PROC every time a message is sent. A *location* flag is set prior to a move and reset after the move completes. The flag is tested prior to the sender acquiring a lock, thus eliminating the need for a lock if the receiver is not undergoing a move. At the same time, a counter indicating the number of outstanding messages prevents a PROC move operation from being executed while the PROC is receiving and handling a message from another PROC.

At the destination host, a locking mechanism synchronizes the PROC initialization and the updates to the engine data structures with the ongoing message send operations addressed to the moving PROC, and with potential subsequent move operations that may be initiated before the data structures corresponding to the moving PROC are updated. We use message sequence numbers and a sliding window protocol to prevent messages potentially originating at different hosts from being delivered out of order to the moving PROC.

The destination engine sends notifications about the PROC move to all other engines that have PROCs connected to the moving PROC. Upon receiving a move notification, an engine updates its local information about the PROC's location, thus avoiding the need for message forwarding in the future. A notification contains the ID of the moving PROC, the new location of the PROC, and the change sequence number that keeps track of the order of changes applied to the PROC (a change can be a move, connect or disconnect). The sequence number is used to prevent old PROC move notifications received out of order from being applied.

4. Persistent Connectivity

In many cases, the temporary failure of a connection between engines can be made transparent to PROCs and applications. When a network connection is broken, the engines will try to re-establish the connection during a timeout interval. The timeout interval can be set on an application basis, regardless of the physical network settings. Assuming that the disconnection is temporary, an engine caches messages addressed to a remote PROC until the connection is re-established. The use of a timeout for re-connection allows an administrator to briefly shut down an application running on a host, and immediately restart it, without other connected applications noticing it. All inter-component connections are transparently re-established, and neither one of the applications loses any state information. We have found this feature useful during the testing, debugging, and upgrading of distributed applications.

A logical connection between PROCs is maintained even if the underlying physical connection changes. PROCs can move between hosts while maintaining persistent connectivity to other PROCs. The structure of the application does not change and the flow of data in the system is not interrupted. Using move notification information, the engines transparently re-establish inter-PROC connections over the physical connections between the hosts of the communicating PROCs. The seamless connectivity between DACIA components offers a great benefit to mobile users, who can move applications from one device to another without having to manually re-establish all connections to other parties. It can also provide transparency of the location of PROCs and users, if so desired.

Through *application parking* [5] (Figure 4), component mobility and persistent connectivity can be used to support the off-line operation of interactive applications. Initially developed in the context of groupware applications, application parking is suitable to any interactive distributed application. Using DACIA, a parked application can continue to maintain state and to participate, on a limited basis, to collaborations on the user's behalf, while the user is disconnected or is not active. When the user reconnects, eventually from a different place, she can take over the control from the parked application.

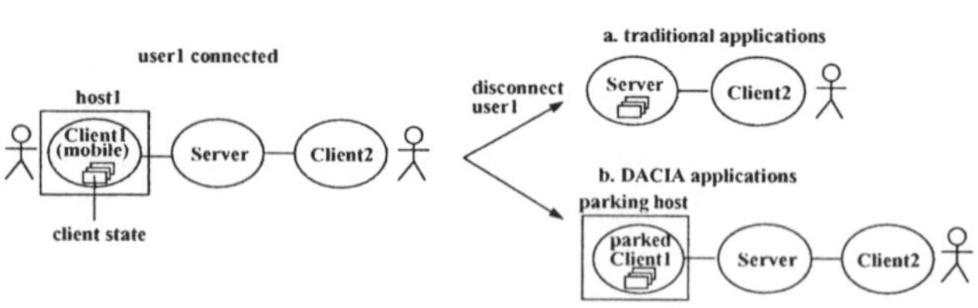

Figure 4. Application parking. (a.) In traditional groupware applications, when a user disconnects, its state has to be saved on a server. If the user later connects to a different host, the state is transferred between servers and between the new server and client. (b.) Using DACIA, the user can park her client to a fixed, connected host. While the user is disconnected, a parked client can continue to maintain state and its connections, and it can interact with collaborative partners.

5. Conclusions

Our work addresses the challenges of using reconfigurable component-based applications to support the specific needs of mobile and wireless environments. Solutions exist for maintaining connectivity between mobile hosts, that attach to different networks and change their IP addresses. However, maintaining logical connectivity between mobile applications has not been thoroughly investigated.

In this paper, we propose a solution for implementing component mobility so as to ensure reliable message delivery among components in the presence of mobility. Our design also provides support for persistent component connectivity when TCP/IP connectivity incurs transient failures or the IP address of the device changes. To ensure that messages are reliably and orderly delivered both during and after a component relocation, component moves are synchronized with message delivery. We discuss the applicability of component mobility and persistent connectivity to supporting off-line operation of interactive applications.

We are currently extending DACIA to address the security concerns of both mobile components and the hosts where they execute.

Acknowledgments

This work has been supported in part by the National Science Foundation under Grant No. ITR 0082851.

References

[1] G. D. Abowd. Software Engineering Issues for Ubiquitous Computing. In *Proceedings of the 21st International Conference on Software Engineering (ICSE '99)*, Los Angeles, CA, May 1999.

[2] A. Fox, S. D. Gribble, E. A. Brewer, and E. Amir. Adapting to Network and Client Variability via On-Demand Dynamic Distillation. In *Proceedings of the Seventh International ACM Conference on Architectural Support for Programming Languages and Operating Systems (ASPLOS '96)*, Cambridge, MA, Oct. 1996.

[3] X. Fu, W. Shi, A. Akkerman, and V. Karamcheti. CANS: Composable, Adaptive Network Services Infrastructure. In *Proceedings of the 3rd USENIX Symposium on Internet Technologies and Systems (USITS)*, San Francisco, CA, March 2001.

[4] T. D. Hodes, R. H. Katz, E. Servan-Schreiber, and L. Rowe. Composable Ad-hoc Mobile Services for Universal Interaction. In *Proceedings of the 3rd ACM/IEEE MobiCom*, Budapest, Hungary, Sep 1997.

[5] R. Litiu and A. Prakash. Developing Adaptive Groupware Applications Using a Mobile Component Framework. In *Proceedings of the 2000 ACM Conference on Computer-Supported Cooperative Work, (CSCW 2000)*, Philadelphia, PA, Dec. 2000.

[6] B. D. Noble, M. Satyanarayanan, D. Narayanan, J. E. Tilton, J. Flinn, and K. R. Walker. Agile Application-Aware Adaptation for Mobility. In *Proceedings of the 16th ACM Symposium on Operating Systems Principles (SOSP '97)*, Saint-Malo, France, Oct. 1997.

[7] C. Perkins. Mobile IP. *IEEE Communications*, 35(5):84–99, 1997.

[8] J. P. Sousa and D. Garlan. Aura: an Architectural Framework for User Mobility in Ubiquitous Computing Environments. In *Proceedings of the 3rd Working IEEE/IFIP Conference on Software Architecture*, Montreal, CA, August 2002.

[9] M. Weiser. Hot Topics: Ubiquitous Computing. *IEEE Computer*, Oct. 1993.

[10] M. Yarvis, P. Reiher, and G. J. Popek. Conductor: A Framework for Distributed Adaptation. In *Proceedings of the Seventh Workshop on Hot Topics in Operating Systems (HotOS '99)*, March 1999.

Service and Personal Mobility in Next Generation Networks

Natalia Kryvinska[1], Salem Lepaja[1], Hoang M. Nguyen[2]

[1]Vienna University of Technology
Favoritenstrasse 9–388, A1040, Vienna, Austria

[2]Networking Department, Communications and Device Division,
Institute for Infocomm Research (Member of A*STAR)
21, Heng Mui Keng Terrace. Singapore 119613

E-mail: {natalia.kryvinska, salem.lepaja}@tuwien.ac.at; hoangnm@i2r.a-star.edu.sg

Abstract: Next Generation Networks (NGN) is a concept for defining and deploying networks, which, due to their formal separation into different layers and planes, and use of open interfaces, offers service providers and operators a platform that can evolve to create, deploy and manage innovative services. In our paper, we present a generic definition of *mobility* in NGN. Then, we address our vision of NGN service architecture to support personal and service mobility. Next, we motivate a session initiation protocol (SIP) signaling protocol usage for service and user mobility. Finally, we discuss concept of Virtual Home Environment (VHE) and its element: the user Personal Service Environment (PSE), as a basis for users' access to personalized services from any network and with any terminal used.

Key Words: NGN, SIP, VHE, PSE, mobility.

1. INTRODUCTION

Mobility refers to the generic concept of being mobile. Referring to the OSI model – *mobility* can be identified on three different levels (Figure 1):

1. Terminal/Host mobility,
2. Personal/User mobility, and
3. Service/Application mobility,

Terminal mobility allows users to move from one physical location to another while having the same set of services available. From an outside perspective the host always seems to be attached to its home network. Hence, sessions are routed transparently to the current attachment point of the mobile node.

Personal mobility is the ability to redirect communication for one user across different devices. It is the concept of having the user as the communication endpoint. By using a single identity for an individual, we can implement a level of indirection to the desired endpoint for communication. Personal mobility is a key motivation for integrating services across heterogeneous devices from diverse networks. Some specific functions are used to support user mobility:

- authentication of the user as a single person,
- authorization of the user to access resources from that network,
- local identity for the user and the host, which is valid on that network,
- maintaining information about the local identity in a central or distributed database, sessions initiation from and towards the user using distributed databases.

Service mobility refers to the seamless mobility across different networks and devices during a service session (for example, switching from a cell-phone to an IP-phone during the conversation). It provides access to services independent of the user's access point, i.e., the user sees the same set of services from all end-points [1, 2].

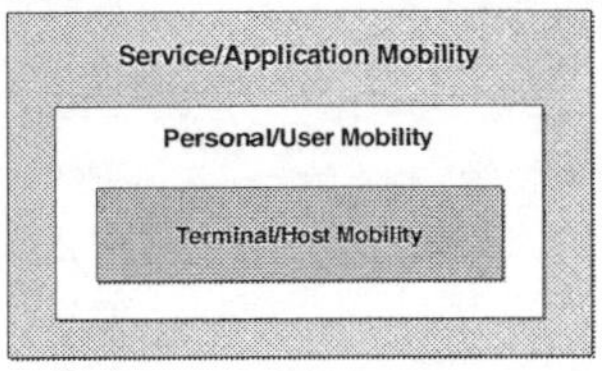

Figure 1. Mobility concept

According to the abovementioned definitions of mobility, service and personal mobility are the most important features of NGN services architecture. And, the goal of the Next Generation Networks is to provide service and personal mobility, e.g., the ability to access customized services anywhere, anytime, and from any end-device. To achieve this, it is necessary to have a new architecture that will be able to support mobility on all three levels described above.

2. NGN SERVICE ARCHITECTURE

In this part we present our vision of NGN service architecture, which can support service and personal mobility. The NGN services should be accessed from different types of terminals in different network environments. The service needs to be adjusted for different types of terminals without affecting the basic service functionality. Same service can be experienced differently in terminals with different capabilities; hence, the *specific service logic* must be kept separate from the *service core logic*. The specific service logic or service user interface resides in the terminal, which is used to access the service.

Figure 2 presents a NGN service architecture, which consists of two main parts – service level and access level. The services part includes *core service logic* with services residing in a server inside the network. Whereas the access part embodies the *specific service logic* that can be inside the terminal if it is capable to host the whole specific service logic or, if terminal is very simple, the access network performs special service logic [3, 4].

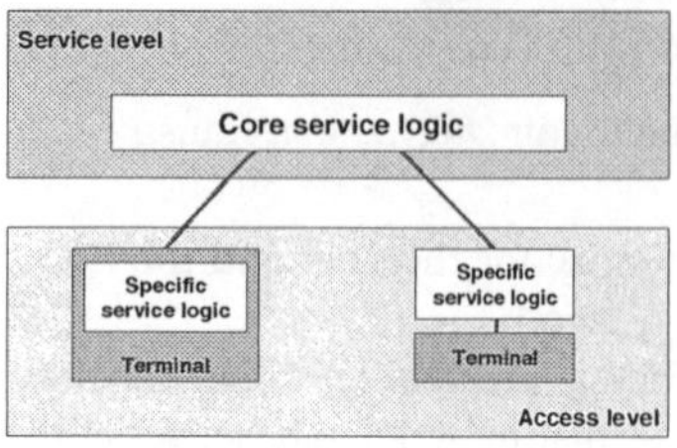

Figure 2. NGN services architecture

Traditional fixed networks and mobile networks provide a relatively complete set of voice services. However, even when enhanced by the use of IN technology, services evolve slowly and the paradigm remains essentially the same.

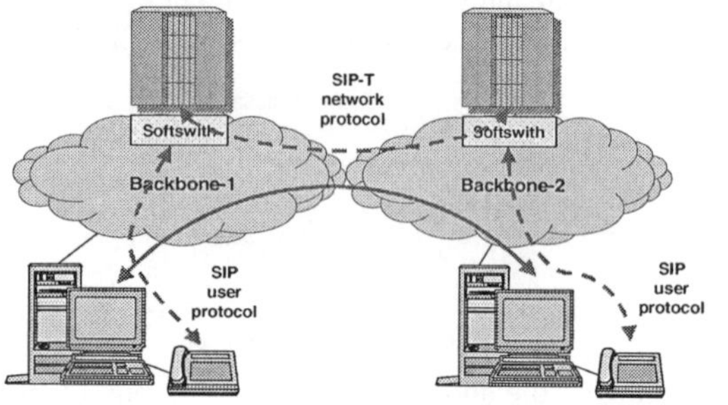

Figure 3. NGN functional architecture

Next Generation Networks, that include 3G mobile networks, will overcome these limitations by reorganizing the network architecture (Figure 3) in order to separate the provision of services from the network, to merge information and telephone technologies, and to introduce open protocols, like session initiation protocol (SIP). These networks include all the ingredients necessary to meet a wide range of user expectations with regard to services. Also, they will be able to provide new and successful added-value services. In general, they have three main strengths concerning their service capabilities:

- bring together the information and telecommunication worlds;
- move away intelligence from the network core to the edge servers and terminals;
- do not need a state machine with triggers for the call and service control [5].

3. SIGNALING PROTOCOLS FOR SERVICE MOBILITY SUPPORT IN NGN

Signaling is a key enabler for supporting Service mobility architecture. It allows delivering new services in NGN. Signaling System No.7 undoubtedly evolves, and new IP based mechanisms are included. Clearly, signaling is the glue that provides a seamless migration towards next generation networks.

The SIP protocol is chosen as a basis for the session/signaling management in NGN since it can support terminal, personal, as well as service mobility for all applications. Following are some advantages of using SIP for supporting mobility:

- Allowing users to depend on their appliances rather than on the network for supporting mobility on an end-to-end basis. This means mobile users can roam across SIP environments without concern about whether they support network layer mobility.
- Providing a means of route optimization and improved performance for real-time services via SIP signaling messages for address binding, registration.
- Allowing to deal with mobility at a semantic level above IP terminals (e.g., moving of a media stream from one terminal to another).

In the short term, supporting mobility with SIP at the application layer complements other approaches that rely on network layer mobility protocols (e.g., mobile IP) though it may replace them in the longer term [5, 6, 7].

3.1 The advantages of SIP usage for NGN

Actually, SIP is a textual protocol for client/server architectures. Its simplicity enables services to be developed easily and rapidly. Like the HyperText Transfer Protocol (HTTP), SIP is designed so that extensions can be added relatively easily. In common with other text-based protocols for the Internet world, it has end-to-end transparency. SIP-T is an extension for communication between softswitches. ISDN User Part (ISUP) messages with parameters can be transported by SIP as MIME-type (Multipurpose Internet Mail Extension) attachments. SIP encompasses more than the strict call process. It not only includes the registration process, but also supports information domain capabilities such as the presence protocol (SIMPLE). As a consequence, the use of SIP is a key factor for achieving consistency between the information and telephone worlds.

SIP can carry the Simple Object Access Protocol (SOAP), enabling one endpoint to access applications at another endpoint. For example, a server application can launch an application on a SIP handset (e.g. to change its ring tone for an incoming call). In the same way, account information relating to a prepaid card can be carried from a SIP application server to a proxy to control the duration of a call.

A key driver behind the introduction of SIP services will be their integration with web applications. SIP services for the web will be described in eXtensible Markup Language (XML) scripts. XML gives the operator or user the power to create and use an abstraction layer in order to easily define service scripts.

A variety of new services will be deployed in the near future. The roll-out of Microsoft Messenger means that each desktop is now a SIP endpoint, and each web developer is equipped with XML and a SIP Service Creation Environment (SCE). This is a major opportunity for the SIP industry, both in the corporate domain (SIP private branch exchanges) and in the carrier domain (SIP proxies and softswitches). These are strong grounds for using SIP services in the enterprise and public Internet domains today, and in the IP mobile domains in the near future.

The value of these new services for the network operator is that they increase either the subscriber base (new voice over IP lines, new corporate virtual private networks) or the network traffic (connectivity), or the network value (e.g. quality of service management, location, billing or presence management). It is not found in cannibalizing voice and leased line services at a lower price, even though there is a market for best effort voice in most countries [5, 6].

3.2 Comparison of HTTP and SIP for 3G and NGN services

In the 2G mobile networks, communication services are almost driven by the control protocol. In NGN and 3G mobile networks, control flows can originate from two sources, since both SIP and HTTP can be used to set up a connection.

The current methodology used to set up a simple connection (look and feel, procedure, behavior) on a mobile terminal is also likely to be used for multimedia calls, so SIP will be the main protocol. However, complex-connectivity services, such as multiparty managed conferences or training applications, involve more complex control and many more semantic control elements. In this case, the obvious approach is to use web technology for the main service control flow, with SIP only handling the simple point-to-point connectivity control. SIP and HTTP can be used separately to control a connectivity service. In addition, they can cooperate to control a service. Furthermore, HTTP is a content carrier protocol. A basic SIP-driven call can turn into an HTTP-enhanced SIP service. As an example, one can:

- Pushing a web page while the connection is being established, in a similar way to the 2G Calling Line Identification Presentation (CLIP) service. However, it can be much broader in concept, extending to all the phases in a call (answer, connection, release, assuming these phases are identified).
- Pushing this web page at invite time can lead to multiple HTTP-driven choices on what and how to proceed [5].

3.3 Service mobility support with SIP – location based services

There are many location-based services that can be implemented within a mobile network that utilizes information about the location of a user. This can range from providing the end user with relevant information on travel services, restaurants, or tourist information. It could also be used to provide a subscriber's location to 3rd parties, which could then push content. This could be in the form of a web page, video clip or instant message. An Application Server embeds media within a SIP message that would be rendered by the SIP enabled handset. The location services can be implemented at all stages of network migration, from GSM to UMTS release 5.

As an example of LBS (Location Base Services), we take an application that sends to the end user content such as web page detailing local attractions (e.g., listing of services such as taxis, restaurants, hotels and so on) providing him with click-to-dial capabilities, while he lands at a local airport. When this mobile user would like to order a taxi, he accesses a directory service that provides information on local taxi firms. He uses a 'click-to-message' capability allowing an instant message to be sent to the taxi firm requesting a taxi immediately, or at a certain time. The taxi firm then uses click-to-dial to speak and confirm to the user that a taxi is on its way. At this point an application server is used to integrate with mobility servers and the location is provided automatically to the taxi firm. The implementation of this type of service requires the development of an application that interacts with the mobility servers and provides a directory service. The application then interacts with the Instant Message and 3rd Party Call Control capabilities, using SIP for the signaling messages exchange [6].

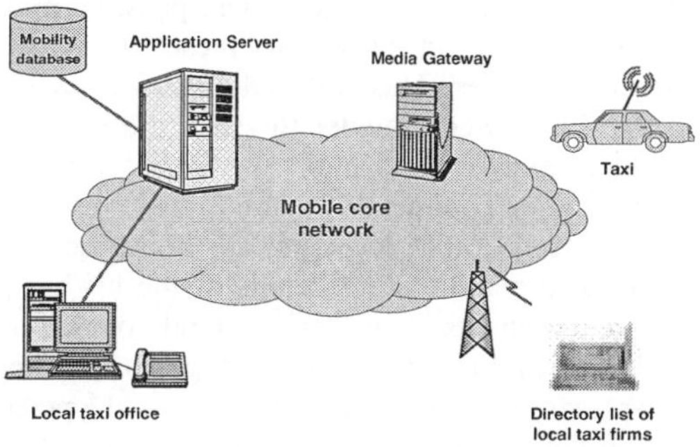

Figure 4. An example of SIP service supporting mobility

4. VHE AND ITS ELEMENT PSE IN NGN

Virtual Home Environment (VHE) is introduced as a fundamental concept in the Universal Mobile Telecommunications System (UMTS). But, it seems to be basis for all Next Generation Networks because it aims to provide users with access to personalized services in any network, with any terminal used, and at any location.

The VHE is also defined as a basic concept for personalized service portability across networks and terminals. This definition addresses three important aspects:

- service personalization, allowing users to customize their services in the way that is most convenient to them;
- service portability across networks, expands the applicability of the system concept across the boundaries of fixed, mobile, data, cable, satellite and other networks;
- service portability between terminals is the most visible aspect to the VHE user.

The broad definition of the VHE system concept, variety of networks and the diversity of terminals, introduces a number of technical obstacles which need to be taken into account.

The serving network and the terminal being used provide certain capabilities, respectively, determine the service set

that can be offered to the served user and the way in which the services within this service set are provided to the user.
NGN users will use different services, provided by different network systems, and user data may be associated with each service. On the application level, VHE provides the integration necessary to offer global personalized services, and points out the need for service providers to manage user service information. Thus an important element of the VHE is the user Personal Service Environment (PSE).
The VHE identifies the provision of a PSE to the users as a key function of it. The PSE is essentially a combination of information on services and on the personalization of their use. From the user's point of view, the PSE is a personalized service portfolio provided by a single service provider. With new regulations that are opening up the market to a variety of competing actors, the service provider can obtain these services from various VASPs (Value-Added Service Providers).
NGN user sees only his/her own PSE, which is provided and controlled by the service provider. The possible multiplicity of VASPs involved in providing these services remains transparent to the user. This makes the services easier to use, and allows the service provider to select the best VASP for each service, and even change it, without modifying the user's subscription contract.
Each user in his/her PSE identifies all the services to which he/she has access and storing details of the his/her personal services requirements (e.g., what kind of services are preferred, time and day dependency, types of terminal). The PSE also stores the characteristics and personalized information defined by the user and service provider. Hence, it can logically be split into:

- user interface part, containing information on settings such as menu structure, preferred access to information, preferred language, and network- related preferences, such as the language used for announcements.
- service part, containing the list of subscribed services (e.g., prepaid, family pack, services portal, sponsored calls) and references to service preferences, such as preferred access (voice or data), personal/family directory, do not disturb mode, incoming and outgoing call screening lists, and time-of-day variations for each of the services, if applicable.

A PSE enables a rich interaction between networked services and their users. We have introduced the concept of a PSE as a solution to many of the challenges for mediation of networked services. The main contribution of the PSE is that it increases its user control of networked services by offering support for spontaneous collaboration between peers [8, 9].

CONCLUSIONS

NGN is a concept for defining and deploying networks which due to their formal separation into different layers, planes and use of open interfaces, offers service providers and operators a platform which can evolve in a step-by-step manner to create, deploy and manage innovative services. This architecture is based on decoupling of services and networks with multiple layers and planes. It provides capabilities create, deploy and manage all kinds of services.
The Service and Personal mobility are the most important features of NGN services architecture. In this paper, we have defined a generic concept of mobility used in the NGN, which is based on service and personal mobility, e.g., ability to access services anywhere, anytime, from any end-terminal. After that, we present our vision of the NGN architecture for its services support. Next, we propose a SIP signaling protocol usage for service and user mobility support in NGN. Finally, we have discussed concept of Virtual Home Environment (VHE) and its element: the user Personal Service Environment (PSE), as a basis for Next Generation Network users access to personalized services in any network and with any terminal used.

REFERENCES

[1] A. Laukkanen, "IP Quality-of-Service in Mobile Networks", Seminar on IP Quality-of-Service, Department of Computer Science, University of Helsinki, November 2000.

[2] H.J. Wang, B. Raman, Chen-nee Chuah, et al., "ICEBERG: An Internet-core Network Architecture for Integrated Communication", IEEE Personal Communications, August 2000.

[3] B. Raman, R.H. Katz, and A.D. Joseph, "Universal Inbox: Providing Extensible Personal Mobility and Service Mobility in an Integrated Communication Network", Workshop on Mobile Computing Systems and Applications (WMSCA'00), December 2000.

[4] M. Jaseemuddin, R. Khuwaja, B. Major, B. Nguyen, M. Loryman, and P. Buckle, "Architecture for Service Mobility across Multiple Provider Domains, ACTS workshop on Advanced Services in Fixed and Mobile Telecommunications Networks", Singapore, September 1999.

[5] G. Bonnet, Y. Shen, "Next generation telecommunication services based on SIP", Alcatel Telecommunications Review, 2nd Quarter, 2002.

[6] White Paper "SIP Enhanced Mobile Network Services", Ubiquity Software Corporation 2002.

[7] B. Raman, R.H. Katz, and A.D. Joseph, "Personal Mobility in the ICEBERG Integrated Communication Network", Report No. UCB/CSD-99-1048, EECS Department, U.C. Berkley, May 1999.

[8] C. Abarca, C. Bazin, Y. Hervé, Draft Report of TSG-S1/S2/S3/T2, Doc-11 (Alcatel), "Managing a Personal Service Environment for the UMTS Virtual Home Environment", Joint ad-hoc Meeting On User Profile Stockholm, June 2001.

[9] M. Peters, R. Roscam, "Mobile or mobility? – Evolution of mobility services", Alcatel Telecommunication Review, 2nd Quarter, 2000

Composite Simulations for B3G Service and Network Management Platform

Guillaume Vivier[1], Karim El-Khazen[1], Panagiotis Demestichas[2], Pierre Kuonen[3]
[1]Motorola Labs (CRM) - European Communications Research Lab
Espace Technologique Saint-Aubin, 91193 Gif-sur-Yvette, France
{vivier, elkhazen}@crm.mot.com
[2]University of Piraeus, Greece
[3]Ecole d'Ingénieurs et d'architectes de Fribourg, Switzerland

Abstract

Assuming that different wireless technologies will be cooperating components of the future heterogeneous context, there is a need for designing new management platform in order to handle efficiently the use of the various access systems. This paper provides a high level description of such a management platform, validated by a prototype. This prototype includes two types of the simulation tools that were developed first for palliating the lack of real networks and secondly for being used within the optimisation process. The first simulator is dedicated to the simulation of the IP core network and is based on the well-known Network Simulator (NS) that have been modified. The second simulator represents the various radio access systems (UMTS, DVB, Hiperlan) and was developed over a radio planning software. Means to interconnect these two simulators between them as well as with the B3G management platform were provided.

Key Words: Simulation, NS, Heterogeneous systems, B3G.

1 Introduction

The wireless world is witnessing the convergence of different radio access networks around the IP technology. This convergence, often called "Beyond 3G" (B3G), is oriented towards the user for whom the access to different and novel services should be easy and transparent.

However, for properly allocating the radio and core network resources, the network operator should take into account several parameters. Because of the heterogeneity of the wireless technologies, management actions should be properly assessed before their real application into the networks. For that purpose, advanced management platforms should be developed. These platforms will necessarily include simulators engines in order to help finding the most appropriate management actions.

This paper presents a management platform prototype developed within the framework of the European project MONASIDRE [1], with an emphasis given on its simulation tools. Two different tools were considered and bridged together: one specific to the radio aspects and another adapted to the core network. It was possible to reuse existing tools, such as the Network Simulator (NS) [2], after modifications, leading to a considerable gain in terms of development and validation time.

The paper is organized as follows. The first section draws the context of the management platform. Then requirements on the simulations of the prototype are given followed by a description of the simulation components. Finally, the last section provides an example with the results that could be obtained with this platform.

2 A Management Platform Adapted to B3G Systems

Wireless communications continue to attract immense research and development effort. From a technology perspective, the main evolutions are the enhancement of mobile cellular infrastructures towards the 2.5G/3G, the introduction of Broadband Radio Access Networks through the development of mobile computing and Wireless LAN, and the advent of Digital Video Broadcasting (DVB). Instead of making these standards compete, the B3G convergence trend encompasses all the wireless standards and leverages on their differences. A B3G system coordinates different radio access techniques to offer advanced experience for users. Through this composite infrastructure, a network operator is able to provide to final users a more efficient wireless access to broadband IP-based services.

This convergence trend leads to an evolution of the requirements from the different wireless world actors: the users, the service providers and the network operators. These requirements have been identified and collected [1] and were the starting point of the development of the service and network management platform described here. A summary of the high level requirements is given below:

- *Final user*. Transparent and easy access to services and content. QoS and availability enhancement.
- *Service provider*. Unified interface with network operators, so as to ease the management of Service Level Agreements. Possibility to adapt the content accordingly to available resources.
- *Network operator*. Simple and efficient exploitation of its heterogeneous infrastructure. Scalability and easy evolution of the platform. Possibility to plug new radio interface.

From these requirements the management platform specifications were derived. Two layers were defined. The first one is independent of a specific radio access technology

(RAT); it allows for instance interaction with the service provider, as well as communication between different platforms. It also acts as a resource broker to offer the possibility to accept or move traffic from/to another network operator. The second layer is technology-dependent and the components have been derived into Cellular, WLAN, DVB-T, and Core IP versions.

In addition, there is two kinds of functions needed in the management platform: the first one is related to communication, i.e. gathering data, enforcing management action, answering to requests, communicating with other platforms. The second kind of function is obviously dedicated to optimisation, providing the most appropriate management configuration in order to serve the services taken into account the current and forecast networks status.

As a result, a B3G management platform should comprise the following elements:

- Inter-platform communication and service provider interworking: MASPI-RATIP (Monitoring and Service Provider Interworking, RAT Independent),
- Communication with network elements: MASPI-RATD (RAT Dependent),
- Optimisation and resource partitioning: RMS-RATIP (Resource Management Strategies),
- Optimisation to find the appropriate configuration of a specific RAT: RMS-RATD.

It is noted that this architecture allows the possibility to introduce later new resource management strategies. The proposed management platform can be seen as an open platform, providing a skeleton from which any network operator can plug its own management strategy. Figure 1 represents the architectural decomposition showing only the functions directly linked to the management

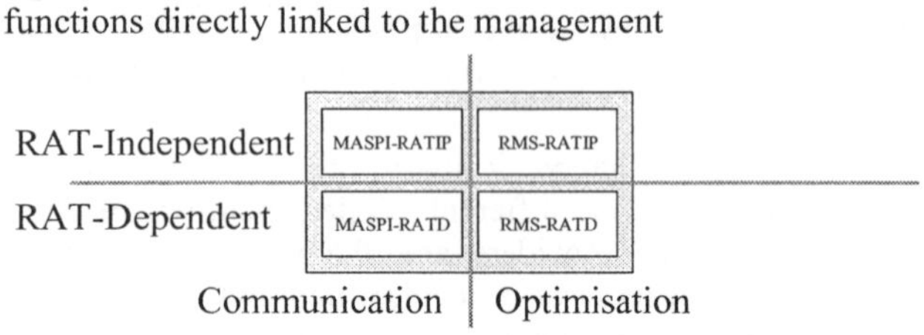

Figure 1. Management platform architecture

3 Platform Prototype

For the prototype development, all these blocks have been developed separately. Of course, for the dependent part, a block per radio technology has been developed, plus one for the IP part. These blocks could be located anywhere in the network operator domain to form a B3G management platform. In fact, these blocks are interconnected thanks to a CORBA bus while the communication between a platform and the outside world uses only standard protocols: HTTP for the communication between the platform and the service provider and COPS for the communication between the platform and the managed network elements, for the configuration flows. Regarding the monitoring, instead of using a real SMNP agent for gathering information coming from the simulators, simple sockets or direct requests to the databases storing the simulated events are used. Figure 2 details the communication flows within a B3G management platform, depicting the CORBA, COPS, SQL requests and sockets as well as the main interactions.

The components run over Linux or Windows indifferently. The motivation of the technology choices for building the prototype (CORBA, COPS, SQL, etc...) are not in the scope of this paper, an interested reader could further consult [1].

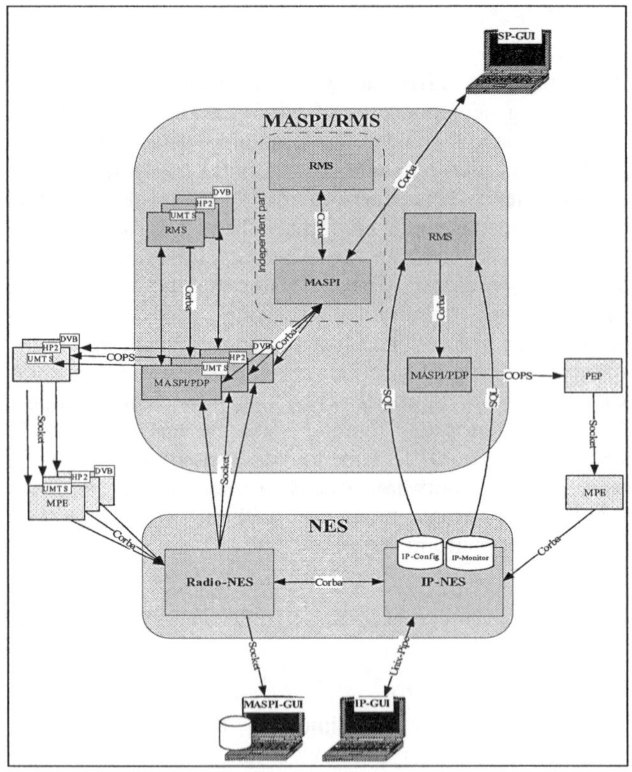

Figure 2. Prototype platform

4 Simulations in the B3G Management Platform

The role of the Network and Environment Simulator (NES) component is:

- To palliate the lack of large-scale experimental testbeds. When the prototype was developed, there were little even no commercial 3G or digital broadcast systems covering a same area. It was necessary to emulate such networks for validating and demonstrating the prototype,
- To allow network operator to simulate the effect of management actions prior to their application to the real networks. The simulators could help the platform finding the most appropriate set of management actions in the context of composite radio systems.

For both applications, the NES must interact with the rest of the platform as close as possible to the way the platform will interact with real networks. This requirement has lead to the use of standard protocols for communication between the management platform and the NES.

Moreover, the NES should be able to represent the various aspects of the considered radio systems: UMTS, DVB and WLAN, as well as the core networks. It is assumed that all these radio systems have a similar core network based on traditional IP networks or are directly connected to a public IP network.

Consequently, four kinds of simulation must be developed: one dedicated to the IP segment and three others simulating the radio part of the cellular, broadcast and WLAN systems. Even if these three systems are different, they could be based on a similar core simulation engine. The IP simulator is more classical and should allow emulating the router management, following different QoS strategies.

Thus, two basic simulator engines were considered: one dedicated to the radio systems (Radio-NES) and the other for the IP network (IP-NES). The radio simulators were based on the same engine, with different parameters and ran as different processes. On the other hand, there was only one simulation process for the various core networks, but with separate management domains.

At last, instead of developing from scratch the engines, some existing software was adapted. NS was considered for the IP part, and radio-planning software inherited from other projects [4] was a starting point. Modifications were necessary to adapt the softwares to the specific needs of the platform. The issue was to make the two discrete event simulators communicating together without loosing their internal time/event representation. Moreover, any simulated data communication must be derived in the two simulators, generating linked events.

For that purpose, a master/slave concept with a pseudo-time synchronization of the two simulators was considered. The IP-NES was the slave of the Radio-NES: the calls were generated/terminated by the Radio-NES, leading to the creation of the corresponding event in the IP-NES, for uplink and downlink calls.

The Figure 3 represents the NES overall architecture, showing the basic components: the IP-NES and the 3 types of Radio-NES, each one corresponding to a particular RAT.

To allow the IP-NES executing on demand request from the Radio-NES and/or the B3G platform, it was mandatory to modify NS that is originally script-based: the simulation to be executed is theoretically described in advance in a script file. Then, the outcome of the simulation is written in another file that can be played back in a visualization software, Network Animator (NAM) [3]. This way of running the simulation implies that the simulator user knows in advance the lists of events to simulate. It also implies that neither modification of the network topology, management action or unscheduled traffic is allowed during the simulation. In order to overcome this limitation, NS was modified to allow external interactions when simulation is running. Several "input pipes" have been added in which commands can be inserted during the simulation, as depicted in the Figure 4.

Similarly, NAM has been modified to allow "real-time" visualisation of the IP-NES. For that purpose, a pipe mechanism has been added to NAM that feed the visualisation tool with the events to be shown.

Regarding the visualization of the Radio-NES part, a specific GUI written in Java was developed, showing in "real-time" what happened in Radio-NES. The synchronization between the visualization interfaces was ensured by the intrinsic synchronization between IP-NES and Radio-NES.

5 Example of Use

The following scenario illustrates how the B3G management platform could work: a service provider (SP) has a commercial agreement with a cellular operator, and this network operator owns WLAN access points in some dense areas of its cellular coverage and made a cooperation agreement with a DVB-T operator.

This sample scenario has been validated with the prototype, leading to capacity improvement when the different radio systems cooperate. Detailed information on the B3G management platform could be found in [5] while performance results are captured in [6].

The prototype, including the simulation tools demonstrated its ability to handle SP requests as well as to react dynamically to congestion events that could occur in any wireless systems. The simulation tools embedded in the platform allows reaction time in average below five minutes. Such reaction time is largely compatible with the objective of semi-dynamic performance management of the B3G platform.

6 Conclusion

Current trend of wireless systems is convergence, or at least cooperation between complementary types of access. As a result, WLAN access is used in urban dense areas to reduce the load of cellular systems. Similarly, digital broadcast is foreseen to provide efficiently broadcast/multicast IP services. In order to manage efficiently the use of these various radio access systems, new management platforms should be designed. This paper provides a high level description of such a platform, validated by a prototype and using simulation tools first for palliating the lack of real networks and secondly for being used within the optimisation process. Instead of developing a proprietary unique tool, the paper presents the interconnection of two simulators, one dedicated to radio environment that could be derived for representing the various radio access systems. The second simulator is dedicated to simulation of IP networks and was built on the public NS software. Our approach leads to efficient simulations that allow dynamic response of the B3G management platform prototype when facing congestion event or simple requests from service provider.

REFERENCES

[1] MONASIDRE project (Management Of Networks And Services In A Diversified Radio Environment), http://www.monasidre.com.

[2] UCB/LBNL/VINT Network Simulator, http://www.isi.edu/nsnam/ns/

[3] Network Animator, http://www.isi.edu/nsnam/nam/

[4] STORMS project (Software Tools for the Optimization of Resources in Mobile Systems), http://www.cordis.lu/infowin/acts/rus/projects/prtits.htm

[5] P. Demestichas and all., Wireless Beyond 3G:Managing Services and Network Resources, IEEE Computer Magazine, Vol. 35. Issue 8, August 2002

[6] P. Demestichas and all., Management of Networks and Services in a Composite Radio Context, to be published in IEEE Wireless Communication Magazine

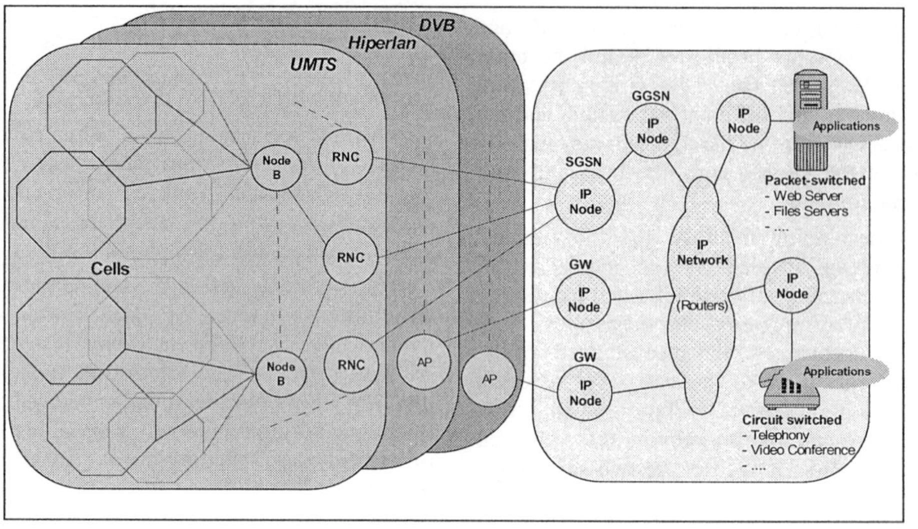

Figure 3. NES overall architecture

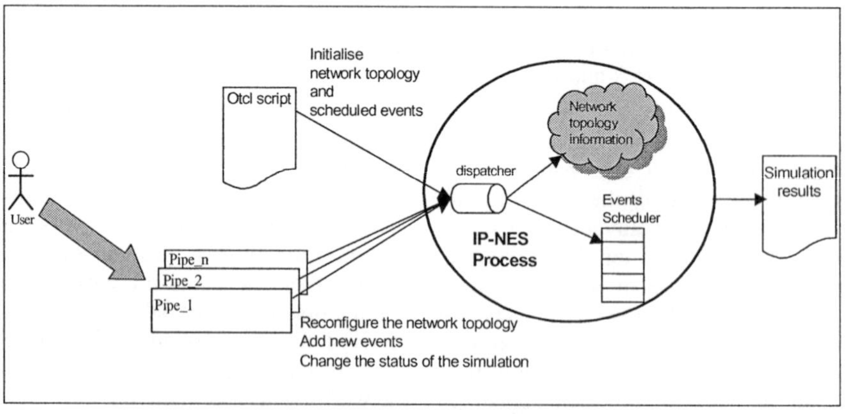

Figure 4. Modified NS architecture

MULTICASTING IN UMTS: EFFECT OF MOBILITY ON TREE MAINTAINANCE

NILANJAN BANERJEE and SAJAL K. DAS
Center for Research in Wireless Mobility and Networking (CReWMaN)
Department of Computer Science and Engineering
The University of Texas at Arlington
Arlington, TX 76019-0015

As the concept of wireless Internet is flourishing and multicast based applications such as teleconferencing, live video broadcasts and stock quotes are targeted for its deployment, a performance modeling of multicasting in wireless domain is needed to develop efficient multicast protocols. This paper presents an analysis of the dynamics and the cost (total length) of the multicast tree in an UMTS based network. A relative cost comparison for Mobile IP based bi-directional tunnelling and remote subscription has been also presented.

1 Introduction

With the tremendous advancement in wireless communication technology, *wireless Internet*, is supposed to provide an end-to-end IP platform supporting real-time and non-real-time multimedia services. Realization of wireless Internet would mean dissemination of any type of information anywhere anytime. With the increase in network capabilities, multimedia applications such as Internet television, videoconferencing, interactive network games, video on demand, etc. are gaining popularity in wired networks. Multicasting has been inherently found to be the most suitable technology for these applications. The advantage of multicasting is to save bandwidth by avoiding duplication of data over the shared links.

UMTS (Universal Mobile Telecommunications System) is a family of third generation mobile networks designed to offer high bandwidth wireless access. UMTS is supposed to converge gradually into the wireless Internet architecture. Since in UMTS network, bandwidth is a limited resource UMTS is likely to incorporate multicasting as a compulsory feature. The UMTS standard specifies that the multicast tree should be terminated at the GGSN (Gateway GPRS Support Node interfacing the UMTS network and the external packet data network) [1] and it is the responsibility of the GGSN to deliver multicast data to the user via unicast tunnels. But this puts a huge load on the already complex GGSN functionality besides being bandwidth inefficient. In order to avoid this, it was proposed in [4] to push the multicast further down into the UMTS network and terminate the multicast at the base stations. The basic multicast architecture for such base station terminated multicasting in UMTS is shown in Figure 1. Of-course there may be more than one GGSN if there are more than one PLMN (Public Land Mobile Network) under consideration. In this paper, we would study multicasting within a single PLMN.

Mobility plays an important role in the overall behavior of the multicast protocol. The current version of Mobile IP [8] proposes two approaches to support user mobility, while maintaining connection during a data delivery session. One is called *bi-directional tunnelling* (BT) and the other is called *remote subscription* (RS) method. Depending on the strategy of handling host mobility, the multicast delivery tree structure and hence the cost or the total length of the treeinvolved, vary.

In BT, user's terminal mobile host (MH) receives the multicast datagram through its home agent (HA) using unicast Mobile IP tunnels. This approach hides the host mobility from all other members of the group. Therefore, there is no need to change the multicast tree because of a change in the member locations. The main drawback of this approach is that the routing paths for multicast delivery to the MHs are sub-optimal, and hence they introduce unwanted delay in data delivery. Moreover, the HA has to duplicate and deliver tunnelled multicast datagrams to all the MHs to their corresponding foreign networks. This is a considerable waste of network resources. Tunnel management is also very cumbersome.

In RS, each MH subscribes again when it enters a foreign network. This makes the local multicast router for that foreign network join the multicast delivery tree, if not already a part of the tree. This re-subscription frequency depends on the rate of handoff that the MHs face. The main advantage of this approach is in the use of optimal multicast

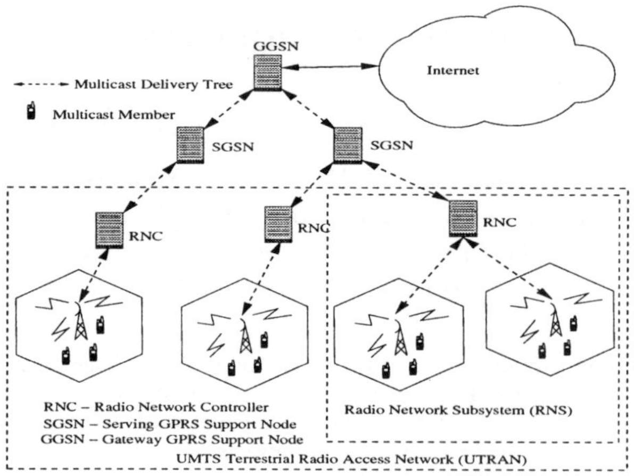

Figure 1. Multicasting in UMTS terminated at the base stations

tree. However, there is an overhead associated with the re-subscription, buffering of datagrams to ensure minimal loss during handoff, and modification of the multicast delivery tree.

A soft handoff technique has also been proposed [4] for handling host mobility. Soft handoff means both the old and new base stations will be in contact with the host during handoff. This allows the base station to join the specified multicast session prior to the actual handoff. In principle, soft handoff is similar to the RS, except in soft handoff mechanism a make before type of connection is done. Both BT and soft handoff ensure minimal packet loss, but the latter also ensures minimal cost of the multicast tree.

In this paper we present an analysis of multicasting in UMTS network, with respect to the multicast tree dynamics and the total cost involved, using a random walk based user mobility model. A cost comparison for the multicast tree cost under BT and RS has also been presented. The organization of the paper is as follows. Section 2 describes the random walk mobility model considered. Analytical results on multicast tree dynamics and the corresponding cost of the tree has been presented in Section 3. Section 4 presents the cost comparison between BT and RS. Section 5 concludes the paper.

2 Random Walk Based Mobility Model

Let us consider the cells (area covered by an individual base station) of an UMTS network form a connected, non-bipartite, undirected graph, $G = (V, E)$, where V is the set of the cells in the network and E is the set pairwise arrangement of adjacent cells. Also let $|V| = n$ and $|E| = m$. If the movement of a mobile user in this cellular network is modelled as a *Random Walk* on the graph, a Markov chain M_G is induced as follows: the states of the M_G are the vertices of G and for any two vertices $u, v \in V$ [6],

$$P = \begin{cases} \frac{1}{d(u)} & \text{if } (u,v) \in E \\ 0 & \text{otherwise} \end{cases} \tag{1}$$

where $d(u)$ is the degree of vertex u.

Now the mean residence time for a mobile user can be obtained as the expected number of time steps to reach from one state (i) to itself in the Markov chain, denoted by h_{ii}. Now for any irreducible, finite Markov chain $h_{ii} = \frac{1}{\pi_i}$, where π_i is the stationary probability of state i. Also for random walk on graph it is known that $\pi_v = \frac{d(v)}{2m}$ for $\forall v \in V$ [6]. Therefore $h_{ii} = \frac{2m}{d(v)}$ [6]. But for a cellular network the cells can be assumed to be of hexagonal cell approximately and hence we can assume that $d(v) = 6$. Therefore, $h_{ii} = \frac{m}{3}$ or in other words the mean residence time for this general mobility model is given as $\frac{m}{3}$, where m is the number of edges in the graph abstraction of the cellular network under consideration.

3 Multicast Tree Dynamics

We assume that the user residence time in a cell follows a general with a specified mean. We also assume that the user population in each cell is modeled using a $M/G/c/c$ queuing model [3] for finite number of servers and finite buffer. Here $c = C$ corresponds to the number of channels in a cell. According to [3], the steady state system size distribution for such a queuing model is given by,

$$p_k = \left\{ \frac{(\frac{\lambda}{\mu})^k/k!}{\sum_{i=0}^{C} (\frac{\lambda}{\mu})^i/i!} \quad k = 0, 1, \ldots, C \right. \tag{2}$$

where p_k is the probability that there are k users in a particular cell and $\lambda = \lambda_N + \lambda_H$ (λ_N and λ_H are the arrival rates for new calls and handoff calls, respectively) and $\mu = \mu_R + \mu_T$ (μ_R and μ_T are the mean residence time of a user and the mean call duration time, respectively), and $\mu_R = \frac{1}{\text{mean residence time}} = \frac{3}{m}$, as derived above. Now a cell belongs to the multicast delivery tree when it has one or more mobile user within its range. So, the probability of a cell to be in the multicast tree is given as:

$$\sum_{k=1}^{M} p_k = 1 - p_0 = 1 - \frac{1}{\sum_{i=0}^{C} (\frac{\lambda}{\mu})^i/i!} \tag{3}$$

The expected number of cells involved in the multicast delivery tree is as follows:

$$\mathcal{G} = n\,[1 - p_0] \tag{4}$$

where N is the total number of cells in the network.

A new route is grafted in a multicast tree, when either a new call or a hand off call arrives at a cell which is not already a member of the multicast delivery tree. So the probability, $P_g(t)$, that a given cell A will be added to the multicast tree at time t is derived as follows: $P_g(t) =$ *Prob* [a new call or a hand off call arrives in cell A at time t | cell A does not have any user].

Let π_j^A be the probability of finding j users in cell A. Therefore, from Equation (2) we can write,

$$\pi_j^A = \frac{(\frac{\lambda}{\mu})^j/j!}{\sum_{i=0}^{C} \frac{(\frac{\lambda}{\mu})^i}{i!}} \tag{5}$$

Now, *Prob* [a new call or a hand off call arrives in cell A at time t] $= \lambda e^{-\lambda t}$, and *Prob* [cell A does not have any user] $= \pi_0^A$. Since these two probabilities are independent,

$$P_g(t) = \lambda e^{-\lambda t}\left[\frac{(\frac{\lambda}{\mu})^0/0!}{\sum_{i=0}^{C} (\frac{\lambda}{\mu})^i/i!}\right] = \frac{\lambda e^{-\lambda t}}{\sum_{i=0}^{C} (\frac{\lambda}{\mu})^i/i!} \tag{6}$$

Therefore, the expected number of new cells grafted to the multicast tree in time T is given as

$$\mathcal{N}_g = \int_0^T n P_g(t)\,dt = \frac{n}{\sum_{i=0}^{C} \frac{(\frac{\lambda}{\mu})^i}{i!}} \left[1 - e^{-\lambda T}\right] \tag{7}$$

3.1 Multicast Tree Cost

Once we have the group size, it is possible to estimate the cost (henceforth used to mean the total length) of the multicast distribution tree using an empirical expression called the Chuang-Sirbu scaling law [2]. The total length of the multicast tree and the average unicast path length are typically expressed in terms of either number of packet hops or bandwidth consumed. Then the total length of the multicast tree (L_m), the average length of unicast routing path ($\bar{L}_u$) and the multicast group size ($\mathcal{G}$) are related by the relation $L_m = \bar{L}_u \mathcal{G}^k$, where k is typically found to be 0.8. $\bar{L}_u$ is typically obtained by taking the average of all the unicast path length to each multicast receiver. Hence, using Equation 4 for $\mathcal{G}$ we can derive L_m as follows :

$$L_m = \bar{L}_u \left(n \left[1 - \frac{1}{\sum_{i=0}^{C} \frac{(\frac{\lambda}{\mu})^i}{i!}} \right] \right)^{0.8} \tag{8}$$

Although this is an approximate estimate only based on an empirical relationship, elaborate experimentation shows that it holds good for the multicast trees which are economic, efficient and fair.

To get a numerical estimate of the multicast tree cost we assume the following values [7] for the parameters considered. (i) Arrival rate for new calls, $\lambda_N = 2.75e^{-04}$ calls/sec, (ii) Arrival rate for handoff calls, $\lambda_H = 5.91e^{-02}$ calls/sec, (iii) Service rate for calls or the call termination rate, $\mu_T = 1.55e^{-02}$ calls/sec. Now assuming that each node has six neighbors, we get the $|E|$, the number of edges as, $|E| = \frac{\sum_{i=1}^{n} d_i}{2} = 3n$, where N is the total number of nodes in the network and d_i is the degree of node i. Therefore μ_R is obtained as $\mu_R = \frac{3}{|E|} = \frac{1}{n}$. Typically in an UMTS network $\bar{L}_u = 3$. With these values of the parameters, the cost of the multicast tree with the number of cells considered in the network, is depicted in Figure 2 for different number of channels in each cell. With lesser number of channels per cell, lesser number of users are served and hence less number of cells are added to the multicast tree. Consequently, the multicast tree lower than that with higher number of channels per cell.

4 Cost Comparison Between BT and Remote Subscription

Considering each hop in the UMTS network of unit length, a branch of length 1 is grafted to the tree if the new cell is not already in the multicast tree, for RS. The length of the grafted branch is 2 or 3 depending on which of the corresponding radio network controller (RNC) and serving GPRS support

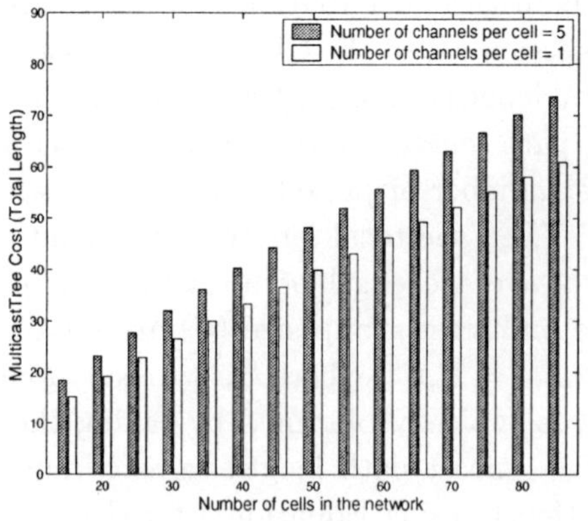

Figure 2. Total length of multicast tree vs. number of cells in the network

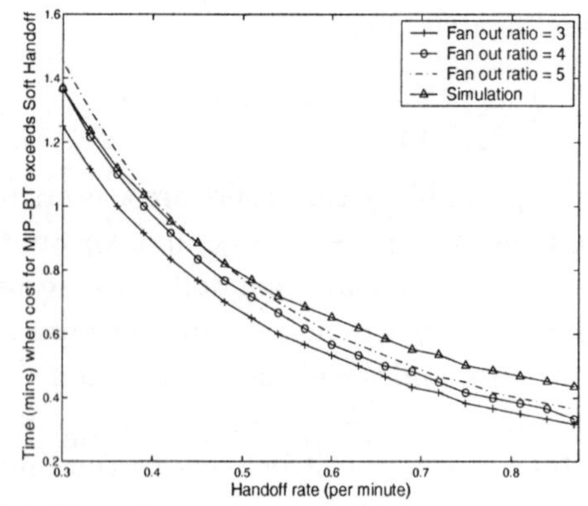

Figure 3. Time required for the total length of the tree in Mobile IP based bi-directional tunnelling to exceed the corresponding length in remote subscription

node (SGSN) is not on the multicast tree. The total expected cost of the tree, however, can be estimated by using Equation 8.

To estimate the cost of Mobile IP based BT we first estimate the tunnel length, which can be of three possible length, 2, 4 and 6 for intra RNC mobility, intra SGSN mobility and intra GGSN mobility, respectively. Let the probability of intra RNC, intra SGSN and intra GGSN mobility be p_1, p_2, p_3, respectively. Therefore the total expected length of the tunnel (T_l) in the tree for BT in time T is given by $T_l = (2p_1 + 4p_2 + 6p_3) \int_0^T nP_H(t)\, dt = n(2p_1 + 4p_2 + 6p_3)\left[1 - e^{-\lambda_H T}\right]$, where $P_H(t) = Prob$[a handoff call arrives in a cell at time t] $= \lambda_H e^{-\lambda_H t}$. The relative fraction of the number of cells, RNC and SGSN in the UMTS network can be used to estimate the probabilities p_1, p_2, p_3, respectively.

The total length of the multicast tree for BT can be derived as $T_l + L_b$, where L_b can be obtained in a similar fashion to that of L_m by replacing λ and μ with λ_N and μ_N (i.e. we are not considering handoff while computing L_b), respectively.

Figure 3 presents the numerical and simulation results on the time when the cost for BT exceeds the cost for RS. The fan out ratio indicates the ratio between the number of base stations to the number of RNC, which has been assumed to be equal to the ratio between the number of RNC and the number of SGSN under the control of a single GGSN. The simulation has been done only for fan out ratio = 5. The cost for BT exceeds the corresponding cost of RS very quickly (a few minutes) and the cross over time decreases with the increase of the handoff rate.

Since BT is costlier than RS but ensures minimal packet loss, neither of them is the best option. In fact a hybrid protocol starting with BT and switching to RS later on when the tree grows out of balance and becomes cumbersome, is a better option. For such hybrid protocols, the results presented in Figure 3 gives an idea of the switch-over times.

5 Conclusion

In this paper we have presented the modelling and subsequent analysis of the cost of multicasting in UMTS based 3G networks. The dynamics of the tree due to user mobility have been derived in terms of the average group size and the number of average nodes added to the multicast delivery tree. Mobile IP based multicasting with BT and RS have been compared for their respective costs.

References

1. 3GPP, "TS 29.061: Internetworking between the Public Land Mobile Network (PLMN)", V5.0.0, 2001.
2. J. Chuang and M. Sirbu, "Pricing multicast communications: A cost-based approach", *Proceedings of the 1998 INET*, 1998.
3. J. W. Cohen, *On regenerative processes in queueing theory*, Springer, Berlin, 1976.
4. M. Hauge and O. Kure, "Multicast in 3G Networks: Employment of Existing IP Multicast Protocols in UMTS", *International Workshop on Wireless Mobile Multimedia, WoWMoM*, Pages: 96-103, 2002.
5. Y. Lin, "Modeling Techniques for Large-Scale PCS Networks", *IEEE Communication Magazine*, Vol. 35, N. 2, Pages: 102-107, February 1997.
6. R. Motwani and P. Raghavan, *Randomized Algorithms*, Cambridge University Press, 1995.
7. P. V. Orlik and S. S. Rappaport, "A Model for Teletraffic Performance and Channel Holding Time Characterization in Wireless Cellular Communication with General Session and Dwell Time Distributions", *IEEE Journal on Selected Areas in Communications*, Vol. 16, No. 5, Pages: 788-803, June 1998.
8. C. Perkins, "IP Mobility Support", *RFC 2002*, October 1996.

SESSION 7A:
PHYSICAL LAYER

A HARDWARE TESTBED FOR THE IMPLEMENTATION AND EVALUATION OF MIMO ALGORITHMS

PATRICK MURPHY, FEIFEI LOU and J. PATRICK FRANTZ

Rice University, 6100 Main Street, Houston, TX, 77005, USA

E-mail: {murphpo, fflou, jpfrantz}@rice.edu

As the demand for higher performance wireless communications continues to grow, novel algorithms have been developed which provide increased performance and efficiency. One such class of algorithms involves the use of multiple antennas on either end of a wireless link. Many of these multiple input, multiple output (MIMO) algorithms offer impressive performance gains over their single antenna counterparts. The practicality of implementing such algorithms in a real system, however, is a topic in need of further exploration. We present in this paper a testbed designed specifically to test such algorithms. This testbed provides hardware for baseband processing, up and downconversion to RF and emulation of multiple wireless channels. It was designed to provide sufficient flexibility to implement a wide range of algorithms while preserving the ability to evaluate an algorithm's resource and power requirements. The testbed hardware and configuration options are presented, along with a basic demonstration of its functionality.

1 Introduction

The wireless communications industry has seen continuous growth in the demand for higher throughput communications. This increasing demand, however, has not been met with the availability of increased spectral resources. As a result, researchers have sought algorithms which provide higher performance without requiring additional allocation of spectrum.

As mentioned above, MIMO techniques encompass one class of algorithms which fit this requirement. There has been a great deal of effort spent developing and analytically evaluating MIMO algorithms. Far less effort, however, has been expended evaluating the challenges and practicality of implementing these algorithms in real hardware. This is due in part to the relative scarcity of hardware testbeds with sufficient resources to realize an end-to-end, multiple antenna wireless link. We present a testbed designed explicitly to fill this void.

The remainder of this paper is organized as follows. Section 2 discusses the testbed design and the hardware chosen to construct it. Section 3 outlines the implementation of a simple MIMO scheme using the testbed's resources. Finally, section 4 offers some concluding remarks.

2 Hardware

The testbed consists of hardware which performs three major functions: baseband processing, conversion to and from radio frequencies and emulation of realistic wireless channels. The components we chose to fill each role are discussed below.

2.1 Baseband Hardware

One of the major challenges in designing a testbed capable of real-time wideband wireless communication is choosing baseband hardware. We first identified the major requirements the baseband hardware must meet. First, it must possess sufficient processing power to implement a wide variety of complex algorithms. Reliable wireless communication alone is a computationally intensive task; the added complexity of multiple transmit and receive antennas can increase the computational requirements significantly. The hardware must also allow reliable measurement of resource utilization and power consumption on a per-algorithm basis. This is an important requirement as the computational requirements of a particular algorithm play a large role in gauging its suitability for real-world use.

We found early on that of all the options for baseband processing, FPGAs provided the greatest flexibility in algorithm design and visibility of resource utilization. We acknowledge that FPGAs may not be ideal for deployment in large scale wireless systems. They are, however, very well suited for use in a rapid prototyping and research environment such as this testbed.

We decided on FPGA-based development systems designed by Nallatech of Glasgow, Scotland as our baseband processing platform. As shown in Figure 1, each board houses a multi-million gate Xilinx

Virtex-II FPGA and two channels each of analog-to-digital and digital-to-analog conversion. The board's ADCs and DACs are fast enough to connect directly to the intermediate frequency interfaces on the testbed's RF hardware. These connections allow each board to potentially act as a dual-antenna transceiver. Each board also contains a smaller FPGA dedicated to providing clocks to the larger FPGA and the external analog interfaces. This feature provides a great deal of flexibility in connecting to hardware from different designers with various interface requirements. Finally, the Nallatech board is also a PCI card, providing a high throughput connection over which a host PC can provide and receive data and monitor system performance.

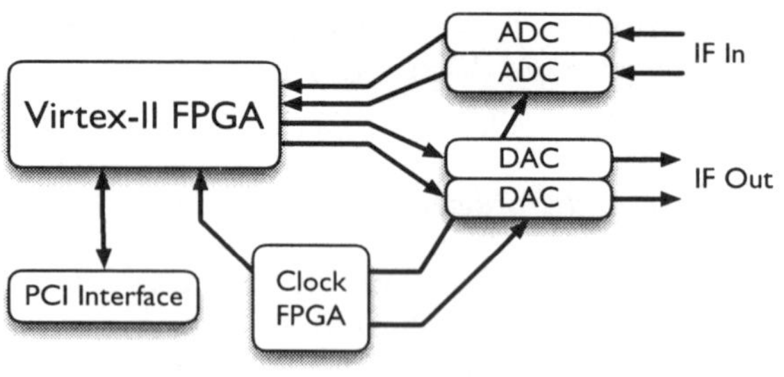

Figure 1. FPGA-based processing board

2.2 *RF Hardware*

This testbed is designed for wideband wireless communication in the 2.4 GHz ISM band. The RF hardware must not be tied to any particular standard and must have intermediate frequencies within the sampling capability of the FPGA baseband platform. We also desired control over the synchronization of the frequency translation stages of the radios. Such control allows variation in the carrier frequency offsets between antennas and will prove useful in characterizing the performance of multi-antenna carrier recovery schemes. It also allows offsets between the carriers of multiple transmit or receive antennas to be eliminated entirely, a useful option when testing algorithms lacking carrier recovery systems.

We decided on 2.4GHz RF hardware from National Instruments. NI has only recently ventured into the RF hardware market, but we found that their hardware fit the needs of our testbed very well. These devices operate with an intermediate frequency of just 15MHz, easily within range of the FPGA baseband's sampling capabilities. They also support signal bandwidths up to 20MHz, accommodating a large variety of wideband schemes.

In order to maximize flexibility in choosing multiple antenna configurations, the testbed also uses a number of RF multiplexers. These devices form the interconnect between the RF transceivers and the channel emulators, described in section 2.3. The multiplexers allow most configurations to be implemented programmatically without any changes to the equipment's physical connections. This both minimizes the risk of damage to the equipment and allows for a great deal of automation in switching between interconnection configurations.

2.3 *Wireless Channel Emulation*

Another significant challenge in designing this wireless testbed was identifying some means by which we could provide a realistic wireless channel in a laboratory setting. We required the ability to simulate a wide variety of channel conditions, including fading, multi-path effects and delay spread. We also needed to assure repeatability of experimental results, a virtually impossible requirement when using real wireless channels.

To fill this need, we chose RF channel emulators from Spirent Communications. Each device is capable of emulating two wireless channels, each with six independently configured paths. Figure 2 illustrates the capabilities of each emulated wireless channel.

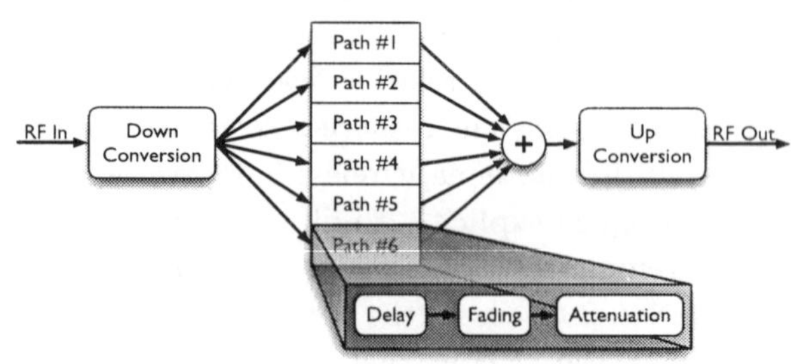

Figure 2. Emulated wireless channel

The emulator can apply one of many fading models, including Rayleigh, Rician and Nakagami, plus programmed delay and attenuation to each path. This flexibility allows the emulation of a large number of realistic line-of-sight and multi-path environments. The emulator can also introduce arbitrary frequency shifts to each path, providing a reliable

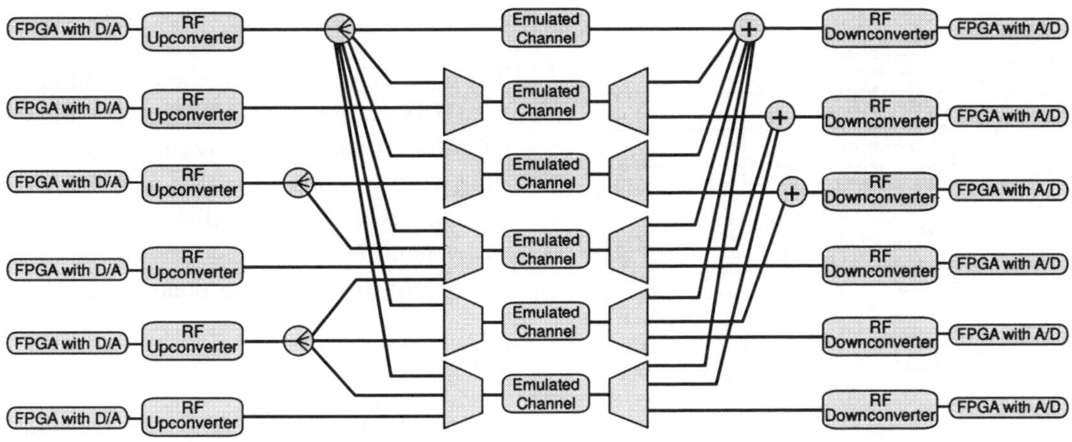

Figure 3. Testbed connectivity for half-duplex testing

means by which to simulate the Doppler effects of various mobile scenarios.

Our testbed currently houses three such emulators, providing a total of six independent emulated channels. This fairly large number of channels provides significant flexibility in designing test configurations. In terms of transmit-to-receive antennas, the testbed can implement half-duplex systems of 6-to-1, 3-to-2, 2-to-3, 1-to-6 or any smaller configuration requiring six or fewer channels. Figure 3 illustrates the full connectivity for multiple-antenna half-duplex testing.

3 Example Application

The final stage of funding for this testbed has only recently been approved. As a result, the first projects which utilize its resources are still under development and will be the focus of future publications. However, in order to demonstrate the testbed's functionality, we offer the implementation of the simple algorithm utilizing multiple transmit antennas first described by Alamouti [1]. This scheme is well suited for such a demonstration as it uses every stage of the testbed but is fairly straightforward to implement.

Alamouti's technique is a transmit diversity scheme utilizing, at a minimum, two transmit antennas and one receive antenna. It is an example of a space-time block code, a multiple antenna technique which requires only linear processing in the encoding and decoding states to achieve its diversity gain [2].

Our implementation effort began with the selection of Xilinx's System Generator for DSP [3]. System Generator is an extension for The Mathworks' Simulink environment which allows models to be rapidly constructed, simulated and synthesized for use in real FPGA hardware.

Our model currently consists of a transmitter and receiver in the same FPGA. The transmitter is fairly simple, consisting of a QPSK modulator, Alamouti encoder, pulse shaping filter and digital upconverter for translation to IF.

The receiver is substantially more complicated. It includes subsystems for carrier recovery, channel estimation, decoding and demodulation. The carrier recovery block assumes the transmit antennas share a reference clock and have no offset. It then corrects for any frequency offset between the transmit and receive radios. The channel estimation system correlates the received data against a known training sequence to estimate each channel's phase offset and attenuation. The channel estimates are used until the training sequence is transmitted again. This approach assumes both channels have slowly varying characteristics, a limitation that will be eliminated with more sophisticated algorithms in future revisions of the design.

Top level views of the System Generator design for the Alamouti transmitter and receiver are shown in Figures 4 and 5. The design has been synthesized and implemented in a two million gate Xilinx Virtex-II FPGA (XC2V2000). The current revisions of the transmitter and receiver together utilize approximately 85% of the logic resources in the FPGA. This utilization should be viewed as an upper bound as there exist a variety of possible optimiza-

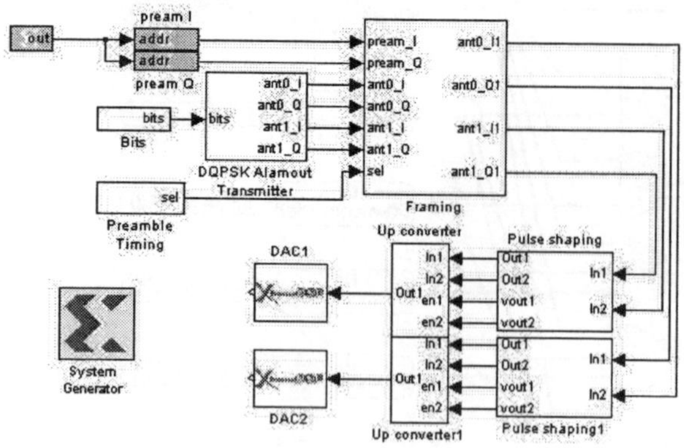

Figure 4. System Generator model of Alamouti transmitter

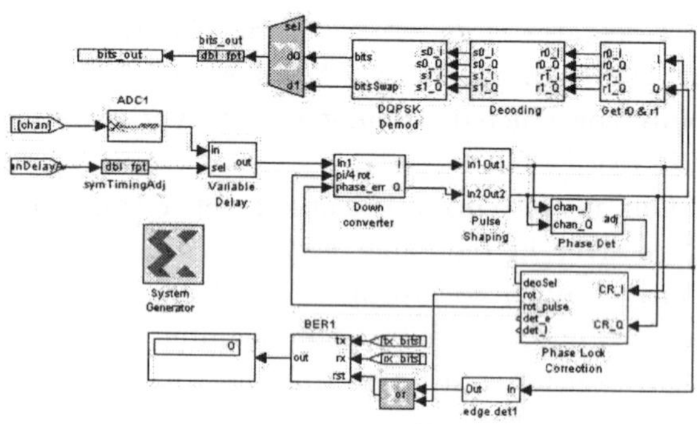

Figure 5. System Generator model of Alamouti receiver

tions not yet applied to the design. The system has been tested as an end-to-end link using the baseband, RF and channel emulation hardware described above with realized throughputs in excess of 1 Mb/sec.

4 Conclusion

There is little doubt that many future wireless communications systems will employ multiple antenna techniques to provide high performance while maximizing spectral efficiency. This expected prevalence of MIMO systems highlights the need for platforms designed to evaluate such algorithms under realistic conditions. The practical value of an algorithm is increased significantly when it is tested in real hardware and evaluated in terms of computational requirements and power consumption. This paper has described in detail the hardware platform we have constructed to perform this analysis.

Acknowledgments

The equipment for this testbed was purchased in part with NSF grants CISE-ANI-0224458 and CISE-MRI-0321266. Authors Murphy and Lou are each partly supported by Texas Instruments Fellowships. Special thanks to Xilinx, Nallatech, Spirent Communications and National Instruments for their generous equipment discounts and donations.

References

1. S. M. Alamouti, "A simple transmit diversity technique for wireless communications," IEEE Journal on Select Areas in Comm., vol. 16, pp. 1451-1458, Oct. 1998.
2. V. Tarokh, H. Jafarkhani, A.R. Calderbank, "Space-time block codes from orthogonal designs," IEEE Transactions on Information Theory, vol. 45, pp. 1456-1467 July 1999.
3. http://www.xilinx.com/system_generator/

A COMPARISON OF MC-CDMA AND SC-CDMA WITH SCC SCHEMES IN RAYLEIGH FADING CHANNELS

Thorben Detert
Darmstadt University of Technology
Institute for Communication Technology
Merckstrasse 25, 64283 Darmstadt, Germany
t.detert@nt.tu-darmstadt.de

Abstract— **In this paper, a performance comparison of SC-CDMA and MC-CDMA systems in Rayleigh fading channels is presented in the asynchronous uplink. Both coded and uncoded systems are considered as the uncoded systems show the behavior of the different signal models, the coded systems provide a realistic scenario for comparison. MC-CDMA is used in the frequency spreading approach. Serial concatenated coding (SCC) of a convolutional, a Walsh-Hadamard and a repetition code is used for the comparison of the coded systems. Iterative decoding is applied.**
The bit error probability for MC-CDMA and SC-CDMA is calculated for the uncoded systems and a union bound for the coded systems is given. Potentials to meliorate the occurring disadvantages of both system designs are presented and differences in multi-user environments are demonstrated.

Keywords - **CDMA, MC-DS-CDMA, error bounds, iterative decoding, concatenated codes.**

1. INTRODUCTION

As is discussed in [1], the spectral efficiency of a SC-CDMA system is comparable to that of a MC-CDMA system when Nyquist pulses are utilized. Otherwise, the bandwidth of a MC-CDMA system is 1.9 times higher than that of a SC-CDMA system. Multi-user detection can be applied in the uplink in order to mitigate the MAI for both CDMA techniques. The MAI stems from the use of pseudo random sequences, as orthogonal scrambling can not be applied in the asynchronous uplink. The limits on system capacity which stem from the MAI can be overcome by channel coding which demonstrates good performance at low signal-to-noise ratios. Asynchronous transmission circumvents the use of orthogonal scrambling sequences. Here, pseudo-random sequences are applied. Spreading can be regarded as repetition coding with successive scrambling. The entire code rate R_c can be set to the reciprocal of the processing gain G_p and the inner repetition code can be partly replaced by a Walsh-Hadamard code. Thus, a more powerful SCC scheme results, which has the potential to yield low BERs at low signal-to-noise ratios and which can be decoded by iterative decoding.

The text at hand is organized as follows: Section 2 describes the single and multi-carrier CDMA systems under consideration. Single user bounds are given for uncoded and coded systems. This is followed up by simulation results and discussion in section 3. Section 4 contains a short performance analysis. Finally, section 5 concludes the paper.

2. SYSTEM DESCRIPTION

The single user receiver used in SC-CDMA is the RAKE receiver, which exploits the diversity of the multipath channel by matched filtering each multipath and applying maximum ratio combining (MRC). In a multi-user environment, the performance of the RAKE receiver is limited by the MAI. Multi-user detectors have been proposed in the last years, but their complexity rises considerably with the number of users. Therefore, the RAKE receiver is still the most common SC-CDMA detector.

MC-CDMA is an alternative to classic SC-CDMA. OFDM is combined with CDMA. Various realization approaches exist, e.g. [2] [3] [4]. Only spreading in frequency domain will be considered. Figure 1 shows the block diagram.

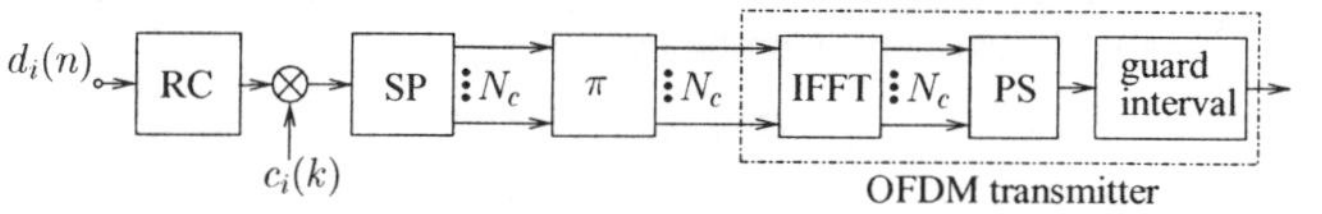

Fig. 1. The block diagram of the MC-CDMA transmitter with N_c subcarriers

A DS-CDMA signal is generated, described by a repetition code, followed by multiplication with a scrambling sequence. After serial-parallel processing and interleaving, the chips are allocated to the subcarriers, whereby one chip has to modulate one subcarrier. This is followed by an ordinary OFDM-transmitter, performing an IFFT and parallel-to-serial processing. Finally, a guard interval is inserted. Figure 2 clarifies the spreading in the frequency domain.

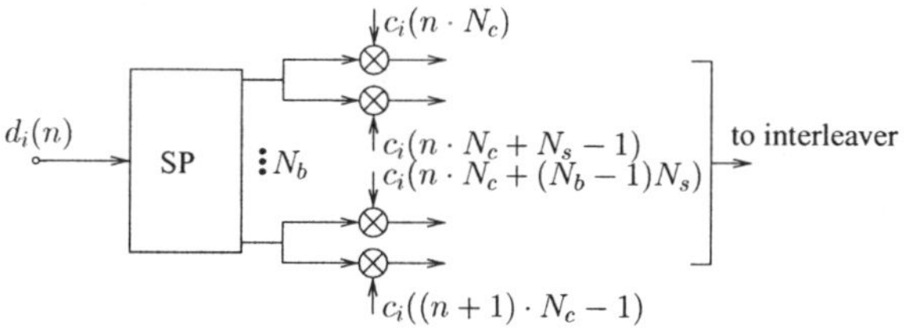

Fig. 2. A different view on the MC-CDMA transmitter in equivalent baseband

A block of N_b bits is mapped onto one OFDM-Symbol. Each bit is then distributed over N_s subcarriers (repetition by N_s times). Finally, N_c chips of the spreading code $c_j(k)$ are multiplied with the symbols on the subcarriers. We set N_c equal to the processing gain G_p to achieve the same conditions for both CDMA systems, therefore, on the average, one information bit is mapped onto one OFDM symbol. Figure 3 shows the related receiver.

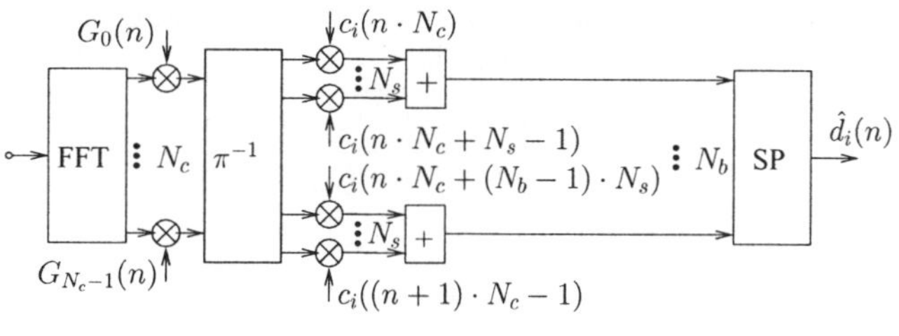

Fig. 3. The block diagram of the MC-CDMA receiver

The N_c subcarriers are equalized by multiplicative correction by factor $G_l(n)$, $0 \leq \mu < N_c - 1$. $G_l(n)$ can be chosen to perform a compromise between restoring the orthogonality of the spreading sequences and an optimization of the signal-to-noise ratio, for the orthogonality of the spreading sequences is destroyed on the channel. The combining schemes considered in general are MRC;

$$G_l^{MRC}(n) = H_l^*(n), \tag{1}$$

leading to the optimum signal-to-noise ratio. This causes heavy MAI by weighting each chip with the squared absolute value of the related channel coefficient, thus disrupting the orthogonality of the spreading sequences.

Orthogonal Restoring Combining (ORC);

$$G_l^{ORC}(n) = \frac{1}{H_l(n)}, \tag{2}$$

is a zero forcing approach which restores the orthogonality of the spreading sequences though amplifying the noise power.

Finally, Equal Gain Combining (EGC);

$$G_l^{EGC}(n) = \frac{H_l^*(n)}{|H_l(n)|}, \tag{3}$$

is a compromise between MRC and ORC, compensating the phase of each chip, leading to a coherent superposition of each chip after despreading. It is well known that that MRC has advantages over ORC at higher SNR, whereas EGC outperforms MRC and ORC for high SNR. Therefore, ORC is dropped. MRC will be considered in the simulations, for it is performed by the RAKE receiver and lends itself to a comparison with the MC-CDMA system.

The analytical formula for the uncoded binary bit error probability in an L-tap diversity mobile channel is given by, see [5],

$$P_2^{SC} = \left[\frac{1}{2}(1-\mu)\right]^L \cdot \sum_{k=0}^{L-1} \binom{L-1+k}{k} \cdot \left[\frac{1}{2}(1+\mu)\right]^k, \tag{4}$$

where by definition

$$\mu = \sqrt{\frac{\bar{\gamma_u}}{1+\bar{\gamma_u}}}, \tag{5}$$

and $\bar{\gamma_u}$ is the average SNR (signal-to-noise ratio) per channel, given by

$$\bar{\gamma_u} = E_b \cdot E\{\alpha_k^2\}/N_0. \tag{6}$$

The attenuation factor of each channel is denoted α_k. This formula applies to SC-CDMA and has to be adapted to MC-CDMA with MRC combining, where the guard interval has to be taken into consideration. The power spent for the guard interval is

$$S_{\text{excess}} = 10lg\left(\frac{G_p}{L-1+G_p}\right). \tag{7}$$

Thus, it can be easily shown that the bit error ratio for the MC-CDMA systems with MRC is

$$\begin{aligned} P_2^{MC} &= \left[\frac{1}{2}\left(1-\sqrt{\frac{E_b \cdot E\{\alpha_k^2\} \cdot \frac{G_p}{L-1+G_p}/N_0}{L+E_b \cdot E\{\alpha_k^2\} \cdot \frac{G_p}{L-1+G_p}/N_0}}\right)\right]^L \\ &\cdot \sum_{k=0}^{L-1} \binom{L-1+k}{k} \\ &\cdot \left[\frac{1}{2}\left(1+\sqrt{\frac{E_b \cdot E\{\alpha_k^2\} \cdot \frac{G_p}{L-1+G_p}/N_0}{L+E_b \cdot E\{\alpha_k^2\} \cdot \frac{G_p}{L-1+G_p}/N_0}}\right)\right]^k. \end{aligned} \tag{8}$$

This result will be verified by simulation.

The union bound for the coded systems is

$$P_{2,\text{coded}} \leq \sum_d w_d P_d \tag{9}$$

with P_d being the pairwise error probability for two code words with Hamming distance d. The weights w_d have to be computed from the distance spectrum of the coding scheme [6]. P_d is given by

$$P_d = \left[\frac{1}{2}(1-\mu)\right]^{L \cdot d} \cdot \sum_{k=0}^{L \cdot d-1} \binom{L \cdot d-1+k}{k} \cdot \left[\frac{1}{2}(1+\theta)\right]^k, \tag{10}$$

where θ is defined as

$$\theta = \sqrt{\frac{\bar{\gamma_c}}{1+\bar{\gamma_c}}}. \tag{11}$$

For SC-CDMA, $\bar{\gamma_c}$ is given by

$$\bar{\gamma_c}^{SC} = \frac{L \cdot R_c \cdot E_b \cdot E\{\alpha_k^2\}/N_0}{L+(J \cdot L-1) \cdot R_c \cdot E_b \cdot E\{\alpha_k^2\}/N_0} \tag{12}$$

and for MC-CDMA,

$$\bar{\gamma_c}^{MC} = \frac{L \cdot R_c \cdot E_b \cdot E\{\alpha_k^2\}\frac{G_p}{L-1+G_p}/N_0}{L+(J \cdot L-1) \cdot R_c \cdot E_b \cdot E\{\alpha_k^2\}\frac{G_p}{L-1+G_p}/N_0}. \tag{13}$$

Here, J is the number of users. The code rate R_c is tuned to be the inverse of the processing gain G_p throughout this paper:

$$R_c = \frac{1}{G_p}. \tag{14}$$

Therefore, one code bit is mapped onto one OFDM symbol in the MC-CDMA case. The SCC scheme is shown in figure 4.

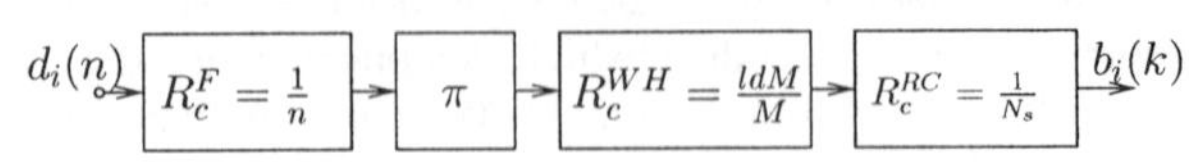

Fig. 4. The underlying SCC scheme

This replaces the inherent repetition code of the CDMA systems. A Walsh Hadamard code is used as an inner spreading code of rate

$$R_c^{WH} = \frac{\text{ld}M}{M}, \tag{15}$$

succeeding a convolutional code of rate

$$R_c^F = \frac{1}{n} \tag{16}$$

and an interleaver. The repetition code has its code rate $R_c^{RC} = \frac{1}{N_s}$ chosen so that the entire code rate is

$$R_c = R_c^{RC} \cdot R_c^{WH} \cdot R_c^{RC} = \frac{1}{G_p}. \tag{17}$$

The decoding of the SCC scheme by iterative processing is given in [7] and can be summed up as follows: after decoding the repetition code by a $1/R_c^{RC}$-tap integrate and dump operation, the inner symbol-by-symbol Max-Log-MAP decoder decodes the inner Walsh-Hadamard code, the outer symbol-by-symbol Max-Log-MAP decoder decodes the convolutional code.

3. SIMULATIONS

For the rest of this paper, the power of the attenuation factor of each channel will be set to unity:

$$E\{\alpha_k^2\} = 1 \quad \forall k. \tag{18}$$

For both CDMA systems, the process gain is $G_p = 64$. A real valued pseudo-noise sequence is applied for scrambling after the repetition code in the SC-CDMA system. For MC-CDMA, the same process gain and number of channel taps are used. A maximum ratio combiner is applied. The number of subcarriers is $N_c = 64$.

Figure 5 shows the results achieved with different combiners.

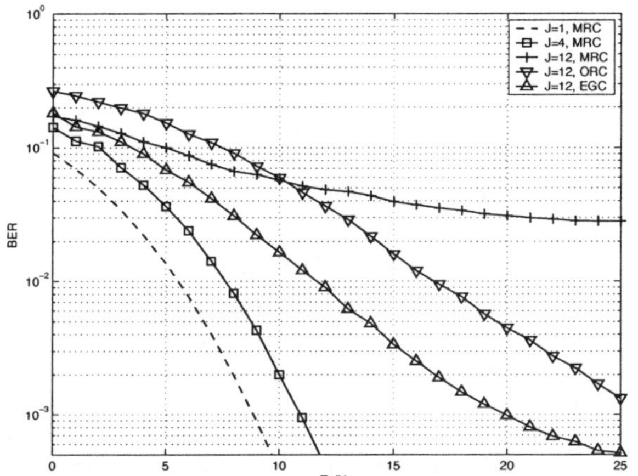

Fig. 5. Performance of MRC, ORC and EGC combiners, $L = 6$, $G_p = 64$

Despite the optimization of the signal-to-noise ratio through MRC equalization, the lack of orthogonality leads to huge MAI, which is increasing with J.

The number of RAKE fingers is 2 and 6. Each RAKE receiver corresponds to the number of channel taps L, so L=2 and L=6 channel taps are used. We consider the multi-user case where J is 12 and 20, as well as the single user case ($J = 1$). Figure 6 shows the simulation results for uncoded SC-CDMA, together with the theoretical and simulated single user bounds.

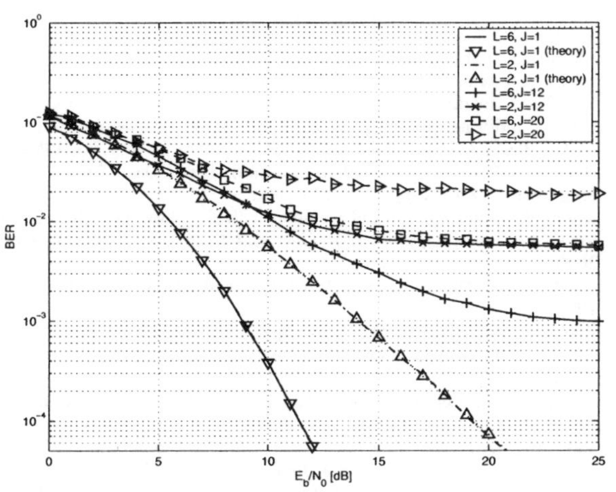

Fig. 6. BER for an SC-CDMA system, also single user bounds for MRC

The diversity gain is approximately 7.5 dB at a BER of 10^{-4}. It is observed that the theoretical and simulated single user bounds match closely in agreement with equation (4). However, a deviation of 0.4 dB for higher diversity is due to the cross talk within the RAKE receiver. Various error floors can be observed for $J = 12$ and 20, but the advantage of diversity is maintained.

Figure 7 shows the simulation results for MC-CDMA, together with the theoretical and simulated single user bounds for J=1.

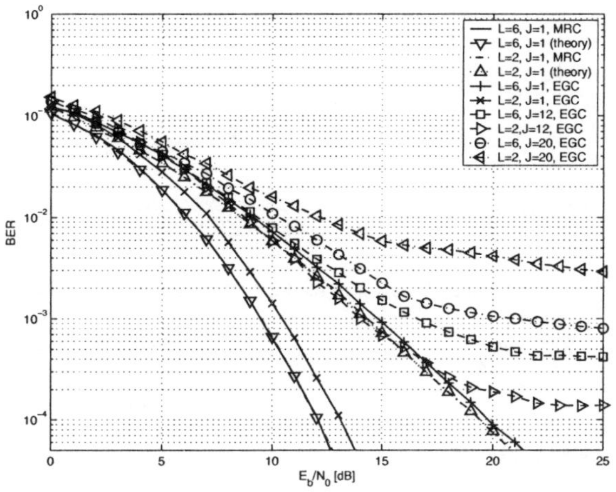

Fig. 7. BER for an MC-CDMA system, also single user bounds for MRC

The simulation results are in agreement with the theoretical performance given by equation (8). In a single user environment, the mismatching is about 0.25 dB with twofold diversity. It decreases with L. Therefore, for $L = 6$ and above, the BER is worse than for SC-CDMA. The mismatching in the multicarrier case is worse than the cross talk in the RAKE receiver. However, the simulations show that in a multi-user environment, uncoded MC-CDMA slightly outperforms uncoded SC-CDMA. The gain achieved in the multi-user case with more diversity is less than for the multi-user case in SC-CDMA, for more diversity means more frequency selectivity, which in turn destroys the orthogonality of the spreading sequences. This effect is to be compensated by EGC, whereas MRC would be better for less diversity. Clearly, the analytical formula for L-fold diversity fading channels perfectly matches the simulation result, in consequence, the MC-CDMA receiver fully utilizes diversity.

In the coded systems, an interleaver of variable length $N = 1000$ or $N = 5000$ is chosen. The convolutional code has a constraint length of $L_c = 7$ and a code rate of $R_c^F = 0.5$. The repetition code has a code rate of $R_c^{RC} = \frac{1}{3}$, M is equal to 64, so that according to equation (17) the entire code rate is $R_c = \frac{1}{64}$. Figure 8 shows the union bounds of the coded systems for the given parameters and $J = 12$ users.

The higher error floor of the SC-CDMA system at higher SNR is caused by the path cross talk in the RAKE due to imperfect scrambling codes, leading to MAI, as in the uncoded case.

In the following, results with 10 decoding iterations are given. The interleaver length is varied between $N = 1000$ and $N = 5000$. Figure 9 shows the resulting BER curves for various numbers of user J in an $L = 6$-path Rayleigh fading channel.

The long interleaver leads to improvements of up to $3.7dB$. The typical curves for SCC schemes occur, with significant coding gains for SNR values above $4dB$. Even with $J = 20$, the highest load simulated here, BERs of 10^{-6} are achievable.

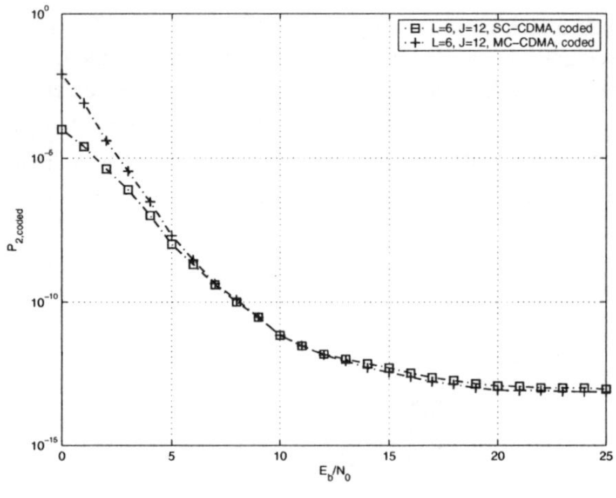

Fig. 8. Union bounds $P_{2,\mathrm{coded}}$ of the SC-CDMA and MC-CDMA systems

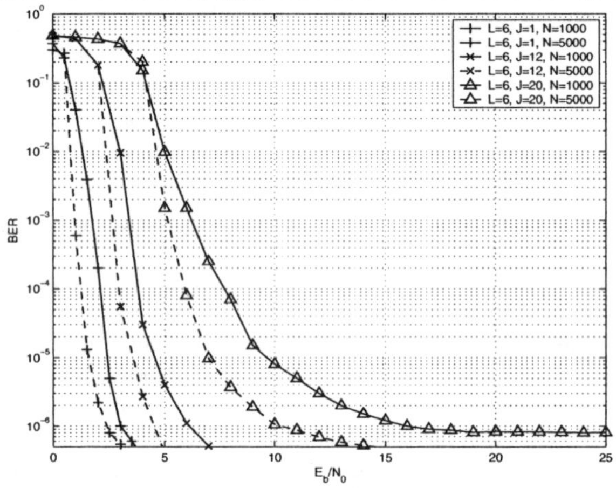

Fig. 9. BER for different J with 10 decoding iterations, $L = 6$

Figure 10 compares MC-CDMA and SC-CDMA for $J = 20$ in a 6-path Rayleigh fading channel, $N = 1000$ and $N = 5000$. A slight advantage of SC-CDMA over MC-CDMA about $0.2dB$ can be observed, due to the guard interval.

4. PERFORMANCE ANALYSIS

As was shown in the simulation results, MC-CDMA is advantageous over SC-CDMA in asynchronous uplink scenarios by nearly restoring the orthogonality of the spreading sequences, given here by Walsh-codes, through low complexity equalization. Additionally, there is less cross-talk compared to the RAKE-receiver. However, the aberrations between SC-CDMA and MC-CDMA in terms of the BER are so small that they should be neglected and implementation aspects should come to the fore.

Not simulated is the problem of envelope variations, leading to spectral regrowth especially in multicarrier systems and necessitating a higher back-off in the power amplifier, being measured by the PAPR. A higher amplifier back-off maps directly to a lower amplifier power conversion efficiency, being proportional to the terminal talk time. SC-CDMA systems have no PAPR problem in the uplink, as only the data for one user is transmitted [8]. Variations of the envelope can be mitigated through the use of PAPR reduction codes, like Schroeder-codes [9], or given by [10], also by clipping and peak windowing; see [1] for an overview.

Due to the repetition code, the data bits are distributed to three

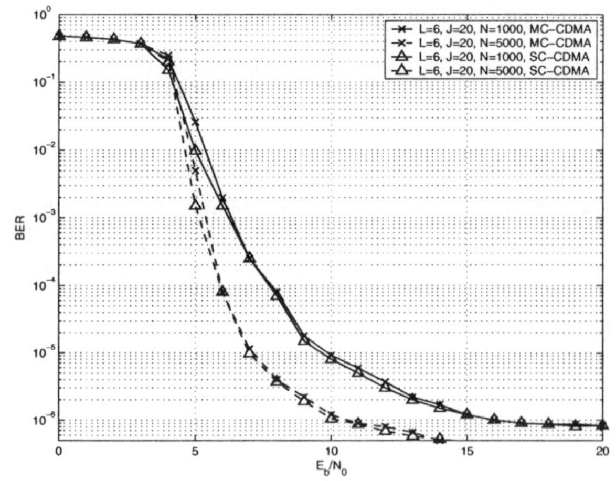

Fig. 10. Comparison between MC-CDMA and SC-CDMA for different N

successive OFDM-symbols that see different channel conditions, leading to relatively high diversity gains.

5. CONCLUSION

We conclude that MC-CDMA and SC-CDMA achieve comparable BER results. A viable solution to mitigate the MAI, even in SC-CDMA, is the replacement of the weak repetition code inherent in CDMA systems by more powerful codes, e.g. convolutional codes with small constraint length and strong inner Walsh-Hadamard codes [4], yielding good performance at low SNR ratios. Thus, with the resulting SCC schemes, the MAI can be overcome without computationally complex multi-user detection, treating the MAI as additional white noise.

The derived formula for the BER was validated and has to be extended to the M-ary and EGC-cases in future work. The simulation results have shown that low rate channel coding is an appropriate means to mitigate MAI.

ACKNOWLEDGEMENTS

I would like to thank K. D. Kammeyer, W. Zschunke and my colleagues, for the fruitful discussion on multicarrier systems we had during the visit of Mr Kammeyer at our institute.

REFERENCES

[1] van Nee, R. and Prasad, R., *OFDM for wireless multimedia communications*, 1st ed. Boston, London: Artech House, 2000.
[2] Yee, N., Linnartz J.-P., and Fettweis, G., "Multicarrier CDMA in indoor wireless radio networks," *Proc. IEEE Int. Symp. on Personal, Indoor and Mobile Radio Communications (PIMRC), pp. D.1.3.1-D.1.3.5*, 1993.
[3] Hara, S. and Prasad, R., "Overview of multicarrier CDMA," *IEEE Commun. Mag., vol. 35 no. 12, pp. 126 -133*, 1997.
[4] Kühn, V., Dekorsky, A., and Kammeyer, K.D., "Low rate channel coding for CDMA systems," *Int. J. of Elec. and Com. (AEÜ), 54, pp. 353-363*, 2000.
[5] Proakis, J. G., *Digital Communications*, 3rd ed. New York: Mc Graw Hill, 1995.
[6] Benedetto, S., Divsalar, D., Montorsi, G., and Polara, F., "Serial concatenation of interleaved codes: Performance analysis, design, and iterative decoding," *TDA Progress Report 126*, 1996.
[7] Hagenauer, J., Herzog, R., and Schmidbauer, A., "Iterative decoding and despreading improves CDMA-Systems using M-ary orthogonal modulation and FEC," *IEEE Trans. Acoust., Speech, Signal Processing*, 1981.
[8] Martoyo, I., Schober, H., and Jondral, F., "CDMA versus OFDM, a performance comparison in selective fading channels," *IEEE 7th Int. Symp. on Spread-Spectrum Tech. & Appl., Prague*, 2002.
[9] M. Schröder, "Synthesis of low-peak-factor signals and binary sequences with low autocorrelations," *IEEE Trans. Inform. Theory, 16, pp. 85-89*, 1970.
[10] M. Friese, "Multitone signals with low crest factor," *IEEE Trans. Commun., vol. 45 no. 10, pp. 1338 -1344*, 1997.

M-ARY PPM UWB RADIO IN FADING CHANNELS*

MOHAMMED ABDEL-HAFEZ,
United Arab Emirates University, Department of Electrical Engineering
P.O.BOX 17555, Al-Ain, United Arab Emirates, Email: mhafez@uaeu.ac.ae,
FATIH ALAGOZ
Harran University, Department of Electrical and Electronics Engineering
, Sanliurfa, TURKEY, Email: alagoz@harran.edu.tr
MATTI HAMALAINEN
University of Oulu, Centre for Wireless Communications
Tutkijantie 2 E, FIN-90014, FINLAND, Email: matti.hamalainen@ee.oulu.fi

This paper investigates the performance of M-ary pulse position modulation (PPM) multiuser ultra wideband (UWB) communication systems in terms of symbol error rate (SER) over fading and additive white Gaussian noise (AWGN) channel. Based on Gaussian approximation for the multiple access interference, an expression for the signal-to-noise ratio is derived for the UWB system. This expression is used to derive exact SER expressions for coherent UWB receivers. The effect of pulse selection on the SER of multiuser UWB system is studied. In addition to rectangular pulse, 2nd derivative Gaussian waveform is considered. We show that the system capacity and/or SER performance can be significantly increased by using the 2nd derivative Gaussian pulse.

1. Introduction

Ultra wideband (UWB) systems were recently proposed as one of the possible solutions for short range wireless networks [1]. The new technology has the potential to deliver high data rates with very low power densities. The UWB systems are characterized by the transmission of series of sub-nanoseconds pulses (monocycles) that spread the energy of the signal from near DC to a few GHz. The pulse train is transmitted without any modulation with sinusoidal carrier. This is a major advantage of UWB since it has high immunity against multipath fading effect as experienced in other wireless systems. In addition, high processing gain and very low power density ensure minimal mutual interference between the UWB and other wireless systems. Pulse position modulation (PPM) has been proposed as a modulation scheme suitable for the UWB communications [1]. With PPM, the data modulates the position of the transmitted pulse within an assigned window in time. UWB radio is the generic term describing radio systems having very large bandwidths; "bandwidth greater than 20% of the center frequency measured at the –10dB points," and "RF bandwidth greater than 500MHz," are the two of the definitions under consideration by the US Federal Communications Commissions (FCC).

Time Hopping (TH) is used in the UWB system as a multiple-access method. The PPM scheme is used in TH-mode with pulse transmission instants defined by a pseudo random code. One data bit is spread over multiple pulses to achieve a processing gain due to the pulse repetition. The processing gain is increased by the low duty cycle. The multiple-access interference (MAI) may be the dominant factor on the bit error rate (BER) performance. Some published work considered the BER of UWB system in nonfading multipath channel [2,3]. The correlation properties and the frequency spectra of UWB pulses are very crucial. The BER of a single user M-ary PPM UWB system was considered for fixed multipath fading channel by [2], extension of this work to multiuser UWB is done by [3]. This paper is different from other previous works by introducing the effect of pulse shape and fading on the performance of multiuser UWB system. Rician channel has been selected for its generalized modelling of short range UWB channel [4].

The remainder of the paper is organized as follows. Section 2 introduces the signal, channel, and receiver models. Section 3 carries out the performance analysis of the UWB M-ary PPM both in Gaussian and Rician fading channels. Section 4 presents the numerical results obtained. Section 5 concludes this study.

2. UWB System Model

The time-hopping M-ary PPM system model examined in this paper is shown in Fig. 1. The j-th data symbol for the v-th user's transmitted signal has the form [1]:

$$S_j^v(t) = A^v P\left(t - jT_f - C_j^v T_C - d_j^v\right) \quad (1)$$

and the v-th user's transmitted signal has the form [1]:

$$S^v(t) = \sum_{j=-\infty}^{\infty} S_j^v(t) = \sum_{j=-\infty}^{\infty} A^v P\left(t - jT_f - C_j^v T_C - d_j^v\right) \quad (2)$$

where $P(t)$ is the UWB pulse of duration T_p, E_p is the energy per pulse. The pulse repetition interval, referred to as frame, is T_f , $A^v = \sqrt{E_p^v}$, C_j^v, and d_j^v are respectively, the amplitude, time-hopping and modulation for user v. The PPM time shift is $d_j^v \in \{\delta_1, \ldots, \delta_M\}$. For a fixed T_f, the symbol rate $R_s = 1/(N_p T_f)$ where N_p is the number of pulses that form a symbol. The symbol duration is then $T_s = N_p T_f$. The spreading ratio is defined by $\beta = T_f / T_p$.

Each user's signal propagates over a signal path channel with attenuation factor α and propagation delay τ as shown in Fig.2. It is assumed that the channel is a single path Rician channel. The channel is random and its amplitude, α, can be described by Rician distribution. The instantaneous power level, $\gamma = \alpha^2$, follows the non-central Chi-square probability distribution function (*pdf*) as:

* This work is supported by Centre for Wireless Communications, University of Oulu, Finland.

$$f(\gamma)=\frac{1+K}{\Omega}e^{-\left(\frac{\gamma(1+K)+K\Omega}{\Omega}\right)}I_0\left(2\sqrt{\frac{(1+K)K\gamma}{\Omega}}\right) \tag{3}$$

where Ω is the mean power level, which is equal to the sum of the powers due to the multipath diffused , γ^0_{diff}, and Line of Sight (LOS) , γ^0_{LOS}, and $I_0(x)$ is the modified Bessel function of the first kind and zero'th order. The Rician distribution is more commonly specified by the Rice factor, K, which is interpreted as the direct to multipath diffused power ratio, $K=\gamma^0_{LOS}/\gamma^0 diff$. The 1st and 2nd moments of α are:

$$E[\alpha]=\frac{\sqrt{\pi\Omega}}{2}e^{-\frac{K}{2}}\left[(1+K)I_0\left(\tfrac{K}{2}\right)+KI_1\left(\tfrac{K}{2}\right)\right] \tag{4}$$

$$E\left[\alpha^2\right]=\Omega=\gamma^0_{LOS}+\gamma^0_{diff} \tag{5}$$

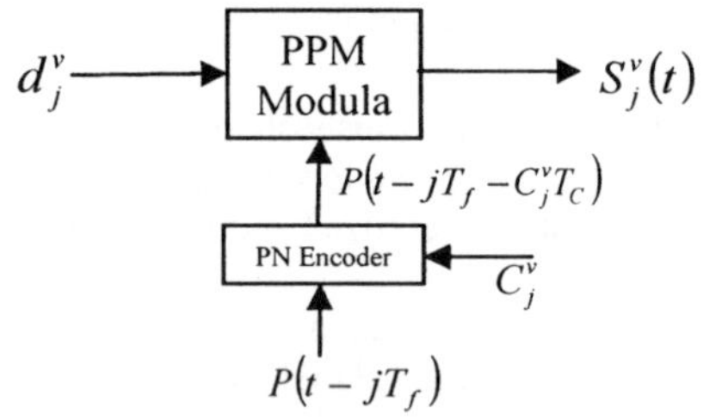

Figure 1. TH-PPM UWB modulator.

The received signal $R(t)$ from all users is given by:

$$R(t)=\sum_{v=1}^{N_u}\alpha_v S^v\left(t-\tau^v\right)+\eta(t) \tag{6}$$

where α_v and τ^v are the channel attenuation and time delay associated with user v and $n(t)$ is zero-mean AWGN with power spectral density $N_0/2$.

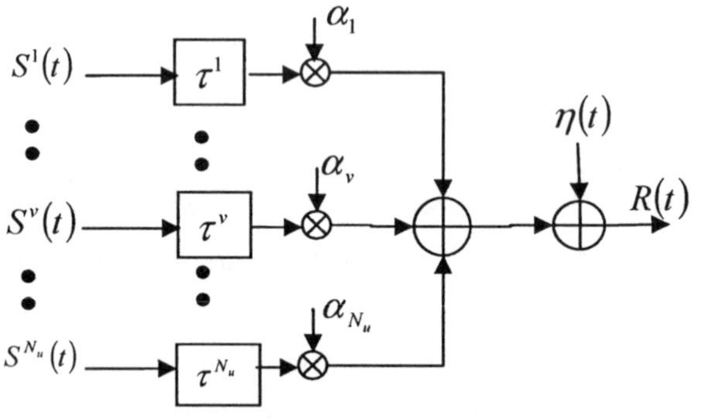

Figure 2. The channel model.

Without loss generality, we assume the desired user is v=1. The single-user optimal receiver is the M-ary pulse correlation receiver followed by a detector. We assume the receiver is perfectly synchronized to user 1, i.e., τ^1 is known. Furthermore, the time hopping sequence C_j^1 is known at the receiver. The M-ary correlation receiver for user 1 consists of M filters matched for any pulse $P(t)$ with duration T_P. We define the correlation (N_u) to the basis functions $\phi_i^1(t)$ defined as:

$$\phi_i^1(t)=P\left(t-d_i^1-\tau^1\right) \qquad i=1..M \tag{7}$$

The detector selects the Max i-th symbols of M possible outputs, Fig. 3.

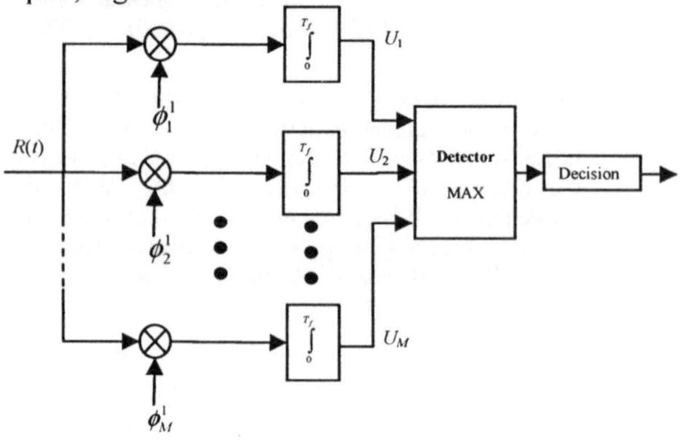

Figure 3. M-ary PPM UWB receiver for the 1st user.

The decision variables at time sample ($t=jT_f$), are now given as:

$$U_i=\int_{(j-1)T_f}^{jT_f}R(t)\phi_j^1\left(t-jT_f-C_j^1T_c\right)dt \tag{8}$$

by substituting for $R(t)$ from Eq.6, the parameter U_i can be represented as:

$$U_i=d+NI+N \tag{9}$$

where, d denotes the desired part of the received signal expressed as:

$$d=\int_{(j-1)T_f}^{jT_f}\alpha_1 S^1\left(t-\tau^1\right)\phi_i^1\left(t-jT_f-C_j^1T_c\right)dt \tag{10}$$

NI is the multiple access interference (MAI) from other users expressed as:

$$NI=\int_{(j-1)T_f}^{jT_f}\sum_{v=2}^{N_u}\alpha_v S^v\left(t-\tau^v\right)\phi_i^1\left(t-jT_f-C_j^1T_c\right)dt \tag{11}$$

N is the noise component at the output of the receiver,

$$N=\int_{(j-1)T_f}^{jT_f}\eta(t)\phi_i^1\left(t-jT_f-C_j^1T_c\right)dt \tag{12}$$

with $P(t)$ being normalized and orthogonal pulses, the desired part of the signal is expressed as:

$$d=\alpha_1 A^1\delta(i-j) \tag{13}$$

The MAI part can be written as:

$$NI = \sum_{v=2}^{N_u} \alpha_v A^v \int_0^{T_f} P(t)P(t-\Delta)dt \quad (14)$$

where is the time difference between the different users expressed as:

$$\Delta = \left(C_j^1 - C_j^v\right) - \left(d_i^1 - d_j^v\right) - \left(\tau^1 - \tau^v\right) \quad (15)$$

If the correlation function of the pulse $P(t)$ is defined by:

$$h(\Delta) = \int_0^{T_f} P(t)P(t-\Delta)dt \quad (16)$$

then the expression for MAI in Eq.14 can be written as:

$$NI = \sum_{v=2}^{N_u} \alpha_v A^v h(\Delta) \quad (17)$$

It is assumed that the time-hopping code elements are random, independent, with uniform distribution over frame interval T_f for all users and frames. Each user has a uniformly distributed data source. The time delays are also assumed random, i.i.d. uniformly distributed over the frame interval. Under the assumptions listed above, and noting that MAI pulses of interest fall within the same UWB frame, the time difference Δ is a uniformly distributed random variable over the interval $[-T_f, T_f]$.

Various waveforms with complex mathematical formats have been proposed for impulse radio including Gaussian pulse, Gaussian monocycle [1], and Rayleigh monocycle [6]. All of these waveforms reflect the high-pass-filtering impact of the transmitter and receiver antennas. To simplify our calculation, we consider two types of waveforms, namely, rectangular and 2nd derivative Gaussian pulses. This analysis can be extended to other waveforms.

We assume the following *rectangular waveform*:

$$P_{rect}(t) = \begin{cases} \sqrt{\dfrac{1}{T_P}} & 0 \le t \le T_P \\ 0 & \text{Otherwise} \end{cases} \quad (18)$$

The correlation function $h(\Delta)$ for $P_{rect}(t)$ in Eq.18 is:

$$h_{rect}(\Delta) = \begin{cases} 1 - \dfrac{|\Delta|}{T_P} & 0 \le |\Delta| \le T_b \\ 0 & \text{Otherwise} \end{cases} \quad (19)$$

Comparing with a rectangular waveform, the main characteristic of monocycle signal is that they have zero DC component to allow them radiate effectively. The normalized *2nd derivative Gaussian pulse* is expressed as:

$$P_{gauss}(t) = \sqrt{\frac{8}{3\varepsilon}}\left[1 - 4\pi\left(\frac{t}{\varepsilon}\right)^2\right]e^{-2\pi\left(\frac{t}{\varepsilon}\right)^2} \quad (20)$$

where ε is a time scale factor and its relation to pulse width T_P is $T_P = 7\varepsilon$ which contains 99.99% of the total energy. The corresponding correlation function is:

$$h_{gauss}(\Delta) = \left[1 - 4\pi\left(\frac{\Delta}{\varepsilon}\right)^2 + \frac{4\pi^2}{3}\left(\frac{\Delta}{\varepsilon}\right)^4\right]e^{-\pi\left(\frac{\Delta}{\varepsilon}\right)^2} \quad (21)$$

3. Performance Analysis

For simplicity, and without loss of generality, we assume that signals are transmitted in the l-th time slot. With perfect synchronization, channel delay τ^1 is known, the time hopping sequence C_j^1 is known, and the decision variables are:

$$U_i = \begin{cases} U_l & : \alpha^1 A^1 + NI + N & i = l \\ U_p & : NI + N & i = 1..M, i \ne l \end{cases} \quad (22)$$

The multiple-access interference is not Gaussian. However, we assume large number of users to justify the Gaussian assumption for the MAI by invoking the central limit theorem. The decision variables are independent Gaussian random variables distributed as follows:

$$\begin{aligned} U_l &: N\left(\alpha^1 A^1 + E[NI] + E[N], \sigma_{NI}^2 + Var(N)\right) \\ U_p &: N\left(E[NI] + E[N], \sigma_{NI}^2 + Var(N)\right) \end{aligned} \quad (23)$$

where $N(0,1)$ denotes the Gaussian distribution with zero mean and unit variance, and $E(x)$ and $Var(x)$ are the mean and variance of x, respectively.

The mean $E(N)$ and variance $Var(N)$ are found to be 0 and $N_0/2$, respectively. The corresponding parameters for MAI are found to be:

$$E[NI] = \sum_{v=2}^{N_u} E[\alpha_v]A^v E[h(\Delta)] \quad (24)$$

and

$$\sigma_{NI}^2 = \sum_{v=2}^{N_u} (A^v)^2 \left[E[\alpha_v^2]E[h^2(\Delta)] - E^2[\alpha_v]E^2[h(\Delta)]\right] \quad (25)$$

The mean of $h(\Delta)$ can be calculated as follows:

For rectangular waveform:

$$\begin{aligned} E[h(\Delta)] &= E[h_{rect}(\Delta)] \\ &= \int_{-T_P}^{T_P} h_{rect}(\Delta)\frac{1}{2T_f}d\Delta = \frac{1}{2\beta} \end{aligned} \quad (26)$$

For 2nd derivative Gaussian waveform:

$$E[h_{gauss}(\Delta)] = \frac{2}{\beta}\left[\frac{T_P}{2\varepsilon}\left(\frac{1}{\pi} - 1\right)erf\left(\sqrt{\pi}\frac{T_P}{\varepsilon}\right) + \left(2 - 2\left(\frac{T_P}{\varepsilon}\right)^2 - \frac{1}{\pi}\right)\exp\left(-\pi\left(\frac{T_P}{\varepsilon}\right)^2\right)\right] \quad (27)$$

The 2nd moment of $h(\Delta)$, $E[h^2(\Delta)]$, is calculated as follows:

For rectangular waveform:

$$E[h^2(\Delta)] = E[h_{rect}^2(\Delta)] = 2\int_0^{T_P} h_{rect}^2(\Delta)\frac{1}{2T_f}d\Delta = \frac{1}{3\beta} \quad (28)$$

For 2nd derivative Gaussian waveform:

$$E\left[h_{gauss}^{2}(\Delta)\right]=$$

$$\frac{2\varepsilon}{T_P\beta}\left[\sqrt{2}\left(\frac{1}{2}-\frac{1}{2\pi}+\frac{35\sqrt{2}}{192\pi^2}\right)erf\left(\sqrt{2\pi}\frac{T_P}{\varepsilon}\right)\right.$$
$$+\left(\left(-1-\frac{35}{48\pi^2}+\frac{2}{\pi}\right)\frac{T_P}{\varepsilon}+\left(\frac{8}{3}-4\pi-\frac{35}{36\pi}\right)\left(\frac{T_P}{\varepsilon}\right)^3\right.$$
$$\left.\left.+\left(-\frac{7}{9}+\frac{8}{3}\pi\right)\left(\frac{T_P}{\varepsilon}\right)^5+-\frac{4}{9}\pi\left(\frac{T_P}{\varepsilon}\right)^7\right)e^{-2\pi\left(\frac{T_P}{\varepsilon}\right)^2}\right] \quad (29)$$

If $T_P/\varepsilon=7$, Eq.27 and Eq.29 can be reduced to,

$E\left[h_{gauss}(\Delta)\right]=\frac{-9.738\times10^{-2}}{\beta}$, $E\left[h_{gauss}^{2}(\Delta)\right]=\frac{.14519}{\beta}$

The *pdf* for U_l can be represented as:

$$p(U_l)=\frac{1}{\sqrt{2\pi}\sigma_l}e^{-\frac{(U_l-\mu_l)^2}{2\sigma_l^2}} \quad (30)$$

where $\mu_l=\alpha^1A^1+E[NI]$ and $\sigma_l=\sqrt{\sigma_{NI}^2+\frac{N_0}{2}}$

$$p(U_p)=\frac{1}{\sqrt{2\pi}\sigma_p}e^{-\frac{(U_p-\mu_p)^2}{2\sigma_p^2}} \quad (31)$$

where $\mu_p=E[NI]$ and $\sigma_p=\sigma_l=\sqrt{\sigma_{NI}^2+\frac{N_0}{2}}$

The conditional probability of correct decision is:

$$P_C|U_l=prob(U_l>U_1,...,U_l>U_i,..,U_l>U_M|U_l,l\neq i) \quad (32)$$

and the average probability of correct decision is:

$$P_C=\int_{-\infty}^{\infty}\left[prob(U_l>U_i|U_l,l\neq i)\right]^{M-1}p(U_l)dU_l \quad (33)$$

by substituting Eq.30 and Eq.31 into Eq.33, and after some calculus with the substitution $x=\frac{U_l-\mu_l}{\sqrt{2}\sigma}$, the probability of correct decision becomes,

$$P_C=\frac{1}{\sqrt{\pi}2^{M-1}}\int_{-\infty}^{\infty}\left[1+erf\left(x+\frac{\alpha^1A^1}{\sqrt{2}\sigma}\right)\right]^{M-1}e^{-x^2}dx \quad (34)$$

where σ is defined as:

$$\sigma=\sqrt{\frac{N_0}{2}+\sum_{v=2}^{N_u}(A^v)^2\left[E\left[\alpha_v^2\right]E\left[h^2(\Delta)\right]-E^2[\alpha_v]E^2[h(\Delta)]\right]} \quad (35)$$

If perfect power control is assumed then the term $\frac{\alpha^1A^1}{\sqrt{2}\sigma}$ in Eq.30 can be expressed as,

$$\frac{\alpha^1A^1}{\sqrt{2}\sigma}=\sqrt{\frac{\alpha^2\frac{E_P}{N_0}}{1+(N_u-1)2\left(\frac{E_P}{N_0}\right)\left[E\left[\alpha_v^2\right]E\left[h^2(\Delta)\right]-E^2[\alpha_v]E^2[h(\Delta)]\right]}} \quad (36)$$

The probability of symbol error is $P_S=1-P_C$.

3.1 *AWGN Channel*

In Gaussian channel, the channel attenuation factor can be assumed as unity, $\alpha=1$, and the average probability of correct decision error can be reduced as:

$$P_{C(AWGN)}=\frac{1}{2^{M-1}}\int_{-\infty}^{\infty}\left[1+erf\left(x+\sqrt{\frac{\frac{E_P}{N_0}}{1+(N_u-1)2\left(\frac{E_P}{N_0}\right)\left[E\left[h^2(\Delta)\right]-E^2[h(\Delta)]\right]}}\right)\right]^{M-1}e^{-x^2}dx \quad (37)$$

this integration is evaluated by Hermite integration method [5].

The term $\left[E\left[h^2(\Delta)\right]-E^2[h(\Delta)]\right]$ in Eq.37 represents the variance of the correlation function denoted by σ_h^2. From Eq.26 and Eq.28 and for $\beta>100$, the parameter σ_h^2 for rectangular pulse can be approximated by:

$$\sigma_h^2(rect)=\frac{1}{3\beta}-\frac{1}{4\beta^2}\approx\frac{0.333}{\beta} \quad (38)$$

and the corresponding parameter for 2nd derivative Gaussian pulse with $T_P/\varepsilon=7$ is:

$$\sigma_h^2(gauss)=\frac{0.14519}{\beta}-\frac{9.48368\times10^{-3}}{\beta^2}\approx\frac{0.14519}{\beta} \quad (39)$$

From Eq.37 and by comparing Eq.39 to Eq.38, the MAI interference using monocycle pulses is more than 50% of the rectangular pulses.

3.2 *Rician Fading Channel*

To evaluate the system performance in terms of probability of correct decision, we substitute Eq.36 into Eq.34, and average P_c over the non-central Chi–square distribution for γ defined by Eq.3. The average probability of correct decision is:

$$\overline{P_C}=\frac{1}{\sqrt{\pi}2^{M-1}}\int_0^{\infty}\int_{-\infty}^{\infty}\left[1+erf\left(x+\sqrt{\frac{\gamma\frac{E_P}{N_0}}{\zeta}}\right)\right]^{M-1}e^{-x^2}f(\gamma)dxd\gamma \quad (40)$$

where $\zeta=1+(N_u-1)2\left(\frac{E_P}{N_0}\right)\left[E\left[\alpha^2\right]E\left[h^2(\Delta)\right]-E^2[\alpha]E^2[h(\Delta)]\right]$

and the integrations in Eq. 40 are evaluated using Laguerre and Hermite integration methods described in [5].

The performance expressions found are very general and can be used for Rayleigh and Rician channels for different pulse shapes.

4. Numerical Results

Figure 4 depicts the effect of pulse selection on the user BER of 2-ary PPM UWB in Rician fading channel. It is obvious that the monocycle pulses represented by the 2nd derivative Gaussian pulse outperform the rectangular pulse regardless of the Rice factor. This is due to the correlation properties of the monocycle pulses.

The effect of system capacity of the BER is presented in Figure 5. Similarly, we observe that practical pulses such as monocycle pulses provide higher capacity than theoretical ones. For the same BER, the number of accommodated users in the system can be significantly increased using the 2nd derivative Gaussian pulse.

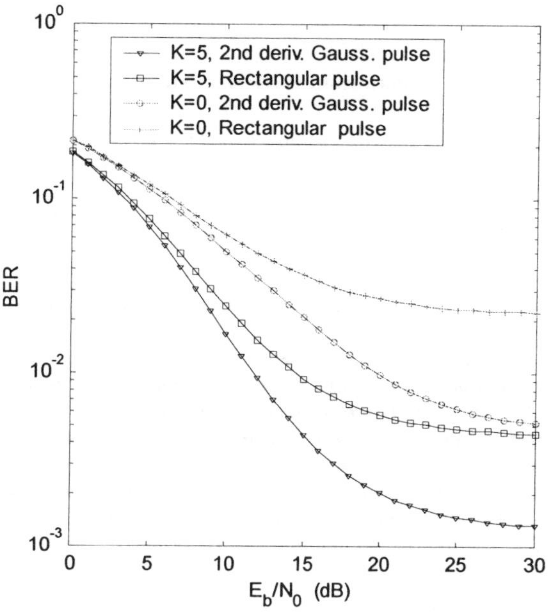

Figure 4. User BER with respect to Rice factor, K, for multiuser 2-ary PPM UWB radio system with N_u=10 users and spreading ratio β =100 in Rician fading channel.

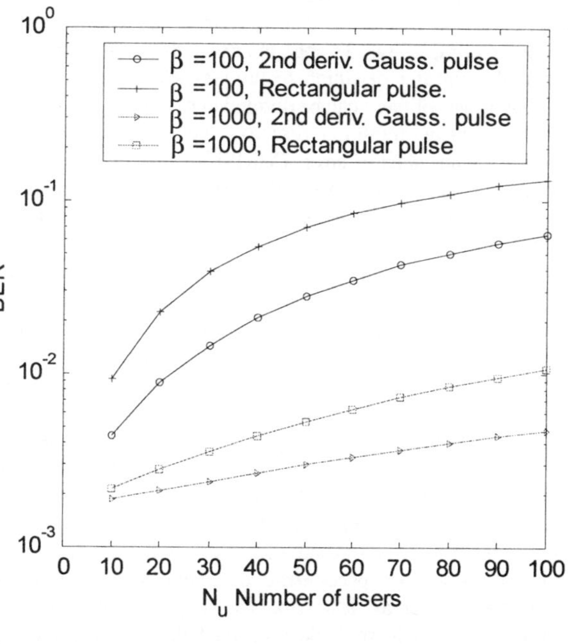

Figure 5. User BER with respect to number of users and spreading ratio, β, for multiuser 2-ary PPM UWB radio system with E_b/N_0=15dB in Rician fading channel with K=5.

5. Conclusion

In this paper, we have investigated the SER performance of M-ary PPM UWB multiple-access system in fading channels. Rician channel was selected for its accurate modeling of indoor propagation.channel. Among many practical pulse waveforms, Rectangular and 2nd derivative Gaussian pulses were considered. Closed form expressions for SER were derived using M-ary PPM UWB system for Rician fading channel. The expressions are generalized in such a way that it can be applied for multiuser UWB environment using various pulse waveforms. This work can be extended to account for multipath fading using a suitable UWB receiver.

References

[1] M. Win, R. Scholtz, "Ultra-Wide Bandwidth Time-Hopping Spread-Spectrum Impulse Radio for Wireless Multiple-Access Communications," *IEE Trans. On Comm.*, Vol. 48, No. 4, pp. 679-691, April 2000.

[2] Foerster, J.R., "The Effects of Multipath Interference on the Performance of UWB Systems in an Indoor Wireless Channel," *IEEE VTS 53rd Vehicular Technology Conference.*, 2001, Spring, vol. 2, pp. 1176-1180.

[3] Lijia Ge, Guangrong Yue, and Sofiene Affes, " On the BER Performance of Pulse-Position-Modulation UWB Radio in Multipath Channels," *IEEE Conference on Ultra Wideband Systems and Technologies*, 2002.

[4] Hovinen, V., Hämäläinen, M., Pätsi, T., "Ultra Wideband Indoor Radio Channel Models: Preliminary Results". *IEEE Conference on Ultra Wideband Systems and Technology*, 2002.

[5] Gryadshteyn I.S. and Ryzhik I.M., Table of Integrals, Series, and Products, 5th ed. San Diego: Academic, 1994.J.T.

[6] Conroy, J.L. LoCicero, and D.R. Ucci, "Communication techniques using monopulse waveforms," *Proceeding of IEEE MILCOM' 99*, vol. 2, pp. 1191–1185, 1999.

Comparison of Several Multiple Antenna Multiuser Detectors in Wireless CDMA Systems

S.Mohammad Razavizadeh[1,3] **Vahid TabaTaba Vakili[1]** **Paeiz Azmi[2]**

[1] Iran University of Science and Technology (IUST), Narmak 16844,Tehran, Iran
[2] Tarbiat Modarres University (TMU), Tehran, Iran
[3] Iran Telecommunications Research Center (ITRC), Tehran, Iran
e-mail: m_razavi@iust.ac.ir

Abstract - *Two main problems in CDMA cellular systems are multiple access interference and multipath fading. To alleviate these problems, two usually used methods are multiuser detection and space-time coding. Many of works on multiuser detection are done for single antenna systems. In this paper, we study the multiuser detection in multiple transmit/receive antenna CDMA systems. We analyze the combination of linear and non-linear multiuser detection techniques with space-time block coding methods. We also compare the performance of various multiuser detection techniques in multiple transmit/receive antennas by computer simulations.*

I. Introduction

One of the main problems in CDMA systems is multiple access interference that limits the capacity of these systems. Multiuser detection is a technique that mitigates this problem [1]. The optimum multiuser detector that was introduced by Verdu had the problem of complexity. To encounter with this inefficiency, many suboptimal multiuser detection techniques are introduced in recent years that most of them can be divided in two groups of linear and non-linear. Linear multiuser detectors include decorrelating and minimum mean square error (MMSE) detectors and parallel interference canceller (PIC), serial interference canceller (SIC) and decision feedback (DF) detector are among the non-linear multiuser detectors [2].

On other hand, in recent years much research has been done on signal processing techniques for transmit diversity over fading channels. For this purpose, many space-time codes are introduced. The first space-time code was introduced by Alamouti [3] that can implement maximum likelihood decoding using linear combination in the receiver. After that, Tarokh et al. extended the Alamouti's work to an arbitrary number of antennas and gave a construction criteria for space-time block codes [4], [5]. Moreover they introduced space-time trellis codes [4].

Many researches on multiuser detection area consider single antenna links between users and base station. On other hand, almost all-original works on space-time coding have been done for single user systems. Whereas, few papers have been published on the multiuser detection in multiple antenna links or using space-time codes in multiuser CDMA systems.

In this paper, we consider space-time block codes in CDMA systems. Our main work is on combining of space-time decoding with multiuser detection in the receiver of CDMA systems. For this reason, we combine various multiuser detection techniques in a multiple input/output CDMA system and propose some new receiver structures.

II. Space-Time Multiuser Detection

In this section, we analyze the receiver structure for synchronous CDMA systems with space-time block codes. The space-time block code that we have selected is Alamouti's code [3]. In this code the number of antennas in transmitter and receiver is two. Our work can also be extended to more than two antennas and the other kinds of space-time codes. In the receiver, the combination of multiuser detection (MUD) techniques, space-time decoding and maximal ratio combining (MRC) are used. We use five MUD schemes, that are decorrelating detector, MMSE detector, PIC, SIC and decision feedback detector. We compare the receivers that use these detectors with conventional (matched filter) receiver. We divide our analyze into four cases:

A) One transmit antenna, Two receive antennas
B) Two transmit antennas, One receive antenna
C) One transmit antenna, Four receive antennas
D) Two transmit antennas, Two receive antennas

In each of the above cases, we use six types of CDMA receivers and study the general receiver structure. In fig.1 the structure of receiver is depicted.

A: One transmit antenna, Two receive antennas

In this case, the received CDMA signal in each receive antenna is as follow:

$$r_p = \sum_{k=1}^{K} A_k b_k h_{k.p} C_k + n_p \;, \qquad p = 1,2 \qquad \text{(A.1)}$$

where $\boldsymbol{r_p}$ is the received signal at the pth antenna, A_k the amplitude of kth user and $b_k \in \{-1,+1\}$ is the data bit of the kth user. $h_{k.p}$ is the fading coefficient between the kth user and the pth antenna, $\boldsymbol{C_k}$ is the spreading sequence of kth user, $\boldsymbol{n_p}$ is AWGN noise on the pth antenna. We can also present the received signal in matrix form as follow:

$$\mathbf{r}_p = \mathbf{b} \cdot \mathbf{A} \cdot \mathbf{h}_p \cdot \mathbf{C}^T \qquad p = 1,2 \qquad \text{(A.2)}$$

where $\mathbf{C} = [C_1 \; C_2 \; \ldots \; C_K]^T$, $\mathbf{b} = [b_1, b_2, \ldots, b_K]$, $\mathbf{A} = diag([A_1, A_2, \ldots, A_K])$ and $\mathbf{h}_p = diag([h_{1.p}, h_{2.p}, \ldots, h_{K.p}])$.

In a conventional receiver, the received signal passes through a bank of matched filters or correlators in which each correlator is related to one of the signature sequences as follow:

$$\mathbf{t}_p = \mathbf{r}_p \mathbf{C}^T \qquad p = 1,2 \qquad \text{(A.3)}$$

Then, we use MRC combining at the outputs of each matched filter and then add them:

$$\mathbf{z} = \mathbf{h}_1^* \cdot \mathbf{t}_1 + \mathbf{h}_2^* \cdot \mathbf{t}_2 \quad \text{(A.4)}$$

Then maximum likelihood (ML) decision for input data can be performed as follow:

$$\hat{\mathbf{b}} = sign(\Re e[\mathbf{z}]) \quad \text{(A.5)}$$

Multiuser detector is applied to the outputs of conventional receivers in two antennas for improving the performance of the receiver.

For linear multiuser detectors (decorrelating and MMSE), before MRC, the outputs of matched filters are multiplied to a matrix, as follow:

$$\mathbf{u_p} = \mathbf{t_p}\, \mathbf{R}_{\mathbf{lin}}^{-1} \qquad p = 1,2 \quad \text{(A.6)}$$

where for decorrelating detector $\mathbf{R}_{lin} = \mathbf{R}_{dec} = \mathbf{C}\,\mathbf{C}^T$ and for MMSE detector $\mathbf{R}_{lin} = \mathbf{R}_{MMSE} = \left(\mathbf{R}_{dec} + \sigma^2/A^2\right)$. σ^2 is the noise power.

For the case of non-linear or multi-stage multiuser detectors (PIC, SIC and DF), in each stage the MAI is computed and subtracted from the received signal.

In PIC detector, the MAI duo to each user in each antenna is computed as follow:

$$\mathbf{MAI}_{\mathbf{p,k}} = \sum_{\substack{m=1 \\ m \neq k}}^{K} (A_m \hat{b}_m h_{p,m}\, \mathrm{C}_{\mathrm{m}}) \;, \quad p = 1,2 \quad \text{(A.7)}$$

and then MRC and interference cancellation by using the outputs of two antennas is performed as follow:

$$\mathbf{z} = \mathbf{h}_1^* \cdot (\mathbf{r}_1 - \mathbf{MAI}_1) + \mathbf{h}_2^* \cdot (\mathbf{r}_2 - \mathbf{MAI}_2) \quad \text{(A.8)}$$

where $\mathbf{MAI}_p = [MAI_{p,1}, MAI_{p,2}, ..., MAI_{p,K}]$. Here, we use MRC between all stages of multiuser detectors and this can enhance the performance of receiver.

The only difference between SIC and PIC is that in the SIC the MAI estimation and subtraction are performed serially. Also, in this detector, at the first, the users are arranged by their received power.

In the DF detector, the estimation and subtraction of MAI are performed serially. If F is a lower triangular matrix that obtained from Cholesky decomposition of correlation matrix R, i.e., $\mathbf{R} = \mathbf{F}^{\mathbf{T}}\,\mathbf{F}$ then at the first stage of detector (user 1),

$$\mathbf{u_p} = \mathbf{t_p} (\mathbf{F}^T)^{-1} \quad , \quad p = 1,2 \quad \text{(A.9)}$$

Then, the MRC and estimation are performed as (A.4) and (A.5). For the other users from 2 to K, the MAI is computed as follow:

$$MAI_{p,k} = \sum_{m=1}^{k-1} A_m \hat{b}_m h_{p,m} F_{m,k} \quad , \quad p = 1,2 \quad \text{(A.10)}$$

and then this MAI is subtracted from the input as follow:

$$u_p^k = u_p^{k-1} - MAI_{p,k} \quad \text{(A.11)}$$

and MRC and ML detection in each stage (each user) is :

$$\hat{b}_k = sign\left(\Re e\left[h_{1,k}^* u_1^k + h_{2,k}^* u_2^k\right]\right) \quad \text{(A.12)}$$

As SIC, in this receiver, users are ranked in order of decreasing received power levels.

B: Two transmit antennas, One receive antenna

In this case, all users transmit their data on their antennas according to Alamouti's code [3]. The received signals in two consecutive time slots in matrix form are:

$$\begin{cases} \mathbf{r}_1 = (\ \mathbf{b}_1 \cdot \mathbf{A} \cdot \mathbf{h}_1 + \mathbf{b}_2 \cdot \mathbf{A} \cdot \mathbf{h}_2)\mathbf{C} + \mathbf{n}_1 \\ \mathbf{r}_2 = (-\mathbf{b}_2 \cdot \mathbf{A} \cdot \mathbf{h}_1 + \mathbf{b}_1 \cdot \mathbf{A} \cdot \mathbf{h}_2)\mathbf{C} + \mathbf{n}_2 \end{cases} \quad \text{(B.1)}$$

where $\mathbf{r}_m$ is the received signals at the mth time slot, $\mathbf{b}_i = [b_{1i}, b_{2i}, ..., b_{Ki}]\ i = 1,2$ are the even and odd data bits of the kth user and $\mathbf{h}_i = diag([h_{1i}, h_{2i}, ..., h_{Ki}])\ i = 1,2$. $h_{m,k}$ is the channel coefficient between the mth transmitting antenna of the kth user and the receiver antenna and $b_{k,m}$ the data bit of the kth user at the mth time slot.

In conventional receiver, the received signals in two time slots are passe through a bank of correlators as follow:

$$\mathbf{t}_m = \mathbf{r}_m\, \mathbf{C}^{\mathbf{T}} \quad , \quad m = 1,2 \quad \text{(B.2)}$$

Then the signals from two conventional receivers are added by space-time combining:

$$\begin{cases} \mathbf{u}_1 = \mathbf{h}_1^* \cdot \mathbf{t}_1 + \mathbf{h}_2 \cdot \mathbf{t}_2^* \\ \mathbf{u}_2 = \mathbf{h}_2^* \cdot \mathbf{t}_1 - \mathbf{h}_1 \cdot \mathbf{t}_2^* \end{cases} \quad \text{(B.3)}$$

The ML decision for input data can be performed as follow:

$$\hat{\mathbf{b}}_m = \mathrm{sign}(\Re e[\mathbf{u}_m]) \quad , \quad m = 1,2 \quad \text{(B.4)}$$

Similar to receive diversity, multiuser detection is applied to the output of conventional receivers in two time slots. For linear multiuser detectors, this is done as follow:

$$\mathbf{v}_m = \mathbf{t}_m\, \mathbf{R}_{lin}^{-1} \quad , \quad m = 1,2 \quad \text{(B.5)}$$

Then, space-time combining and ML detection are done same as (B.3) and (B.4).

For the case of non-linear PIC detector the MAI in each stage must be computed as follow:

$$\begin{cases} \mathbf{MAI}_{1,k} = \sum_{\substack{i=1 \\ i \neq k}}^{K} (\ A_i \hat{b}_{i,1} h_{i,1} + A_i \hat{b}_{i,2} h_{i,2}) \mathrm{C}_i \\ \mathbf{MAI}_{2,k} = \sum_{\substack{i=1 \\ i \neq k}}^{K} (-A_i \hat{b}_{i,2} h_{i,1} + A_i \hat{b}_{i,1} h_{i,2}) \mathrm{C}_i \end{cases} \quad \text{(B.6)}$$

These MAI are subtracted from the two received signals.

The SIC detector is also similar to PIC. The only difference is that the MAI estimation and cancellation is performed serially.

In DF detector, in the first stage, we compute the estimate of first (strongest) user as follow:

$$\mathbf{v}_m = \mathbf{t}_m (\mathbf{F})^{-1} \quad , \quad m = 1,2 \quad \text{(B.7)}$$

$$\begin{cases} u_{1,1} = h_{1,1}^* . v_{1,1} + h_{1,2} . v_{1,2}^* \\ u_{1,2} = h_{1,2}^* . v_{1,1} - h_{1,1} . v_{1,2}^* \end{cases} \quad \text{(B.8)}$$

$$\hat{b}_{1,m} = sign(\Re e[u_{1,m}]) \qquad m = 1,2 \quad \text{(B.9)}$$

For the other users from 2 to K, the following procedure is done:

$$v_{k,m} = v_{k,m} - MAI_{k,m} \quad , \quad m = 1,2 \quad \text{(B.10)}$$

$$\begin{cases} MAI_{k,1} = \sum_{m=1}^{k-1} (\ A_m \hat{b}_{m,1} h_{m,1} + A_m \hat{b}_{m,2} h_{m,2}) F_{m,k} \\ MAI_{k,2} = \sum_{m=1}^{k-1} (-A_m \hat{b}_{m,2} h_{m,1} + A_m \hat{b}_{m,1} h_{m,2}) F_{m,k} \end{cases} \quad \text{(B.11)}$$

$$v_{k,p} = v_{k,p} - MAI_{k,p} \qquad p=1,2 \tag{B.12}$$

$$\begin{cases} u_{k,1} = h^*_{k,1} . v_{k,1} + h_{k,2} . v^*_{k,2} \\ u_{k,2} = h^*_{k,2} . v_{k,1} - h_{k,1} . v^*_{k,2} \end{cases} \tag{B.13}$$

$$\hat{b}_{1,m} = sign(\Re e[u_{1,m}]) \qquad m=1,2 \tag{B.14}$$

C: One Transmit Antenna, Four Receive Antennas

This case is similar to case A (two receive antennas) but the only difference is that p=1,2,3,4.

D: Two Transmit Antennas, Two Receive Antennas

In this case, the signal is transmitted from two antennas and in two consecutive time slots and is also received by two antennas. In transmitter the user's data are transmitted via two antennas according to methods in equation (1). In the receiver in each antenna there are two signals in two time slots. Hence:

$$\begin{cases} \mathbf{r}_{11} = (\ \mathbf{b}_1 \cdot \mathbf{A} \cdot \mathbf{h}_{11} + \mathbf{b}_2 \cdot \mathbf{A} \cdot \mathbf{h}_{21}) \cdot \mathbf{C} + \mathbf{n}_{11} \\ \mathbf{r}_{12} = (-\mathbf{b}_2 \cdot \mathbf{A} \cdot \mathbf{h}_{11} + \mathbf{b}_1 \cdot \mathbf{A} \cdot \mathbf{h}_{21}) \cdot \mathbf{C} + \mathbf{n}_{12} \\ \mathbf{r}_{21} = (\ \mathbf{b}_1 \cdot \mathbf{A} \cdot \mathbf{h}_{12} + \mathbf{b}_2 \cdot \mathbf{A} \cdot \mathbf{h}_{22}) \cdot \mathbf{C} + \mathbf{n}_{21} \\ \mathbf{r}_{22} = (-\mathbf{b}_2 \cdot \mathbf{A} \cdot \mathbf{h}_{12} + \mathbf{b}_1 \cdot \mathbf{A} \cdot \mathbf{h}_{22}) \cdot \mathbf{C} + \mathbf{n}_{22} \end{cases} \tag{D.1}$$

Where $\mathbf{r}_{pm}$ the received signals at the *p*th receive antenna and the *m*th time slot and

$\mathbf{h}_{11} = diag([h_{11,1}, h_{11,2}, \ldots, h_{11,K}])$, $\mathbf{h}_{12} = diag([h_{12,1}, h_{12,2}, \ldots, h_{12,K}])$,
$\mathbf{h}_{21} = diag([h_{21,1}, h_{21,2}, \ldots, h_{21,K}])$, $\mathbf{h}_{22} = diag([h_{22,1}, h_{22,2}, \ldots, h_{22,K}])$.

$h_{mn,k}$ is the channel coefficient between the *m*th transmitting antenna of the *k*th user and the *n*th receiver antenna. The structure of receiver is depicted in fig.1.

In the receiver, first the signals of two antennas in two time slots are passed through a bank of correlators:

$$\mathbf{t}_{pm} = \mathbf{r}_{pm}\ \mathbf{C}^{\mathbf{T}} \quad , \qquad p,m=1,2 \tag{D.2}$$

Then the signals from two receivers are added by space-time decoding and MRC combining as follow:

$$\begin{cases} \mathbf{u}_1 = \mathbf{t}_{11} \cdot \mathbf{h}^*_{11} + \mathbf{t}^*_{12} \cdot \mathbf{h}_{21} + \mathbf{t}_{21} \cdot \mathbf{h}^*_{12} + \mathbf{t}^*_{22} \cdot \mathbf{h}_{22} \\ \mathbf{u}_2 = \mathbf{t}_{11} \cdot \mathbf{h}^*_{21} - \mathbf{t}^*_{12} \cdot \mathbf{h}_{11} + \mathbf{t}_{21} \cdot \mathbf{h}^*_{22} + \mathbf{t}^*_{22} \cdot \mathbf{h}_{12} \end{cases} \tag{D.3}$$

$$\hat{\mathbf{b}}_m = \mathrm{sign}(\Re e[\mathbf{u}_m]) \quad , \quad m=1,2 \tag{D.4}$$

For linear multiuser detectors (decorrelating and MMSE), the outputs of matched filter are multiplied to a matrix as follow:

$$\mathbf{v}_{pm} = \mathbf{t}_{pm}\ \mathbf{R}^{-1}_{lin} \quad , \quad p,m=1,2$$

Then, space-time combining and ML detection are done same as (D.3) and (D.4).

For the case of non-linear PIC detector the MAI in each stage must be computed as follow:

$$\begin{cases} \mathbf{MAI}_{11,k} = \sum_{\substack{i=1 \\ i\neq k}}^{K} (\ A_i \hat{b}_{i,1} h_{11,i} + A_i \hat{b}_{i,2} h_{21,i}) \mathbf{C}_i \\ \mathbf{MAI}_{12,k} = \sum_{\substack{i=1 \\ i\neq k}}^{K} (-A_i \hat{b}_{i,2} h_{11,i} + A_i \hat{b}_{i,1} h_{21,i}) \mathbf{C}_i \\ \mathbf{MAI}_{21,k} = \sum_{\substack{i=1 \\ i\neq k}}^{K} (\ A_i \hat{b}_{i,1} h_{12,i} + A_i \hat{b}_{i,2} h_{22,i}) \mathbf{C}_i \\ \mathbf{MAI}_{22,k} = \sum_{\substack{i=1 \\ i\neq k}}^{K} (-A_i \hat{b}_{i,2} h_{12,i} + A_i \hat{b}_{i,1} h_{22,i}) \mathbf{C}_i \end{cases} \tag{D.5}$$

In DF detector, in the first stage, we compute the estimate of first (strongest) user as follow:

$$\mathbf{v}_{pm} = \mathbf{t}_{pm} (\mathbf{F})^{-1} \quad , \quad p,m=1,2 \tag{D.6}$$

After the first stage:

$$\begin{cases} u_{1,1} = h^*_{11,1} v_{11,1} + h_{21,1} v^*_{12,1} + h^*_{12,1} v_{21,1} + h_{22,1} v^*_{22,1} \\ u_{2,1} = h^*_{21,1} v_{11,1} - h_{11,1} v^*_{12,1} + h^*_{22,1} v_{21,1} + h_{12,1} v^*_{22,1} \end{cases} \tag{D.7}$$

$$\hat{b}_{m,1} = sign(\Re e[u_{m,1}]) \quad , \quad m=1,2 \tag{D.8}$$

For the other users from 2 to K, the following process is repeated:

$$\begin{cases} \mathbf{MAI}_{11,k} = \sum_{i=1}^{k-1} (\ A_i \hat{b}_{1,i} h_{11,i} + A_i \hat{b}_{2,i} h_{21,i}) F_{ik} \\ \mathbf{MAI}_{12,k} = \sum_{i=1}^{k-1} (-A_i \hat{b}_{2,i} h_{11,i} + A_i \hat{b}_{1,i} h_{21,i}) F_{ik} \\ \mathbf{MAI}_{21,k} = \sum_{i=1}^{k-1} (\ A_i \hat{b}_{1,i} h_{12,i} + A_i \hat{b}_{2,i} h_{22,i}) F_{ik} \\ \mathbf{MAI}_{22,k} = \sum_{i=1}^{k-1} (-A_i \hat{b}_{2,i} h_{12,i} + A_i \hat{b}_{1,i} h_{22,i}) F_{ik} \end{cases} \tag{D.9}$$

$$v_{pm,k} = v_{pm,k} - MAI_{pm,k} \quad , \quad p,m=1,2 \tag{D.10}$$

$$\begin{cases} u_{1,k} = h^*_{11,k} v_{11,k} + h_{21,k} v^*_{12,k} + h^*_{12,k} v_{21,k} + h_{22,k} v^*_{22,k} \\ u_{2,k} = h^*_{21,k} v_{11,k} - h_{11,k} v^*_{12,k} + h^*_{22,k} v_{21,k} + h_{12,k} v^*_{22,k} \end{cases} \tag{D.11}$$

$$\hat{b}_{m,k} = sign(\Re e[u_{m,k}]) \quad , \quad m=1,2 \tag{D.12}$$

III. Simulation Results

In this section, we present simulation results, that demonstrate the performance of combined multiuser detection and space-time block coding in a CDMA system. In the following simulations, we have assumed a K=10-user synchronous CDMA system with random sequences of length N=31. We have assumed that users have equal powers but they are scales such that the total transmitted power is constant in all antenna configurations. Each user transmits its signal through a fading channel using one or two antenna. In the base station, the transmitted signals are received by one or two antenna. Then one of the above described methods is used in the receiver to improve the BER performance of the system. The fading channel is assumed to be a flat Rayleigh fading channel. This fading generated by a complex gaussian distributed random variable with variance .5 per real dimension.

Fig.2 demonstrates the performance of various MUD receivers in fading channels using two antennas in receiver (receive diversity). Fig.3 also demonstrates the same performances but in the case of Transmit diversity. It is seen that the performance of transmit diversity is similar to the receive diversity. In both cases, the PIC and DF detectors work better than others. But the performance of each detector is not exactly the same as its performance in the receive diversity.

Fig.4 demonstrates the performance of various MUD receivers using four antennas receive diversity and fig.5 demonstrates those performances with two antennas in both side of channel. Here again, the overall performance is the same. But the performance of various multiuser detectors is slightly different in two cases. In contrast to two antenna cases, four antenna diversity (receive or transmit/receive) has more diversity gain.

IV. Conclusion

In this paper, we investigated the combination of space-time block coding method with multiuser detection techniques in CDMA systems. We considered a multiple transmitter/receiver antennas CDMA system and derived the structure of receivers that using space-time block coding and various multiuser detection techniques. We also compared these MUD techniques in the same number of transmitter/receiver antennas by computer simulations. The results show that combining MUD and STC can improve the performance of CDMA systems respect to single antenna CDMA systems or STC decoding of multiple antenna CDMA systems.

References

[1] Verdu, Multiuser Detection, Cambridge, Cambridge University Press, 1998.

[2] S.Moshavi,"Multi-user detection for DS-CDMA communications", IEEE Communications Magazine, 1996

[3] S.M. Alamouti, "A Simple transmit diversity technique for wireless communications," IEEE Journal on Select Areas in Communications, vol. 16, pp. 1451-1458, 1998.

[4] V. Tarokh, N. Seshadri, and A.R. Calderbank, "Space-time codes for high data rates wireless communications: Performance criterion and code construction," IEEE Trans. Inform. Theory, vol. 44, 1998.

[5] V. Tarokh, H. Jafarkhani and A. R. Calderbank, "Space-time block codes from orthogonal designs," IEEE Trans. Inform. Theory, vol. 45, Jul. 1999.

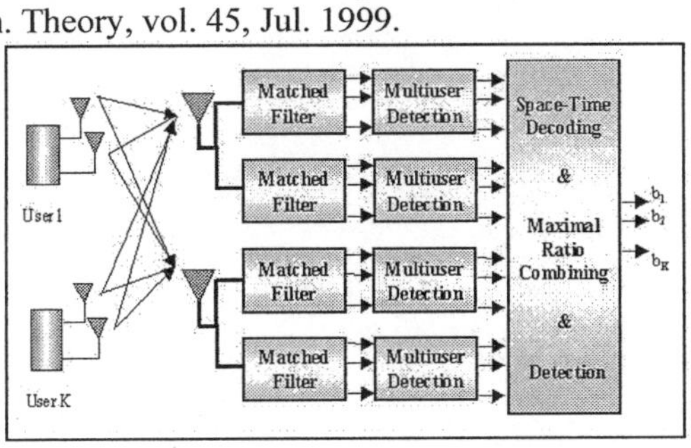

Fig. 1. The receiver structure

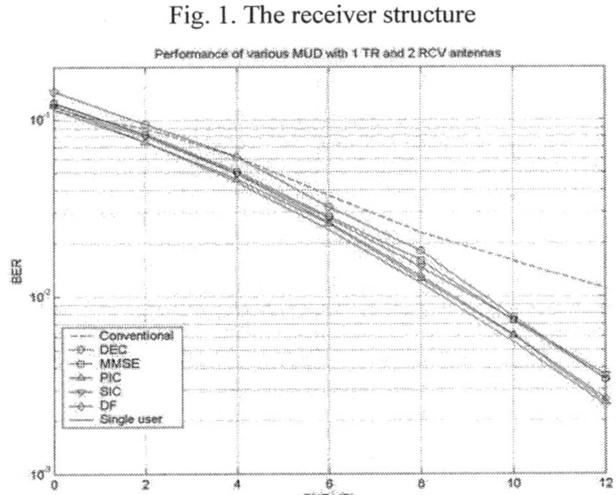

Fig. 2. Performance of various MUD using Receive Diversity (2 antennas)

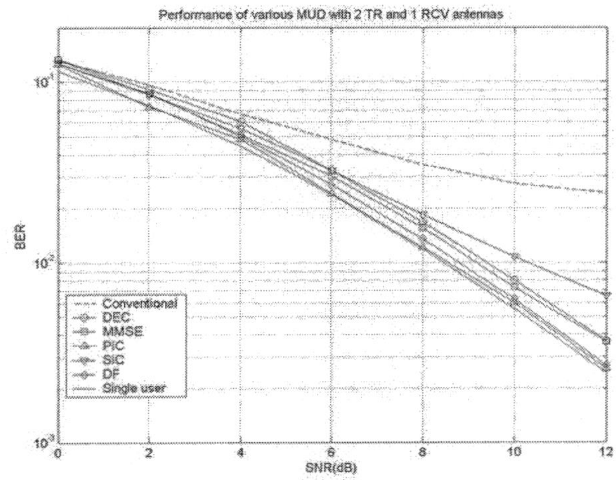

Fig. 3. Performance of various MUD using Transmit Diversity (2 antennas)

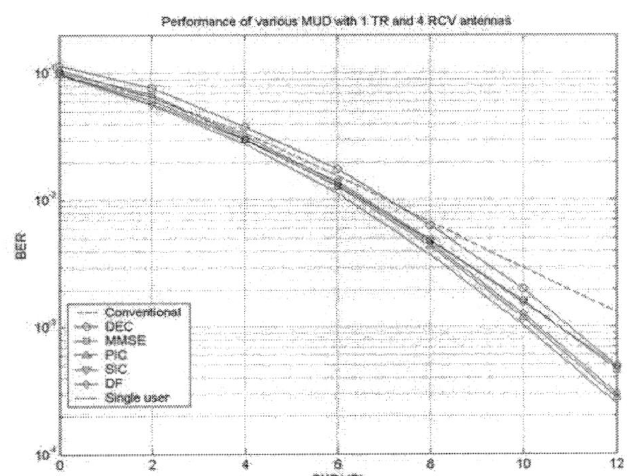

Fig. 4. Performance of various MUD using Receive Diversity (4 antennas)

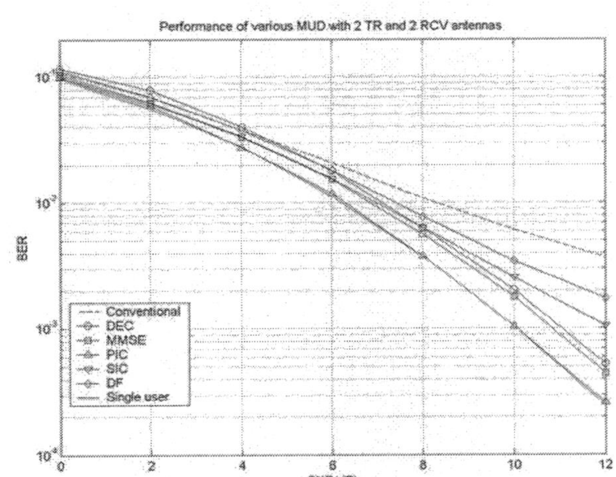

Fig. 5. Performance of various MUD using Transmit and Receive Diversity (2 x 2 antennas)

SESSION 7B:
SECURITY II

HETEROGENEOUS NETWORK ESTABLISHMENT ASSISTED BY CELLULAR OPERATORS

Marc Danzeisen[(1)(2)], Torsten Braun[(1)], Daniel Rodellar[(2)], Simon Winiker[(1)(2)]
[(1)] University of Bern, Computer Networks and Distributed Systems, Bern, Switzerland
[(2)] Swisscom AG, Innovations, CH-3050 Bern, Switzerland

Abstract: In this paper, we describe a novel architecture to enable a secure communication among mobile devices using different wireless technologies like wireless LAN, Bluetooth, cellular systems or even infrared. Making use of the combination of these technologies for the data transmission and for the signaling of the communication, we analyze several scenarios with increasing complexity. The complete picture appears in the last scenario where all technologies are involved and the network is composed of heterogeneous mobile nodes. The paper also presents a solution for the setup of a secured communication channel (i.e. a Virtual Private Network connection) between several heterogeneous mobile nodes controlled by the cellular network operator. The mobile nodes can be either cellular aware or non-cellular aware in this framework. We propose to setup the heterogeneous network communications via the cellular network using the cellular aware nodes.

Key Words: WPAN, Ad-hoc networks, Cellular, Wireless LAN, Bluetooth

1. INTRODUCTION

Today, mobile wireless networking combines data connectivity with user mobility. Using technologies like wireless LAN [1], Bluetooth [2], or GPRS [3] the users access their data and are mobile simultaneously. In this paper we are concentrating on networking applications that these technologies can enable. We are especially interested in the establishment of Virtual Private Networks (VPNs) among mobile nodes, where a given set of mobile devices use one or several of those communication technologies to establish a secured common networking area to share their data. The paper is organized as follows: First we discuss the current situation and the problems users have to face when they want to establish such secured communication channels among their mobile devices. In section 3 we present our vision of future heterogeneous networking. Then the central issue of this paper is exposed in section 4: we propose to use the cellular network to enable the ad-hoc networking setup. The discussion on different case studies in section 5 makes this proposal more concrete and allows better understanding and positioning of the whole framework for this study. In section 6, we conclude and give some remarks on future work.

2. SECURED MOBILE COMMUNICATIONS

When users want to establish a secured connection among their mobile nodes, several actions have to be taken. The users have to be aware of the capabilities of their devices and often also require certain knowledge about the communication technologies supported by these devices. For example, in the case of WLAN, different parameters like SSID and the used channel have to be set, before the devices become visible to each other. If the user takes this hurdle the proper identification and authentication of the correspondent node has to be managed. The mostly used mechanisms for that are based on shared secrets; the involved mobile nodes have to know a certain value to proof their authorization to establish the secured communication.

After establishing the connectivity among the authenticated nodes, the links have to be secured. Therefore, the nodes have to agree on security parameters. The smallest set of security credentials includes an encryption algorithm and an encryption key. For more sophisticated security measures further information has to be negotiated like authentication mechanism and key for authenticating the transmitted data, etc. In other words, there has to be a primary handshake procedure to establish a secure communication link between two or more mobile nodes. This exchange of information should also happen in a secure way. In most of today's situations, this happens either verbally or by writing down the credentials (Figure 1).

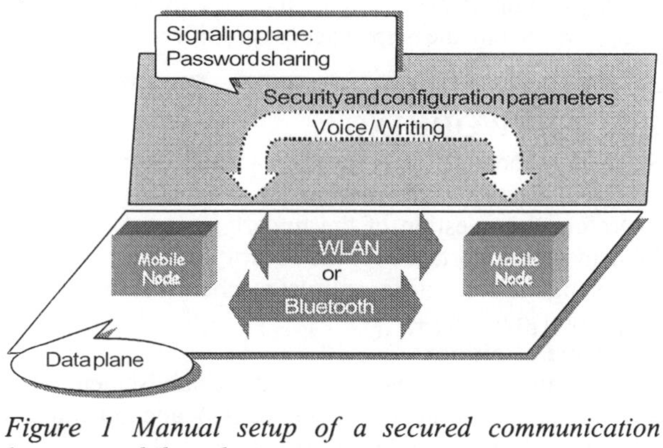

Figure 1 Manual setup of a secured communication between mobile nodes

In this paper we assume that these negotiations of parameters for communication and security form the main part of the signaling needed to build up a secured private connection between mobile nodes. In the case of a self-organizing technology like Bluetooth, where the nodes are able to discover their neighborhood and automatically establish a communication link, the signaling information is reduced to authentication process and the exchange of securityparameters.

3. A VISION OF CONVENIENT HETEROGENEOUS WIRELESS NETWORKING

We believe that users do not want to be aware of all these different configuration and security issues; end users just want to successfully transfer a file from one of their devices to a device of a certain person. Furthermore, we think that most users even do not care to which device that file is transferred to, as long as the right person finally has access to that file. This brings up two main requirements for our vision on heterogeneous wireless networking. First, mobile devices belonging to one person should also be addressable by an individual identifier of that person, for instance his mobile subscriber phone number (MSISDN). Often the sender does not know what type of device the receiver has. Hence, it is up to the receiver to decide which of his devices should be involved in the specific application. This abstraction of the destination node to one existing personal identifier helps to overcome the problem of temporary identifier of the different destination nodes. Most of the broadband wireless connection are set up on demand and have therefore only temporary valid identifiers like a leased IP address for instance. So if the sender can reach the receiver by a static, personal identifier, the receiver can trigger the appropriate device to get temporary connected to the broadband access and fulfill the requested transaction.

The second main requirement to make heterogeneous networking convenient is probably the automatic choice of the most suitable communication technology to cover the needs of a certain service at the lowest possible costs. This is of high importance when the involved nodes are interconnected by the help of access providers.

4. CELLULAR OPERATOR ENABLES AD-HOC NETWORKING

Since the authentication is based on a verification process, the involved entities and mechanisms have to be trusted. Therefore the question of the underlying trust model is of fundamental importance for the security model. In this paper we distinguish three different trust mechanisms following the suggestion given in [4]

Trust by Definition

This is the easiest way of trust establishment. Legal authorities, big companies (like banks) and operators are often trusted a priori, or by definition.

Trust based on Heuristics

This is the most natural trust establishment, which happens often in the social life. Entities make the level of trust they have in other entities dependent on their experience with the latter. As this model is based on experience, it can be a very time consuming and hard to deploy approach.

Trust by Delegation

In this model, trust is established by the help of entities, with which a trust relationship already exists, and which are guaranteeing the trustworthiness of a third entity.

Trust delegations are made through third parties, with which a trust relationship exists. This relation again is based on one of the three trust models, mentioned above. Therefore a trust relation chain has to begin somewhere, with either a trust based on heuristics or with a 'Trust by definition' relation. In pure ad-hoc networks [5] [9] where no fixed infrastructure is available, the only possibility to anchor a trust relation chain is the application of the 'Trust by Heuristics' model. As mentioned, this model can be hard to deploy. Only with the help of fixed infrastructure it is possible to apply the 'Trust by Definition' model, which is in our view a must for a platform offering commercial services. Consequently, there are several advantages in reusing the cellular infrastructure to manage heterogeneous wireless networking.

In the mobile as well as in the fixed communications there is a strong need of signaling channels to establish and control data communication. In the same way as it is done in the fixed networks we propose to separate the control plane from the actual data plane. We propose to use the cellular network to support an ad-hoc communication, from the setup phase until the end of the data transfer. The large coverage of the cellular system makes it very valuable as a signaling system for wireless broadband connections.

The cellular operator will play a role of signaling and configuration provider for heterogeneous ad-hoc networks. The mobile operator will locate the related peers, provide the security mechanisms including the authentication. Furthermore the operator's billing system can be reused to charge the establishment of the networking capabilities.

To establish a secured link or a Virtual Private Network (VPN) among mobile nodes that do not have any prior security relationship with each other, one has to be able to interact with the infrastructure of an operator. This is a very hard constraint since most of today's portable devices, like notebooks or PDAs, do not directly access the cellular networks. A device that does not explicitly support the cellular network is defined as "**non-cellular aware mobile node**" (NCAN), and the device that supports the cellular network is defined as "**cellular aware mobile node**" (CAN).

An interconnection of devices belonging to the same person is often treated as a Wireless Personal Area Network (WPAN). The IEEE is putting standardization effort for WPANs in their 802.15.x workgroup [7]. There is a special focus on the specification and standardization of new physical and link layers to provide short range radio with different bandwidth utilizations and power consumption. The most known outcome of these efforts up to date is the definition of the Bluetooth stack. This wireless communication stack includes mechanisms for device and service detection in the neighborhood of Bluetooth enabled nodes.

For the following analysis of the different use cases, we assume, that the devices belonging to the same WPAN are preconfigured to form one private network and therefore we suppose the NCAN to be able to access the cellular services offered by a CAN within the same WPAN.

5. ANALYSIS OF THE KEY SCENARIOS

In the following section we discuss the required actions to successfully establish a secured connection between mobile nodes. Therefore we focus on the information exchange needed to establish a VPN connection between the nodes. The use cases gradually increase in complexity while allowing more general applicability and less user interaction.

To overcome the limitations of a restricted reachability, we propose to use a signaling plane with high availability. In our view, a cellular network like GSM meets very well the requirements of such a signaling medium because of its high availability, because of the built-in security mechanisms and because of the provided always-on feature.

The ongoing trends towards trusted operators [8] are encouraging the reuse of the cellular network infrastructure to provide authentication and billing services. In the context of heterogeneous networking we focus on the authentication and location of mobile nodes and the secured delivery of the needed signaling information over the cellular network to enable the establishment of secured broadband channels. Figure 2 depicts the separation of the signaling and data channel in the case of short range and peer-to-peer broadband wireless technology like WLAN or Bluetooth.

In such a deployment, the mobile nodes are always connected to the cellular network and therefore always reachable for signaling messages.

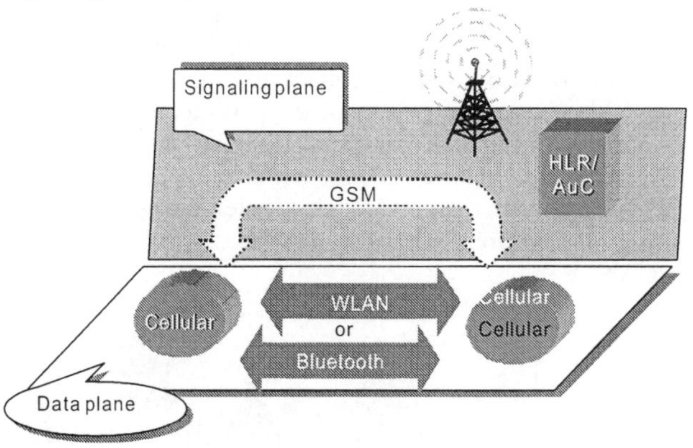

Figure 2 Signaling via cellular operator and data channel via wireless LAN or Bluetooth

There are different possibilities to use the cellular infrastructure to provide the authentication service and to transfer signaling information between the mobile nodes. Probably, the simplest way to do this would be the use of the standard cellular messaging subsystem service (SMS) to distribute information among the mobile nodes involved in the ad-hoc network. In this case, the operator's contribution would be limited to the secured distribution of the signaling information between the participants (the CANs are authenticated by the operator based on the user's PIN when accessing the cellular network). The main advantage of this scenario has its biggest effect when the mobile nodes do not have any prior security relation between each other. Then the trust chain is built up via the cellular operator having a security association with each of his customers [6].

For example one Mobile Node (MN1) trusts the operator and sends security information like a session key for a broadband link via SMS to a second Mobile Node (MN2). The only information MN1 needs is the mobile subscriber phone number (MSISDN) of the MN2 to successfully send the SMS. The SMS can be secured with the security mechanisms provided by the cellular network like the shared Key stored on the SIM card and on the authentication server (AuC).

In the scenario where the data channels used between the nodes are not peer-to-peer links, the interconnection takes place via broadband access providers. This might be the normal case when the nodes are not situated in the same location. For the setup of a secured broadband communication channel between the mobile nodes this situation is pretty similar to the former scenario. The main difference is the additionally needed information about the location and/or access used by the peers. This can include the IP addresses obtained by the broadband access provider, but also the characteristics of this data channel.

In the case, where one mobile node is not connected to a broadband access network, it is even imaginable that the initiating mobile node sends a request to its peer to setup broadband access and report the obtained IP address.

The main drawback of the architecture described in the previous sections is probably the limitation to cellular supporting nodes (CAN). The cellular awareness is required for the exchange on signaling information that is needed to setup the secured communication between mobile nodes. In other words, there has to be a mean to securely reach a given node without having more information than a cellular identifier of that node (i.e. the MSISDN).

In the prior deployments where only CANs were involved, every node had direct access to the authentication and message delivery service of the cellular operator by definition. In the mixed environment where also NCANs are available, they need some support to access these cellular services. To do so the CAN has to relay the signaling information between the cellular network and the NCANs. This requires a secured channel between the CANs and the NCANs. For simplicity reasons, we assume that there is a prior security relationship between the NCANs and their CAN. This is acceptable, especially in the case of wireless personal area networks (WPAN), where the CAN and the NCANs form a short range wireless private network and the security association between the devices is established by one administrative authority. In the following figure, this hierarchical approach to extend the signaling to NCANs is shown.

To clarify the steps involved in a setup of a secured VPN among NCANs, an example is presented based on the scenario depicted in Figure 3 with a closer look at the required infrastructure and the exchanged messages.

This use case involves two people having at least one NCAN and one CAN each. Namely the first person owns NCAN1 and CAN1 and the second NCAN2 and CAN2. The CANs could be standard mobile phones that support Bluetooth for local communication. The NCANs are supposed to support Bluetooth as well, and at least one broadband wireless communication technology.

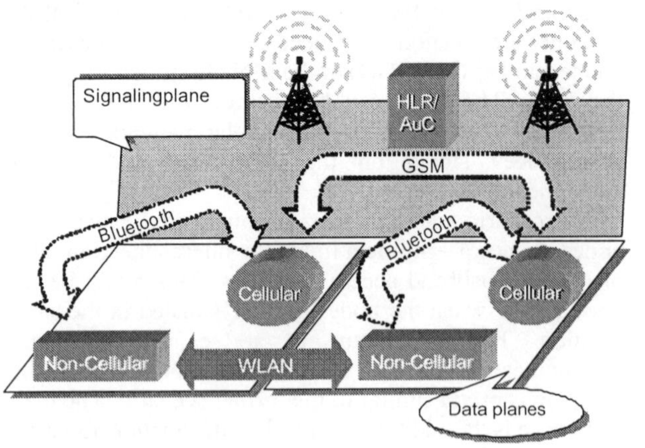

Figure 3 Data transfer between non-cellular aware nodes in the differentlocations scenario

The first person would like to send a file from his NCAN1 to the second person in a secured manner. Therefore he triggers his NCAN1 to acquire broadband wireless access. He does not have any further information than the MSISDN of the second person (MSISDN2); he sends a file send request (*fs_req*) from his NCAN1 via CAN1 to CAN2. This *fs_req* includes the IP configuration of the wireless broadband access of NCAN1, his MSISDN, the type and size of the file to transfer. The transfer of this request is made in two steps. The whole message exchange is depicted in Figure 4.

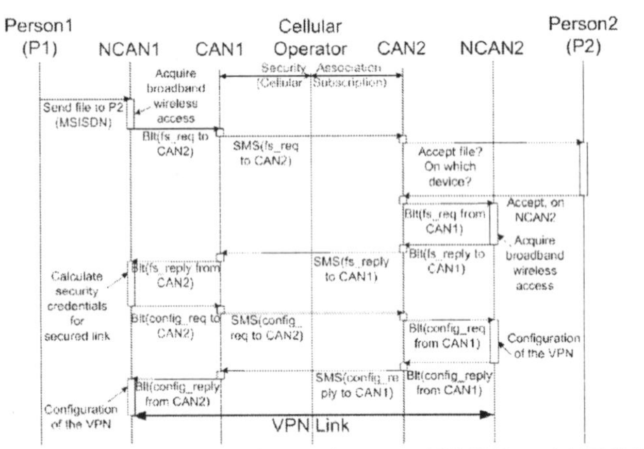

Figure 4 Message exchange between NCAN1 and NCAN2 using CAN1 and CAN2 for the VPN

First, NCAN1 sends the request through a prior secured Bluetooth channel, which is part of the WPAN setup, to CAN1. Then it is relayed via the cellular network to CAN2. As mentioned earlier, this relaying might happen by using the short messaging service subsystem (SMS) of the cellular network. Upon receiving the *fs_req* on CAN2, the second person can decide if he wants to accept or reject. In the case of acceptance he can also choose the device that should receive the file (in the case of having multiple NCANs in his WPAN). The mobile phone of the second person then forwards the *fs_req* to the selected device (NCAN2), which in turn tries to connect to a broadband wireless network and returns the resulting configuration to CAN2 (*fs_reply*). In the case of an IP based access network, for example, the device reports back the resulting IP configuration. This information is then included into *fs_reply* back to the originator of the file send request (NCAN1). Finally, NCAN1 has enough information collected to calculate security credentials to establish a VPN connection towards NCAN2. These credentials can again be delivered via the cellularnetwork.

HETEROGENEOUS TECHNOLOGIES NETWORKING ASSISTED BY CELLULAR NETWORKS

The previous use case becomes even more interesting, when the participating nodes have different communication interfaces. For example, several business people meet in a conference room and want to share some data among different mobile nodes. This circumstance seems to be fairly similar to the previous scenario except for the use of direct broadband connections between the participating nodes. These direct connections are built on a combination of different technologies like infrared (IR), Bluetooth, WLAN, and even wired links like Ethernet.

Because of the increasing variety and complexity of these communication technologies, it becomes more and more difficult for ordinary users to interconnect different devices. Our proposed solution can help to overcome this problem. If the signaling channel offered by the cellular network is also used to collect information about the available network interfaces of the participating devices it is possible to calculate a network design to interconnect all the different mobile nodes within an heterogeneous network. Figure 5 shows such a heterogeneous network using the different network interfaces available on the NCANs.

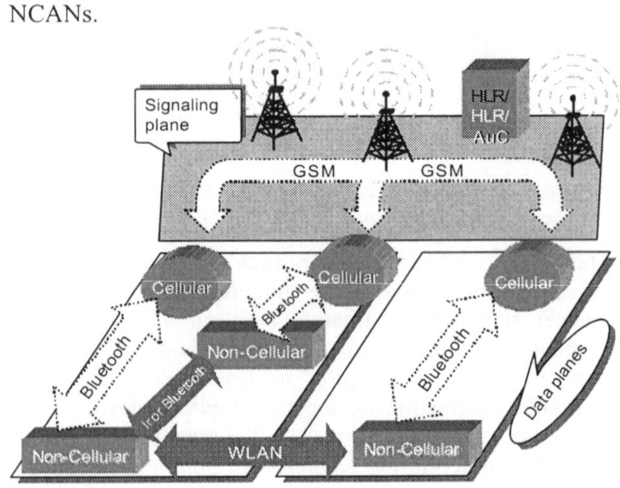

Figure 5 Heterogeneous technologies networking assisted bycellularnetworks

Note that devices having more than one network interface could be configured as gateways to interconnect devices that do not support the same communication technologies or that are not within the range of each other.

For all these different scenarios the degree of integration of the operator can vary. The operator's role can be adjusted from just an enabler, by the use of the SMS subsystem, up to a full integration, where the operator calculates the needed configuration and security information and distributes it to the appropriate participants.

6. CONCLUSION

In this paper we propose to reuse the cellular network for the signaling plane to enable the networking infrastructure of an heterogeneous networking environment. We first studied the secured mobile communications topic, with the signaling channel for authentication and for exchanging the security parameters. We have presented our vision of a convenient wireless networking where a group of devices with different communication technologies (like WLAN, Bluetooth, Infrared, GPRS, etc) can communicate together, without the explicit configuration by the end user. This vision leads to the statement of the cellular operator as the key player in the signaling for the ad-hoc networking configurations. This proposal is quite aggressive for a follower of pure ad-hoc networking, where by definition there should be no managed infrastructure. Our genuine approach is based on the user's trust in the cellular operator, and it only inserts the cellular operator for the signaling part, leaving the data transaction as the same procedure as in ad-hoc networking.

We have analyzed in detail and with examples different scenarios where a secure connection for communications is an established.

We have developed the first system architecture for a cellular assisted heterogeneous networking platform that will allow us to build a first implementation of the signaling messaging, and to deploy the first services onto this platform.

REFERENCES

[1] WLAN IEEE 802.11, http://grouper.ieee.org/groups/802/11/main.html

[2] Bluetooth SIG, http://www.bluetooth.org/

[3] General Packet Radio Service in GSM. Jian Cai and David J.Goodman. IEEE Communications Magazine, Oct.1997.

[4] Deployment of Jini Services in an insecure environment, Susanna Mäkinen, Roger Kehr, Roland Schmitz, Frederico Vieira, Tom Wall, Peter Windirsch, http://www.eurescom.de/public/projectresults/P1000-series/1005D6ti5.asp

[5] "Cooperation of Nodes: The CONFIDANT Approach - Abstract." Sonja Buchegger and Jean-Yves Le Boudec, Levente Buttyan and Jean-Pierre Hubaux (eds.), Report on a Working Session on Security in Wireless Ad Hoc Networks, ACM Mobile Computing and Communications Review (MC2R), Vol. 6., No. 4., 2002.

[6] Internet Security Association and Key Management Protocol (ISAKMP), http://www.ietf.org/rfc/rfc2408.txt

[7] IEEE WPAN Workgroup, http://grouper.ieee.org/groups/802/15/

[8] "Trusted Operator" Sten Lannerström, SmartTrust, Revision: B, September 19, 2002, http://www.smarttrust.com/trustedoperator/trustedoperator.pdf

[9] Mobile Ad-hoc networks (manet), http://www.ietf.org/html.charters/manet-charter.html

AN EFFICIENT AUTHENTICATION AND KEY-EXCHANGE PROTOCOL FOR LOW POWER CLIENTS

AMOL DABHOLKAR

School of Computer Engineering, Nanyang Technological University
Singapore, 639798, email: amold@pmail.ntu.edu.sg

KIN CHOONG YOW

Divison of Computer Communications, SCE, Nanyang Technological University
Singapore, 639798, email: askcyow@ntu.edu.sg

Wireless devices are characterized by low computational power and memory. Hence security protocols dealing with these devices have to be designed to give minimal computational and memory load. We present an efficient authentication and key exchange protocol for low-end wireless clients and high-end servers, which is overall nearly thrice as fast as comparable protocols. The basic idea of our protocol is to use symmetric key encryption in place of public key encryption wherever possible.

1. Introduction

Wireless environments are insecure by their very nature and wireless devices are inherently less secure than their wired counterparts [6, 7]. This is partly due to their low computational power, memory and limited bandwidth. Wireless transmission implies anyone with similar technology can intercept it. Thus wireless technology by its very nature violates fundamental security principles. So we are faced with two major mutually problematic issues in dealing with wireless devices and environments

1 A high level of security due to the insecure nature of the wireless media (which generally needs correspondingly more powerful computational devices)
2 Limited computational and memory power of wireless devices (which implies greater delays in computing protocols providing higher security).

The first phase of a typical secure transaction or communication is a secure key exchange where both parties decide on a common secret-key known only to them. The security of the rest of the transaction then depends on this first step and if there is a compromise at this phase, the rest of the communication maybe compromised.

This paper deals with the issue of designing a secure and efficient mutual authentication and key exchange protocol (MAKEP) for wireless devices. We consider the wireless devices as clients which interact with high-powered servers, typically desktop PCs. We will first look into the design strategies for a typical MAKEP and study two recently proposed protocols that deal with similar scenarios. We will then present our protocol and discus its design and security. Finally we will compare implementations of the protocols based on elliptic curve cryptography using PDA clients.

2. Design Strategies

In this section we will briefly analyze the designs of two MAKEPs designed for low-powered clients and discus their advantages and disadvantages. For a more in-depth analysis refer [9].

2.1. *Server-Specific MAKEP*

This protocol eliminates <u>all</u> public key cryptography on the client side [1]. Symmetric key cryptography is much faster than public key cryptography [7]. Hence the protocol executes much faster on the client side, than it would have if public key operations were involved.

Protocol Analysis

The protocol relies on server specific certificates of the form given by $Cert_A^B$ = <ID_A, $\{K_A\}_{PKB}$, $Sig_{TA}(ID_A, \{K_A\}_{PKB})$> which is a certificate from the low powered wireless client *Alice* to the high powered server *Bob*.

K_A is *Alice*'s long-lived symmetric secret key. PK_B is *Bob's* well known public key. $\{K_A\}_{PKB}$ means *Alice*'s secret key K_A is encrypted with *Bob*'s public key PK_B. Sig_{TA} is a signature from a trusted authority TA and ID_A is *Alice*'s identity.

We will briefly describe the idea behind the protocol. First *Alice* sends her certificate to *Bob*, along with an encrypted random number. *Bob* verifies the certificate and uses *Alice*'s secret key to encrypt another random number which is sent back to *Alice*. Now they both have each others random numbers and can calculate the same shared secret from them.

A major advantage here is that the client uses only symmetric-key cryptography with her secret key K_A. This increases the speed and efficiency as compared to other comparable MAKEPs that use public-key cryptography. The freshness of the session key is maintained by using the random numbers contributed by both, the client and server, which also increases the strength of the key.

However there are some disadvantages to this protocol

1. The protocol needs server specific certificates. For *n* servers it will need *n* distinct certificates. This creates scalability issues.
2. The client's long lived secret key K_A is known to the server. If the server is malicious it can create non existent sessions and there is no way for *Alice* to deny these occurred.
3. Even if the server is trustworthy, it has access to the client's secret key K_A. If the server's security is compromised in anyway it implies the client's security is also compromised since a hacker can get K_A from *Bob*'s memory.

These problems are more or less taken care of by the protocol we will look at in the next section.

2.2. *Client-Server MAKEP*

This protocol solves the scalability and other problems of the Server-Specific protocol by using public-key cryptography on the client side [2]. However the public-key computations are kept to a minimum and most of the costly operations are done on the server side.

We use the same notations to describe this protocol except that the certificates are no longer server-specific. *Alice*'s certificate will now be $Cert_A$= <ID_A, PK_A, $Sig_{TA}(ID_A, PK_A)$>, where PK_A is *Alice's* public key.

Protocol Analysis

The protocol has been shown to be secure using the Bellare and Rogaway model. Since it uses general certificates there is no scaling problem like the one in section 2.1. Most of the computationally expensive operations like Certificate verification are carried out on the server-side.

We observed that the protocol used the public/private keys of the server only for encrypting and decrypting data. It does not make use of the unique mathematical relation between the public and private key of the server for authentication. We also observed that a third step is required to authenticate the client *Alice* to the server *Bob*, by binding *Alice's* private key '*a*' in an equation involving some value from the first step. By this the protocol has effectively bound the first and the third step to each other, thus preventing interleaving and replay attacks. If the protocol had authenticated *Alice* to *Bob* in, say, the first step, the third step would not be required. Consequently the binding between these two steps would not be an issue and the computations on the client side for calculating extra binding values could be removed. We have used this idea in our protocol which we describe in the next section.

3. Our Protocol

Our design strategy is based on the following ideas that we have drawn from the protocols described in section 2

1. By authenticating the client to the server, in the first step, we can prevent computations that are required to thwart interleaving attacks.
2. By using public-key cryptography on the client side, we can prevent server-specific certificates which lead to scalability issues.
3. By replacing public-key *encryption* with symmetric key *encryption* on the client side, we can have significant increase in speed.

3.1. *Description*

The protocol is shown in Figure 1. Let '*g*' be the generator of a field Z_p. (a, g^a) and (y, g^y) are the private-key public-key pairs of *Alice* (the client) and *Bob* (the server) respectively; where $a, y \in Z_p$. *Alice* initially computes a random value 'r_A' which is its key-contribution. She then computes a Diffie-Hellman key using *Bob's* well known public key and her own secret key [4]. The DH Key will be $(g^y)^a = g^{(ya)}$.

Using point 3 from our design strategy described in the start of Section 3, we will now use symmetric key encryption using this DH Key instead of public key encryption as is the normal case. We convert this DH key into an AES key and encrypt Alice's random value 'r_A' with it.

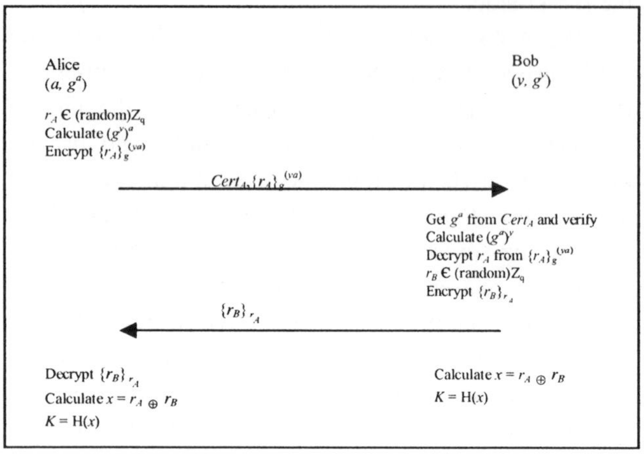

Figure 1 Our Protocol

Alice sends this encrypted value along with her public key certificate $Cert_A$, (which is similar to the certificate described in Section 2) to *Bob*. *Bob* will verify the certificate and get Alice's public key g^a. Now using the Diffie Hellman algorithm, he will calculate the DH Key using his private key and *Alice's* public key. The DH Key will be $(g^a)^y = g^{(ya)}$. This is the same as the one *Alice* had computed. Now *Bob* will convert this into an AES key and use it to decrypt the encrypted 'r_A'. In this step, if the decryption is successful, *Bob* has also effectively authenticated *Alice*, since only *Alice* can calculate the DH key, as it requires the use of her private key. Thus we have incorporated points 1 and 2 of our design strategy.

In the second step, *Bob* will encrypt his random value contribution r_B with an AES key derived from r_A and send it to *Alice*. *Alice* will decrypt this value using r_A, which she already knows. This effectively binds the two steps of the protocol together, preventing interleaving and replay attacks.

Now both sides have r_A and r_B and can calculate a new session key by taking the XOR of the values. By using random contributions from both the client and the server we have strengthened the security of the session key and also ensured forward secrecy i.e. security of the data encrypted in this session, even if some other session keys are compromised.

3.2. *ECC Version*

As we have mentioned in Section 1, using elliptic curve cryptography is a good idea for wireless devices, since we have to deal with much smaller key sizes while ensuring the same level of security as other contemporary systems like RSA. E.g. a 1024 bit RSA key is comparable to a 164 bit ECC key [5]. A 1024 bit RSA encryption is considered to be intractable by today's standards and if we have to maintain the same level of security for our wireless clients using ECC is much more sensible as we will be dealing with 164 bits only. We present an ECC version of our protocol in Figure 2.

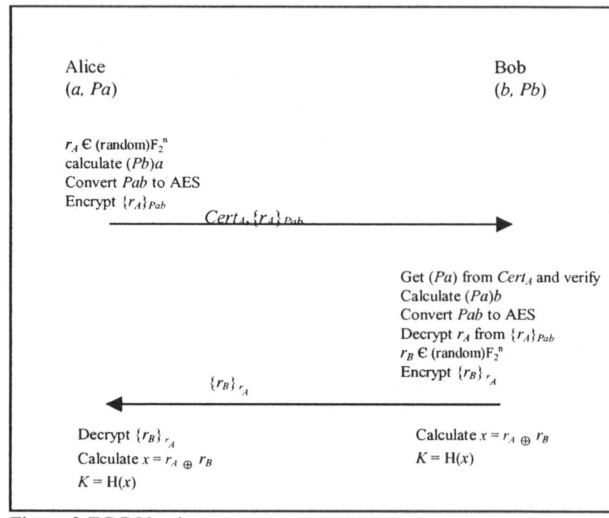

Figure 2 ECC Version

The private-key public-key pairs of *Alice* and *Bob* are (a,Pa) and (b,Pb) respectively, where P is a point on the elliptic curve E. P is called a base point and is publicly known. Pb refers to the point obtained by multiplying the point P with the value 'b' where b is an element of the field over which the curve E is defined. Point multiplication in ECC is analogous to exponentiation in RSA and is the costliest operation computationally. So in step 1, we calculate the DH key for *Alice* by multiplying her secret key 'a' and Bob's public key point Pb to give the point Pab. Similarly Bob obtains the same DH key by multiplying his private key 'b' with *Alice's* public key point Pa to obtain the point Pab. The rest of the protocol is exactly the same as described in Section 3.1.

4. Implementation

We have implemented the ECC version of our protocol and an ECC version of the Client-Server protocol described in Section 2. The ECC version of the Client-Server protocol has been done by replacing the exponentiation operations with point multiplication operations, and by replacing the field generator g with the base point P. By doing this we have a level playing field on which to compare both the protocols for timing efficiency using the same key sizes and the same ECC implementation.

Our implementation of ECC is based on the one described by Rosing [8]. We have used a 206 MHz 64 MB RAM Hp-Compaq iPAQ PDA as the client and a 1.7Hz P4 PC as the server.

4.1. *Results*

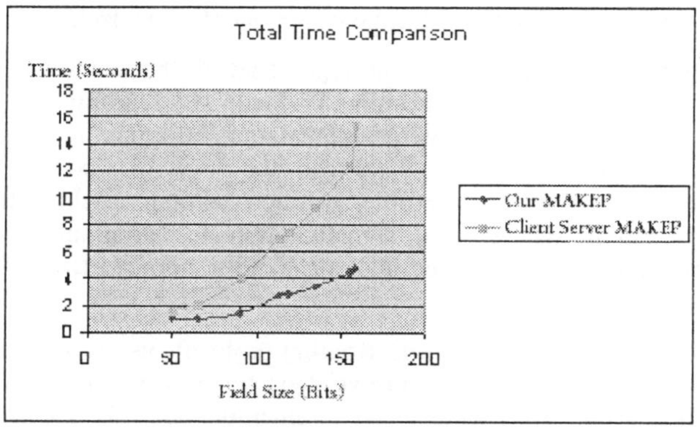

Figure 3 Total Time Comparison

As we can see from Figure 3, our protocol is almost 1.5 times faster at low field sizes of around 50 bits (1.04 seconds vs. 1.63 seconds) and more than 3 times faster for higher bits (at 158 bits our protocol runs at around 4.81 seconds as opposed to 15.3 seconds by the Client-Server MAKEP).

The biggest improvement we have obtained in our protocol is from using symmetric key encryption instead of public key encryption. We can see this from Table 1.

Table 1 Encryption Timings

Field Size (Bits)	**Our Protocol** Using AES Encryption in Step 1 (Milli Secs.)	**Client-Server MAKEP** Using El Gamal Encryption in Step 1 (Secs.)
158	13	5.6
155	12	5.09
134	12	3.5
119	12	2.7
113	13	2.03
90	12	1.04
65	12	0.4
50	12	0.2

As shown in Table 1, we have used ElGamal public key encryption for doing encryption in our ECC implementation of the Client-Server MAKEP which is quite fast as public-key encryptions go [8]. But it is still around 430 times slower than symmetric key encryption like AES at higher key sizes of around 158 bits. For a more detailed evaluation refer [9].

5. Conclusion

We have proposed a new and efficient authentication and key exchange protocol for low powered clients like PDAs in wireless environments. We have analyzed two recently proposed protocols and have implemented one of them using ECC for comparison with our protocol. We have seen that our protocol is around thrice as fast at key sizes of around 160 bits. Furthermore our protocol is scalable as it does not require server specific certificates. We use random contributions from both the client and the server to produce a session key that ensures forward secrecy.

References

1. Wong D, Chan A. *'Mutual Authentication and Key exchange for low power wireless communications'*, IEEE MILCOM 2001 Conference Proceedings.
2. Wong D, Chan A (2001). *'Efficient and Mutually Authenticated Key Exchange for Low Power Computing Devices'*; ASIACRYPT (2001).
3. Jakobsson M & Pointcheval D. *'Mutual Authentication for low-power mobile devices.'* Proceedings of Financial Cryptography, Soringer-Verlag (February 2001).
4. Diffie W. & Hellman M. *'New Directions in Cryptography'*, IEEE Transactions on Information Theory Pg. 644-654. (1976).
5. Lenstra A. & Verheul E. *'Selecting Cryptographic Key Sizes'*, Journal of Cryptology: the journal of the International Association for Cryptologic Research. Pg. 255-293. (1999).
6. Nichols R. & Lekkas P. '*Wireless Security: Models, Threats and Solutions*' McGraw-Hill. (2002)
7. Schneider B. *'Applied Cryptography'* Wiley. (1996)
8. Rosing M. '*Implementing Elliptic Curve Cryptography*' Manning. (1998).
9. Dabholkar A. & Yow K.C. *'Security in Wireless Networks'*. Technical Report: SCE-CEMNET-TR03-01. Nanyang Technological University (2003).

A Secure Location Service in Mobile Ad Hoc Networks

Zhi Zhou, Kin Choong Yow, and Sukumar Nandi
School of Computer Engineering, Nanyang Technological University, Singapore 639798
Email: zhouzhi@pmail.ntu.edu.sg, {askcyow, assukumar}@ntu.edu.sg

Abstract: A mobile ad hoc network consists of a collection of wireless mobile nodes that cooperate to forward packets via multi-hops without any fixed infrastructure. However, this type of network is particularly vulnerable due to its underlying characteristics. Routing issues has received most of the prior research on security while the focus is on some prominent reactive and proactive ad hoc routing protocols. In this paper, we examine attacks and threats against the location-based routing in ad hoc networks, and present a secure location service based on DLM against those discussed attacks. The proposed secure location service can also tolerate to some extent system failures and compromised nodes to achieve enhanced availability and reliability of the location service. Simulation results and analysis presented verified the performance of our design.

Key Words: ad hoc, security, location service, MVFLQ

1. INTRODUCTION

Mobile Ad Hoc Network (MANET) represents a new wireless networking paradigm, in which nodes have the symmetric capabilities and responsibilities, and cooperate to forward packets without any fixed infrastructure, such as access points or base stations. Routing is an important issue in MANET, which attracts lots of research. Most of them, however, consider a trusted environment which may not be true in the realistic setting or applications.

With the increasing attention to security, more and more valuable results are turning out, though published prior work is still quite limited on routing security issues in ad hoc networks so far. Sanzgiri et al. [8] examined some security flaws in AODV [6] and DSR [2], and proposed ARAN, a secure routing protocol for ad hoc networks based on AODV-like route discovery and maintenance. Hu et al. [1] proposed Ariadne, which is another secure on-demand routing protocol based on DSR and TESLA. TESLA [7] is the preferred authentication mechanism in Ariadne because of its computation efficiency compared with public-key encryption. But it requires clock synchronization which is not easy to realize in an ad hoc network. More works on routing security could be found in [5,10,11].

Security issues in AODV and DSR have been widely studied. However, no work about security issues in the location-based routing has been reported so far to the knowledge of the authors, though location-based routing shows its superiority of scalability.

In this paper, we first describe possible attacks and threats encountered in the location-based routing. Based on the analysis, we propose a secure location service that a location-based routing protocol should heavily depend upon.

2. LOCATION-BASED ROUTING SECURITY

In this section, we discuss general attacks and threats in the location-based routing when no security mechanism is applied. From an application level perspective, most of the attacks discussed next are Denial of Service (DoS) attacks.

Black hole Black hole is one type of attack that could easily happen in location-based routing protocols. By sending a reply to the location query with false location information, an attacker can make data packets from the requesting node destined for itself or some other position without reaching the destination. What's more, a malicious node can simply suppress all packets received, looking like the normal effect of a dead-end in topology.

Location Cache poisoning Routing protocols use some optimization techniques to improve their performance. However, optimization itself tends to be a vulnerable point for security. Location cache is usually adopted as an optimization technique in location-based routing protocols. In the case of GLS [3], the content of the location cache is learned from update packets or other packets containing location information. But an attacker can easily fool nodes along the forwarding path with a modified or fabricated packet including erroneous location information. Naturally, location cache poisoning may contribute to a black hole attack as well.

Threats in Geographic Forwarding Generally, geographic forwarding [4] allows routers to be nearly stateless, and requires the local propagation of the topology information: each node only needs to know its neighbors' positions. Usually, such a neighbor list with their positions can be realized by a simple beaconing algorithm that each node periodically sends a broadcast hello message with its location information to its neighbors. Consequently, the functioning of geographic forwarding will depend on the trustable neighborhood. An attacker, however, may not behave in an expected manner. It can continuously disseminate hello messages including false location information or even inform some fabricated neighbors. Specifically in greedy forwarding, a selfish node prefers to claim a very close position to its neighbor in order to avoid forwarding packets whereas a malicious node prefers to claim as far a location to its neighbor as possible in order to attract data traffic. In addition, enough number of deliberately-fabricated nodes around a victim node can lead to a denial of service attack which blocks or misdirects all outgoing packets of that node.

Threats in Location Service Location service aims at helping a sender determine the exact location of the receiver. Typically, it functions based on two parts: *location update* and *location query*, which will be naturally the target for an adversary to launch attacks. In location update, a malicious node can spoof as an updating node to send forged update packets to corresponding servers, or modify the location information in update packets while forwarding. Specifically in DLM [9], a sophisticated attacker may spoof as different updating nodes to send update packets to a specific partition, which put that partition under great pressure of storing false location information. In location query, query packets may be modified or directed to other place without reaching the correct location server. A malicious node can send false reply claiming itself to be a location server or simply modify the reply packet with a false location, which leads to a black hole attack. Moreover, an attacker can simply initiate a query for the location of any node it wants to know.

3. SECURE LOCATION SERVICE

As described in the last section, location-based routing is subjected to different kinds of attacks. To thwart these attacks, we propose in this section a secure location service based on DLM [9] with the full length address policy.

3.1 Assumptions

The proposed secure location service employs greedy forwarding to deliver packets and uses the digital signature scheme as the authentication mechanism due to its superiority in authentication and its popularity in real life security systems. We assume an existing key management service, which allows every legitimate node in the network to have a digital certificate typically signed by a trusted third party such as certification authority (CA). A digital certificate mainly contains the ID of the node, public key of the node, time of issuing, time of expiration, and the signature of the trusted party.

3.2 Neighbor Verification

A reliable greedy forwarding system requires a trustable neighbor list. Neighbor verification is proposed to reach this goal. To maintain a one-hop neighbor list, every node periodically broadcasts a hello message that includes its current location. We require that a hello message be attached with the sender's signature. For example, node A sends a hello message which should be constructed as follows: <[HELLO,id(A),location(A)]signature(A)>. By using the sender's public key available in its certificate any receiver can verify the authenticity of the sender as well as the message integrity. If the certificate is not available, the receiver can ask for it from the sender. Each time a node only selects the verified neighbor to forward packets. Thus, unauthorized nodes are isolated from the network.

If a node receives a hello message with a valid signature but invalid location information, i.e. a location out of the range of the receiver, it sends a two-hop broadcast alarm packet with its signature indicating that either the node or its certificate may be compromised and should never be trusted any more. Since the alerter itself may be compromised also, there leaves the decision for the alarm receiver to trust it or not in order to avoid a blackmail attack. Moreover, digital signature mechanism provides non-repudiation that is an effective deterrent against maliciously distributing rumors.

3.3 Authenticated Location Update

As discussed in Section 3, location update is vulnerable to masquerading attack and modification. So, it is desirable for a location server to verify the authenticity of the source as well as the integrity of the update packet. For this purpose, each node attaches its signature to every update packet before sending it out. An update packet includes the following fields: <[UPDATE, dest_grid, id, location, timestamp]signature> When the location server receives an update packet, it first checks if the source should choose it as a location server according to the server selection algorithm, and then verifies the validity of the signature. The location server will accept the update packet only if both verifications pass.

3.4 Authenticated Location Query

A node initiates a location query when it does not have the location of the node it wants to communicate with. However, query packet may not reach the location server properly and reply packet may be maliciously modified during propagation or comes from an attacker. So, end-to-end authentication for both query packet and reply packet is required for location query. Each query packet will carry the querying node's signature to enable the location server to authenticate the query request. Similarly, the corresponding reply packet includes the location server's signature and its grid information which ensures the querying node that the received location information is provided by the legitimate server in the expected grid. Figure 1 illustrates the procedure of one such location query.

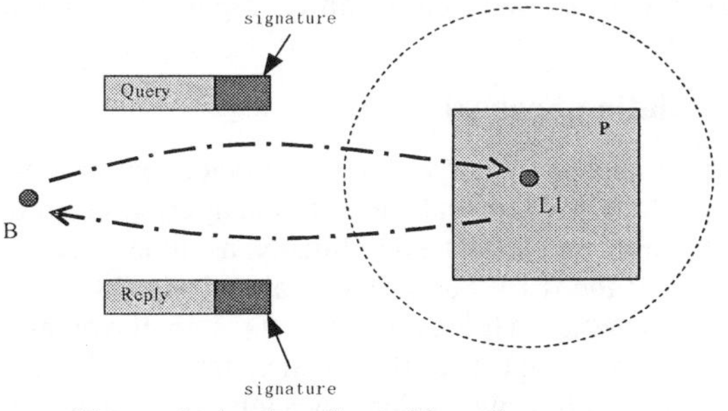

Figure 1. Authenticated location query.

3.5 Majority Voting for Location Query

In fact, failure of location query may result from many reasons. For example, the location server leaves a grid and other candidates have not been updated yet. Packet loss and even packet dropping from some misbehaving nodes may also lead to query failures. A much more severe case is that compromised location server may even provide false location information. In order to tolerate different kinds of failures and compromised nodes, we propose Majority Voting for Location Query (MVFLQ), which achieves a much more reliable query system. In MVFLQ, a node initiates *m* queries to *m* location servers instead of querying for only one server. The querying node discards all replies from non-server grids since they are definitely false replies that can be used to charge for their senders. Then, the querying node performs a majority vote on locations in the remaining replies to obtain the final location. In Figure 2, although the reply from location server *S2* is dropped due to the network congestion or malicious packet dropping, and the compromised server *Si* does not reply, MVFLQ still works well because other loyal location servers, such as *S1* and *Sm*, will provide correct locations. In a worse scenario, *Si* may send a false location to the querying node *B*. But it will not stop *B* obtaining the correct location because enough loyal servers provide the correct location and the false location will be ignored after applying a majority vote on location.

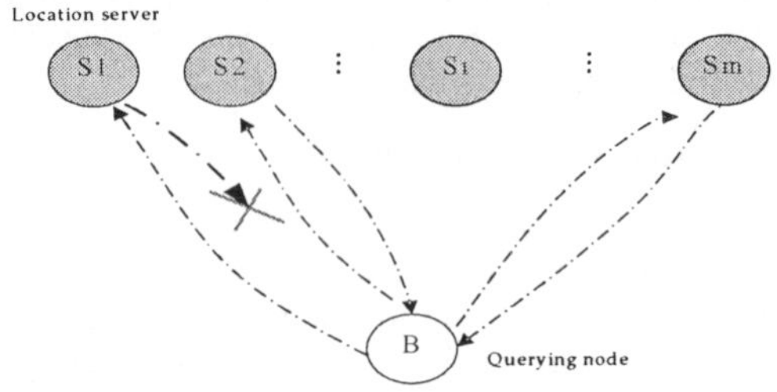

Figure 2. Majority voting for location query.

4. PERFORMANCE EVALUATION

To examine the performance of the proposed secure location service, we conducted a set of preliminary simulations. For the convenience of comparison with the original DLM, we name the proposed scheme SDLM. Due to the limited space, we present some representative results in the following discussion. We refer our technical report [12] as a full discussion.

4.1 Simulation Scenario

We choose *ns-2* simulator with CMU wireless extensions as our simulation environment since it is actively used in network research and offers high fidelity as it includes full simulation of the IEEE 802.11 physical and MAC layers. In our implementation, GPSR greedy mode is employed as the packet forwarding agent for the location service. The nodes are placed at uniformly random locations in a square universe 700 m × 700 m.

The smallest DLM grid has 175 meters on one side. The total number of location servers for a node is 4. Each node moves using a random waypoint model and initiates a location query to a random destination every 15 seconds over the course of the 300 seconds simulation. In the simulations, no data traffic was involved.

4.2 Simulation Results

Figure 3 shows the query success ratio as a function of the total number of nodes in the network. MVFLQ takes the advantage of redundant servers and effectively mitigates the effect of different failures discussed earlier, presenting a prominently higher success ratio than DLM. Furthermore, as indicated in the figure, there is a slight increasing of success ratio with the increase of nodes since the high node density might experience less "void regions" and dead-end topology.

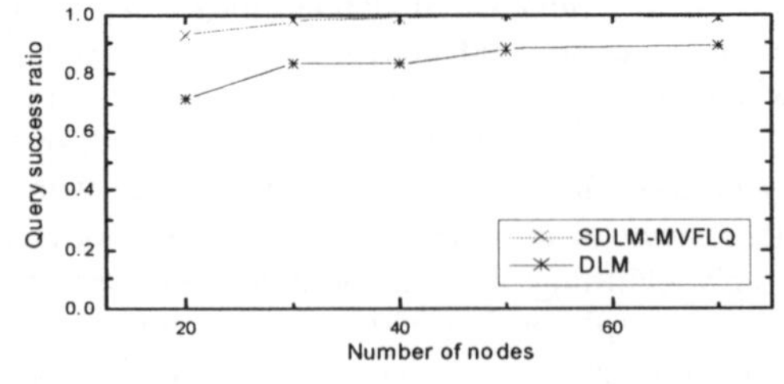

Figure 3. Query success ratio as a function of the number of nodes. Nodes move at speeds up to 10 m/s.

Figure 4 shows the normalized protocol packet overhead for each query success. In DLM, high mobility causes more frequent update traffic, and hence the packet overhead increases. It is always true that some overheads have to be paid for mitigating the mobility effect. An interesting result is that the packet overhead is almost the same in both schemes when node mobility is beyond 30 m/s. It indicates that in order to achieve the same success ratio, DLM needs to pay nearly the same packet overhead as SDLM with MVFLQ due to more query failures caused by high mobility.

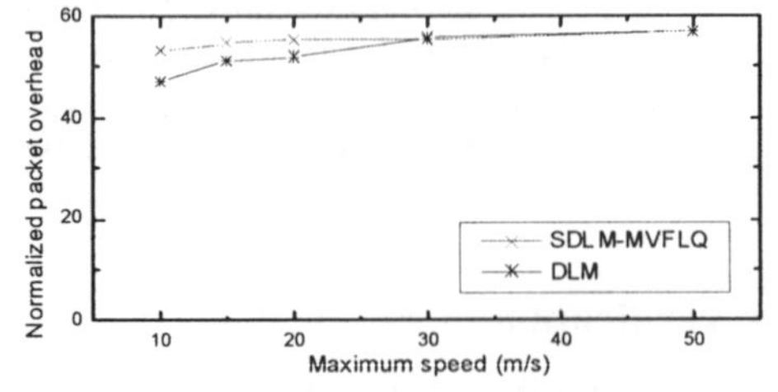

Figure 4. Normalized packet overhead as a function of the node mobility in the network of 50 nodes.

We conducted a particular set of simulations to examine the performance in the malicious network environment by introducing certain percent of misbehaving nodes. These nodes tend to randomly drop packets or not respond at all. Figure 5 presents a prominent sliding of the success ratio for both DLM and SDLM with MVFLQ. However, even when half of nodes in the network are misbehaving, SDLM with MVFLQ still has a satisfactory success ratio, around 80%. Instead, half of location queries will fail in DLM.

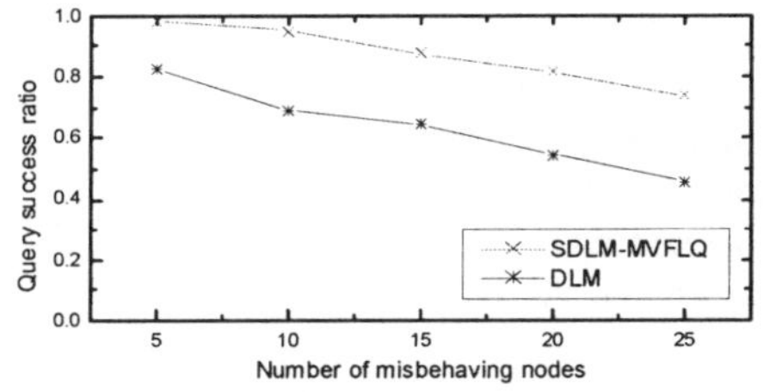

Figure 5. Query success ratio as a function of the number of misbehaving nodes in the network of 50 nodes. Nodes move at speeds up to 10 m/s.

4.3 Security Analysis

Assuming the well-keeping of personal private key and strong enough security of digital signature scheme, the proposed secure location service resists all attacks that we have presented in Section 2. The authenticated location update guarantees that the location server gets genuine location information of the node, which serves a valid location update for that node. Black hole attack using forged location is not possible since any modified or forged replies will be detected by the querying node through authentication. The proposed secure location service does not make use of location cache to avoid cache poisoning. Neighbor verification mechanism maintains a trustable neighbor list, which guarantees a reliable end-to-end packet delivery. With the proposed mechanism, unauthorized nodes are isolated from the network.

In some hostile environment, such as a battlefield, compromising a node is possible. Furthermore, some legitimate nodes could be selfish or even malicious due to some reasons and begin to drop or selectively drop packets. The proposed MVFLQ enhances the availability of location service based on the idea that it is much more difficult for an adversary to block all location queries or replies than to block one. In addition, MVFLQ provides a much higher probability of obtaining the correct location information under the same hostile situation because enough correct locations from loyal servers eliminate the effect of false ones while in single query the querying node may unconditionally accept the false location.

The effectiveness of MVFLQ is based on two assumptions:

A1. All loyal servers have the same location information L1 for a specific node.

A2. Loyal servers are more than compromised nodes that can cooperate to send same location L2, where $L1 \neq L2$.

Under a strong attacker model, certain number of compromised nodes can cooperate to send the same location in order to corrupt the majority voting. *A2* gives the idea about how many such strong attackers the system can tolerate. *A1* is equivalent to the condition that all loyal servers are correctly updated. Practically, it is not easy to strictly maintain *A1* and *A2*. The number of compromised nodes that can cooperate is difficult to predict, and some of loyal servers may not be correctly updated. However, MVFLQ still succeeds when enough loyal servers are correctly updated, and provides the acceptable reliability in a highly attacked environment.

5. CONCLUSION

In this paper, we have explored possible attacks and threats encountered in location-based routing and proposed a secure location service, which isolates unauthorized nodes and provides reliable packet forwarding. Considering the existence of system failures and compromised nodes, we proposed Majority Voting for Location Query to enhance the availability and reliability of obtaining a correct location. A secure location service is essential to secure a location-based routing protocol. The focus of this piece of work is to provide a secure location service. Security issues about data delivery will be reported separately. Moreover, secure location service is not limited to the use in location-based routing. It can be used in any application that relies heavily on the available and accurate location information.

REFERENCES

[1] Y. C. Hu, A. Perrig, and D. B. Johnson, "Ariadne: A Secure On-Demand Routing Protocol for Ad Hoc Networks", In Proceedings of the Eighth Annual International Conference on Mobile Computing and Networking (MobiCom 2002), September 2002.

[2] D. B. Johnson and D. A. Maltz, Dynamic Source Routing in Ad Hoc Wireless Networks, Imielinski and Korth, editors, Mobile Computing, vol. 353, Kluwer Academic Publishers, 1996.

[3] J. Li, J. Jannotti, D. De Couto, D. Karger, and R. Morris, "A scalable location service for geographic ad-hoc routing", In Proceedings of the 6th ACM International Conference on Mobile Computing and Networking (MobiCom '00), pp 120–130, 2000.

[4] M. Mauve, J. Widmer, and H. Hartenstein, A survey on position-based routing in mobile ad-hoc networks, IEEE Network, vol. 15, no. 6, page 30–39, 2001.

[5] P. Papadimitratos and Z. J. Haas, "Secure Routing for Mobile Ad Hoc Networks", In Proceedings of the SCS Communication Networks and Distributed Systems Modeling and Simulation onference (CNDS 2002), 2002.

[6] C.E. Perkins and E.M. Royer, "Ad-hoc on-demand distance vector routing", In Proceedings of the 2nd IEEE Workshop on Mobile Computing Systems and Applications, Feb 1999.

[7] A. Perrig, R. Canetti, D. Tygar, and D. Song, The TESLA Broadcast Authentication Protocol, RSA CryptoBytes, vol. 5, no. 2, page 2–13, 2002.

[8] K. Sanzgiri, B. Dahill, B. N. Levine, C. Shields, and E. M. Belding-Royer, "A Secure Routing Protocol for Ad Hoc Networks"; In Proceedings of IEEE International Conference on Network Protocols(ICNP), November 2002.

[9] Y. Xue, B. Li, and K. Nahrstedt. "A scalable location management scheme in mobile ad-hoc networks", In 26th Annual IEEE Conference on Local Computer Networks (LCN'01), 2001.

[10] S. Yi, P. Naldurg, and R. Kravets, "Security-Aware Ad Hoc Routing for Wireless Networks", In Proceedings of the ACM Symposium on Mobile Ad Hoc Networking and Computing (MobiHOC 2001), Long Beach, CA, 2001.

[11] M. G. Zapata and N. Asokan, "Securing Ad Hoc Routing Protocol", In Proceedings of the 2002 ACM Workshop on Wireless Security (WiSe 2002), page 1–10, September 2002.

[12] Z. Zhou, K. C. Yow, and S. Nandi, "Securing Location-based Routing in Mobile Ad Hoc Networks", Technical Report SCE-CEMNET-TR03-02, July, 2003.

Certification and Authentication Services for Securing MANET Routing Protocols

Ricardo Staciarini Puttini[1], Ludovic Me[2], Rafael Timóteo de Sousa[1]
[1]Department of Electrical Engineering - University of Brasília
Campus Universitário Darcy Ribeiro - Asa Norte - 70910-900 - Brasília - DF - Brazil
[2]École Supérieure d'Électricité - Supélec - Campus de Rennes
Avenue de la Boulaie - BP 81127 - 35511 Cesson-Sévigné Cedex - France
puttini@unb.br, desousa@unb.br, ludovic.me@supelec.fr

Abstract: In this paper, we describe a new authentication service for securing mobile ad hoc network (MANET) routing protocols. A MANET authentication extension (MAE) is appended to each routing protocol message. Our design includes a self-organized certification service, adapted from [6] with the following improvements: appropriated initialization, correcting vulnerabilities in the original design; local certificate management; and support to multiple Certification Authorities. Our design is policy-configurable in the sense that certification and authentication services may be adjusted according to specific security requirements. We discuss the use of MAE and certification service for securing DSR, AODV, OLSR and TBRPF MANET routing protocols. We have implemented the MAE securing OLSR and we evaluate computational and network requirements for this case.

....

Key Words: MANET, routing protocol security, authentication, certification services.

1. INTRODUCTION

In this paper, we tackle the problem of securing ad hoc network routing protocols. An authentication service is considered as a basic preventive protection for the routing protocol. Our proposal consists in the definition of a MANET authentication extension (MAE) to be appended to each routing protocol message or packet, providing the authentication service. The salient features of our design, differentiating the work presented in this paper from prior published results [1,2,3,4,5] are: (1) protocol-independent authentication extension design, (2) policy configurable authentication service definition and, (3) design of a certification service adapted to the MANET context.

Concerning the protocol-independent design, we aim to provide a security solution that could be easily adapted and extended to provide authentication service to different MANET routing protocols. This is different from previous work, in the sense that, instead of focusing on a specific protocol and even specifying changes to the original protocol, we design a general authentication protocol that is accurate in form and procedures, while guarding direct extensibility (by simple instantiation of more specialized authentication objects) and semantic flexibility, allowing adaptation with the protocol-specific needs.

Regarding the configuration of the authentication service, our objective is to allow the service to be configured according to the particular requirements of a specific security policy. Our design is different from most previous work by giving post-design configuration options that could map different security requirements. This feature is particularly related to the diversity on the MANET application, which results in different scenarios with different security requirements.

The design of a simple authentication extension mechanism is not enough to provide a complete routing protocol security solution, as such design usually rely on some key distribution scheme. Moreover, in the MANET context, this scheme must be provided in a completely distributed manner and should be, as much as possible, self-organized[1]. The design of a conjugate key distribution scheme is an important part of the overall security solution, although in previous work this is presumed [4] or delegated to classical solutions [3], which do not meet the specific MANETs requirements.

Therefore, we have designed a certification service that has distribution and self-organization features. The rationale of our proposal is adapted from [6], with the following improvements: (1) appropriated initialization conditions (correcting some flaws in [6], which is vulnerable to sybil attacks [7]); (2) certificate distribution and management services; and (3) support to multiple Certification Authorities (CA) and trust path construction. This last feature is mandatory if secure communications are needed between nodes originating from different MANETs (merging of two MANETs).

---

[1] Indeed, this is the reason for not using IPSec directly, as the security model adopt in the IPSec architecture is based on peer-to-peer security associations. It would be impractical to have a security association between each pair of Manet nodes. Also, as discussed further ahead in the paper, routing protocol authentication isn't restricted to a single authentication object, as it is the case for the the IPSec Authentication Header (AH).

In this paper, We show how to adopt the proposed solution for securing the four MANET routing protocol standard candidates: DSR [8], AODV [9], OLSR [10] and TBRPF[11]. To validate our design, we have implemented both authentication and certification services for securing the OLSR routing protocol; we present here the implementation and some experimental results.

2. MANET CERTIFICATION SERVICE

The design of self-organized certification services in MANET has been discussed in a few recent papers [6,13,14,15]. The approaches in [6,13,14] are all based on a distributed certification authority (DCA) trust model. The CA private key (K_{CA}) is used to sign certificates for the nodes in the MANET. A certificate signed with K_{CA} can be readily verified with the well-known system public key. The distribution of the CA capabilities is achieved by sharing K_{CA} among network nodes by means of threshold cryptography [16]. Each node holds a private-key-share (SK_{CA}) and any K (a system wide constant, usually related to the average number of neighbors) of such SK_{CA} holders can collectively function as a CA. The K_{CA}, however, is not recoverable by any node. Counter-certificate issuing does certificate revocation, which must also be signed with K_{CA}. All nodes keep a local Certificate Revocation List (CRL) with all non-expired counter-certificates. Our proposal is based on [6,13], with the following improvements:

(1) Certification issuing and renewal in [6,13] follows some pre-established rule. In our design, we propose a set of policy configurable options for certificate issuing and renewal, and private-key-share issuing and updating. This policy may be used to enforce that new certificate and SK_{CA} issuing in the DCA follow strictly the same identity verification that an equivalent centralized CA, avoiding multiple identity tampering attacks (Sybil attacks) [7].

(2) Approaches described in [6,13] do not specify how certificates are acquired in each node. Also, the local CRL is progressively built by accumulating counter-certificates that are flooded when they are signed, but the synchronization of arriving nodes is left unspecified. We propose the localized maintenance of a certificate cache and CRL.

(3) Designs from [6,13] are based on a single DCA. We extend the model to support multiple DCA and trust path constructions.

2.1. Certificate Issuing and Renewal

A node without certificate or needing to renew his certificate must ask to other nodes in the MANET for a certificate issuing. The authentication policy must specify how the nodes receiving a certificate request serve such request. Different policies can be specified for certificate issuing and renewal, which includes: (1) serve according to some policy-specified identity check, such as [12]; (2) serve manually (the user is prompted to decide if certificate request can be served); (3) deny (certificate request are rejected and an error message is returned to requester); (4) reject (certificate request is silently discarded); and (5) other (user defined).

2.2. Local Certificate Cache and CRL

In our design, building, synchronization and management of local certificate databases are setup depending on the type of the routing protocol being used. Options for certificate management includes: choice for proactive or on-demand construction of local certificate cache and CRL; time for keeping an unused certificate in the cache; and maximum size for the certificate cache.

Proactive certificate distribution consists in requiring the sender to attach his certificate along with the messages being transmitted. If certificates are transmitted with all messages, we have maximum certificate availability. However, it also implies considerable network and processing overhead, as a certificate is transmitted and must be validated for each message. Some optimization can be introduced by caching recently validated certificates and requiring the sender to attach his certificate only in some of the messages he originates.

Alternatively, nodes can cache recently validated certificates and explicitly ask other nodes for certificates that are needed but not available in the cache. In this reactive certificate distribution, sender does not attach his certificate in the message but only provides it when explicitly queried. The certificate owner is not the only one who can serve a certificate query, as other nodes having the needed certificate in their cache can also answer the query. Finally, a hybrid proactive-reactive policy is also possible.

A synchronization strategy is needed to allow roaming/arriving nodes to have their local CRL updated. Reactive and proactive approaches are possible. In the reactive approach, a roaming/arriving node explicitly query his neighbors for a copy of the up-to-date CRL, while in the proactive approach nodes identifying messages from some node with a revoked certificate should broadcast the counter-certificate in its neighborhood.

2.3. Usage with Multiple DCA

Nodes trusting and being trusted by more than one CA should apply for certificate and private-key-shares from each CA. A cross trust relation between the nodes trusting in two different CAs (e.g. CA1 and CA2) can be established if there are, at least, k nodes trusting both CAs. This is done by generating a certificate for CA1 and a certificate for CA2 that are signed by k private-key-shares from CA2 and CA1, respectively. The CAs certificates generated in this processes should be flooded in the network, finalizing the cross trust relation establishing. Establishment of cross trust relation is also controlled by the certification policy.

3. MANET AUTHENTICATION SERVICE

The authentication service considered in our model is provided by a MANET authentication extension (MAE), which is appended to each routing protocol message or packet. This MAE contains all the authentication information required to correctly assure authenticity and integrity to the message or packet being protected. Our objective is to design such extension in a flexible and adaptable way, so that it could be used to secure different MANET routing protocols. We leave the routing protocol unchanged, differently from [2,3,4].

3.1. Authentication Objects

MAE is composed by authentication objects. At least one (mandatory) authentication object must be present in the MAE and contains a message authentication code (MAC), which is computed as a hash-function applied to the data being authenticated keyed with a secret shared team key, or a digital signature (DS). MAC/DS authenticates all the non-mutable in the packet/message. Additional authentication objects are used to provide optional services. Currently defined options are signer certificate (certificate distribution), hash chains (mutable fields protection) and sequence number (reply protection).

3.2. Mutable Fields

In DSR and AODV there are messages that progressively change while they are forwarded by intermediate nodes in the path between message source and destination [8,9]. While it is desirable that this mutable information should also be protected, such protection usually implies in increasing the authentication information size each time the message is forwarded. This is not surprising as the information contained in the message is due to all nodes that have previously forwarded it and, each of them should be authenticated. Methods to protect typical mutable information (e.g. hop-count, IP address routing trace, etc.) have been proposed [2,3] and are adapted here.

3.3. MAE for DSR, AODV, OLSR and TBRPF

OLSR and TBRPF are proactive link-state routing protocols whose messages don't have mutable fields in the routing messages that are actually used by the routing protocol algorithm. MAE for securing OLSR and TBRPF is simply built with a single MAC or DS.

AODV have mutable fields in *route request* (RREQ) and *route reply* (RREP). These fields are hop-count metrics that changes every time the packet is forwarded by nodes between the message source and destination. A hash chain object (HC) [4] is included and updated each time these fields change. Such protection avoids that an attacker could decrement the hop count. *Route error* (RERR) messages are signed only by the node forwarding them. Route reply acknowledgments (RREP-ACK) have no mutable fields and are only authenticated by the message originator.

Securing DSR is quite more complex, although limited security can be achieved by combining the mandatory authentication object with a hash chain object implementing a per-hop hashing schema in RREQ messages. This will avoid an attacker from faking of the initiator node and from removing correct IP address in the route list [2]. RREP messages could be simply signed by the target of the route discovery (e.g. the node originating the RREP message).

Table 1 illustrates the main features of each MANET routing protocol and MAE requirements for each of them.

Table 1 – MAE for MANET routing protocols

Routing Protocol	Routing Discovery	Routing Algorithm	Relevant Messages	Auth. Objects
DSR[8]	on-demand	source-routing	RREQ	DS+HC
			RREP	DS
AODV[9]	on-demand	distance-vector	RREQ	DS+HC
			RREP	DS+HC
			RERR	DS
			RREP-ACK	DS
OLSR[10]	proactive	link-state	Hello, TC	DS
TBRPF[11]	proactive	link-state	Hello, TU	DS

3.4. Authentication and Certification Policy Issues

MAE and certification service specifications are provided with a set of configurable options that should be used to define a security context. Figure 1 below summarizes these authentication and certification policy parameters.

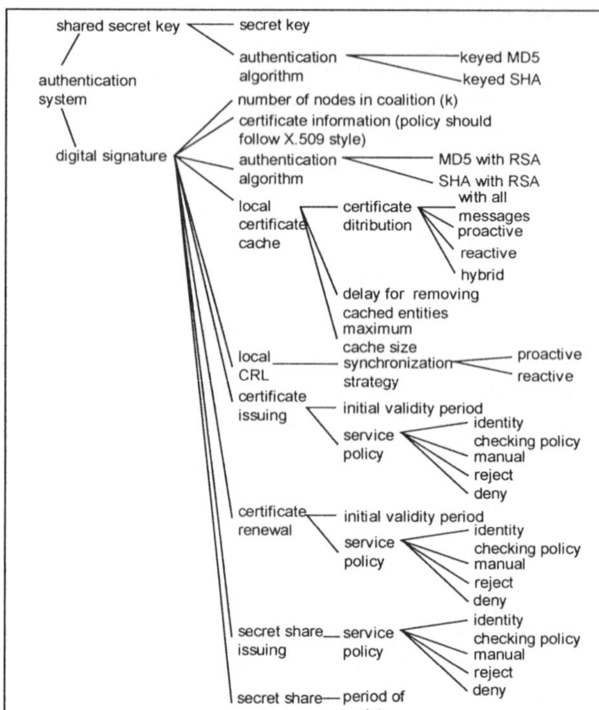

Figure 1 – Authentication Policy Parameters

4. EXPERIMENTS AND RESULTS

MAE authentication services where implemented for the available implementation of OLSR (v.3). The *openssl* library was used for cryptography routines. Tests were executed with both shared secret key and digital signatures authentication systems. In the second case, a "proactive" certificate distribution and CRL synchronization were configured. A 512 bits RSA key was defined. Both certificate and MAE digital signatures were using an MD5 with RSA signature algorithm. Certificates were issued in the X.509 format. Minimal information was included in the certificate, aiming to reduce its size. Timeout for excluding a valid certificate cache from the local certificate database was set to 10 times the HELLO cycle. All certificates were previously issued out-of-band. Issuing policy was set to "reject" in the case of new certificates and to "always" in the case of certificate renewal. Number of nodes in certification service coalition was fixed to K = 3.

Computational and Network Performance Considerations: Overhead of the proposed protocols has been preliminarily evaluated, through our experiments with the OLSR implementation. Considering the network overhead, MAE transmitted (without certificates) have a fixed size of 72 bytes. Average size of OLSR messages depends on the network size and density. The average size of a HELLO message is 62 bytes, each node having an average neighborhood of 12 nodes. Computational overhead were indirectly evaluate. The most expensive operation in MAE generation and verification is RSA signature and verification, respectively. Time for executing a RSA signature generation and verification (512-bits key) were averaged in a Pentium III (900MHz, 128Mbytes de RAM, running Linux with kernel 2.4.7) to 9ms and 2.6ms, respectively. Storage requirements of our proposal are mainly related to certificate cache storage (as CRL can not oversize K, a small constant). If all certificates in the 100 nodes MANET being simulated were locally cached, a 26kbyte cache is due, which is perfectly reasonable.

5. CONCLUSION

We have designed certification and authentication services to secure MANET routing protocols. Our design is protocol-independent and post-design options allow accomplishing of different security contexts/policies. An implementation has been realized for the OLSR protocol to demonstrate the feasibility and provide preliminary evaluation of computational and network aspects of the designed protocols. The important key management problem is treated as part of security solution and not simply presumed as in many of the prior published works.

REFERENCES

[1] Y.C. Hu, D. Johnson, and A. Perrig, "SEAD: Secure eficient distance vector routing for mobile wireless ad hoc networks," Proc. 4th IEEE Workshop on Mobile Computing Systems and Applications (WMCSA '02), pp. 3-13, 2002.

[2] Y.C. Hu, A. Perrig, and D. Johnson, "Ariadne: A secure On-demand routing protocol for ad hoc networks," Proc. of ACM Int. Conf. on Mobile Computing and Networking (MobiCom 2002), pp. 12-23, 2002.

[3] B. Dahill, K. Sanzgiri, B. N. Levine, C. Shields and E. Royer, "A secure routing protocol for ad hoc networks," Proc. 10th IEEE International Conference on Network Protocols (ICNP 2002), pp. 78-87, 2002.

[4] M. Guerrero and N. Asokan, "Securing Ad Hoc Routing Protocols, " Proc. 2002 ACM Workshop on Wireless Security (WiSe 2002) pp. 1-10, 2002.

[5] P. Papadimitratos and Z.J. Haas, "Secure routing for mobile ad hoc networks," SCS Communication Networks and Distributed Systems Modeling and Simulation Conf. (CNDS 2002), 2002.

[6] J. Kong, P. Zerfos, H. Luo, S. Lu and L. Zhang, "Providing robust and ubiquitous security support for MANET," 9th IEEE International Conference on Network Protocols (ICNP 2001), pp.251-260, 2001.

[7] J. Douceur, "The Sybil Attack," 1st Int. Workshop on Peer-to-Peer Systems (IPTPS'02), 2002.

[8] D.B. Johnson, D.A. Maltz and Y.C. Hu, "The dynamic source routing protocol for mobile ad hoc networks (DSR)," IETF Internet Draft, MANET working group, version 9, Apr. 2003.

[9] C.E. Perkins, E.M. Royer, and S.R. Das, "Ad hoc on-demand distance vector (AODV) routing," IETF Experimental RFC, MANET working group, RFC 3561, Jul. 2003.

[10] T. Clausen and P. Jacquet, "Optimized Link State Routing Protocol," - IETF Internet Draft, MANET working group, version 11, Jul. 2003.

[11] R. Ogier, M. Lewis and F. Templin, "Topology Dissemination Based on Reverse-path Forwarding (TBRPF)," Internet Draft, MANET working group, version 9, Jun. 2002.

[12] L Feeney, B. Ahlgren, A. Westerlund, "Spontaneous Networking: An Application-oriented Approach to Ad Hoc Networking. IEEE Communications Magazine, pp. 176-181, June 2001.

[13] H. Luo, P. Zerfos, J. Kong, S. Lu and L. Zhang, "Self-securing Ad Hoc Wireless Networks, " Proc 7th IEEE International Symposium on Computer and Communications (ISCC'02), pp. 567-576, 2002.

[14] L. Zhou and Z. J. Haas, "Securing ad hoc networks," IEEE Network Magazine, vol. 13, issue 6, pp. 24-30, Nov./Dec. 1999.

[15] J. Hubaux, L. Buttyan, and S. Capkun, "The quest for security in mobile ad hoc networks," Proc. 2001 ACM Symposium on Mobile Ad Hoc Networking and Computing (MobiHoc 2001), pp. 146-155, 2001.

[16] A. Shamir, "How to Share a Secret," Communications of the ACM, vol. 22 issue 11, pp. 612-613, 1979.

POSTER SESSION P1

PERFORMANCE ANALYSIS OF ADMISSION CONTROL ALGORITHMS FOR CDMA NETWORKS

KUO-CHUNG CHU*, FRANK YEONG-SUNG LIN, and SHIN-FU WANG

Department of Information Management, National Taiwan University

50, Lane 144, Keelung Rd., Sec. 4, Taipei, Taiwan, R.O.C.

This paper proposes a modified mathematical formulation of revenue optimization problem in terms of CDMA admission control. The problem solution is based upon proposed algorithm as well as Lagrangean relaxation approach. Performance analysis on three algorithms with respect to voice activity factor (VAF) is considered, where 9 base stations, 50 existing mobile stations, and new mobile stations in Poisson arrival process (λ=100) on 500 test cases are given. Computational results illustrate that no matter which value of VAF is given; proposed algorithm always is with an outstanding performance on solution optimality, blocking rate, as well as revenue contribution.

1. Introduction

Generally, CDMA (code division multiple access) system provides no upper limit of available channels, but it is an interference-based channel assignment. For example, in the reverse-link, received signal-to-interference ratio (SIR) at the base station affects the connection quality. Interferences base station incurred must be lower than pre-defined acceptable interference threshold to ensuring communication quality of service (QoS) [4] [6] [7]. Thus, to preserve the whole system QoS, a number of interferences sources, including existing connections and other interferences propagated from cells, must be taken into account. Besides, mobility of new, handover, and outbound calls should be effectively managed [8]. To manage system capacity, call admission control (CAC) is a prevalent mechanism to allocating channel resources. The more users are admitted, the more revenue is contributed. However, less research discusses revenue optimization in terms of admission control [5] [9]. Even though previous research [9] has been illustrated the revenue analysis in terms of admission control, but no consideration of homing blocked mobile stations into another base station. It would be a poor performance. This paper not only modifies the formulation in [9], but also proposes a solution algorithm with dropping and homing mechanisms to enhance total system revenue. In order to proof the performance of proposed algorithm, analyses including solution optimality, blocking rate, revenue contribution are presented. The remainder of this paper is organized as follows. In Section 2, a modified mathematical formulation is expressed. Section 3 applies Lagrangean relaxation as a solution approach. Besides, the algorithm is also developed here. Section 4 illustrates the computational experiments and related analyses. Finally, Section 5 concludes this paper.

2. Admission Control Problem

In previous research [9], for simplicity of modeling admission control problem, only new mobile stations are considered. Besides, new mobile stations in previous model can either be homed to the controlling base station or blocked. However, to optimally contribute system revenue, the mechanism that will home blocked mobile stations to another base station is taken into account. Accordingly, we modify the previous formulation as problem (IP).

$$Z_{IP} = \max \sum_{t\in T''} a_t \sum_{j\in B} z_{jt} = \min(-\sum_{t\in T''} a_t \sum_{j\in B} z_{jt}) \qquad \text{(IP)}$$

subject to:

$$\left(\frac{E_b}{N_{total}}\right)_{req} \le \frac{\frac{S}{N_0}}{1+\frac{1}{G}\alpha\frac{S}{N_0}\left(\sum_{t\in T'}\delta_{jt}+\sum_{t\in T''}z_{jt}-1\right)+\frac{1}{G}\alpha\frac{S}{N_0}\sum_{\substack{j'\in B\\ j'\neq j}}\left(\sum_{t\in T'}\left(\frac{D_{j't}}{D_{jt}}\right)^{\tau}\delta_{j't}+\sum_{t\in T''}\left(\frac{D_{j't}}{D_{jt}}\right)^{\tau}z_{j't}\right)} \qquad \forall j \in B \ (1)$$

$$\sum_{t\in T'}\delta_{jt}+\sum_{t\in T''}z_{jt} \le M_j \qquad \forall j \in B \ (2)$$

$$D_{jt}z_{jt} \le R_j\mu_{jt} \qquad \forall j \in B, \forall t \in T'' \ (3)$$

$$z_{jt} \le \mu_{jt} \qquad \forall j \in B, \forall t \in T'' \ (4)$$

$$\sum_{j\in\{b',b_t\}} z_{jt} = 1 \qquad \forall t \in T'' \ (5)$$

$$z_{jt} = 0 \text{ or } 1 \qquad \forall j \in B', \forall t \in T'' \ (6)$$

$$\delta_{jt} = 0 \text{ or } 1 \qquad \forall j \in B, \forall t \in T' \ (7)$$

The objective function is to minimize the total revenue loss in the process of admitting new mobile stations. A lot of constraints are the same in [9] except additional constraint (3) and (4) are added. Constraint (1) requires that every one mobile station is served with its homing base station in the required QoS. The left hand side of (1) is the threshold of acceptable SIR for each connection. The right hand side means the real SIR. The denominator of the right hand side is the total interference value, including white noise, the intra-cell interference as well as inter-cell interference. Constraint (2) is to ensure that the number of users who can be active at the same time in a base station would not exceed the base station's upper bound. Constraint (3) ensures that a base station can only serve the mobile stations inside its coverage of effective transmission power radius, where R_j is upper bound of power transmission

* Correspondence author, E-mail: d5725003@im.ntu.edu.tw, who is also a lecturer of the Department of Information Management, Jin-Wen Institute of Technology, Taipei, Taiwan.

Table 1. Description of Notations

Notation	Description
B	the set of candidate locations for base stations
b'	the artificial base station to carry the rejected call when admission control function decides to reject the call
B'	the set of $B \cup \{b'\}$
b_t	the controlling base station of mobile station t
T	the set of mobile stations
T'	the set of existing mobile stations
T''	the set of new mobile stations whose admittance into the cell is to be determined
G	the processing gain
S	the power that a base station received from a mobile station that is homed to the base station with perfect power control
E_b	the energy that BS received
N_{total}	total noise
N_0	the background noise
α	voice activity factor
τ	attenuation factor
D_{jt}	distance between base station j and mobile station t
M_j	upper bound on the number of users that can active at the same time in base station j
μ_{jt}	indicator function which is 1 if mobile station t can be served by base station j and 0 otherwise
a_t	the revenue from admitting mobile station $t \in T''$ into the system
R_j	upper bound of power transmission radius of base station j
δ_{jt}	indicator function which is 1 if mobile station t is homed to base station j and 0 otherwise
z_{jt}	decision variable which is 1 if mobile station t is served by base station j and 0 otherwise

radius of base station j. Constraint (4) guarantees that if a base station does not provide service to a mobile station, then the decision variable z_{jt} must be equal to 0. Constraint (5) ensures that each mobile station can be homed to only one physical base station or rejected. Constraint (6) and (7) guarantee the integer property of decision variables and indicator functions. Notations used to modeling the problem are described in Table 1.

3. Solution Procedure

3.1. *Lagrangean Relaxation Approach*

The solution approach applied to solving admission control problem is Lagrangean relaxation that is originally designed to solve large-scale linear as well as integer programming problems [1] [2]. Modified revenue optimization problem (IP) is transformed into the following Lagrangean relaxation problem (LR) where Constraints (1) (2) (3) are relaxed.

$$Z_D\left(v_j^1, v_j^2, v_{jt}^3\right) = \min - \sum_{t \in T''} a_t \sum_{j \in B} z_{jt}$$

$$+\sum_{j\in B} v_j^1 \left((\frac{E_b}{N_{total}})_{req} + (\frac{E_b}{N_{total}})_{req} \frac{1}{G} \alpha \frac{S}{N_0} \left(\left(\sum_{t\in T'} \delta_{jt} + \sum_{t \in T''} z_{jt} - 1 \right) + \sum_{\substack{j' \in B \\ j' \neq j}} \left(\sum_{t \in T'} \left(\frac{D_{j't}}{D_{jt}} \right)^{\tau} \delta_{j't} + \sum_{t \in T''} \left(\frac{D_{j't}}{D_{jt}} \right)^{\tau} z_{j't} \right) \right) - \frac{S}{N_0} \right)$$

$$+ \sum_{j \in B} v_j^2 \left(\sum_{t \in T'} \delta_{jt} + \sum_{t \in T''} z_{jt} - M_j \right) + \sum_{t \in T''} \sum_{j \in B} v_{jt}^3 \left(D_{jt} z_{jt} - R_j \mu_{jt} \right) \qquad \text{(LR)}$$

subject to: (4) (5) (6) (7).

Lagrangean relaxation problem can further be decomposed into a lot of independent subproblems that could be optimally solved with respect to decision variables. To getting primal optimal solutions, we should iteratively adjust Lagrangean multipliers by subgradient method to optimally solve Lagrangean dual problem. Here, we express (LR) into subproblem 1 related to decision variables z_{jt}.

Subproblem 1: for z_{jt}

$$= \sum_{t \in T''} \left(\sum_{j \in B} \left(z_{jt} \left(-a_t + (\frac{E_b}{N_{total}})_{req} \frac{1}{G} \alpha \frac{S}{N_0} \left(v_j^1 + \sum_{\substack{j' \in B \\ j' \neq j}} v_{j'}^1 (\frac{D_{jt}}{D_{j't}})^{\tau} \right) + v_j^2 + v_{jt}^3 D_{jt} \right) - v_{jt}^3 R_j \mu_{jt} \right) \right)$$

$$+ \sum_{t \in T'} \left(\sum_{j \in B} \delta_{jt} \left((\frac{E_b}{N_{total}})_{req} \frac{1}{G} \alpha \frac{S}{N_0} \left(v_j^1 + \sum_{\substack{j' \in B \\ j' \neq j}} v_j^1 (\frac{D_{jt}}{D_{j't}})^{\tau} \right) + v_j^2 \right) \right)$$

$$+ \sum_{j \in B} \left(v_j^1 \left((\frac{E_b}{N_{total}})_{req} - \frac{S}{N_0} - (\frac{E_b}{N_{total}})_{req} \frac{1}{G} \alpha \frac{S}{N_0} \right) - v_j^2 M_j \right) \qquad \text{(SUB)}$$

subject to: (4) (5) (6) (7).

In (SUB), indication function δ_{jt} and decision variable z_{jt} are used to track existing and new mobile stations, respectively. From which, δ_{jt} just indicates the homing status of existing mobile stations since existing ones would not be blocked at all. The second and the third term of (SUB) are constant, because of all variables are constant that can be calculated. Finally, the first term of (SUB) is what we intent to treat it. Let q_{jt} as follows, the first term of (SUB) can be decomposed into |T''| sub-problems for treatment of new mobile stations whether to be admitted or not in terms of revenue optimization. If q_{jt} is less than 0, assign z_{jt} to 1 or 0 otherwise.

$$q_{jt} = -a_t + (\frac{E_b}{N_{total}})_{req} \frac{1}{G} \alpha \frac{S}{N_0} \left(v_j^1 + \sum_{\substack{j' \in B \\ j' \neq j}} v_j^1 (\frac{D_{jt}}{D_{j't}})^{\tau} \right) + v_j^2 + v_{jt}^3 D_{jt}$$

According to the weak Lagrangean duality theorem [1], for any $(v_j^1, v_j^2, v_{jt}^3) \geq 0$, the objective value of $Z_D(v_j^1, v_j^2, v_{jt}^3)$ is a lower bound of Z_{IP}. Based on problem (LR), the following dual problem (D) is constructed to calculate the tightest lower bound.

$$Z_D = \max Z_D\left(v_j^1, v_j^2, v_{jt}^3 \right) \qquad \text{(D)}$$

subject to: $(v_j^1, v_j^2, v_{jt}^3) \geq 0$.

Then, subgradient method [3] is applied to solving the dual problem. Let the vector S is a subgradient of $Z_D\left(v_j^1, v_j^2, v_{jt}^3 \right)$ at (v_j^1, v_j^2, v_{jt}^3). In iteration k of subgradient optimization procedure, the multiplier vector π is updated by $\pi^{k+1} = \pi^k + t^k S^k$, where t^k is a step size determined by $t^k = \delta\left(Z_{IP}^* - Z_D(\pi^k)\right)/\|S^k\|^2$, where Z_{IP}^* is an upper bound on the primal objective function value after iteration k, and δ is a constant where $0 \le \delta \le 2$.

3.2. *Getting Primal Feasible Solutions*

Based on decision variables solved in Lagrangean dual problem (D), we must adjust them for getting primal feasible solutions of problem (IP). Actually, the primal feasible solution is an upper bound (UB) of (IP) while the solution of Lagrangean dual problem guarantees the lower bound (LB) of (IP). Iteratively, both solving Lagrangean dual problem and getting primal feasible solution, we get the LB and UB, respectively. The gap between UB and LB, expressed by (UB-LB)/LB*100%, illustrates the optimality of problem [1] [2]. In getting primal feasible solutions, there are two mechanisms, say call dropping and call homing, described as follows. 1) Call dropping mechanism (*CDM*): No surprisingly, some of new users would be dropped due to capacity and QoS constraints. However, selection of dropped users has something to do with the homing mechanism. If system properly chooses a new mobile station to be dropped, then it can be homed (added back) into another base station (system) later. For each mobile station, the more number that mobile station is covered by base stations, the more possibility that it can be homed into another base station in the homing stage. This implies total system revenue is enhanced. In this paper, we propose *CDM* with number it is covered by base stations (*CDM-NCBS*), from which the system will pick up the mobile station in the descending number that it is covered by base stations; 2) Call homing mechanism (*CHM*): the system tries to home each dropped new mobile station to another/candidate base station. Selection of candidate base station is also a key issue. Here we propose *CHM* with randomly selected candidate (*CHM-RSC*). Combining to *CDM-NCBS* and *CHM-RSC*, an algorithm, denoted algorithm AA, shown in the following is our solution for problem (IP).

[Algorithm AA]

Step 1. Check capacity constraint (2), for each base station $j \in B$, based upon decision variables z_{jt} solved in Lagrangean dual problem (D). Drop the new mobile station, i.e. set z_{jt} =0, with *CDM-NCBS*, if violates the constraint (2), or go to Step 2 otherwise.

Step 2. Make sure QoS constraint (1) for each base station j is satisfied. Drop the new mobile station, i.e. set z_{jt} =0, with *CDM-NCBS*, if violates the constraint (1), or go to Step 3 otherwise.

Step 3. Try re-adding back all dropped new mobile stations in Step 1 & 2 into system.

- 3-1) sequentially picks up a dropped new mobile station.
- 3-2) home to another base station based upon *CHM-RSC*, i.e. set z_{jt}=1 again, if this setting satisfies constraint (1) as well as capacity constraint (2) for each base station, or go to Step 4 otherwise.

Step 4. End algorithm AA.

4. Computational Experiments

4.1. *Other Primal Heuristics*

As mentioned in section 3.2, selection of candidate base station is a key issue. For the purpose of comparison with algorithm AA, other primal heuristic based upon another homing mechanism, denoted algorithm AB, is proposed that is *CHM* with rank of interference incurred at base station (*CHM-RII*). For *CHM-RII*, let SIR_j equals to the right side of constraint (1), based upon SIR_j in terms of ascending interference loading, candidate is selected. Algorithm AB is following up *CDM-NCBS* with *CHM-RII*, the system will re-add dropped mobile stations into candidate base station, i.e. set z_{jt} =1, if it is not violate constraints. Additional heuristic, denoted algorithm AC, applied in [9] is also implemented for performance analysis.

4.2. *Experiments Scenario*

A few of constants used in the formulation, including S/N_0, E_b/N_{total}, M_j, τ, G, and a_t are the same as Table II in [9]. Number of base stations (|B|), existing mobile stations (|T'|) are given to 9 and 50, respectively. Number of new mobile stations is generated in Poisson arrival process with λ=100. More generically, all locations of base stations, existing as well as new mobile stations are randomized, even thought a few of number of new mobile stations generated in Poisson process may be the same. For the purpose of statistic analysis, 500 test cases with Poisson arrival are generated. Three algorithms, say AA, AB, and AC, conjunction with voice activity factor (VAF) are analyzed.

4.3. *Performance Analysis*

4.3.1. Optimality of solution

Table 2 summaries a statistics of error gaps on best and average cases for three algorithms. Because of the problem is with a strong integrality property, no matter which algorithm applied, the more traffic is penetrated in the system (heavy VAF), the loose gaps the solutions are incurred. Even thought proposed algorithm AA is with an average gap up to 14.57%, it still is a best solution among three algorithms in all four VAF cases.

4.3.2. Blocking rate

Based on the solution the total call blocking rate is also analyzed, which is defined by ratio of blocked users (new mobile stations only) to total users (including existing and new mobile stations) in the system. Figure 1 illustrates the experiment results on blocking rate with respect to four VAF values. Inevitably, the bigger VAF is applied, the more blocking rate is incurred. Proposed algorithm AA is with an outstanding performance.

4.3.3. Revenue contribution

Table 3 summaries the aggregate revenue of problem (IP) on 500 test cases. Besides, improvement on algorithm AC is also shown. Proposed algorithm AA always contributes better revenue. This result is consistent with our assumption that CHM is an important issue for revenue optimization in terms of admission control policy. Another interesting finding is that the revenue improvement is monotonically increasing from 0.55% to 3.60% and from 0.49% to 3.13% for algorithm AA and AB, respectively, when VAF is increasing from 0.3 to 0.375. Especially, the more VAF is applied, the more improvement is calculated.

5. Conclusions

This paper analyzes the performance of three admission control algorithms for CDMA networks. First of all, a modified admission control problem is formulated as a mathematical revenue optimization problem and solution algorithm is proposed. The solution algorithm is based upon call dropping mechanism (*CDM*) and call homing mechanism (*CHM*). We propose an algorithm integrating *CDM* with number it is covered by base stations (*CDM-NCBS*) and *CHM* with randomly selected candidate (*CHM-RSC*) to handle mobile users. For proving the effectiveness of the proposed algorithm, other primal heuristic and previous algorithm in literature is implemented. Three algorithms for solving admission control problem are jointly considered. The analysis of experiments include optimality of solution, blocking rate, and total system revenue with respective to voice activity factor (VAF). Computational results illustrate that no matter which value of VAF is given, proposed algorithm AA always is with an outstanding performance on solution optimality, blocking rate, as well as revenue contribution.

References

1. A. M. Geoffrion, "Lagrangean relaxation and its use in integer programming", *Math. Programming Study*, vol. 2, pp. 82-114, 1974.
2. M. L. Fisher, "The Lagrangian relaxation method for solving integer programming problems", *Management Science*, vol. 27, pp. 1-18, 1981.
3. M. Held, P. Wolfe, and H. D. Crowder, "Validation of subgradient optimization", *Math. Programming*, vol. 6, pp. 62-88, 1974.
4. S. M. Shin and D. K. Sung, "DS-CDMA Reverse Link Channel Assignment Based on Interference Measurements," *Electronics Letters*, vol. 31, pp. 1897-1899, Oct. 1995.
5. H.-F. Wang, "Admission control and homing in DS-CDMA wireless communications networks," M.S. thesis, Department of Information Management, National Taiwan University, Taipei, Taiwan, 2001.
6. K. Kim and Y. Han, "A call admission control with thresholds for multi-rate traffic in CDMA systems," in *Proc. IEEE 51st VTC-Spring*, pp. 830–834, vol. 2, Tokyo, Japan, 2000.
7. S. M. Shin, C. -H. Cho, and D. K. Sung, "Interference-based Channel Assignment for DS-CDMA Cellular Systems," *IEEE Trans. Veh. Technol.*, vol. 48, pp. 233-239, Jan. 1999.
8. N. Dimitriou, G. Sfikas, and R. Tafazolli, "Call admission policies for UMTS," in *Proc. IEEE 51st VTC-Spring*, vol. 2, pp. 1420-1424, Tokyo, Japan , 2000.
9. Kuo-Chung Chu, Frank Yeong-Sung Lin, and Shin-Fu Wang, "An admission control algorithm for CDMA networks," accepted for publication, *IEEE 11th IEEE ICON*, Sydney, Australia, 2003.

Table 2. Summary of error gaps percentage (%) for three algorithms based on 500 test cases of new mobile users with Poisson arrival process (λ=100) with respect to VAF.

VAF	0.3			0.325		
Algorithm	AA	AB	AC	AA	AB	AC
Best	0.00	0.00	0.00	0.00	0.00	0.00
Average	1.60	1.65	2.12	4.27	4.43	5.48
VAF	0.35			0.375		
Algorithm	AA	AB	AC	AA	AB	AC
Best	0.00	0.00	0.00	0.00	0.00	0.00
Average	8.50	8.78	10.52	14.57	14.98	17.59

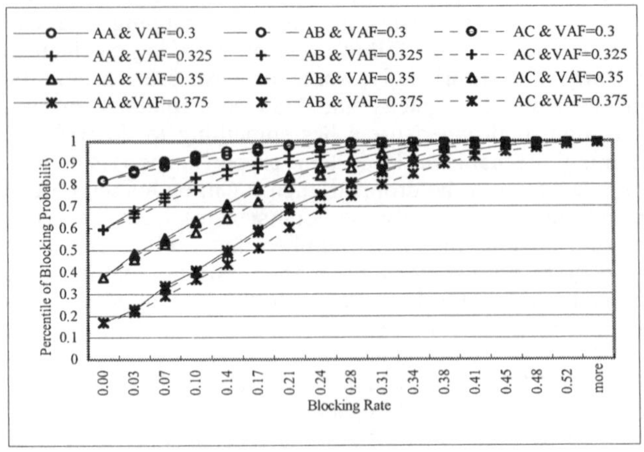

Figure 1. Percentile of blocking rate for three algorithms based on 500 test cases of new mobile users with Poisson arrival process (λ=100) with respect to VAF.

Table 3. Revenue aggregation and improvement on 500 test cases of new mobile users with Poisson arrival process (λ=100) with respect to VAF.

VAF	0.3		
Algorithm	AA	AB	AC
Revenue Aggregation	490960	490680	488270
Revenue Improvement on AC	0.55 %	0.49 %	N/A
VAF	0.325		
Algorithm	AA	AB	AC
Aggregate Revenue	476370	475560	470220
Improvement on AC	1.31 %	1.14 %	N/A
VAF	0.35		
Algorithm	AA	AB	AC
Aggregate Revenue	453580	452170	443640
Improvement on AC	2.24 %	1.92 %	N/A
VAF	0.375		
Algorithm	AA	AB	AC
Aggregate Revenue	421780	419830	407080
Improvement on AC	3.61 %	3.13 %	N/A

A PERFORMANCE ENHANCING PROXY FOR MOBILE SATELLITE INTERNET

Giovanni Ciccarese ° Mario De Blasi ° Luigi Patrono °
Sebastiano Elia ° Giuseppe Tomasicchio °°

° Department of Innovation Engineering, University of Lecce
Via Monteroni, 73100 Lecce, Italy
E-mail: {mario.deblasi, gianni.ciccarese, luigi.patrono, sebastiano.elia}@unile.it

°° Alenia Spazio, Business Multimedia Unit
Via G.V. Bona, n.85, 00156 Rome, Italy
E-mail: {g.tomasicchio@roma.alespazio.it}

Abstract: Integration between satellite and terrestrial networks is to play a very important role in the creation of a Global Internet which supports user mobility. Unfortunately, in these scenarios, TCP protocol tends to perform poorly because of high delay-bandwidth product, high bit error rate (BER) and burst errors mainly due to shadowing. The use of a reliable data link protocol, well suited to mobile geo-stationary satellite channel characteristics, can improve TCP performance appreciably. However, introducing a more variable end-to-end delay causes, as a side effect, a new problem: the competition between retransmission mechanisms at transport layer and data link layer. In order to solve this problem a TCP Performance Enhancing Proxy has been defined and validated. The simulation results, in terms of web-page download mean time, have shown that a local approach results in substantial improvements in HTTP performance, solving the retransmission competition problem.

Key Words: Satellite, Ka-band, HTTP, Performance, Mobility, PEP.

1. INTRODUCTION

Satellite networks will play a key role in supporting multimedia mobile communication services in tomorrow's Global Internet. Unfortunately, Satellite Internet raises some issues, mainly related to the poor performances TCP protocol exhibits in a network including a mobile geo-stationary satellite channel. In fact, TCP mechanisms have not been designed for wireless channels with high bit error rate (BER) and burst errors. More specifically, the TCP interprets all packet losses, even those due to transmission errors, as signs of network congestion and reacts by unnecessarily slowing down the traffic sent over the connection.

A number of schemes have been proposed to improve TCP performance in a wireless environment [1, 2, 3]. They usually fall into one of two main approaches. The first, well-known as end-to-end, aims at adapting TCP to the wireless link characteristics through protocol enhancements. The second approach, a local approach, on the contrary, requires no changes to existing sender implementations as it attempts to hide any non-congestion related losses from the TCP sender and to solve the wireless problems locally in order to improve the lossy link quality. Techniques based on the local approach are often called Performance Enhancing Proxies (PEP) [4].

A PEP is a network entity that is used to improve the performance of Internet protocols on network paths where native performance suffers due to characteristics of a link or sub-network on the path. There are many types of Performance Enhancing Proxies. Different types of PEPs are used in different environments to overcome different link characteristics which affect protocol performance.

Recently, some research activities [5, 6, 7] included in the EuroSkyWay (ESW) program showed that, by using an ARQ data link protocol well suited to mobile geo-stationary satellite channel characteristics, appreciable TCP performance improvements are achievable. However, the more variable end-to-end delay due to the adoption of this approach introduces, as a side effect, a new problem resulting in useless retransmissions of TCP segments: the competition [7] between retransmission mechanisms at transport layer and data link layer.

The activities summarized in this work are also included in the EuroSkyWay framework and have addressed the above competition problem by focusing on Web interaction in the terrestrial-satellite integrated scenario. The research output consists of an efficient TCP PEP (T-PEP) scheme that is based on the possibility of forcing a TCP sender to go into *persistent mode* by setting the WINDOW field in a segment to zero.

Thus, through computer simulation, the performance of the HTTP 1.1 protocol, in terms of mean time per web-page downloaded has been evaluated in an integrated satellite-terrestrial scenario affected by both congestion- and non-congestion-related losses.

The simulation results have shown how using the defined T-PEP, in addition to ARQ data link protocol well suited to mobile geo-stationary satellite channel characteristics, enhances TCP performance to the extent that access to Web services in a satellite mobility scenario become very attractive, by virtue of a significant reduction in TCP timeouts even in urban environments.

The rest of the paper is organised as follows. Section 2 reports a description of the TCP PEP defined. In Section 3 the network model under consideration is described. The simulation results are discussed in Section 4. Finally, conclusions and future works are provided in Section 5.

2. PEP DESCRIPTION

In order to mitigate the competition problem between retransmission mechanisms at transport layer and data link layer, the T-PEP has been defined.

The main key elements of the defined T-PEP are:

- *Distribution*: it is integrated, i.e. it comprises a single PEP component implemented within the satellite gateway, well known as Base Station (BS);
- *Symmetry*: it is asymmetric, i.e. the actions taken by the PEP depend on the interface which a packet is received from;
- *Split Connection*: it does not split TCP connection and it preserves the end-to-end semantic;
- *Transparency*: it operates totally transparently to the end systems, thus requiring no modification to TCP implementation.

As above said, T-PEP consists of an efficient scheme that is based on the possibility of forcing a TCP sender to go into *persistent mode* by setting the WINDOW field in a segment to zero. The scheme exploits this opportunity in order to freeze TCP timers whenever the time needed by the reliable data link protocol on satellite channel to deliver data can cause a RTT (Round Trip Time) longer than the TCP RTO (Retransmission Timeout).

More specifically, T-PEP implementation runs on the BS and works as follows:

- It monitors all TCP segments, stores in a local buffer the last ACK segment received from TCP receiver on the mobile host and activates, when this occurs, a local timer.
- When a new ACK is buffered, the previous ACK buffered is forwarded unaltered to the TCP sender on the fixed host. On the contrary, when the local timer expires, the ACK buffered is forwarded to the fixed host-side TCP sender with the WINDOW field set to zero.

The local timer is set to a value close to the round trip propagation delay between the fixed host and the mobile host in order to avoid useless TCP retransmissions.

The T-PEP agent stops its activities when some segments received from TCP sender on the fixed host are out-order. In this case, the inactivity of the T-PEP scheme is justified because this event represents a clear sign of congestion losses and ACKs forwarding permits an immediate reaction of the TCP sender to recover congestion losses.

It is worth noting that the proposed scheme requires no changes in TCP and also maintains TCP semantics. However, it needs snooping (read and re-write) TCP segments and no longer works if network encryption schemes, such as IPSEC, are applied end-to-end. One solution possible solution, non transparent to IPSeC, is represented by Multi-Layer IPSEC [8], which divides the IP datagram into zones and applies different protection schemes to each one. This way, for example, a TCP flow that desires support from T-PEP at a satellite gateway can divide the IP datagram payload into two zones, TCP header and TCP data. The TCP data part can use an end-to-end protection with keys shared only between the source node and the destination node (hosts or security gateways). The TCP header part can use a separate protection scheme with keys shared among the source node, the destination node and certain trusted intermediate nodes. This way, no one in the public Internet other than the source node, destination node and the trusted intermediate nodes has access to TCP header or TCP data, and no one other than source node and destination node (not even the trusted intermediate nodes) has access to TCP data.

Finally, use of T-PEP does not introduce scalability issues on the BS because the amount of T-PEP processing is limited to reading the sequence number and the acknowledgement number of packets and to buffering no more than one packet at time.

3. SIMULATION MODEL

The integrated network model considered in this work is illustrated in Fig. 1: a mobile host, equipped with a satellite interface operating in the Ka-band, accesses an Internet server reachable through a satellite fixed earth station. The bit rate between client and server is supposed symmetric and equal to 304 Kbps. By focusing on transactional traffic applications, such as the Web [9, 10], a HTTP/1.1 session, persistent and with pipelining, over one TCP connection between a client and a server has been modelled. The reference ESW architecture is a connection-oriented network, which operates according to circuit-switching mode. It is based upon a constellation of geo-stationary, Ka-band, re-generative satellites. Both FDMA and TDMA are used as access schemes.

With regard to TDMA, the time axis for both up-link and down-link is subdivided into *Frames*, whose duration is constant and equal to 26.5 ms; a *Frame* consists of *N Frame Units* (*FU*), with *N* depending on the terminal type. Each *FU* contains the basic data unit transferred across the ESW network, called an *ESW Cell*. An ESW cell is 60-bytes long: 7 bytes are reserved for the header, the remaining 53 bytes constitute the payload.

In order to evaluate the efficiency of the proposed T-PEP scheme, the reliable ESW data link layer (called ESW-LL-2I) has been considered. It is characterised by an error recovery procedure based on a particular Selective Repeat ARQ mechanism [5] which has achieved

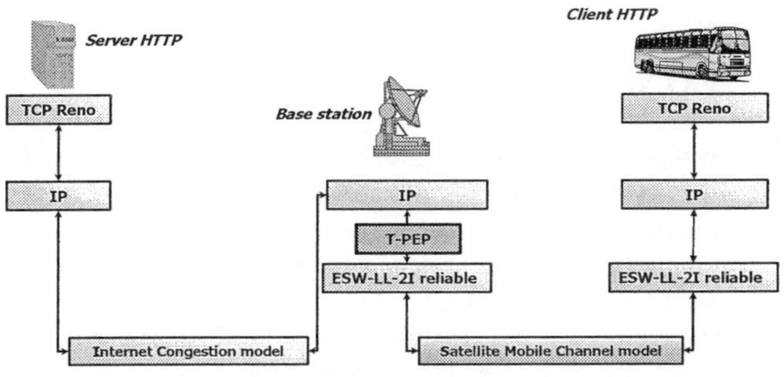

Figure 1. Network Model.

high efficiency through coding, in an ESW-ACK cell, of blocks of contiguous cells received and queued (or the gaps between blocks, as ESW is connection-oriented) in the receiver window. In particular, each block is coded by a pair of values, the first value representing the left edge of the block, that is, the sequence number of the first cell in the block and the second one the block length (number of cells in the block). The maximum number of coded blocks in the payload of one ESW-ACK cell is equal to 14. The transmission of one ESW-ACK cell is triggered by one of the following events: transfer of a complete SDU to higher layer either expiration of the timer *Tack*, whose timeout value is set to 636 ms in ESW architecture. This value represents the maximum time interval between two consecutive ESW-ACK cell transmissions.

In order to evaluate the impact of both congestion and non-congestion related losses on the HTTP performances in a satellite connectivity scenario, the following analysis was carried out, taking into account that transmissions errors on wired portion are due to congestion and can occur in both directions. The characteristics of the congestion losses in the terrestrial Internet are described by a *uniform* error model with the following congestion loss rates: 0.0 and 0.01 [7].

The mobile satellite channel, affected by fading due to mobility, has been modelled using the Lutz model [11, 12], which uses a two-state Markov chain, whose states, "good" and "bad", represent, respectively, the unshadowed and the shadowed periods. Applying ESW system parameters to the Lutz model, the cell error probability in the two states has been calculated: the result is that during shadowing periods everything is irreparably corrupted, whereas, during clear intervals, data is received perfectly. The errors due to satellite link (atmospheric attenuation) were assumed to be almost totally resolved by the physical layer.

The Maximum Segment Size (MSS) for TCP has been set to 536 bytes, the value used by most TCP implementations. Each IP packet is fragmented according to the payload length of an ESW cell, 53 bytes.

The main performance metric considered in this work is the *Web-page download mean time*. The web-page download mean time has been calculated as a function of the mean sojourn time in good state, *Tgood*, the mean sojourn time in bad state, *Tbad*, and the congestion loss rate.

Maximum Segment Size (MSS)	536 bytes
ESW cell size	60 bytes
ESW LL-2 Tx Window on client/server side	1003.82 KB
Bit Rate of client/server	304 Kbps
Data Link timeout	31.8 s
Mean of the main object size	10 KBytes
Std of the main object size	25 KBytes
Mean of the Number of in-line objects	5.5
Std of the Number of in-line objects	11.4
Mean of the in-line object size	7.7 KBytes
Std of the in-line object size	126 KBytes
Mean of the Think time	39.5 s
Std of the Think time	92.6 s
Tack	636 ms
Mobile terminal speed	10 m/s
Frame time	26.5 ms
Round Trip Time	710 ms
Propagation time on wireless	265 ms
Propagation time on wired	50 ms

Table 1 - Simulation model parameters.

All simulations results, obtained using the Network Simulator v2 (ns2) tool, are characterised by a 95% confidence interval whose maximum relative error is equal to 5%. Finally, Table 1 summarises the values related to the main parameters of simulation model.

4. SIMULATION RESULTS

A performance evaluation, in terms of web-page download mean time, of the T-PEP scheme efficiency has been performed. In particular, a comparison between two following configurations has been evaluated: TCP Reno running over a reliable data link protocol and the same configuration with addition of the T-PEP scheme. Fig. 2 gives us a clearer idea showing the web-page download mean time versus T_{good}, for a number of values of T_{bad}; the range 0.5 ÷ 8.5 seconds has been chosen for both T_{good} and T_{bad} because it covers all the environments (urban, suburban and rural) of interest. In this case, the wired portion of the end-to-end connection has been supposed not affected by congestion losses.

These curves clearly show that the T-PEP scheme is able to improve TCP performance considerably,

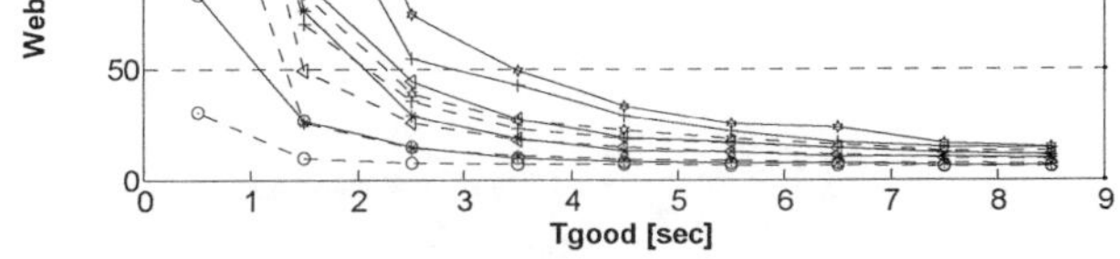

Figure 2. Impact of the T-PEP scheme on the Web page download time without congestion losses.

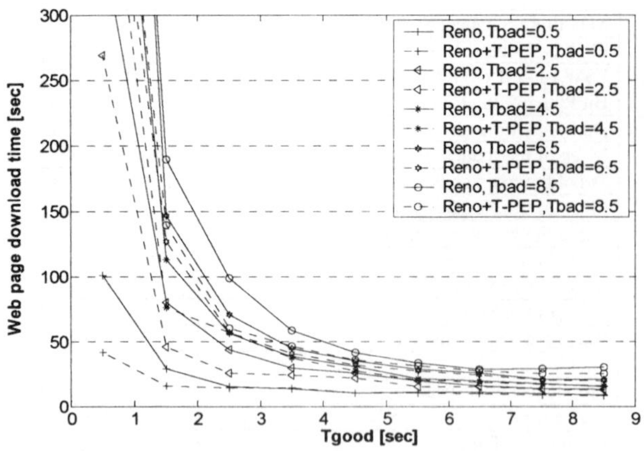

Figure 3. Impact of the T-PEP scheme on the Web page download time with congestion loss rate equal to 0.01.

successfully mitigating the retransmissions competition problem. So, the mean time needed to complete an entire web-page download is low enough to render access to web services attractive in this mobility scenario.

Fig. 3 shows the impact of Internet congestion, characterised by a congestion loss rate equal to 0.01, on the HTTP performance. The plot shows a slight increase in web-page download time for both configurations because of the Internet congestion which slows down the data flow put into the network. In any case, the T-PEP scheme continues to be very efficient, confirming that the addition of the T-PEP represents a rational choice when a reliable data link layer, well suited to mobile ka-band satellite channel characteristics, is adopted to solve the burst error problem.

Finally, Fig. 4 shows a quantitative measure of the TCP timeouts reduction as consequence of the introduction of the T-PEP scheme. A comparison in terms of the TCP timeouts in any environment (urban, sub-urban, rural, etc.) can be seen on this plot.

These curves clearly show that the T-PEP scheme is able to mitigate the retransmissions competition problem considerably, thus rendering the extension of access of Web services to an urban environment attractive.

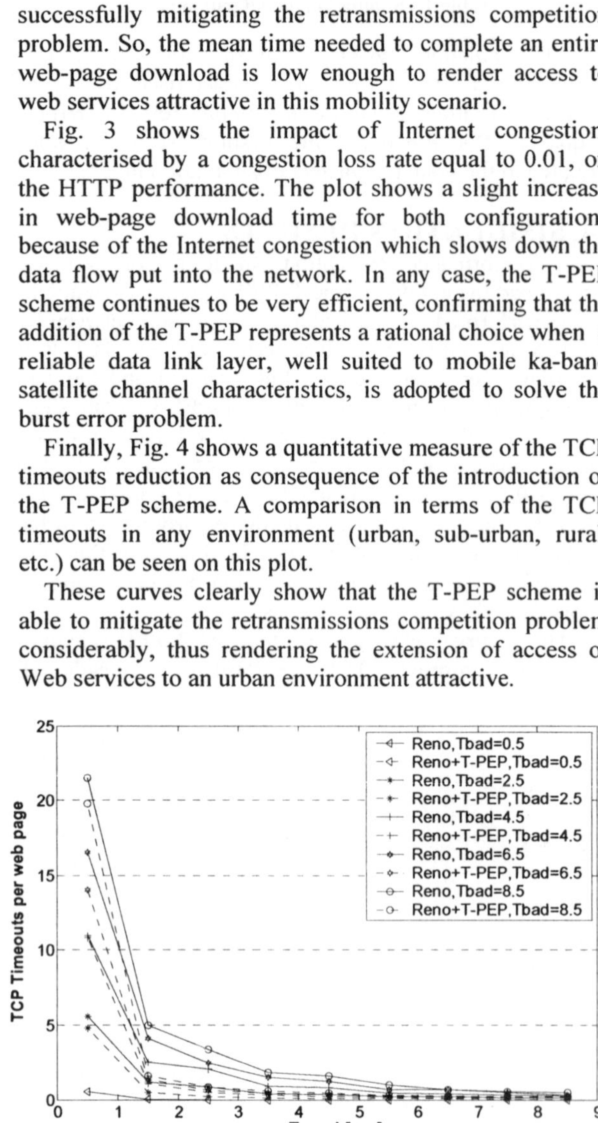

Figure 4. Impact of the T-PEP scheme on the TCP timeouts without congestion losses.

5. CONCLUSION

In this paper a performance evaluation of the proposed T-PEP scheme, aimed at solving the retransmission competition problem, has been performed through simulation in an integrated terrestrial-satellite scenario supporting user mobility.

The results have clearly shown that the T-PEP is able to substantially improve TCP performance, succeeding in mitigating the competition between retransmission mechanisms at transport layer and data link layer. The main advantage obtained by introducing the T-PEP scheme is that the mean time needed to complete an entire web-page download is low enough to render access to web services attractive in this mobility scenario, even in the presence of congestion on the wired portion.

Finally, it is noteworthy that the proposed scheme requires no changes in TCP and also maintains TCP semantics. However, it needs snooping TCP segments and no longer works if network encryption schemes, such as IPSEC, are applied end-to-end. For this reason, research activities will now be directed towards evaluating some extensions which are able to address the competition problem transparently to IPSEC protocol.

REFERENCES

Proceedings Papers:

[4] J. Border, et al., Performance Enhancing Proxies Intended to Mitigate Link-Related Degradation, Internet Society, RFC 3135, June 2001.

[5] L. Casone et al., An Efficient ARQ Protocol for a Mobile Geo-stationary Satellite Channel, Proc. GLOBECOM 2001, Nov. 2001.

[6] L. Casone et al., Throughput of TCP using a data link ARQ protocol over a mobile satellite channel, Proc. SOFTCOM 2001, Oct. 2001.

[7] G. Ciccarese et al., Improving HTTP Performance in a mobile terrestrial-satellite network, 9th Ka Band and Broadband Communications Conference, Italy, November 2003.

[8] Y. Zhang and B. Singh, A Multi-Layer IPSec Protocol, Proc. 9th USENIX Security Symposium Denvor, Colorado, Aug 2000.

[9] H-K. Choi and J. O. Limb, A Behavioral Model of Web Traffic, Proc. of the 7th IEEE International Conference on Network Protocols, November 1999.

[10] A. J. McGregor, M. V. Pearson and J. G. Cleary, Improving the Performance of HTTP over High Bandwidth-Delay Product circuits, Proc. of the Comm. Networks and Distributed Systems Modeling and Simulation Conference, January 2001.

Journal Papers:

[1] Chunlei Liu and Raj Jain, Leading Causes of Wireless TCP Performance Degradation, ACM Wireless Networks Journal, Aug. 2001.

[2] H. Balakrishnan, V. N. Padmanabhan, S. Seshan and R. H. Katz, A Comparison on Mechanisms for Improving TCP Performance over Wireless Links, IEEE/ACM Transactions on Networking, vol. 5, pp. 756-769, Dec. 1997.

[3] N. Ghani and S. Dixit, TCP/IP Enhancements for Satellite Networks, IEEE Commun. Magazine, July 1999, pp.64-72.

[11] E. Lutz, D. Cygan, M. Dippold, F. Dolainsky and W. Papke, The Land Mobile Satellite Communication – Recording, Statistics and Channel Model, IEEE Trans. Veh. Technol., vol. 40, pp. 375-386, May 1991.

Books:

[12] E. Lutz, M. Werner and A. Jahn, *Satellite Systems for Personal and Broadband Communications*, Springer Verlag, 2000.

Mobile Firewall

Guangrui Fu, Daichi Funato, Jonathan Wood, Toshiro Kawahara
DoCoMo Communications Labs, USA Inc.
181 Metro Drive, Suite 300 San Jose, CA, 95110
U.S.A.
{fu, funato, wood, kawahara@docomolabs-usa.com}

Abstract: The main problems with using existing firewall approaches in wireless and mobile environments are these methods cannot prevent internal attacks and at the same time preserve wireless bandwidth, have little or no packet filtering capability for mobile users, and inconvenient for service providers and end users to change firewall policies dynamically. We propose mobile firewall architecture to solve these problems. The paper presents the design and implementation of the mobile firewall, and shows the packet-filtering performance results and the maintenance cost in the mobile firewall.

Key Words: Firewall, Mobility, Network Access Control

1. INTRODUCTION

Wireless service providers are incorporating Internet technologies into their service networks. As this trend progresses, they are facing with gaps in an Internet protocol suite that was not designed with wireless service and mobile networks in mind. One of these gaps is the lack of adequate packet filtering mechanisms to protect mobile users of small mobile devices running IP over precious wireless links. The most common firewall is perimeter firewall [6]. Other existing and proposed firewall architectures include personal firewall [11, 12], distributed firewall [1, 2], and embedded firewall [4, 5]. The problems of these approaches are:

- Perimeter firewalls can not prevent attacks in which the malicious nodes and the victims are in the same network [3].
- Personal firewalls, distributed firewalls, and embedded firewalls are end-point based filter approaches. The main problem of these methods is that even though malicious packets can be filtered out at end hosts, they consume both the scarce access network bandwidth and the limited battery power and computing resources of end hosts. These approaches make resource-constrained end hosts such as cell phones and PDAs more vulnerable to the Denial-of-Service (DoS) attacks.
- Perimeter firewalls, distributed firewalls, and embedded firewalls all have very little or no support of host mobility.

In this paper, we propose the mobile firewall architecture. The main contributions are:

- By using a firewall filter transfer protocol, the firewall rules can be transferred among different network access points. It provides mobile users continuous packet filtering services as users move among networks.
- By having a layered firewall rule set and an access notification mechanism, end users and service providers can define and change firewall rules dynamically.
- By enforcing the policy at network edge point, attacks coming from the same subnet or different subnets can be prevented while preserving access network bandwidth and end-host computing and power resources.

2. DESIGN

A network firewall splits the network into two parts: the internal network which is behind the firewall, and the external network which is outside of the firewall. Traditionally, an enterprise relies on its physical security mechanism (e.g. security gate) to screen out attackers, and depends on the firewall to protect internal hosts from external malicious attacks. However, in a wireless network, attackers can access to the network from anywhere within radio range. The physical security methods employed by enterprises can no longer prevent attackers from reaching their internal networks. The objectives of our mobile firewall design are to protect users from both internal and external attacks, save scarce wireless bandwidth and resources on end hosts, enable packet filtering protection as user moves around, and change firewall rules dynamically.

A. Architecture

The Mobile Firewall architecture consists of three main network entities: end host (EHOST), network edge point (NEP), and service administrative server such as an AAA server, as shown in Figure 1. The NEP is the first network entity that connects the end-host to the network, such as an 802.11 WLAN access point or a Bluetooth access point, or an Ethernet hub. It is the NEP that enforces policies, while all three types of entities can define policies. The EHOST and SAS upload their rules to the NEP by using the filter transfer protocol. The NEP merges all rules into a layered rule table for each legitimate EHOST and filters all packets sent to or come from the EHOST based on this table.

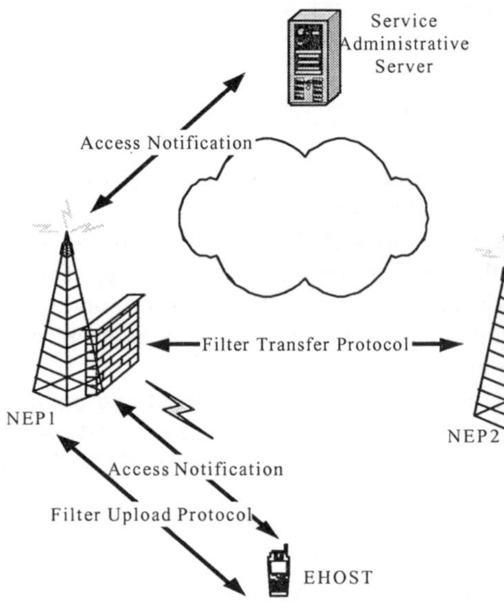

Figure 1 Mobile Firewall Architecture

B. Layered Firewall Rule Management

An EHOST can define firewall rules (personal rules) based on their own preferences. For instance, an EHOST may allow its friends to ping it, but disallow ping packets from unknown hosts. An EHOST uploads its personal rules to the NEP when it is first authorized to access the network. After the EHOST moves to a new network, its personal rules can be transferred from the previous NEP to the new one, by using the filter transfer protocol. The benefit of transferring personal rules between the two NEPs, rather than from the EHOST to the new NEP, is that it preserves both access network bandwidth and the battery and computing resources of the EHOST. It becomes especially beneficial when the access network bandwidth or available resources at the EHOST is very limited while the personal rule set of the EHOST is large.

A NEP can define firewall rules (network rules) based on the security and network management needs from network operators' perspective. For instance, it may deny all TCP SYN attack packets. Ports are disabled by default for each EHOST, except for some ports needed for initial network access control and other basic network management purposes.

An SAS can define firewall rules (service rules) based on service subscriptions. For example, an SAS may enable a voice-over-IP port if the user has paid for this service.

A NEP creates a rule table for each authorized EHOST. After collecting personal rules and service rules, the NEP is responsible for integrating these rules with the network rules into one table for each individual EHOST. In the order of priority to reject packets, a packet traverses network rules first, then service rules, and finally personal rules. The NEP also needs to ensure that an EHOST can only change its personal rules, but not service rules or network rules in its table or any rules in other EHOSTs' rule tables.

Having layered firewall rules with prioritized ownerships, together with filter transfer protocol, it enables networks to filter packets not only for network administrative purposes but also for personal network management preference and service-specific needs.

C. Filter Transfer Protocol

The protocol includes three parts: rule transfer from EHOST to NEP, between two NEPs, and from SAS to NEP. It assumes that mutual authentication between these peers exist. Since the size of rules to be transferred can be larger than the network maximum transmission unit (MTU) size, the filter transfer protocol also assumes that the lower layer protocol handles fragmentation and provides reliable transport. For example, TLS over TCP can be used to set up mutual authentications between the peers and transfer filters reliably and securely.

The NEP needs to contact an SAS to authenticate an EHOST and retrieve the service subscription information of the EHOST. After the EHOST is authorized to access the network, the SAS uploads the service rules for this EHOST to the NEP. The NEP generates a shared session key for securing the communication between the authorized EHOST and the NEP. If it is the first time that the EHOST has accessed the network, the EHOST encrypts its personal rules with the session key and upload them to the NEP. The NEP ignores any filter upload request from unauthorized EHOSTs. The service rules and personal rules can be uploaded in any order or at the same time.

If an EHOST moves from one access network to another, its personal filters can be transferred from previous NEP to current one, in order to preserve access network bandwidth and limited computing and power resources in the EHOST. Figure 2 shows the protocol sequence for transferring filters between two NEPs. Upon authorized to access the new network, the EHOST sends to the new NEP (NEP2 in Figure 2) a Redirection Request message (step a) which has the previous NEP (NEP1 in Figure 2) address. If NEP2 and NEP1 trust each other, NEP2 sends Context Transfer Request (step b) which contains the EHOST identification information to NEP1, and NEP1 replies back with the EHOST filter information if it has it (step c). The new NEP sends a message (step d) back to the previous one to acknowledge whether the rules are successfully transferred. Upon receiving the acknowledgement message, NEP1 can clean up its local copy of the filter table for the EHOST. The previous NEP also can clean up the filter table for the EHOST after the EHOST has left the network for certain period of time.

By utilizing link layer pre-triggers defined in [10], the filter transfer process can be initiated prior to the arrival of the EHOST to the new network. It helps to decrease the packet filtering blackout time during the handoff.

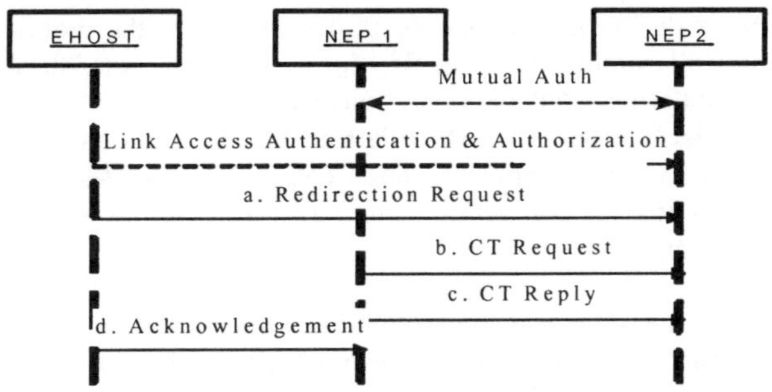

Figure 2 Filter Transfer Protocol Sequence

D. Dynamical Filter Control with Access Notification
Service providers and end users may have different packet filtering requirements in different network environments. For end users and 3rd party service providers, usually it is difficult to change firewall rules dynamically in centralize-managed firewalls such as perimeter firewalls, distributed firewalls, and embedded firewalls. So, one of the key objectives in mobile firewall architecture design is to enable end users and service providers to change firewall rules dynamically.

Other than DENY, REJECT, and ACCEPT etc. common actions, we define a new action PEND in the mobile firewall architecture. If a packet matches with a PEND filter, the NEP temporarily buffers the packet and sends an access notification message to the owner of this firewall rule. Only the EHOST and SAS can specify PEND filter, but not the NEP. When an EHOST or a SAS set PEND filter, other than specifying the traffic signature as for other filters, the EHOST or SAS also needs to specify a call-back function to be triggered when access notification message is received. After receiving the access notification, the mobile firewall client program in the EHOST or SAS calls the registered call-back function and let the application to decide whether to allow or deny such traffic at run-time.

E. Mobile Firewall Application Programming Interface
The mobile firewall application programming interface (API) enables application programs to configure the firewall in a programmatic way. The API has two functions, a registration function and a rule-control function. The registration function allows an application process to register the call-back function for the PEND filter with the mobile firewall client program on the EHOST and SAS. The rule-control function is for application processes to manage firewall rules such as add, delete, change, or search a rule. The API design is still at its preliminary stage, and currently we are investigating how to assure one application cannot change rules defined by other applications.

3. PERFORMANCE EVALUATION

We have prototyped the mobile firewall in C and Java on Red Hat Linux 7.2. Firewall rules are described and transferred in XML format. NEPs need to translate XML into its native firewall implementation language, for example, IP Chain in Linux in our prototype. We have implemented the firewall transfer protocol, the firewall table management program, and modified IP Chain [7] to generate access notification message for PEND filters. We use openSSL [12] to transfer filters securely, and use Open Diameter [8] to authenticate and authorize the EHOST to access the network.

We measure the mobile firewall performance in two aspects: packet filtering throughput and the overhead of filter transfer in NEPs. In the experiment, the NEP runs on a 700MHz Intel Pentium III desktop, and the EHOST is a 1.8GHz Intel Pentium IV IBM A30 Laptop. The link between the EHOST and the NEP is 100Mbps Ethernet.

Figure 3 compares the throughput of IP filters in two implementations: original Linux IP Table module and our Mobile Firewall implementations. It shows both have similar throughput performance as rule table size increases. Note that the throughput suddenly drops as the table size increases to around 2500 rules. It is due to the table management program is optimized for certain size of the table.

The filter transfer delay includes the packet transmission delay and the NEP local operation overhead, and the NEP local operations include the TLS session encryption or decryption, translation rules between XML description and IP Chain description, and rule table management. Figure 4 shows that the delay time of transferring filters between a NEP and an EHOST increases non-linearly as the rule size (size in byte in XML format) increases. However, the total delay is not significant; for example, it takes 20 ms to transfer 50 rules at the size of 6736 bytes. The timer in the EHOST starts when the Filter Upload Request sends out, and the timer stops when the EHOST receives the Acknowledgement message. The timer in the NEP starts when receiving the Upload Request message and stops when sending out the Acknowledgement message. The time duration difference between the two timers, shown as the gap between the two lines in Figure 4, is the transmission delay. One of the possible reasons for the non-linear increase in the delay time is that fragmentations may happen when the packet size is larger than the Ethernet MTU size (1500 bytes). In Figure 5, it shows that in a 1000-rule table it takes less than 5 micro-second for the new NEP to add a rule and less than 10micro-second for the old NEP to delete a rule. The rule table manager needs to search the policy table to delete a rule, while always adds a rule at the beginning of the table; so it takes a little longer to delete a rule than add a rule, as shown in Figure 5.

Figure 4 and 5 show that TLS encryption and XML rule translation are two leading factors in the transfer delay time; the transmission delay becomes less significant as the packet size increases; and the table management overhead is trivial.

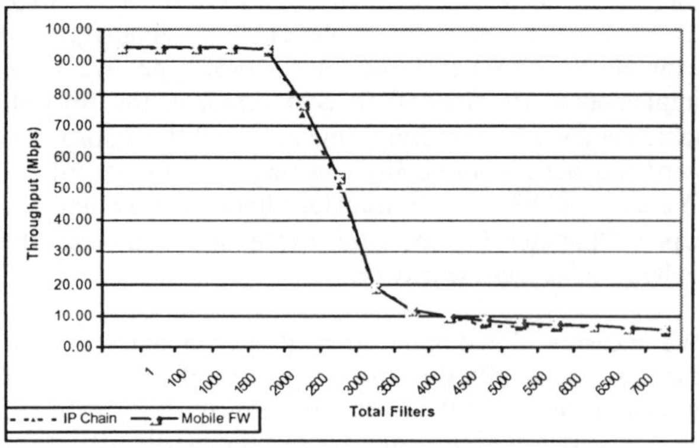

Figure 3 Filter Throughput

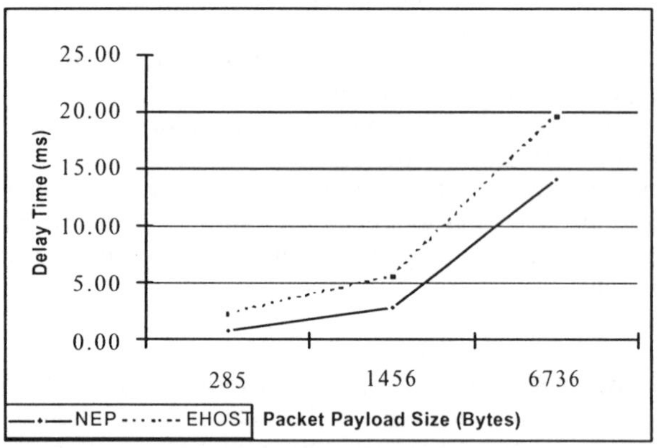

Figure 4 Filter Transfer Delay

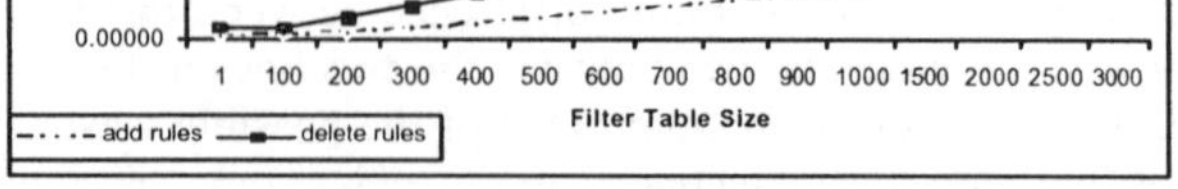

Figure 5 Filter Table Management Cost

CONCLUSION

Existing firewall architectures such as perimeter firewall, personal firewall, and distributed firewall, all have various problems when applied in mobile and wireless environments. The proposed mobile firewall architecture provides packet filtering protection anywhere mobile users go, and offers flexible firewall policy control for end users and service providers. The performance measurement shows that the mobile firewall can archive similar throughout as other firewalls, but may require some computation power and delay time for transfer filters during mobile user handoffs. However, the mobile firewall architecture itself does not rely on these computation intensive operations such as TLS encryption and decryption and XML rule translations. These are the implementation choices and could be speed-up by offloading the computations to hardware co-processors or ASIC chips.

There are several avenues for the further improvements. In particular, we like to extend the mobile firewall architecture into the ad-hoc networks. For example, using filter transfer protocol to protect end hosts in the host-host direct communications and preserve the bandwidth of the intermediate hops in multi-hop wireless networks. Ultimately, we like to explore how to use firewalls to control the usages of all the network-accessible resources and services.

REFERENCES

[1] S. M. Bellovin, "Distributed Firewalls", *;login: magazine, special issue on security*, November 1999.

[2] S. Ioannidis, A. D. Keromytis, S. M. Bellovin, J. M. Smith, "Implementing a Distributed Firewall", ACM Conference on Computer and Communications Security, Athens, Greece, 2000.

[3] D. Keromytis, "Some Problems with Perimeter Firewalls", IETF Draft, work in progress, http://www.ietf.org/internet-drafts/draft-keromytis-disfi-compare-00.txt

[4] C. Payne and T. Markham, "Architecture and Application and of a Distributed Embedded Firewall", 17th Annual Computer Security Applications Conference, December 10-14, 2001, New Orleans, Louisiana

[5] David Friedman and David Nagle, "Building Firewalls with Intelligent Network Interface Cards", CMU SCS Technical Report CMU-CS-00-173, May 2001

[6] W. R. Cheswisk, and S. M. Bellovin, *Firewall and Internet Security: Repelling the Wily Hacker*, Addison-Wesley, 1994

[7] Netfilter, http://www.netfilter.org/, 2003

[8] Open Diameter, http://www.opendiameter.org/, 2003

[9] A. Yegin, et al., "Supporting Optimized Handover for IP Mobility - Requirements for Underlying Systems", IETF Internet Draft, work-in-progress, December, 2002.

[10] Zone Labs Personal Firewall, http://www.zonelabs.com/store/content/home.jsp

[11] Norton Personal Firewall, http://www.symantec.com/sabu/nis/npf/

[12] openSSL, http://www.openssl.org/

Macrodiversity Evaluation of WCDMA with Radio over Fibre Access Networks

Nazem Khashjori and H.S. Al-Raweshidy*
University of Kent , Canterbury, UK
E-mail: nlk1@ukc.ac.uk
*E-mail: H.Al-Raweshidy@ukc.ac.uk

Abstract - In this paper, we propose a novel Macrodiversity architecture in the uplink, based on Maximum Ratio Combining (MRC) of the received symbols by different BS receivers of a WCDMA mobile cellular system with Radio over Fibre (RoF). We evaluated the Macrodiversity performance through link and system level simulations. Our simulation results demonstrate that, for a given service, the coverage efficiency increased by 19% - 23% for the uplink and the capacity improved by 14 - 16%. The simulation results of the system level demonstrate that, for a given service, the transmit power reduction is up to 50% in the case of the RoF technique as compared to the wireless link.

Keywords – Macrodiversity, WCDMA, Radio over Fibre, MRC.

1. INTRODUCTION

Radio over Fibre (RoF) technology had been proposed in the past as a delivery technique for 1st and 2nd generation mobile cellular systems [1-3]. It was argued that with such an arrangement the cost of deploying and maintaining a mobile cellular network can be substantially reduced with the benefit of higher flexibility for radio network planning and operation. Similar arguments can be raised in the case of 3rd generation (3G) systems given their complexity and the wide range of services they are expected to provide and support.

In 3G systems, Wideband Code Division Multiple Access (WCDMA) has been selected as the primary multiple access technique. A soft handover algorithm is employed to switch radio communication from one Node B to another as the mobile stations move amongst the cells. During soft handover more than one Node B is connected simultaneously to the mobile station allowing for coverage and/or capacity improvement through Macrodiversity.

This paper proposes a novel Macrodiversity scheme in the uplink based on Maximum Ratio Combining (MRC) of the received symbols by different BS receivers of a WCDMA mobile cellular system with RoF. In conventional WCDMA systems (systems without RoF), this is not practically feasible because of the spatial separation between the BTSs. Selection Combining (SC) of soft decoded data sequences arriving at the Radio Network Controller (RNC) from different BTSs is usually utilised. It is well known in the literature that the MRC technique is superior to all other combining techniques [4-5]; therefore, it is expected to deliver the maximum possible Macrodiversity gain. This is extremely beneficial to improving the performance of the WCDMA uplink given that the mobile stations would be located near cell boundaries when they are in a state of soft handover.

This paper also investigates through computer simulations the amount of the Macrodiversity gain achievable in vehicular and indoor to outdoor pedestrian test environments with different mobile speeds.

2. MACRODIVERSITY MODEL

Macrodiversity, and its implementation using MRC as suggested in this paper, can be modelled as a MS communicating with two BSs. The block diagram of the uplink chain is shown in figure (1). The transmitted signal of each link can be written as:

$$S_0(t) = A_{(0)} \cos(\omega c t) \sum_n d_n c_n R(t - nTc) \qquad (1)$$

where ωc and $A_{(0)}$ are the angular frequency and amplitude of the carrier respectively, d_n is the data value, c_n is the PN spreading code, R(t) is the impulse response of the root raised cosine filter and Tc is the chip rate.

The received despread and descrambled signal may be represented by:

$$r_l(t) = s_l(t) + n_l(t) + \mathrm{O}(t) \qquad (2)$$

The received $s_l(t)$ is modelled as:

$$S_l(t) = A_{(l)} \cos(\omega_o t + \phi_l(t)) \sum_n d_n c_n R(t - nTc - \tau_l) \qquad (3)$$

where $A_{(l)}$ represents the amplitude and ϕ_l represents the phase of the channel on the ith path, the term $n_l(t)$ represents the additive noise and interference on the same path, and O(t) is the optical link noise.

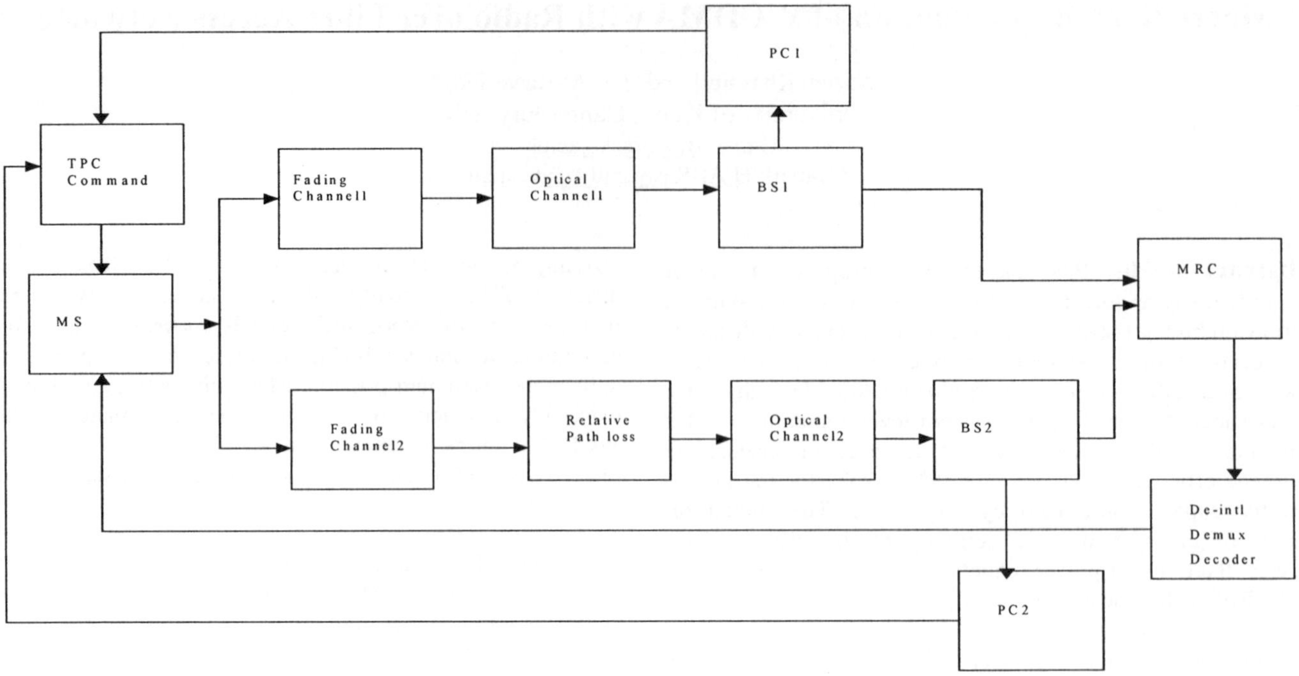

Figure (1) Macrodiversity Model on the Uplink with RoF

The signals from the various fingers are brought into time alignment and then combined coherently using the channel estimates. For MRC, the received signal in each finger is multiplied by the conjugate of the channel estimate (including the amplitude) before the signals are combined. Hence

$$r_{MRC} = \sum_{i=l}^{M} W_i S_l \qquad (4)$$

where M is the number of branches.

The weighting factor w_i can be shown to be proportional to the square root of the SNR of channel l [4]. By using a characteristics function, the pdf of the SNR can be obtained from the statistics of s_l. Finally the Bit Error Probability P_b for each branch can be found.

The signal given by (3) modulates the Laser Diode (LD). The LD nonlinearity can be modelled by a third-order polynomial without memory to give the output optical signal shown in equation (5):

$$P(t) = P_t[1 + I'(t) + a_2 I'^2(t) + a_3 I'^3(t)] + \eta(t) \qquad (5)$$

where P(t) is the average transmitted optical power, $I'(t)$ is the LD current, a_2 and a_3 are the second and third-order coefficients respectively, and $\eta(t)$ is the background noise [6].

3. SIMULATION MODEL

The components of figure (1) are chained together to form the simulation model for the Macrodiversity on the uplink and downlink. The information from the RoF link-level simulations is brought to the system level as look-up tables. The main aim of the system-level simulations is to help to perform more realistic link-level simulations, by providing information about the effect of the environment in a certain link.

The scenario represents a first step in the radio access planning in an urban environment. The networks used in this scenario consisted of a central site based on regular hexagonal cell with omni-direction antennas. The user profile used in the simulations was homogeneous distribution of speech (12.2 kps) with different mobile speeds.

The system level of WCDMA has been investigated within the Vehicular environment described by [7] and characterized by the channel model. The performance of RoF was estimated with a static WCDMA simulator [8].

4. SIMULATION RESULTS

Figure (2) shows the results of the uplink simulation. The figure shows the BER as a function of E_c/N_o for the Vehicular channel with 3dB path loss, with 3km/h mobile speed.

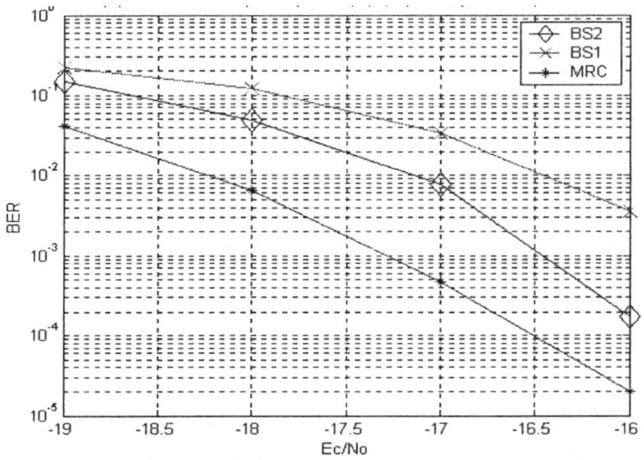

Figure (2) Uplink BER for Vehicular (A) channel with 3km/h mobile speed with 3dB path loss difference.

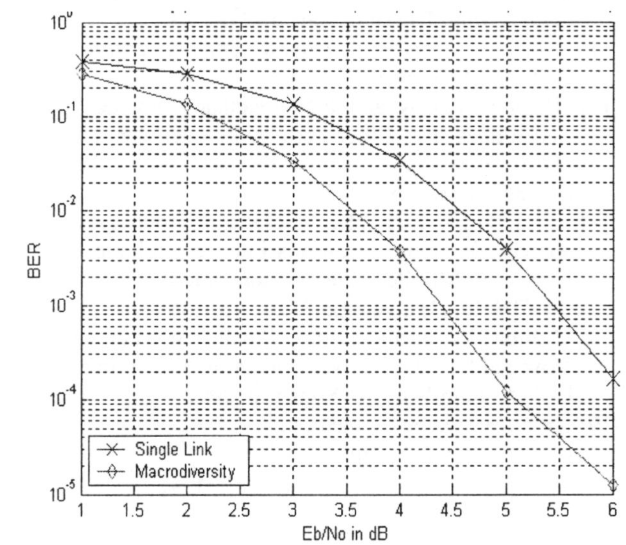

Figure (3) Downlink BER for Pedestrian (A) with 3km/h mobile speed.

From the figure, at the system working point (E_c/N_o @ 10^{-3} BER), the E_b/N_o is calculated by adding the overhead (processing gain and voice activity factor) to E_c/N_o. The results of our simulation show that using MRC in the uplink of a RoF network could result in gain of more than 2dB. The Macrodiversity gain has been calculated between the two values of E_b/N_o of the SC and MRC techniques at the system working point.

Figure (3) shows the results of the downlink simulation. The graph compares the BER performance for the single and Macrodiversity links for different environments and mobile speeds. The behaviour of the downlink simulation is found to be similar to that of the uplink with respect to pathloss and power control estimation. The gain of the

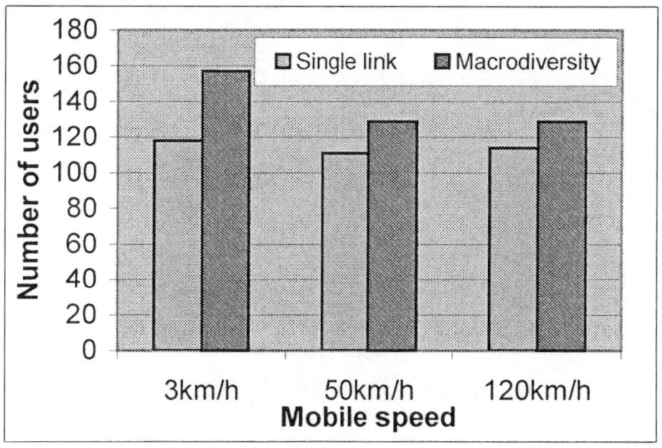

Figure (4) Downlink capacity for Vehicular (A) channel.

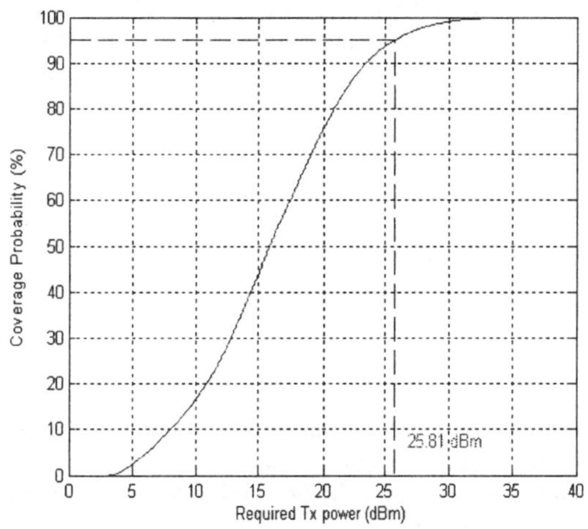

Figure (5) The required power for 95% target coverage with RoF.

Macordiversity in the downlink is found to be less than that in the uplink.

The system-level simulation results are shown in figures (5,6). The figures show the cumulative distribution of the required downlink transmission powers per link. The required power was measured at a 95% coverage probability target. The figures show that using RoF could result in a reduction of the required TX power of up to 50% compared with wireless access.

5. CAPACITY CALCULATIONS

System Capacity is determined by the amount of the interference level and the accepted level of Quality of Service (QoS) [9]. In the downlink capacity, the lower E_b/N_o necessitates lower transmitted power, which translates into more capacity.

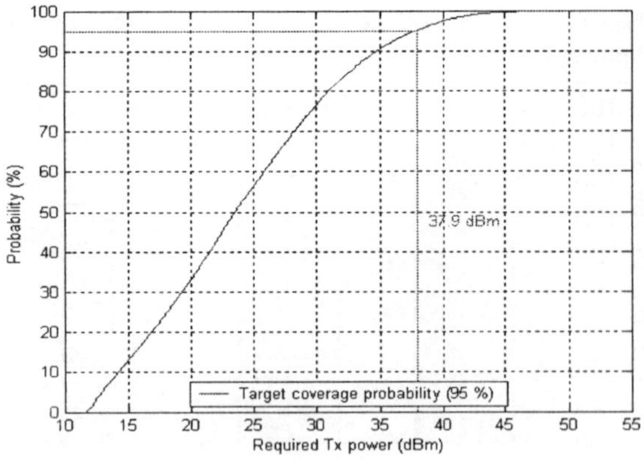

Figure (6) The required power for 95% target coverage with wireless access.

The pole capacity is the theoretical maximum capacity if the mobiles have infinite transmits power available. In practice the capacity is a fraction of the pole capacity. Typical values are 50-60% of the pole capacity. The sustainable capacity is proportional to the processing gain, reduced by the required SNR [10].

$$E_b / (N_o + I_o) = \frac{W / R}{(N_o W / P_s) + N - 1} \quad (6)$$

For maximum capacity $P_S \rightarrow \infty$. Therefore, the basic capacity equation can be reduced to:

$$N \approx \frac{W / R}{E_b / (N_o + I_o)} \quad (7)$$

By considering the effect of additional interference and the voice activity factor, the pole capacity equation is: mobile speeds. The capacity gain is higher in the Pedestrian channel with low mobile speed.

$$N_{pole} = \frac{W / R}{E_b / (N_o + I_o).(F + 1)v} \quad (8)$$

The capacity of the WCDMA system can be represented by the L_f and the total number of traffic transmitted by the network. Figure 4 shows the capacity of Vehicular A channel for different mobile speeds.

6. CONCLUSION

This paper has presented the idea of using Macrodiversity in WCDMA systems with RoF. We computed the E_b/N_o for different cases applying Macrodiversity with SC and MRC. From the uplink we were able to calculate the coverage and our simulation results show that a significant coverage gain, up to 26%, can be realised through the use of MRC in the uplink.

We have also investigated the Macrodiversity performance in the downlink of WCDMA with RoF. The results of our simulations show that using Macrodiversity in the downlink results in a significant capacity gain (up to 46%). The results for system level show that there is considerable power reduction available for each link (up to 50%).

REFERENCES

[1] K. Morita and H. Ohtsuka, "The new generation of wireless communication based on fibre-radio technologies," *IEICE Trans. Commun.*, vol. E76-B, no.9, pp. 1061-1068, Sept. 1993.

[2] Hamed Al-Raweshidy and Shozo Komaki, Radio over Fiber Technology for Mobile Communications Networks, Artech House, 2002.

[3] S. Ariyavisitakul, T.E. Darcice, L.J. Greenstein, M.R. Philips, and N.K. Shankanarayanan, "Performance of simulcast wireless techniques for personal communication systems," *IEEE J. Sel. Areas Commun.*, vol. 1, no.4, pp. 632-643, May 1996.

[4] The Members of Technical staff, Bell labs "*Handbook of CDMA System Design Engineering and Optimization*," Prentic Hall, 1999.

[5] John G. Proakis " *Digital Communications*" Third Edition, McGraw-Hill 1995.

[6] B. John Koshy and P.M Shankar.: 'Spread-Spectrum Techniques for Fiber-Fed Microcellular Networks', *IEEE Trans. on Vehicular Techn.*, Vol. 48, No. 3, pp. 847-857, May 1999.

[7] ETSI TR 101 112, "Selection procedures for the choice of radio transmission technologies of the Universal Mobile Telecommunications System UMTS (UMTS 30.03)", version 3.2.0, April 1998.

[8] Edited by Jaana Laiho, Achim Wacker and Toas Novosad " Radio Network planning and optimisation for UMTS, John Wiley & Sons Ltd, 2002.

[9] http://www.umtsworld.com

[10] http://www.cdg.org/

The Performance Evaluation of Handoff Methods Using H.323

Young-Sin Lee
IP System Gr, Enterprise Communication Lab, System Company
LG R&D Complex Hogye-1dong, Dongan-gu, Anyang-shi, Kyongki-do, 431-749, Korea

Abstract: The purpose of this paper is to evaluate the performance of H.323 handoff methods that use ad hoc multipoint conference, transfer, and third party rerouting signaling respectively. This performance is evaluated in terms of inter-operability and signaling delay. The inter-operability specification includes multipoint conference messages, H.450.2 messages, and null TerminalCapabilitySet message. The delay of each signaling is analyzed into number of TCP connections and number of logical channel setup procedures. The results of our experiment show that third party rerouting signaling has better result than other signalings in inter-operating with terminals of four vendors, since it uses only H.323 mandatory messages. Also it requires fewer number of TCP connections and logical channel setup procedures than the others do, so it can reduce the signaling delay by 1.4 sec in average.

Key Words: VoIP, H.323 Gatekeeper, Mobility, Handoff

1. INTRODUCTION

With an increase in the demand of the wireless LAN environment, many works for IP terminal mobility are being taken in voice over IP (VoIP) technology. The efforts made by ETSI TiPHON Working Group (WG) 7, and ITU-T Study Group (SG) 16 H.323 Mobile Annex are included by the existing activity in the international standard bodies regarding VoIP. TiPHONE addresses user and service mobility for VoIP services, and supports the concept of global multimedia mobility, which assumes that future terminals may connect to several types of access networks and that the choice of access should be made dynamically according to individual needs [4]. H.323 mobile annex addresses mobility issues in layer 2, timers, header compression, the H.323 system architecture, and all three aspects of mobility which are terminal, user and service [5].

In order to provide this function, IP terminal handoff which enables roaming between IP subnets during conversations should be provided. This function can be realized by Mobile IP because Mobile IP's Home Agent and Foreign Agent technology can provide a transparent terminal mobility to the application program. But particularly for this delay-sensitive multimedia application, mobile IP has some limitations, including triangle routing, triangle registration, encapsulation overhead and the need for home address [6]. So this method is needed to be complemented with application level protocols such as H.323.

Some works for IP terminal handoff using H.323 have been taken. H.323 annex H views handoff in the aspect that the mobile terminal dynamically participates in H.323 ad hoc multipoint conference [5]. And since the conference signaling can be used for call transfer function, H.323 call transfer signaling is applied to handoff [1-2]. And since call transfer is a function of rerouting the media channels of a call for one area to another area where a mobile terminal has been moved, H.323 Gatekeeper-initiated-pause-and-rerouting (Third party rerouting) is applied to handoff [3].

In this paper we evaluate the performance of these three H.323 handoff methods. We regard inter-operability and signaling delay as major factors affecting on the performance. The inter-operability means that handoff service can be provided transparently to the existing terminals when mobile terminals are added on an existing network. The signaling delay is time taken to reroute the media channels of a call to a mobile terminal.

The content of this paper has been organized as follows. In section II, H.323 handoff signalings are investigated. In section III, H.323 terminals of five vendors are taken to evaluate the inter-operability. And we analyze the delay of each handoff signaling in terms of call setup delay and the number of signaling messages. In section IV, a conclusion is drawn from the experimental results.

2. H.323 HANDOFF SIGNALINGS

Figure 1 shows terminal mobility architecture in this paper. Mobile IP's mobility entities use an advertisement message (Agent Advertisement) and a solicitation message (Agent Solicitation) for mobility management. Agent Advertisement is transmitted periodically in limited multicast (224.0.0.1) or broadcast (255.255.255.255) address, thus only mobile nodes which locate in same subnet with their home agent can receive this message. When MN1 roams to another subnet, it discovers that its own network connection point has been changed through mobility entities. Then MN2 obtains CareOfAddress (COA), and requests the registration of COA to GK. H.225 RegistrationRequest (RRQ) message which includes terminal's aliases the same as MN1's is used for this registration. Then GK notices the registered endpoint MN1 was changed to MN2, and performs mobility management for MN1. If a call involving MN1 is still exist after the roaming is taken place, then GK performs handoff signaling.

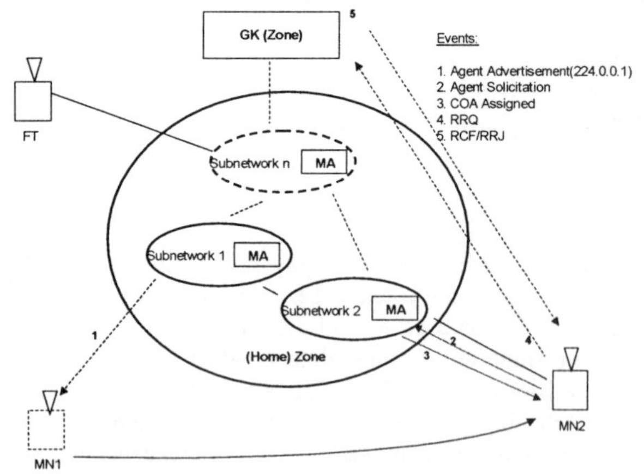

Figure 1. Terminal mobility architecture

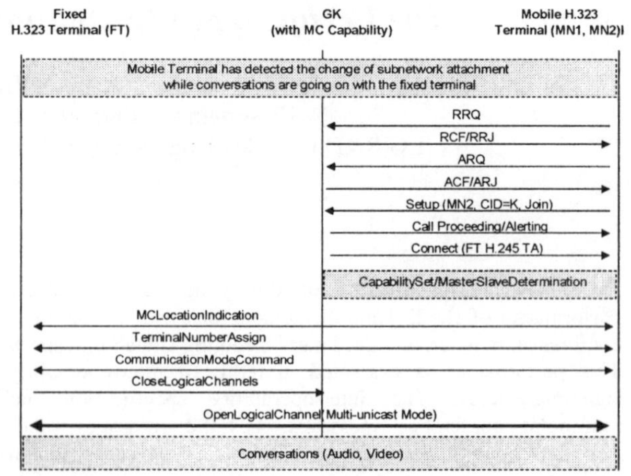

Figure 2. Handoff using ad hoc multipoint conference

2.1 Handoff using Ad hoc Conference

H.323 annex H views handoff from the aspect that mobile terminal dynamically participates in H.323 conference (Figure 2). After MN2 is registered, MN2 expands its call with MN1 to a multipoint conference. To do this, MN2 transmits Setup (Goal = Join) message to GK. Then GK performs a mode switch to relate call information of the peer-to-peer call to the multipoint call. After this mode change, GK takes a role of multipoint controller (MC) controlling the conference. MC transmits McLocationIndication (MLI) message and TerminalNumberAssign (TNA) message to inform the participants that the corresponding call has been changed to multi-point call. In addition, GK transmits CommunicationModeCommand (CMC) message to inform the participants of a selected common mode. Then all the terminals in conference should comply to CMC message. If media processing mode currently opened is different from the mode of CMC message, the terminals should transmit CloseLogicalChannel (CLC) message and then new OpenLogicalChannel (OLC) message to GK. Handoff is accomplished after all the terminals perform logical channel setup signaling (OLC/OLCACK) according to CMC.

An environment in which GK supports only MC and no multipoint processor (MP) mixing several media streams, brings about more complicated signaling. That is, every terminal attending conference should transmit its own audio data in multi-unicast and mix every audio stream coming to itself. Accordingly, the terminal performs the logical channel setup signaling several times.

2.2 Handoff using Call transfer

Since H.323 conference signaling can be used for call transfer function, H.323 call transfer signaling which uses H.450.2 can be also applied to handoff. H.450.2 protocol considers many options and error conditions which can take place in H450.2 signaling entities[7].

Figure 3 shows an handoff using H450.2. GK having received RRQ message from MN2 transmits TransferRequest (CT_init) message to FT. Then handoff is performed by FT which functions as H.450.2 transferred signaling entity (TRDSE). In other words, in this handoff architecture, FT is required to implement H.450.2 stack fully. The reason for this is that FT should consider implementing routines to encode and decode all H450.2 messages, to relate original call with transfer call, and to generate a timer handling abnormal states which take place when an anticipated call state is not reached within a predetermined time. However it is very difficult for IP Phone to implement H.450.2 stack considering all these cases, since IP phone has a severe memory restriction.

2.3 Handoff using Third Party Rerouting

H.323 is a dynamic protocol that can change the logical channel information of a call by using H.245 control channel messages. Due to this property, H.323 can change RTP session information of a call by means of using mandatory H.245 messages. For example, GK transmits null TerminalCapabilitySet (TCS = 0) message to a terminal to close its outgoing channels (Paused state). Then GK transmits TCS message of a different terminal to the paused terminal so that the paused terminal can

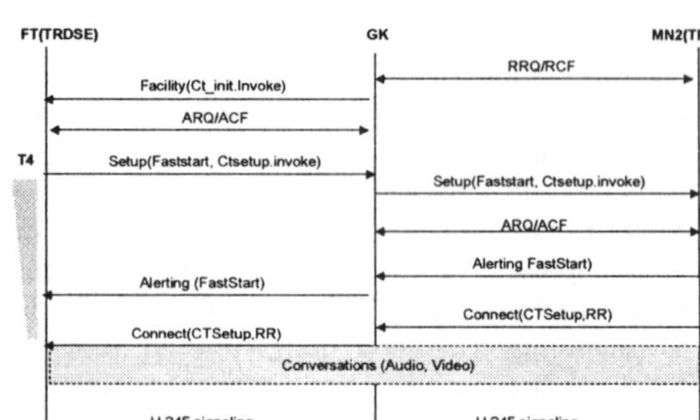

Figure 3. Handoff using FT's TRDSE

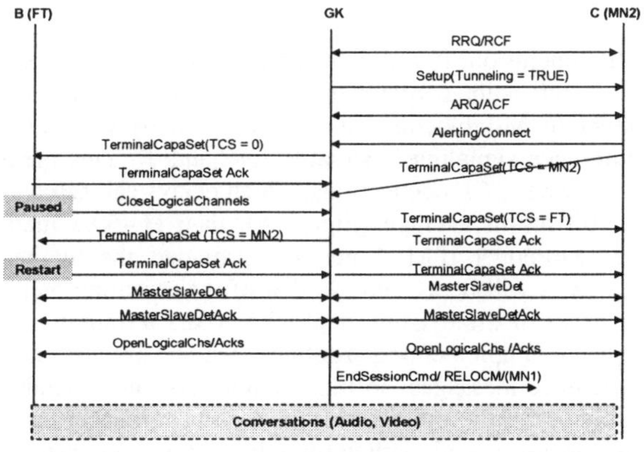

Figure 4. Handoff using third party rerouting

transmit its RTP data changed according to the newly received TCS, to the different terminal (Restart State). Thus if this rerouting function is implemented in GK, the terminal without H.450.2 stack can be provided with handoff service. Figure 4 shows an handoff using third party rerouting signaling. Receiving RRQ message from MN2, GK transmits null TCS message to FT to close its outgoing channels. Then GK transmits MN2's TCS message to FT so that FT can transmit its RTP data to MN2's address.

3. EXPERIMENT

In order to evaluate the performance of handoff methods, we have conducted an experiment that transfers call by the three investigated handoff signalings because this function can simulate handoff function. In particular, we have our GK stressed by a call generator which generates a call per 100ms and measure the conference signaling in the environment where the GK supports only MC.

3.1 Inter-Operability Result

We have taken H.323 terminals of five vendors for inter-operability. Table 1 shows the inter-operability result of each handoff signaling. The row tells inter-operability specifications and the column tells terminals of five vendors.

Table 1. Inter-operability result

Signaling \ Terminal	A	B	C	D	E
Basic H.245	O	O	O	O	O
Tunneling	X	O	O	O	O
H.450.2	X	X	X	X	O
Conference H.245	X	X	X	X	O
Rerouting H.245	X	O	O	O	O

A : Netmeeting.
B : SoundPoint IP 500.
C : AS5300.
D : Teliphone 2000.
E : LG SoftPhone.

O : Interoperable with GK on the spec.
X : Not Interopable with GK on the spec.

For the conference signaling, terminals of four vendors are not inter-operated with our GK in the following reasons. First, they do not transmit Setup message whose conference goal is INVITE or JOIN. Second, they do not conform to CMC message received from MC as shown in Figure 2. Third, MC needs to be complemented by a specific error handling routine. It should transmit TerminalLeftConference (TLC) messages to terminals because MN1 does not exist at the time when handoff occurs, then MN2 and FT should close the logical channel setup signaling which has been tried with MN1. For H.450.2 signaling, terminals of four vendors are not inter-operated in the following reasons. First, GK needs to be complemented by a specific routine that transmits CT_init message on behalf of MN1 which cannot transmit the message until it is reconnected with GK. Second, FT does not signal as shown in Figure 3 because the terminal has implemented only H.450.2 transferring signaling entity (TRRSE). On the other hand, only terminals of one vendor are not interoperated with GK in third party rerouting because they do not comply to null TCS as shown in Figure 4. And because the rerouting signaling for call transfer can be used for handoff without modification, GK does not need to be complemented by additional routine.

3.2 Signaling Delay Result

To estimate an average delay time of each handoff method, we compare call transfer signaling delays. As we see figure 5, the delay of third party rerouting signaling is smaller by 1.8 sec in average than that of call transfer and by 1.4 sec in average than that of ad hoc conference signaling. To look into these differences, we measure the performance of H.323 call setup procedure and take two factors as major contributors to the call setup delay –the number of TCP connections and the number of message exchanges.

Table 2 shows the number of TCP connections, H.225 and H.245 signaling round trip(RT)s for each transfer signaling. Third party rerouting and ad hoc conference require TCP connections as half as those of call transfer. It requires new Q.931 and H.245 TCP channels to be made not only between MN2 and GK but also between FT and GK, whereas the others require Q.931 and H.245 TCP channels only between MN2 and GK. To investigate TCP connection time, we compare the call setup time in tunneling with the call setup time in non-tunneling since H.245 tunneling needs only one TCP connection for both Q.931 channel and H.245 channel. Figure 6 shows that call setup time is faster by 0.23 sec in average when both the caller and the callee use tunneling than when they do not use tunneling.

In GK routed signaling model, uni-direction H.245 request signaling is done in 2 RT. So the TCS, MSD, OLC, CLC signaling of H.450.2 method requires 4 RT respectively because they should be taken over in bi-direction. By the conference method, TCS and MSD signalings are done by 2 RT because they are required only between GK and MN2. And OLC signaling and CLC signaling are done by 3 * 4 RT, 2 * 4 RT respectively because they are taken over in multi-unicast mode in this experiment. On the other hand, in comparison with the H.450.2 method, the rerouting method only requires additional TCS signaling of 2 RT and CLC signaling of 4 RT to pause FT and MN1. This feature requiring fewer OLC and CLC RTs than the conference method cause third party rerouting to save time for opening new RTP resources and releasing used RTP resources.

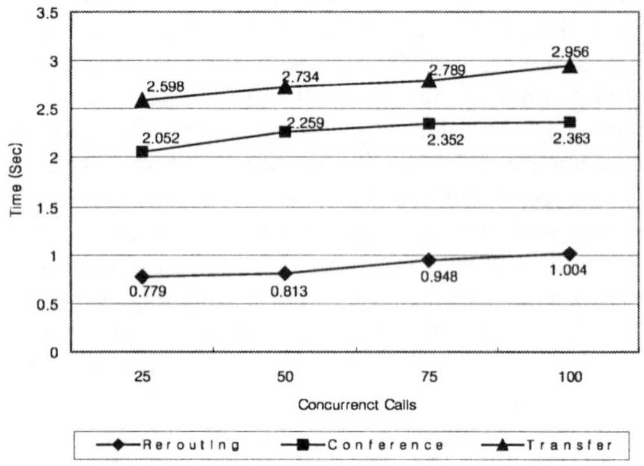

Figure 5. Delay differences of transfer signalings

Table 2. Number of the signaling RTs for each method

	TCP # Connections	ARQ # Round Trips	Q.931 # Round Trips	TCS # Round Trips	MSD # Round Trips	OLC # Round Trips
Conference	1 (tunneling)	1	1	2	2	3 * 4
H.450.2	2 (tunneling)	2	2	4	4	4
Rerouting	1 (tunneling)	1	1	2 + 4	4	4
	CLC # Round Trips	MLI # Round Trips	TNA # Round Trips	TLC # Round Trips	CMC # Round Trips	
Conference	2 * 4	3	3	2	3	
H.450.2	4	0	0	0	0	
Rerouting	4 + 4	0	0	0	0	

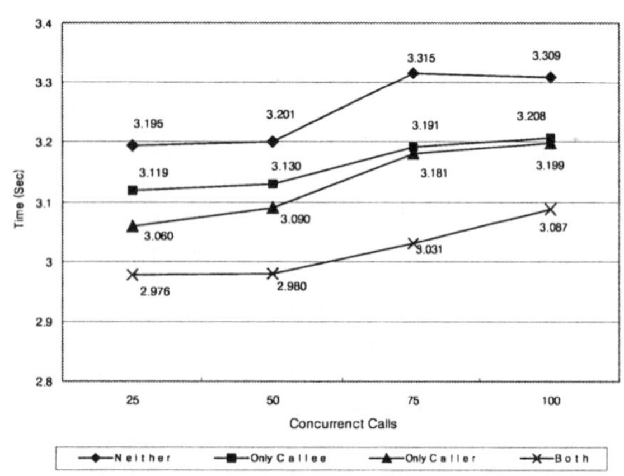

Figure 6. Call setup time differences

4. CONCLUSION

In conclusion, if a mobile terminal is already connected to a multi-point conference before it roams to the other subnet or MP function can be provided by GK, ad hoc conference signaling is suitable for handoff. This method needs fewer TCP connections, which delays call setup time by 100 ms in average (Figure 6), than H.450.2 method does. Moreover, if MP function is provided, the number of OLC RTs will be as same as that of the rerouting method. But most of the existing deployed H.323 terminals are in short of implementing routines which process H.245 conference messages, so this method has inter-operability problem. On the other hand, third party rerouting method inter-operates well with terminals of four vendors since most of H.323 terminals can process null TCS which is one of H.323v2 mandatory requirements. Also this method needs fewer OLC RTs than ad hoc conference multipoint method, so it is faster by 1.4 sec in average.

REFERENCES

Proceedings Papers:

[1] Wanjiun Liao, "Mobile Internet Telephony: Mobile Extension to H.323", Proc. IEEE INFOCOM '99, New York, Mar. 1999

[2] W. Liao and J.-C. Liu, "VoIP mobility in the internet/cellular networks interworking", IEEE Commun, Mag., pp. 2-8, Apr. 2000.

Journal Papers:

[3] Young-Sin Lee, Gi-Moo Choi, "Handoff signaling implementation using H.323 Third party pause initiated and re-routing in IP-based network", The KIPS Transactions VOL. 8-C, pp. 821-830, Dec. 2001

Books:

[4] ETSI TiPHON, *Analysis of existing roaming techniques applicable to TiPHON mobility services*, TR 101 338 v1.1.2, May 1999

[5] ITU-T Rec H.323 Annex H, *H.323 Mobility Architecture and Protocol for Terminal, User, Service Mobility*, Feb 2000

[6] C. E. Perkins, *Mobile IP: Design Principles and Practice*, Addison-Wesley, 1998.

[7] ITU-T Rec. H.450.2, *Call transfer supplementary service for H.323*, Feb 1998

A CONTEXT INFORMATION SERVICE FOR MOBILE APPLICATIONS

FENG WANG, PADDY NIXON
Department of Computer and Information Sciences
University of Strathclyde, Glasgow, Scotland

Context-aware mobile applications take user's context as implicit input and change their behaviors accordingly to reduce the interruptions to user's attention. One significant challenge for context-aware computing is the collection, processing, organization and dissemination of context on a global scale. In this paper we present an approach to the specification of context flows that encapsulate the data flow characteristics of the context provision to applications. Within the flow description we capture 3 core concepts: constraints over basic context sources; the structure of the context flow; and constraints over the deployment of the components. We also outline the underlying architectural assumptions and implementation details.

1. Introduction

Mobile, and more specifically mobile pervasive computing, can broadly be defined as calm technology that delivers the correct service to the correct user, at the correct place and time, and in the correct format for the environment. Context, viewed alongside this definition, is all the information necessary to make a useful decision in the face of real-world complexity. More specifically, Contextual Computing refers to the intersection of several related trends in computing: the increasing pervasiveness of computational devices in the environment, the mobility of users, the connectivity of mobile users' portable devices and the availability to applications of relevant information about the situation of use, especially that based on data from physical sensors. This paper explicitly addresses questions of context in the setting of mobile pervasive devices, within a ubiquitous wireless computing infrastructure.

1.1. *Context*

Historically, the use of "context" grew from roots in linguistics [6]. More recently, in the setting of mobile computing, "context aware" was at first defined by example, with an emphasis on location, identity and spatial relationships [7,8,9,10,11]. This has since been elaborated to incorporate more general elements of the environment or situation. Such definitions are, however, difficult to apply operationally and modern definitions [12,13] generalize the term to cover "any information that can be used to characterize situation". Current work in the field addresses issues including:

- developing new technologies and infrastructure elements, such as sensors, middleware, communication infrastructures to support the capture, storage, management and use of context;
- increasing our understanding of form, structure and representation of context;
- increasing our understanding of the societal impact of these new technologies and approaches and directing their application.

A more detailed retrospect of the academic history of context can be found in [2,4]. While there remains great need for pure research work on the theoretical underpinnings of context, there is also an immediate problem of providing to application developers ways in which they can describe the context needs of their applications in manner that is orthogonal to the application or business logic of the application. We call this *orthogonal context* and define it as the information needed to adapt the application in a given environment.

1.2. *Requirements of Orthogonal Context*

In previous work we, and others, have reviewed the requirement of infrastructures that deliver such orthogonal context in general manner [2,3,4]. The response to these requirements can be seen in various projects, but most notably in projects Aura[14], Solar[16], iQueue[15], Context Toolkit[1], and Contextors[17]. In reviewing these projects, we have observed the following open issues in the management of mobile pervasive computing environments that we wish to address with our infrastructure [3]:

- Dynamic composition of context entities;
- Control over the quality and structure of context composition;
- Adaptivity to environmental changes (e.g. component failure);

- Flexible and extensible representation and retrieval of contextual information; and
- A scalable global infrastructure.

In the following sections, we briefly outline our architectural view of mobile pervasive systems focusing on application infrastructure. We then outline a context flow language for orthogonal context provision. We conclude with a summary of the current implementation status.

2. The SCINET Architecture

The Strathclyde Context Infrastructure (SCI) is organised into two distinct layers. The upper layer of the infrastructure is a network overlay of partially connected nodes and is referred to as the SCINET. The lower layer of the infrastructure concerns the contents of each node, which consists of entities (People, Software, Places, Devices and Artefacts) responsible for producing, managing and using contextual information, and is referred to as a range.

A Range is defined as an area that can be described in logical and/or physical terms. A Range can be can be bounded by a physical area such as a building or by the effective operating range of a particular network type (e.g. a wireless network). Each Range is governed by it's own individual Context Service. More details of the general architecture can be found in [3].

2.1. *The Detailed Design of the Context Information Service*

Within a given range an application dynamically composes elements of context (context entities). An application's context requirement can be specified as a data flow in which data from multiple sources flow through a sequence of components to the application. We call this data flow style specification of context requirement "context flow". The context flow captures 3 core elements:

1. The constraints over basic context sources (sensor input or other primitive context sources) that capture the quality requirements of the context information.
2. The structure of context flow. That is the components and the connections necessary to transform the basic context information into application usable data.
3. The constraints over the deployment of the components.

An application uses the context flow language to specify the context processing steps in a procedural manner. Compared to our procedural approach, a declarative approach just asks applications to describe the context data they need in term of data description, and then the environment will try to get the context data and provides it to the applications. Orthogonal context is user and task specific and different applications have different meaning for the same context. Considering the diversities of user and their tasks, we do not believe that it is feasible for a context service to try to predict all the possible context of roaming applications. Instead we assume a minimum core of basic context information exists in every environment that the user will visit. Here basic context means the physical context such location, temperature. We also assume the environments share a set of metadata about the set of basic context, that is, a minimum ontology of basic context. Thereafter we assume a lifecycle of the context processing flow as follows:

1. The application submits the context flow specification to the run-time environment and waits.
2. The run-time parses the specification and tries to deploy a flow to provide the required context to the application. If successful, a reference of a flow instance will be returned to the application.
3. The application subscribes to the output of the flow instance using the reference.
4. The flow instance disseminates the context information to the application.
5. The flow instance adapts to changes such as sources failure in the environment. If it couldn't recover from the failure, it will notify the application.
6. After the application finishes its work or it receives failure notification from the run-time environment, it requests the run-time to tear up the flow.

2.2. *The Component Model*

In our system, both the sources and the context processing steps are components. The sources and context processing components implement similar interfaces in manner similar to Contextors[17]. The communication between components is through events and supports event subscribers that implement data listener interfaces. The notify method in the interface is called when event publishers publish events. Event publishers implement subscription interface to accept subscriptions and remove subscriptions. They further implement a control interface to modify the parameters at run time.

2.3. *The Context Flow Language*

We base our view of a context flow on the principals embodied in web service flow languages such as WSFL [18] and the context flow language is consequently realised as XML. As we see above, the context specification of one application consists of three parts. The features of our context flow language includes are outlined below and explain by example.

2.4. *The Context Entities*

The entities include sources or context processing components. For a source, the language allows the application developer to specify the multiplicity of the source and constraints over the source. Multiplicity means that one source or all sources that satisfy the constraints. An XML fragment of constraints over a location source is shown below:

```
<source name="LocationSource1">
  <multiplicity>One</multiplicity>
  <contextConstraint type="IndoorLocationSource">
    <predicate compareType="GREATERTHANEQUAL">
      <metaData>UpdateRate</metaData>
      <value>2</value>
    </predicate>
    <predicate compareType="EQUAL">
      <metaData>Resolution</metaData>
      <value>Room</value>
    </predicate>
    <predicate compareType="EQUAL">
      <metaData>PhysicalCoverage</metaData>
      <value>Floor10</value>
    </predicate>
  </contextConstraint>
</source>
```

Figure 1: Location constraints

Depending on the type of binding, there are two kinds of components: one type is a user-provided component with known code, but unknown deployed host. For this kind of component, the language allows the application developer to specify the code base and the constraints over the deployed hosts. An XML fragment for a component with known code is shown in figure 2. The second kind of component is component-by-type. The language must allow an application builder to specify the class name or interface name in manner similar to type lookup in CORBA trading services. An XML fragment for a component-by-type is shown in figure 3. We can also include deploy host constraints to both of the above component constraints although these are not outlined here to aid clarity.

```
<userComponent name="LocationFilter1">
<codeBaseURL>http://www.abcdef.com/codebase/
</codeBaseURL>
  <className>LocationFilter</className>
    <parameter name="username" type="String" value="Bob"/>
</userComponent>
```

Figure 2: a user provided location filter component

```
<systemComponent name="NearbyPrinterList1">
  <interfaceName>NearbyPrinterList</interfaceName>
</systemComponent>
```

Figure 3: a component-by-type example

2.5. *The Connections between Components*

A connection usually connects the data output of one component to the data input of another component. It may happen that the data output of a control component connects the control input of another component. An example is as follows. In this example the input of a location filter is connected to output of a location source.

```
<dataLink source="LocationSource1" target=" LocationFilter1">
</dataLink>
<dataLink source="LocationFilter1" target ="flowOutput">
</dataLink>
```

Figure 4: Connections and constraints on flows.

2.6. *The Run-time Environment*

We assume there is a lookup service in each environment. The lookup service will maintain the status of available basic context sources and context components. The quality of information for a basic context in the environment is decided by the user's application. In our design, there is a source factory server in each environment. It takes the constraints over the source and creates an adaptive source component that will dynamically choose the source available in the environment and which acts as a proxy for the basic source guaranteeing the context information satisfies the users constraint.

For the component with known codebase, the run-time downloads the code and deploy the component. For component-by-type, the run-time searches the lookup service and get a reference to it. After the run-time environment gets the references to all the components in the flow. It will connect the components together according to the interconnection part of the flow. The run-time environment will create a flow component that will monitor the status of the components and manage the dynamic adaptation to changes in the environment. The flow component acts as a proxy between the flow and the application. It receives the output from components inside the flow and forwards it to the application. It also manages the subscribe/unsubscribe and teardown flow request from the application.

3. The Implementation

We choose JINI as the platform to implement the work described above because it provides many facilities we need. The adaptive source has been implemented using the JINI lookup service that supports searching for services based on the type of service and attributes of services. Basic context sources are registered in the JINI lookup service with the quality of context information encoded in the attributes of the service. Whenever the quality of the context information changes, the context source will notify the lookup service. The run-time environment uses the constraints over the context source as parameter and calls a factory server to create an adaptive source object. This adaptive source object dynamically chooses available sources in the environment that satisfies the constraints.

4. Conclusion

We propose a context information service to support context-awareness where mobility on a global scale is our central concern. Our solution is based on workflow-like language and the corresponding run-time environment. We are currently refining the context flow language and implementing the extended run-time environment for the context flow language.

References

1. D. Salber, A.K. Dey, G. Abowd. The Context Toolkit: Aiding the Development of Context-Enabled Applications. In Proc. CHI99, ACM Publ., 1999, pp. 434-441.
2. Dey, A. K. "Understanding and Using Context", Personal and Ubiquitous Computing, Vol 5, No. 1, pp 4-7, 2001.
3. Richard Glassey, Graeme Stevenson, Matthew Richmond, Paddy Nixon, Sotirios Terzis, Feng Wang, Ian Ferguson, Towards a Middleware for Generalised Context Management, to appear in Proceedings of Middleware for Pervasive and Ad Hoc Computing, 2003.
4. J. L. Crowley, J. Coutaz, G. Rey and P. Reignier, "Perceptual Components for Context Aware Computing", Proceedings of Ubicomp 2002, Sweden, September 2002.
5. Schilit, B, N. Adams and R. Want, "Context-Aware Computing Applications", in First international workshop on mobile computing systems and applications, pp 85 - 90, 1994.
6. T. Winograd, "Architecture for Context", Human Computer Interaction, Vol. 16, pp401-419, 2001.
7. B. Schilit, and M. Theimer, "Disseminating Active Map Information to Mobile Hosts", IEEE Network, Vol 8 pp 22-32, 1994.
8. P. J. Brown, "The Stick-E Document: A Framework for Creating Context Aware Applications", in Proceedings of Electronic Publishing, '96, pp 259-272.
9. T. Rodden, K.Cheverest, K. Davies and A. Dix, "Exploiting Context in HCI Design for Mobile Systems", Workshop on Human Computer Interaction with Mobile Devices 1998.
10. A. Ward, A. Jones and A. Hopper, "A New Location Technique for the Active Office", IEEE Personal Comunications 1997. Vol 4. pp 42-47.
11. K. Cheverest, N. Davies and K. Mitchel, "Developing a Context-aware Electronic Tourist Guide: Some Issues and Experiences", in Proceedings of ACM CHI '00, pp 17-24, ACM Press, New York, 2000.
12. J. Pascoe "Adding Generic Contextual Capabilities to Wearable Computers", in Proceedings of the 2nd International Symposium on Wearable Computers, pp 92-99, 1998.
13. Dey, A. K. "Understanding and Using Context", Personal and Ubiquitous Computing, Vol 5, No. 1, pp 4-7, 2001.
14. David Garlan, Daniel P. Siewiorek, Asim Smailagic and Peter Steenkistie. Project Aura: Toward Distraction-Free Pervasive Computing. IEEE Pervasive Computing Magazine, April-June 2002.
15. N. H. Cohen, A. Purakayastha, L. Wong, and D. L. Yeh. iQueue: A Pervasive Data Composition Framework. The 3rd International Conference on Data Management, Singapore, Janauary 8-11, 2002.
16. G. Chen and D. Kotz. Supporting Adaptive Ubiquitous Applications with the Solar System. Technical Report TR2001-397, May, 2001.
17. J. Coutaz and G. Rey, Foundations for a Theory of Contextors. In Computer-Aided Design of User Interfaces III. Kluwer Academic Publishing, 2002, pp 13-32
18. Leymann, F. Web Services Flow Language (WSFL 1.0). IBM (2002)

POSTER SESSION P2

A Simple and Efficient MAC for Implementation in a Wireless Sensor Networks

KL Chee, JG Lim, SV Rao, P K Sivaprasad
Institute for Infocomm Research
Singapore Science Park II
Singapore
Fax: +65 67795441
{cheekl, limjg, raosv, pksiva}@i2r.a-star.edu.sg

Abstract: We present Simple MAC with Relative Time-slot referencing (SMART) for wireless sensors networks. The features of the SMART are relative time referencing and very short time slots which are different from standard fixed allocation TDMA-based MAC. These allow the SMART to operate under loose synchronisation. The SMART also has the capabilities to maintain sufficient guard time between itself and its neighbours, reducing effect of the timing drift. SMART has a software solution for mobility detection. SMART is simple enough to be successfully implemented on a low-power microcontroller.

Key Words: MAC, TDMA, Wireless Sensor Network

1 INTRODUCTION

Wireless Sensor Networks are gaining the attention of researchers worldwide. Currently, there are various projects working on the different aspects of Wireless Sensor Networks, ranging from the hardware implementation to applications like location estimation and data dissemination. Some of these projects employ existing standards and components in their simulations, implementations and testing. The most common one is the use of 802.11 Wireless LAN as the Medium Access Control (MAC) layer.

In-depth analysis [6] has shown that the 802.11 Wireless LAN or other existing MAC protocols, like the BlueTooth, are unsuitable for Wireless Sensor Networks. These protocols focus on providing Quality of Service and bandwidth efficiency. Power conservation is of secondary importance. However, power conservation is an important requirement in Wireless Sensor Networks. A sensor node is equipped with very limited battery, memory and processing power, in order to achieve low cost and small form factor. Additionally, replenishment of resources is almost impossible, because a node often has to operate with minimum or no human intervention.

It is noted that the RF transceiver is usually the component that consumes the most energy on a basic sensor node. The most efficient method for energy conservation is to switch off the RF transceiver when it is not needed. Hence, an indicator of a good MAC protocol will be the percentage of the switch-off-time of the RF Transceiver.

Furthermore, the MAC protocol has to be simple to complement the node's limited memory and processing power. The MAC protocol also has to be operationally reliable even in devices containing low cost components with high tolerance deviations. It is also helpful if the MAC protocol can perform some of the sensory tasks, such as movement detection, as it helps to reduce the number of components on the device.

There are various schemes available to address the energy issue. The more common ones are contention-based [2-5] and fixed allocation TDMA based [1] protocols. Each has its merits and drawbacks. Generally, fixed allocation TDMA protocols have the advantage of having predictable transmission slots to schedule the switch-off-times of RF transceivers and avoid data collision. However, such protocols require some form of timing synchronisation and the process can be taxing for the resource-constrained sensor nodes. Furthermore, variations in components' tolerance result in timing drift. In order to maintain the timing alignment, the design has to be more complex.

In view of these issues, we propose a Simple MAC with Relative Time-slot Referencing, known as SMART, for Wireless Sensor Networks. This is based on fixed allocation TDMA, which provides the predictability of the transmission slot to achieve energy conservation. SMART has simple but effective solutions to combat the timing synchronisation and timing drift problems. Furthermore, SMART has the capability to detect mobility without any additional hardware.

2 DESIGN

The SMART is based on fixed-allocation time division multi-access (TDMA). It is known that the main disadvantages of TDMA systems are the requirement for time synchronisation and time slot allocation. This is especially so for wireless networking environment where the nodes may spread over a large geographical location, making it almost impossible for any form of centralised control for synchronisation or time slot allocation.

The followings are the key features of SMART.

2.1 Network Forming and Joining

SMART does not have a centralised control for time slot allocation. Instead, it uses a semi-randomised approach for time slot allocation. The process is based on a mutual agreement of all the nodes within the communication range of the new node.

In SMART, the node broadcasts only one message per frame. All the neighbours will process the message based on the destination and message type.

The basic operation of the SMART is, upon powering up, the node monitors the neighbourhood for a random period of time. If there is no network present, the node initiates an invitation message at a random time slot and forms a network with any node responding to its invitation. However, if there is a network present, the node monitors the surrounding for a longer period of time. Then it broadcasts a request message in a random free time slot to join the existing network. If the node receives no objection from its neighbours, the node considers that it has joined the network. Otherwise, the node will try again in a different time slot.

Once the node has formed or joined the network, it will consistently broadcast its message in the same time slot.

2.2 Relative Time Slot Referencing

With the nodes spread over a large geographical location and without centralised control, it is almost impossible to have global time slot reference. On top of that, when a node powers up between two unconnected networks, it has to resolve the timing difference between two networks. Furthermore, the time slot referencing is used only within a neighbourhood and not over the entire network. Having a global or absolute time slot referencing is impractical.

Instead, relative time slot referencing is employed as shown in Figure 1. Each node marks its own transmission slot as slot 0. The node keeps track of its neighbours' transmission slots relative to its own. Since this information is used internally in the node, there is no requirement to have global time synchronisation. Relative time slot referencing also makes it easy for the nodes to shift its time slot, simply by changing a counter value. This is particularly useful in maintaining sufficient guard time between itself and the neighbours to reduce the effect of timing drift.

There may be occasions when the node needs to pass time slot information to its neighbours. Relative Time Slot Referencing can handle such occasions with ease, despite that there is no global time slot referencing. Referring to Figure 1, using an example of having 256 slots in a frame, node *x* broadcasts some information at slot 0 (its transmission slot) regarding its time slot 10. The recipient, node *y* receives the message at slot 246. Node *y* does a simple add 10 and 246 which gives 256. Node *y* knows that node *x* is referring to its slot 0 because 256 divided by 256 gives a remainder of 0.

Using such simple technique, SMART eliminates the need for a global time slot reference or frame alignment.

2.3 Short Time Slot

To conserve energy, only useful information is transmitted. Therefore wireless sensor network does not have a fixed message length and it depends on the message type. In our time-allocation scheme, we implemented time-slots with duration much shorter than a typical message length. Hence each message spans over a few time slots. One example is 2 to 4 time slots as shown in Figure 2.

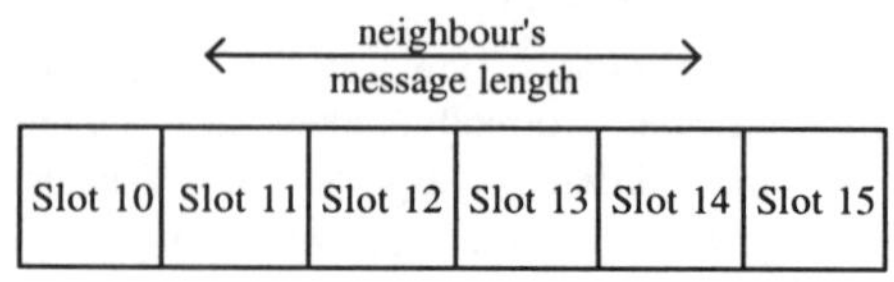

Figure 2 Short Time Slot

Having shorter time slot means that, the sensor node does not have to contain the variable-sized message within a time slot. This simplifies the requirement to align the time slot boundaries to minimise time slot wastage. This not only provides additional flexibility in the time slot allocation process, the synchronization scheme is much simplified for implementation in the resource limited sensor node. This also means that the node can switch on and off the RF transceiver with more precise timings, leading to less energy wastage when the node has to wait for transmission.

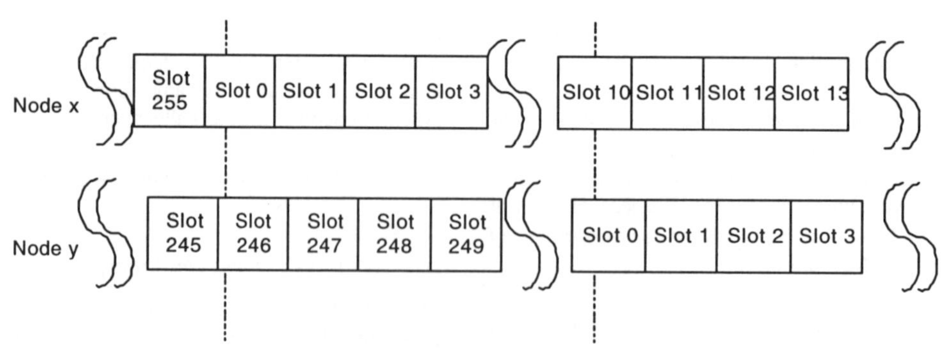

Figure 1 Time Referencing

2.4 *Software-Based Mobility Detection Mechanism*

Referring to Figure 3, when a mobile node moves to a new location, it finds new neighbours. Within a short period of time, it also finds that it is losing some of the existing neighbours.

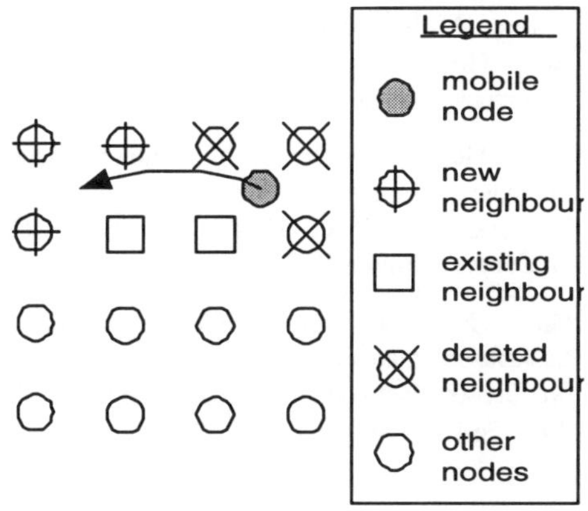

Figure 3 Mobile Nodes

As in all TDMA-based MAC schemes, SMART broadcasts its messages in pre-determined time slots. All neighbours are able to receive the messages sent out by a node. The neighbours will process (or drop) the received message depending on the message source, destination and type header, reducing the complexity of MAC.

In order to conserve energy during the idle slots, each node will keep a record of nodes within its neighbourhood that it is able to "hear". This ability to keep track of neighbours allows a node to monitor the number of new and lost neighbours it has over a short period of time. A node will deem that it has moved from its original location if there is a sudden change in this number.

This provides a simple software solution for mobility detection. With such a capability incorporated in the MAC layer itself, there is no need for a movement detector on the node, when the network has more stationary nodes than mobile nodes.

2.5 *Simplicity and Low Resource Requirement*

SMART has been successfully implemented on a very low power microcontroller running on 2.4MHz clock. The MAC only uses less than 5Kbyte of ROM, about 200 bytes of RAM and a timer interrupt. This gives a lot of allowance for other applications to be added into the node.

3 SIMULATION RESULT

A good indicator of a efficient MAC is the percentage of off time of the radio transceiver. It is a known fact that the transceiver is the most power-consuming device on a wireless sensor node. Switching off the radio transceiver when not needed is an effective way to conserve energy.

In the simulation, we consider two scenarios. The first is the case where the time slot is longer than the message duration (64 Time Slot Frame). The second case is where the message spans over 4 time slots (256 Time Slot Frame).

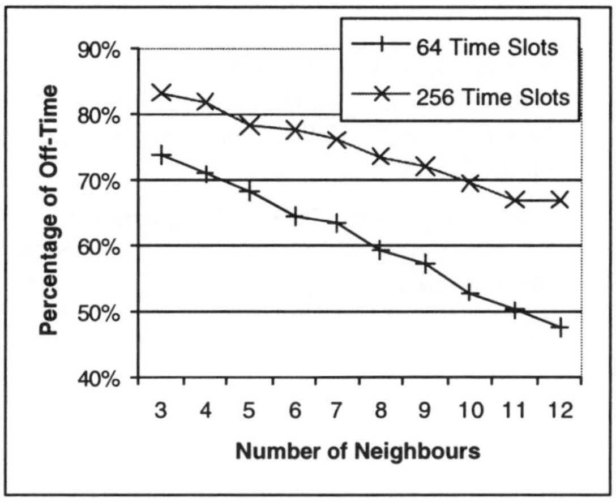

Figure 4 Percentage of Off-Time

From Figure 4, we can clearly see that, a much shorter time than the message duration can help to increase time per frame for the transceiver to be switched off. This is especially visible when the node density increases. When the node is having only 3 neighbours, the improvement is only about 10%. But when the node has 12 neighbours, the improvement is almost doubled.

This means that, this implementation can greatly extend the battery life of the node.

4 CONCLUSION

We present SMART protocol for Wireless Sensor Networks. Unlike similar MAC protocols, SMART employs relative time referencing and very short time slot duration. Through practical implementation, these prove to be a simple solution to the problem of time synchronisation and timing drift commonly found in fixed-allocation TDMA protocol. Through simulations, we also observed that shorter time slot duration could help to improve energy conservation.

The SMART also has the feature of detecting movement simply by monitoring the number of new or lost neighbours within a certain period. Through simulation and practical implementation, we observed that the MAC was able to detect movement using a combination of changes in both the number of new and missing neighbours that it monitors over a short period of time. The node was able to use this information to trigger the necessary actions related to node movement.

Because of its simplicity and low resource requirement, the SMART can be easily implemented in most low-power microcontrollers.

REFERENCES

[1] K Sohrabi, J Gao, V Ailawadhi and G J Pottie, "Protocols for Self-Organization of a Wireless Sensor Network", IEEE Pers. Commun., Oct 2000, Pg 16-27

[2] Alec Woo, David Culler. "A Transmission Control Scheme for Media Access in Sensor Networks". Mobicom 2001.

[3] Wei Ye, John Heidemann and Deborah Estrin, "An Energy-Efficient MAC Protocol for Wireless Sensor Networks", In Proc of the 21st International Annual Joint Conf of the IEEE Comp and Commun Soc (INFOCOM 2002), New York, USA, June, 2002.

[4] S. Singh and C. Raghavendra. "Pamas - Power Aware Multi-Access Protocol with Signalling for Ad Hoc Networks",. In ACM Computer Communication Review, 1998

[5] A. El-Hoiydi, "Aloha with Preamble Sampling for Sporadic Traffic in Ad Hoc Wireless Sensor Networks", in Proc. IEEE International Conference on Communications ICC 2002.

[6] Ian F. Akyildiz, Weilain Su, Yogesh Sankarasubramaniam and Erdal Cayirci, "A Survey on Sensor Networks", IEEE Commun. Mag., Aug 2002, Pg 102-114

OLSR ROUTING FOR MILITARY AD-HOC NETWORKS

Thierry Plesse[1], Jerome Lecomte[1], Cedric Adjih[2], Marc Badel[2], Philippe Jacquet[2],
Anis Laouiti[2], Pascale Minet[2], Paul Muhlethaler[2], Adokoe Plakoo[2]
[1] DGA/CELAR, BP 7419, 35174 Bruz Cedex, France.
[2] INRIA, BP 105, Rocquencourt, 78153 Le Chesnay Cedex, France.

Abstract: Wireless ad-hoc networks do not require any preexisting infrastructure and support mobility. Thus, they are excellent candidates for military tactical networks. In such networks, multihop routing is often needed. In this paper, we report measures made on an OLSR platform consisting of eigtheen nodes implementing OLSR, a proactive routing protocol proposed as experimental RFC. Some nodes are embedded in vehicles. This platform is representative of real military tactical configurations in urban areas.

Key Words: Mobile ad-hoc network, MANET routing, military network, performance evaluation.

Wireless ad-hoc networks do not require any preexisting infrastructure and support mobility. Thus, they are excellent candidates for military tactical networks, where their ability to be operational rapidly and without any centralized entity is essential. As radio coverage is limited, multihop routing is often needed. OLSR (Optimized Link State Routing) is a routing protocol presented at the IETF MANET (Mobile Ad-hoc NETwork) working group for dynamic routing in ad-hoc networks. OLSR is a proactive protocol: periodic control packets are in charge of monitoring the network topology. Proactive protocols are particularly well suited for military applications since all nodes know the network topology, allowing monitoring of the connexity of the network, and also they can be used without modification in the network protocol stack.

This paper is organized as follows. In section 1, we express the requirements of military applications with regard to networks. In section 2, we describe OLSR, a proactive routing protocol. OLSR inherits the stability of link state protocols but at a cheaper cost thanks to its optimization. In section 3, we describe the OLSR platform we have deployed in the CELAR site. Finally, some performance measurements are given.

1 TACTICAL NETWORKS

For several years, the concept of "battlefield numerisation" has made information control a determining stake for the armed forces. This concept requires an architecture of meshed network, adapting automatically to variable topologies. Such a network performs auto-configuration, and dynamic routing. It detects the presence of a new node, the absence of another node and supports nodes mobility. Networks meeting, wholly or partly, these requirements are under development in the civil world: Mobile Ad hoc NETworks (MANETs).

The convergence of the Internet Revolution, which has brought the IP protocols to maturity on the civilian market, and the increasing needs of interoperability for data transmission architectures on the battlefield for joint and allied forces has made the use of COTS technologies and products in Telecommunications protocols a major issue for Defense C3I systems. The effort known as MANET (Mobile Ad-hoc NETwork), led by the Internet Engineering Task Force (IETF), to bring new promising answers to technical issues raised by mobile context about IP network, specifically relative to availability and reliability, is of great interest for military applications. The tactical environment is characterised mainly by the need for:
- Mobile nodes,
- Delay constraints with priorities,
- Reliability and availability,
- Security,
- Unicast and multicast traffics,
- Different kinds of traffic : data, voice, video,
- Interoperability between forces and with allied units.

In military scenarios, nothing can ensure that all communicating nodes will be one hop away due to the inherent nature of the radio propagation. In that case, ad-hoc multi-hop architecture is one of the best approaches to solve the connectivity problem. An architecture with a distributed control is essential. A network should work (even in a downgraded mode) without any centralised function. No node should be indispensable and centralised functions should be limited to optimisation capabilities. Autonomous packet radio networks are essential because of their ability to be operational rapidly and without any infrastructure. Requirements and constraints make MANET more pertinent relatively to the Below of Brigade echelon.

2 OLSR ROUTING PROTOCOL

2.1 MANET routing protocols

As radio coverage is usually limited, multihop routing is often needed. All routing protocols proposed in the IETF

MANET group address unicast routing, while taking into account the features of wireless, multi-hop, mobile ad-hoc networks. They are proactive or reactive.

With reactive protocols, a node discovers routes on-demand and maintains only active ones. Thus, a route is discovered whenever a node needs to communicate with a node for which a route is not available. This discovery mechanism is based on pure flooding [1] in the network. The advantage of reactive protocols is that no control message is needed for non-active routes. The drawback is the latency when establishing a route. Two reactive protocols are on the standard track: AODV [2] and DSR [3].

With proactive protocols, each node maintains the routes to all other nodes in the network by periodic exchange of control messages. When a node needs to send a packet to any other node in the network, the route is immediately available. The main advantage of proactive protocols is that they do not introduce a delay before sending data. Furthermore, these protocols are useful for traffic patterns where a large subset of nodes is communicating with another large subset of nodes, and where the [source, destination] pairs are changing over time. Two proactive protocols are on the standard track: OLSR [6] and TBRPF [4]. This class of protocol is particularly well suited for military applications since all nodes know the network topology and also it can be used without modification in the network protocol stack.

2.2 OLSR

OLSR (Optimized Link State Routing) is an optimization of a pure link state routing protocol. It is based on the concept of *multipoint relays* (*MPRs*) [7]. First, using *multipoint relays* reduces the size of the control messages: rather than declaring all links, a node declares only the set of links with its neighbors that are its "*multipoint relays*". The use of *MPRs* also minimizes flooding of control traffic. Indeed only *multipoint relays* forward control messages. This technique significantly reduces the number of retransmissions of broadcast control messages [5, 7]. The two main OLSR functionalities, Neighbor Discovery and Topology Dissemination, are now detailed.

2.2.1 Neighbor Discovery

Each node must detect the neighbor nodes with which it has a direct link. For this, each node periodically broadcasts *Hello* messages, containing the list of neighbors known to the node and their link status. The link status can be either *symmetric* (if communication is possible in both directions), *asymmetric* (if communication is only possible in one direction), *multipoint relay* (if the link is symmetric and the sender of the *Hello* message has selected this node as a *multipoint relay*), or *lost* (if the link has been lost). The *Hello* messages are received by all one-hop neighbors, but are not forwarded. They are broadcast once per refreshing period "*Hello_interval*". Thus, *Hello* messages enable each node to discover its one-hop neighbors, as well as its two-hop neighbors. This neighborhood and two-hop neighborhood information has an associated holding time "*Neighbor_hold_time*", after which it is no longer valid.

On the basis of this information, each node independently selects its own set of *multipoint relays* among its one-hop neighbors in such a way that the *multipoint relays* cover (in terms of radio range) all two-hop neighbors (see [7] for an algorithm example). The *multipoint relay* set is computed whenever a change in the one-hop or two-hop neighborhood is detected. In addition, each node m maintains its "*MPR selector set*". This set contains the nodes which have selected m as a *multipoint relay*. Node m only forwards broadcast messages received from one of its *MPR selectors*.

2.2.2 Topology Dissemination

Each node of the network maintains topological information about the network obtained by means of *TC (Topology control)* messages. Each node m selected as a *multipoint relay*, broadcasts a *TC* message at least every "*TC_interval*". The *TC* message originated from node m declares the *MPR selectors* of m. If a change occurs in the *MPR selector* set, the next *TC* can be sent earlier. The *TC* messages are flooded to all nodes in the network and take advantage of *MPRs* to reduce the number of retransmissions. Thus, a node is reachable either directly or via its *MPRs*. This topological information collected in each node has an associated holding time "*Top_hold_time*", after which it is no longer valid.

The neighbor information and the topology information are refreshed periodically, and they enable each node to compute the routes to all known destinations. These routes are computed with Dijkstra's shortest path algorithm [1]. Hence, they are optimal as concerns the number of hops. The routing table is computed whenever there is a change in neighborhood or topology information.

3 OLSR PLATFORM

CELAR (French MoD / DGA) works on the concept of ad hoc networks for the preparation of the future. These studies interest military programs like "Soldier of the Future", and could be fully integrated in RHD project (High Data Bit Rate Radio), for Navy and Land tactical networks. The objective of MANET/OLSR CELAR tesbed is to evaluate and demonstrate the potential benefits of MANET advances in military tactical applications, with a special focus on performances and reliability.

3.1 Platform description

CELAR MANET (Mobile Ad-hoc NETwork) demonstrator implements OLSR routing protocol (version 7). This version has been implemented on CELAR testbed at the end of December 2002. CELAR MANET demonstrator features 18 nodes including 4 PDAs, 4 laptops, and 10 OLSR routers. This experimental network is based on a WLAN / 802.11b radio technology (WiFi). A functionnal overview of this demonstrator is depicted on Figure 1.

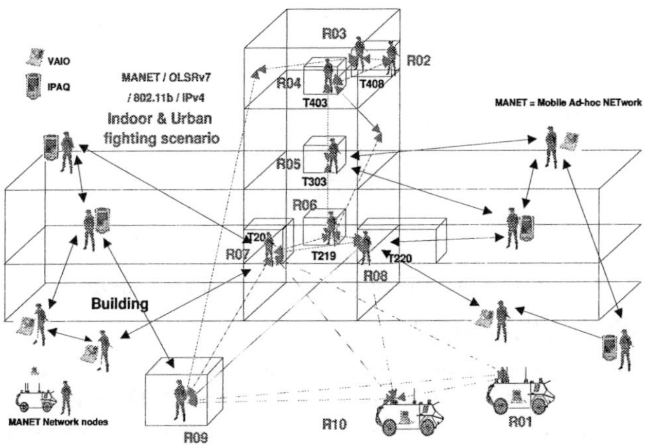

Figure 1. CELAR MANET platform.

3.2 Bandwidth sharing measurements

We now focus on bandwidth sharing between UDP traffics. All the experiments are made for a duration of 60 seconds (based on initial tests ranging from 10 to 180 seconds, 60 seconds found out to be enough for obtaining reproductible results). In each experiment, we progressively increase the load generated by UDP traffic, from medium load to overload, and we study the fairness.

We first consider UDP independent traffics sent by a source (here router r02) to two destinations. In the first experiment, one is one-hop away (here router r03) and the other is three-hop away (here router r08). In such a case, we observe that the unfairness between both traffics does not exist for a source rate of 100 kbps, is about 10% for a source rate of 300Kbps and reaches 1600% for a source rate of 5 Mbps. The unfairness between UDP traffics sent by the same source increases with the source rate: the nearest destination is favored. In the second experiment, both destinations are three-hop away (here routers r08 and r07). As one might expect, we observe a fair bandwidth sharing between UDP traffics sent by the same source, when both destinations are at the same distance from the source and both traffics visit the same intermediate nodes.

	First Exper.		Second Exper.	
src	r02	r02	r02	r02
dest	r03	r08	r08	r07
intermediate	-	r04 r06	r04 r06	r04 r06
hop	1	3	3	3
throughput for 5Mbps	2.3 Mbps	138 kbps	177 kbps	185 kbps
for 500kbps	439 kbps	356 kbps	175 kbps	184 kbps
for 300kbps	300 kbps	291 kbps	182 kbps	172 kbps
for 100kbps	100 kbps	100 kbps	100 kbps	100 kbps

Table 1. UDP throughput: distance influence.

On the other hand, in the third experiment, destinations are at the same distance from the source, but the intermediate nodes differ. For instance, the traffic with destination r09 visits r03, whereas traffic destinated to r05 visit r04. This time, we observe that traffic destinated to r05 is favored as soon as there is overload in the network. Then, in the fourth experiment, one destination (here r09) is two-hop away from the source, whereas the other destination (here r08) is three-hop away. We observe that the traffic destinated to the nearest destination is still favored, but however it receives a smaller bandwidth than in the previous case (first experiment), although it then coexisted with another traffic destinated to a one-hop destination, thus with an heavier bias. In conclusion, the results confirm that when the source rate is overloading the network, the bandwidth sharing becomes unfair: the destination with the smallest hop number is naturally favored, but there are other factors which play a part, likely linked to the physical and MAC layers, and which are subject of further research.

	Third Exper.		Fourth Exper.	
src	r02	r02	r02	r02
dest	r09	r05	r09	r08
intermediate	r03	r04	r03	r04 r06
hop	2	2	2	3
throughput for 5Mbps	264 kbps	353 kbps	152 kbps	111 kbps
for 500kbps	280 kbps	368 kbps	166 kbps	189 kbps
for 300kbps	263 kbps	293 kbps	269 kbps	265 kbps
for 100kbps	100 kbps	100 kbps	100 kbps	100 kbps

Table 2. UDP throughput: intermediate nodes influence.

In the fifth experiment, we consider two nodes exchanging UDP traffics. (the packet size is maximum, i.e.1470 bytes) to each other. Notice that the routes are different, that is, the intermediate router nodes are different in one direction to from another. At a source rate of 300 kbps, both UDP traffics receive approximatively the same bandwidth (289 and 295 kbps), which is fair. However to establish a comparison with TCP, in experiments with the same topology but with TCP flows replacing UDP flows, TCP traffic destinated to router r02 received 281 kbps, whereas TCP traffic in the reverse direction only receives 160 kbps. This point out some unfairness, which is made apparent with UDP, when the source rate increases: the discrepancy between both UDP traffics increases to reach 133 kbps vs 89.3 for a source rate of 5 Mbps. In conclusion, the unfairness is higher with TCP than with UDP, and the unfairness can only be explained by analysis at the level of lower layers (MAC and below). In the sixth experiment, we compare these results with the bandwidth obtained when each UDP traffic runs alone independently, for reference. Measures show that each UDP traffic, if taken alone, approximatively receives the same part of the bandwidth (e.g. 360 kbps for a source rate of 5 Mbps), excluding this potential explanation of unfairness. The addition of another traffic in the reverse direction leads to a smaller bandwidth about 30 % of the bandwidth obtained when the traffic is alone (i.e. goodput is 70%).

	Fifth Exper.		Sixth Exper.	
src	r02	r08	r02	r08
dest	r08	r02	r08	r02
intermediate	r03 r09	r06 r04	r03 r09	r06 r04
hop	3	3	3	3
throughput for 5Mbps	133 kbps	89.3 kbps	395 kbps	360 kbps
for 500kbps	199 kbps	135 kbps	500 kbps	451 kbps
for 300kbps	289 kbps	295 kbps		

Table 3. UDP throughput: reverse traffic influence.

3.3 Mobility measurements

In this scenario, the three VAIOs are embedded in vehicles moving around the tower hosting the platform. The goal here is to demonstrate the dynamic reconfiguration of the OLSR protocol in presence of mobility. The VAIOs are configurated as terminals: their OLSR willingness parameter is set to 0, hence, they do not route packets for other nodes. Router r02 is the source of three UDP flows at 128 kbps, each flow being destinated to one mobile VAIO. The mobiles successively visit points denoted D, T1, T2, R1, R2, R3, R4, R5 and A. We now focus on VAIO2. The next-hop for the route from VAIO2 to r02, along with the distance, gives much of the information about the routes computed and updated by the OLSR routing protocol.

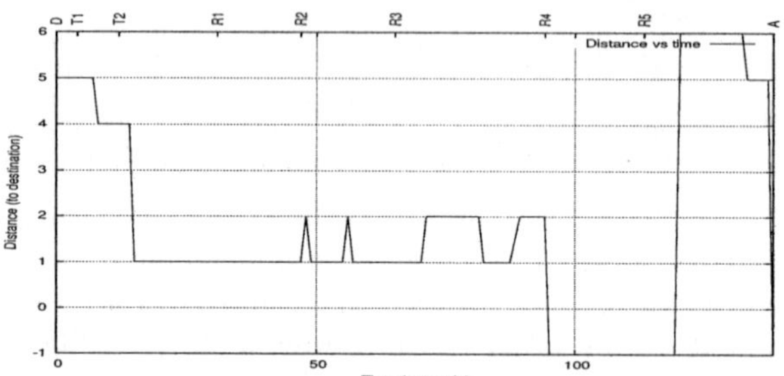

Figure 2. Distance of VAIO2 to r02.

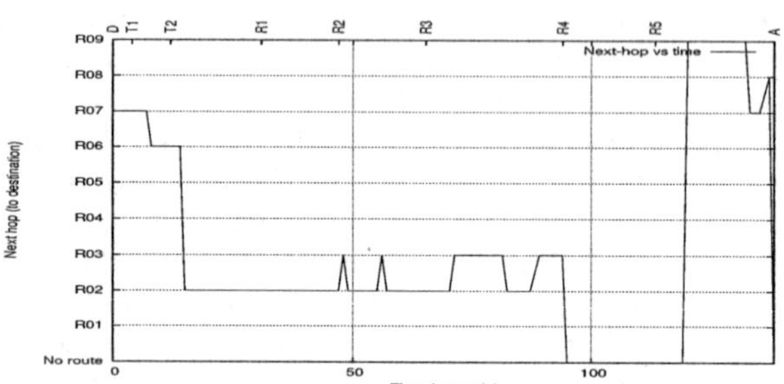

Figure 3. Next hop from VAIO2 to r02.

Figure 2 depicts the distance of VAIO2 to router r02, a distance of -1 means that there is no route. At time $t = 118$, VAIO2 is 6 hops away from r02. Figure 3 shows the next hop selected by VAIO2 to reach router r02. We observe a time interval $[94, 118)$ where there is no route. Before time 94, r03 was selected as the next hop; after time 118, r09 is selected. On Figure 4, we see that VAIO2 receives neither r02, nor r03, nor r04, nor r05 after time 94. Furthermore VAIO2 receives neither r06, nor r07, nor r08, nor r09 in the interval $[42, 103)$, and it receives r09 with a power higher than -85 dBm (low threshold for signal acceptance) only after time 110. Hence, the absence of route is due to a network connectivity problem.

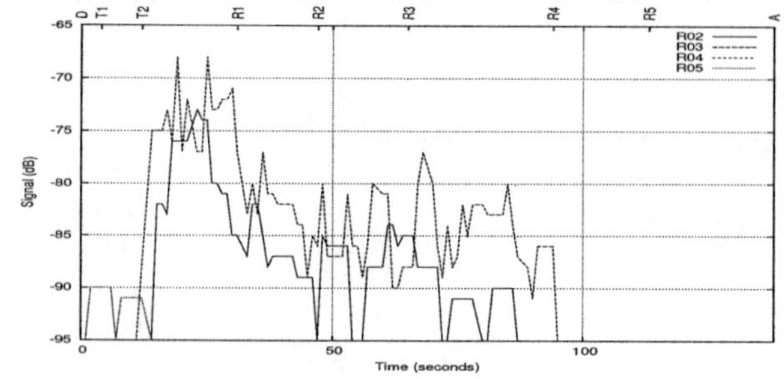

Figure 4. Signal power received by VAIO2.

CONCLUSION

Mobile ad-hoc networks are autonomous, self-adaptive and support mobility. These qualities make them essential for military applications. To succeed, these networks must achieve good performance. That is the reason why we have built a MANET platform to demonstrate and quantitatively evaluate the behavior of such networks. The measurements performed on this platform allow to conclude that OLSR routing seems to be well adapted to military mobile ad-hoc networks: the routing itself has been demonstrated to be dynamically adaptative, and there is some sharing of bandwidth between flows, along with observed UDP bandwidth satisfying many military usages. Still a finer-grained lower-level analysis is required to understand fully the behavior of the network ; and further work is needed to introduce quality of service in such networks to allow first for more fairness, and then for introducing service differentiation.

References

[1] A. S. Tanenbaum, *Computer Networks*, Prentice Hall, 1996.

[2] C. Perkins, E. Belding-Royer, S. Das, *AODV Ad hoc On-Demand Distance Vector Routing* , IETF February 2003, draft-ietf-manet-aodv-13.txt, work in progress.

[3] D. Johnson, D. Maltz, Y-C. Hu, *The Dynamic Source Routing Protocol for Mobile Ad Hoc Networks (DSR)* IETF April 2003, draft-ietf-manet-dsr-09.txt, work in progress.

[4] R. G. Ogier, M. Lewis, F. L. Templin, *Topology Dissemination Based on Reverse-Path Forwarding (TBRPF)*, IETF April 2003, draft-ietf-manet-tbrpf-08.txt, work in progress.

[5] P. Jacquet, P. Muhletaler, A. Qayyum, A. Laouiti, T. Clausen, L. Viennot, *Optimized Link State Routing Protocol*, IEEE INMIC, December 2001, Pakistan.

[6] C. Adjih, T. Clausen, P. Jacquet, A. Laouiti, P. Minet, P. Muhlethaler, A. Qayyum, L. Viennot, *Optimized Link State Routing Protocol*, IETF April 2003, draft-ietf-manet-olsr-09.txt, work in progress.

[7] A. Qayyum, A. Laouiti, L. Viennot, *Multipoint relaying technique for flooding broadcast messages in mobile wireless networks*, HICSS: Hawai Int. Conference on System Sciences, January 2002, Hawai, USA.

Bluetooth Network Performance in Nakagami-m Fading Channels

João Henrique Kleinschmidt, Marcelo E. Pellenz and Edgard Jamhour
Graduate Program in Computer Science
Pontifical Catholic University of Paraná
R. Imaculada Conceição, 1155 – Prado Velho – CEP 80215-901
Curitiba – PR – Brazil – Phone/Fax: (+ 55 41) 330 1669
{joaohk, marcelo, jamhour}@ppgia.pucpr.br

Abstract: The Bluetooth standard is considered a promising technology for implementing ad-hoc networks. This paper addresses issues related to Bluetooth performance in Nakagami-*m* fading channels. This distribution allows a better channel characterization for mobile and wireless communications. The throughput of Bluetooth links using asynchronous packets is investigated. Useful relations between packet error rate and node distances are derived. These results can be used as benchmarks for performance analysis or incorporated into Bluetooth network simulation tools. As an example, we evaluate by computer simulation the performance of a scatternet in a Nakagami fading scenario.

Key Words: Bluetooth, Nakagami-m fading, wireless networks.

1. INTRODUCTION

Bluetooth is emerging as an important standard [1] for short range and low-power wireless communication. It operates in the 2.4 GHz ISM band employing a frequency-hopping spread spectrum technique. The transmission rate is up to 1 Mbps, using GFSK (Gaussian Frequency Shift Keying) modulation. The Bluetooth MAC protocol is designed to facilitate the construction of ad-hoc networks. The devices can communicate with each other forming a network with up to eight nodes, called *piconet*. Within a piconet, one device is assigned as a master node and the others devices acts as slaves nodes. Devices in different piconets can communicate using a structure called *scatternet*. The channel is divided in time slots of 625 μs. A time-division duplex (TDD) scheme is used for full-duplex operation.

For data transmission Bluetooth employs seven asynchronous packets types [1]. All packets use cyclic redundancy check (CRC) codes and retransmission strategy, except the packet type AUX1, where specific error control strategies can be applied [2]. The packet structure is formed by an access code (72 bits), a header (18 bits) and a variable length payload. The header is protected by a (3,1) repetition code, which is capable of correcting only one error. The payload employs a CRC code for error detection and is also protected by a Hamming code (15,10) in packets DM1, DM3 and DM5. The Bluetooth devices are classified in three power classes [1]. The class 3 is a typical Bluetooth device with an output power of 0 dBm. The class 1 operates with a maximum output power of 20 dBm and can cover ranges up to 100 meters. The class 1 makes possible the formation of scatternets for application in telemetry systems and sensor networks.

The throughput of Bluetooth links using asynchronous packets was investigated in [3] for the additive white Gaussian noise (AWGN) channel and for the Rayleigh fading channel. In this work we extend the results presented in [3] in order to investigate the performance of Bluetooth links in Nakagami fading [4] channels. This distribution allows a better characterization of real channels. Through the parameter *m* we can adjust different fading conditions. For $m=1$ we get the Rayleigh distribution. Using $m<1$ or $m>1$ we obtain fading intensities more and less severe than Rayleigh, respectively. In the proposed channel model we also consider the log-distance path loss model [5]. We evaluate the packet error rates in the forward and reverse channels as a function of the distance between devices in the network. These results are incorporated into the simulation software Blueware [6] in order to evaluate scatternet performance.

2. THROUGHPUT EVALUATION IN NAKAGAMI-M FADING CHANNELS

In order to investigate the throughput of Bluetooth links in the presence of Nakagami-*m* fading, we follow the derivation used in [3]. The throughput can be computed using the packet retransmission probability. There are five packet retransmission events:

A: the destination fails to synchronize with the access code of the forward packet.

B: the header of the forward packet is corrupted after the triple redundancy code is decoded.

C: the payload of the forward packet is corrupted after the Hamming code is decoded, causing the CRC check to fail.

D: the source is unable to synchronize with the access code of the reverse packet.

E: the header of the reverse packet is corrupted.

Denoting $\overline{A}$ the complement of event A, $P[A]$ the probability of event A, and so on, the packet retransmission probability is given by

$$P_r(\overline{\gamma}_f, \overline{\gamma}_r) = 1 - \int_0^{\infty} f(\gamma_f) P[\overline{A}] P[\overline{B}] P[\overline{C}]\, d\gamma_f \cdot \int_0^{\infty} f(\gamma_r) P[\overline{D}] P[\overline{E}]\, d\gamma_r \tag{1}$$

where $\overline{\gamma}_f$ and $\overline{\gamma}_r$ are the average received signal-to-noise ratios (SNR) in the forward and reverse channels and $f(\gamma_f)$ and $f(\gamma_r)$ are the probability density functions (pdf) of γ_f e γ_r. The average received SNR is a function of the path loss model or distance between nodes. The analytical expressions for the retransmission probabilities, $P[\cdot]$, were derived in [3]. From equation (1) we can define the packet error rate (PER) of the forward channel, $PER_f(\overline{\gamma}_f)$, and reverse channel, $PER_r(\overline{\gamma}_r)$,

$$PER_f(\overline{\gamma}_f) = 1 - \int_0^{\infty} f(\gamma_f) P[\overline{A}] P[\overline{B}] P[\overline{C}]\, d\gamma_f, \tag{2}$$

$$PER_r(\overline{\gamma}_r) = 1 - \int_0^{\infty} f(\gamma_r) P[\overline{D}] P[\overline{E}]\, d\gamma_r. \tag{3}$$

The data rate R is given by

$$R = \frac{K}{D \cdot N \cdot 625 \cdot 10^{-6}}, \tag{4}$$

where D is the number of occupied slots per transmission including the reverse packet, K is the number of data bits in the packet and N is a random variable representing the total number of times a particular packet must be transmitted. The average throughput is the expected value of R with respect to N,

$$\overline{R} = E\{R\} = \sum_{N=1}^{\infty} \left[1 - P_r(\overline{\gamma}_f, \overline{\gamma}_r)\right] \cdot \left[P_r(\overline{\gamma}_f, \overline{\gamma}_r)\right]^{N-1} \cdot R \tag{5}$$

The Nakagami-*m* probability density function [4] is given by

$$f(\gamma) = \frac{m^m \gamma^{m-1}}{\Gamma(m)\overline{\gamma}^m} \cdot \exp(-\frac{m\gamma}{\overline{\gamma}}), \quad m \geq 0{,}5 \tag{6}$$

Substituting equation (6) in (1), we can evaluate the throughput using (5) for different values of m. It was assumed a GFSK modulation index h=0.32 and a threshold T=65. The parameter T is the minimum number of bits from the access code that must be properly demodulated for synchronization.

Figures 1 and 2 show the throughput for packets DM1 and DH5 for m=1, m=0.5, m=0.75 and m=1.5. Figures 3 and 4 show the throughput for packets DM1, DH1, DM5 and DH5 as a function of PER_f and PER_r for m=0.5. For a given throughput value we read from Figures 3 and 4 the packet error rates associated with the forward and reverse links.

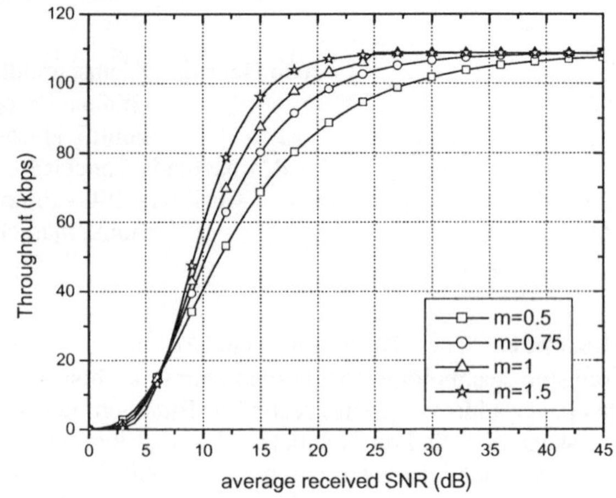

Figure 1 - Performance of DM1 packet

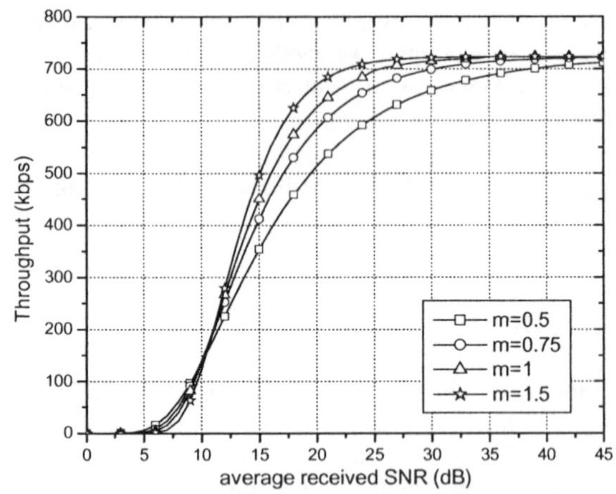

Figure 2 - Performance of DH5 packet

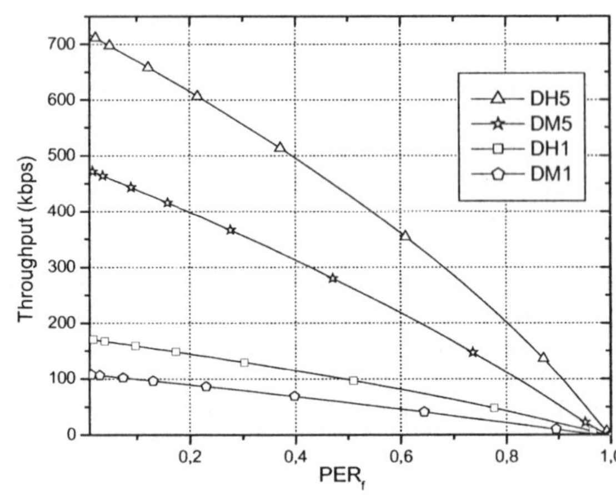

Figure 3. Throughput versus PER$_f$ for $m = 0.5$

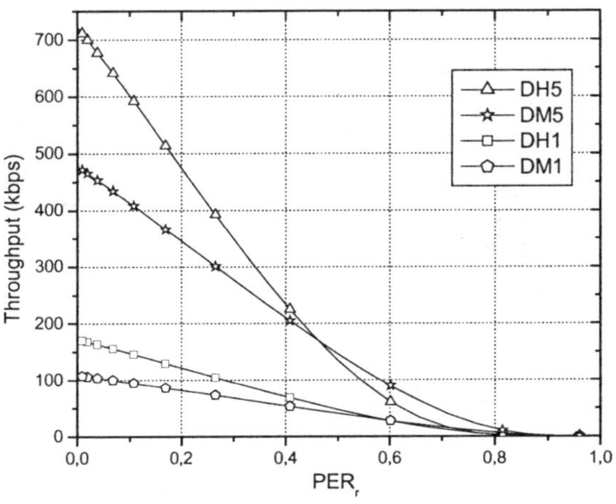

Figure 4. Throughput versus PER$_r$ for $m = 0.5$

3. PATH LOSS MODEL

In our study we apply the log-distance path loss model [5] to predict the mean received power for class one devices, which can cover theoretically a distance range up to 100 meters. A power control is required for class one devices. However, in the path loss computation it was always assumed the maximum transmitted power. The Bluetooth transceiver requires a minimum received power of –70 dBm. The log-distance model [5] is given by

$$\overline{PL}(\mathrm{dB}) = \overline{PL}(d_0) + 10 \cdot n \cdot \log\left(\frac{d}{d_0}\right), \tag{7}$$

where n is the path loss exponent which indicates the rate the path loss increases with distance, d is the transmitter-receiver separation and d_o is a reference distance. The value of n can be selected based on the environment where the devices will operate. The reference path loss $\overline{PL(d_0)}$ is computed using the free space path loss model [5],

$$P_r(d) = P_t \frac{\lambda^2}{(4\pi)^2 d^2}, \tag{8}$$

where P_t is the transmitted power, d is the distance and λ is the wavelength. The wavelength can be evaluated by $\lambda = c/f$, where c is the speed of light and f is the carrier frequency (2.4 GHz for Bluetooth). The received power, P_r, is given by $P_r = P_t - \overline{PL}$. Thus, the average received signal-to-noise ratio is $SNR = E_b / N_0 = P_r(R \cdot N_0)$, where R is the transmission rate (1 Mbps for Bluetooth) and N_0 is the one-sided noise power spectral density. The thermal noise floor for a 1 MHz bandwidth is –114 dBm. Assuming a Bluetooth receiver with noise figure of 23 dB [7] we get a noise power $N = N_0 B = -91\,\mathrm{dBm}$.

Applying the log-distance model we evaluated the average received SNR for different transmitter-receiver separations. This result is shown in Figure 5 for a path loss exponent n = 3, n = 3.5 and n = 4. The dashed line indicates the minimum required received power of -70 dBm (receiver sensitivity). The intersection of this line with the SNR curves represents the maximum transmitter-receiver separation.

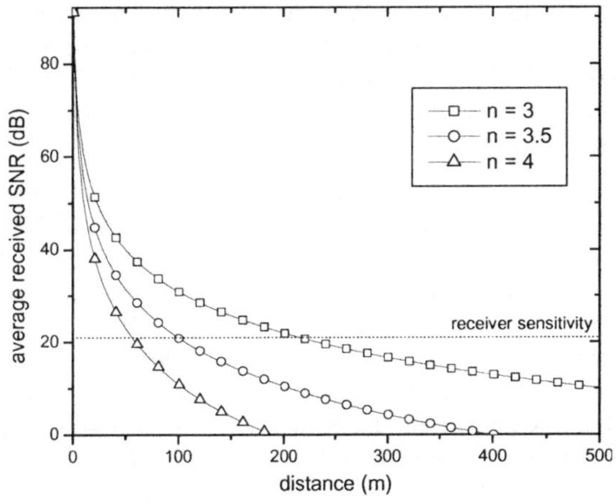

Figure 5. SNR versus distance

Figures 6 and 7 also relate the transmitter-receiver separation with the PER for the asynchronous packets. For the reverse packet (ACK) the PER axis represents the PER_r probability. For packets DM1, DM5, DH1 and DH5 the PER axis represents the PER_f probability.

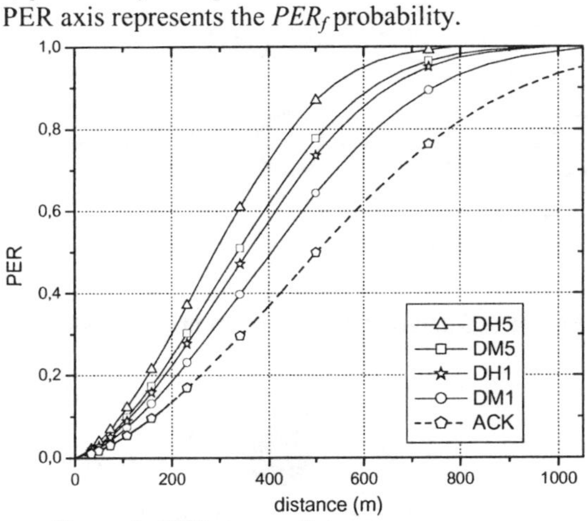

Figure 6. PER versus distance, $m = 0.5$, $n = 3$

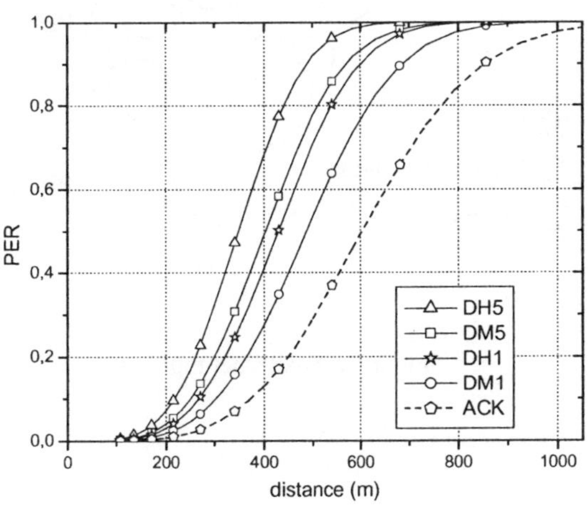

Figure 7. PER versus distance, $m = 1.5$, $n = 3$

4. NETWORK PERFORMANCE

The results obtained in Section 3 were incorporated into Blueware [6] simulator. Originally, Blueware only supports class 3 devices. We also implemented class 1 devices, to allow the formation of a scatternet with different power classes devices.

As an example, we present in Figure 8 a simulation scenario with six Bluetooth nodes to analyze the effects of Nakagami fading in the network performance. In general, near devices (class 3) will have better channel conditions ($m=m_1$) than devices separated by a long distance (class 1), that will experience severe fading conditions ($m=m_2$). The scatternet is formed using the *Tree Scatternet Formation (TSF)* algorithm [8] implemented in the simulator. The scatternet topology constructed is shown in Figure 9. Throughput simulation results for different values of m_1 and m_2 are presented in Table 1. It was used a path loss exponent n=3 and DH5 packets for CBR traffic of 720 kbps. The results are dependent of the intra-piconet and inter-piconet scheduling schemes. The inter-piconet scheduling algorithm utilized was the *Locally Coordinated Scheduling (LCS)* [9].

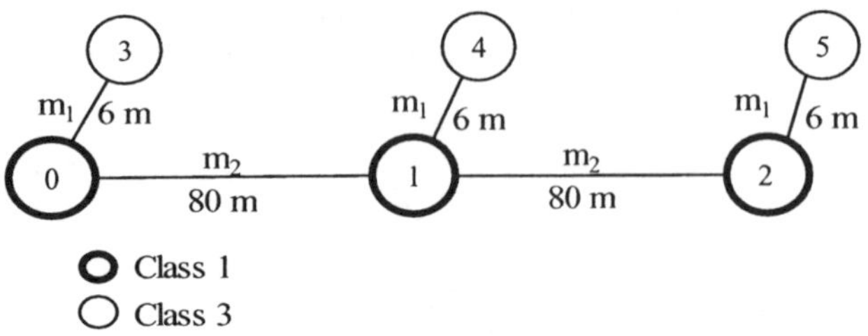

Figure 8. Scatternet with different power classes devices

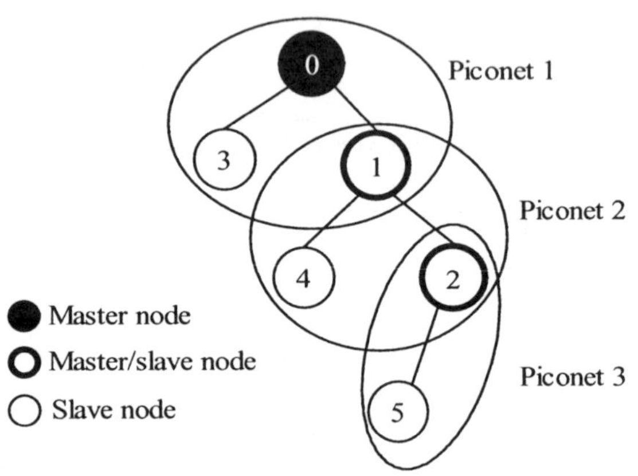

Figure 9. Scatternet topology

Nodes	Throughput (kbps)		
Source - Destiny	$m_1 = 1$ $m_2 = 0.5$	$m_1 = 1$ $m_2 = 0.75$	$m_1 = 1.5$ $m_2 = 0.75$
3-0	648	648	668
3-1	197	248	262
3-4	188	233	243
3-2	40.9	224	214
3-5	32.1	206	209

Table 1 – Throughput for different values of *m*

5. FINAL REMARKS

In this paper is analyzed the Bluetooth performance in Nakagami-*m* fading channels. This distribution allows a better channel characterization for mobile and wireless communications. The throughput of Bluetooth links using asynchronous packets was investigated.

Others scenarios with different topologies, number of nodes and piconets must also be considered. Using devices of class 1 and class 3 in the same network can facilitate the formation of scatternets for applications that have to cover great distances. For future works, the parameter *m* of each link can be estimated and used as a metric for scatternet formation protocols and routing strategies.

REFERENCES

[1] Bluetooth SIG, "Specifications of the Bluetooth System", *Core Version 1.1*, February 2001. http://www.bluetooth.com

[2] M. Valenti and M. Robert, "Improving the QoS of Bluetooth through turbo coding", Proc. IEEE Military Communications Conference, Los Angeles, CA, pp. 1057-1061, October 2002.

[3] M. C. Valenti, M. Robert and J.H. Reed, "On the throughput of Bluetooth data transmissions", IEEE Wireless Communications and Networking Conference, Orlando, FL, pp. 119-123, March 2002.

[4] J. Proakis, *Digital Communications*, New York, NY: McGraw-Hill, 4th edition, 2001.

[5] T. S. Rappaport, *Wireless Communications: Principles and Practice*, Prentice-Hall PTR, 1999.

[6] Blueware: Bluetooth simulator for ns. http://nms.lcs.mit.edu./project/blueware.

[7] B. Koupal, M. Wang and C. Edelman, Case History: Development of a Bluetooth Transceiver in RF CMOS, Signia and Agilent Technologies, April 2002.

[8] G. Tan, A. Miu, H. Balakrishnan and J. Guttag, "An efficient Scatternet formation algorithm for dynamic environments", IASTED Intern. Conf. on Comm. Networks (CCN02), Cambridge, MA, Nov. 2002.

[9] G. Tan and J. Guttag, "A Locally Coordinated Scatternet Scheduling Algorithm", IEEE Conference on Local Computer Networks, Tampa, FL, Nov. 2002.

QoS Support in Wireless LANs Using Dynamic Adaptation of Channel Access Parameters

Liqiang Zhang and Sherali Zeadally
Department of Computer Science, Wayne State University
Detroit, Michigan 48202, USA

Abstract: *Service differentiation can be achieved at the MAC layer of IEEE 802.11 wireless LANs by using different channel access parameters (CAPs) for each traffic class. These parameters can be chosen either a priori (using fixed CAPs) or dynamically. However, the a priori approach cannot provide assured QoS and optimal performance. We propose a dynamic approach called Smart Enhanced Distribution Coordination Function (SEDCF), which is based on the EDCF introduced in the upcoming IEEE 802.11e standard. With SEDCF, the Access Point exploits a novel algorithm to dynamically find optimal CAPs based on traffic load in each class, and then periodically advertises these CAPs to wireless stations through beacon frames. Our performance evaluation of SEDCF demonstrates that it can satisfactorily support assured QoS for real-time applications.*

Key Words: Wireless LANs, Quality of Service, MAC, Channel Access Parameters, Adaptation.

1. Introduction

Over the last few years, we have witnessed a wide proliferation of IEEE 802.11 wireless LANs (WLANs). In most scenarios, they are taken as a wireless extension of Internet. Today, this extension supports services mostly on a best-effort basis. However, emerging QoS sensitive applications such as voice and video requires end-to-end QoS guarantees, which motivates the need for QoS support over both wired and wireless links. The IEEE 802.11 task group E is currently defining a new supplement to the existing legacy 802.11 MAC (Medium Access Control) layer in order to enhance the QoS support for WLANs [5].

In the upcoming standard 802.11e, the mandatory operation mode of legacy 802.11- Distributed Coordination Function (DCF) [3] is replaced by Enhanced DCF (EDCF) [5]. Service differentiation can be achieved in EDCF by assigning different channel access parameters (CAPs) to different traffic classes. However, we currently do not have a proper mapping of application level QoS requirements to MAC layer differentiation parameters. In practice, appropriate distribution of CAPs among different traffic classes to achieve target QoS is a challenging issue. Dynamic load in traffic classes also make it difficult to provide assured QoS when fixed CAPs are used (we refer assured QoS services as those with a lower bound for performance in terms of throughput and latency).

In this work, we propose a novel approach called Smart EDCF (SEDCF), which is based on EDCF, designed to provide assured QoS in WLANs. Our proposed design exploits a MAC layer quality indicator-MACQI (a combination value of drop rate and collision rate) to dynamically adjust CAPs to ensure the QoS of applications. By periodically sampling MACQI for each traffic class and utilizing our proposed Z-scan algorithm, an Access Point (AP) selects the CAPs that best match the QoS requirements of each traffic class. SEDCF's dynamic adaptation of CAPs works over two time scales, one time scale reacting to short-time traffic bursts and the other time scale corresponding to long-time load variations. It is worthwhile noting that our approach works only in BSS mode, and for each traffic class, there exists at least one downlink flow (from AP to a mobile station).

The rest of the paper is organized as follows. In the following section, we briefly review some related work. In section 3, we present the design of SEDCF. Section 4 gives the performance evaluations. Finally, in section 5, we make some concluding remarks and present future work.

2. Related Works

The IEEE 802.11 defined the DCF (Distributed Coordination Function) and PCF (Point Coordination Function) in the MAC layer [3]. DCF does not provide any QoS guarantee. Although PCF is designed to provide QoS for real-time applications, evaluations have shown its inefficiency and uncertainty [6, 7]. Several research efforts [1, 2] have investigated service differentiation mechanisms for the DCF of IEEE 802.11; in parallel, EDCF has been introduced in the upcoming IEEE 802.11e standard [5]. However no adaptive scheme has been investigated in these approaches, which makes it hard to guarantee assured QoS. Adaptive EDCF is proposed in [9]. The goal of this approach is to enhance the overall network performance by dynamically adjusting the size of Contention Window (CW) according to the network conditions. However, since relative priorities among traffic classes rely on a priori selection of CAPs, assured QoS cannot be guaranteed.

3. The design of SEDCF

3.1. QoS mapping

To enable dynamic adaptation for assured QoS, a QoS controller needs feedback indicating the measured QoS to determine any adjustments needed. This feedback can be explicitly generated by the receiver of each flow and sent to the QoS controller. However, this approach needs support from each host and also leads to considerable overheads.

In our design, we propose an alternative approach to avoid explicit feedback. The AP uses MAC layer Quality Indicators (MACQIs) to map the observed QoS at the application layer to the MAC layer. From the simulation experiments we conducted, we demonstrated the effectiveness of MACQI based on the following observations: (i) the flows in the same traffic class experience closely similar QoS; (ii) MACQIs are closely tied to the QoS observed at the application layer. The computation of MACQI is given by the following formulas:

$$MACQI[i] = dr[i] + cr[i] * F_{macqi}$$
$$dr[i] = drop[i]/sent[i]$$
$$cr[i] = collision[i]/sent[i]$$

where i is the index of traffic class; $dr[i]$ is the drop rate experienced at the MAC layer and the interfaces between LLC and MAC layer; $cr[i]$ is the collision rate experienced at MAC layer; and F_{macqi} is the factor to balance the effects of drop rate and collision rate on $MACQI$. $MACQI[i]$ is periodically sampled by the AP during each beacon interval.

To support adaptation for two time scales, we define two estimators of Exponentially Weighted Moving Average (EWMA), namely $MACQIS[i]$ and $MACQIL[i]$, for the adaptation reacting to short-time traffic burst and long-time load changing respectively. These two estimators are expressed as following formulas:

$$MACQIS[i] = MACQISold[i] * (1 - \alpha) + MACQI[i] * \alpha$$
$$MACQIL[i] = MACQILold[i] * (1 - F_{sl} * \alpha) + MACQI[i] * F_{sl} * \alpha$$

where α is the smoothing factor deciding the agility/stability of adaptation regarding to short-term burst, $F_{sl}(< 1)$ is the factor deciding the distance between the two time scales.

3.2. Proposed Z-scan algorithm

The Z-scan algorithm is shown briefly in the Table 1. In our design, the allocation of wireless channel resources is adapted by adjusting three CAPs, namely, CWMin, CWMax, and AIFS [5]. To control the adaptation, for each estimator (MACQIS and MACQIL) in each non best-effert class, we define two thresholds, a low one and a high one.

The adaptation runs at two time scales, namely, short-time scale adaptation and long-time scale adaptation. These two adaptations share the same Z-scan algorithm, however, they use different adaptation scalars. The long-time scale adaptation use a larger adaptation scalar so that CAPs can be quickly adjusted once load changes are perceived. Before we explained the Z-scan algorithm, let us briefly explain some notations used in table 1. All the array variables are indexed by i or ii, which indicates the priority of traffic class. $plevels$ is the total number of the traffic classes, and class $plevels$ has the lowest priority (best-effort traffic). The Z-scan al-

```
/* The code executed by AP */
if beacon timer expired {
    collect data and do calculations for MACQIs
    if the boundary of thresholds for long time scale of any class is crossed
        adaptation_scalar = 5; //for long time scale adaptation.
    else
        adaptation_scalar = 1; //for short time scale adaptation.
    adaptation();
    send updated CAPs through beacon frames;
    update_access_parameters(); //Assign the CWMin_update[i] to CWMin[i], etc.
}

/* The code for adaptation(), only executed by AP */
adaptation() {
  for (int i = 1; i < plevels; i++) {
    if (macqis_curr_avg[i] > macqis_threshold_high[i]) {
        for (int ii = plevels-1; ii > i; ii–) {
            if (macqis_curr_avg[ii] < macqis_threshold_low[ii]) {
                increase_parameters(ii);
                return;
            }
        }
        increase_parameters(plevels);
        return;
    }
    else if (macqis_curr_avg[i] < macqis_threshold_low[i]) {
        for (int ii = i+1; ii < plevels; ii++) {
            if (macqis_curr_avg[ii] > macqis_threshold_high[ii]) {
                decrease_parameters(ii);
                return;
            }
        }
        decrease_parameters(plevels);
        return;
    }
  }
}

/* The code for increase_parameters(), only executed by AP */
increase_parameters(int priority_level){
  int i = priority_level;
    CWMin_update[i] =
        min(255, CWMin_update[i] + adaptation_scalar * i * 8);
    CWMax_update[i] =
        min(1023, CWMax_update[i] + adaptation_scalar * i * 8);
    AIFS_update[i] =
        min(sifs_ + 129 * SlotTime, AIFS_update[i]
            + adaptation_scalar * i * SlotTime);
}

/* The code for decrease_parameters(), only executed by AP */
decrease_parameters(int priority_level){
  int i = priority_level;
    AIFS_update[i] -=
        min(adaptation_scalar * i * SlotTime,
            AIFS_update[i] - AIFS_update[i-1]);
    CWMin_update[i] -=
        min(adaptation_scalar * i * 8,
            CWMin_update[i] - CWMin_update[i-1]);
    CWMax_update[i] -=
        min(adaptation_scalar * i * 8,
            CWMax_update[i] - max(CWMax_update[i-1],CWMin_update[i]));
}

/* The code only executed by mobile nodes */
if get updated access parameters through beacon frames
    update_access_parameters();
```

Table 1. The proposed Z-scan algorithm

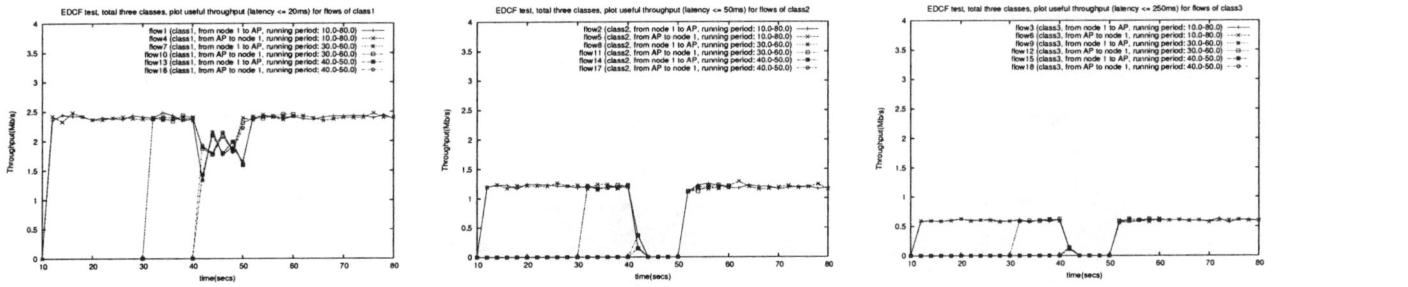

(a) **(b)** **(c)**

Figure 1. The useful throughput of the flows in three classes for EDCF.

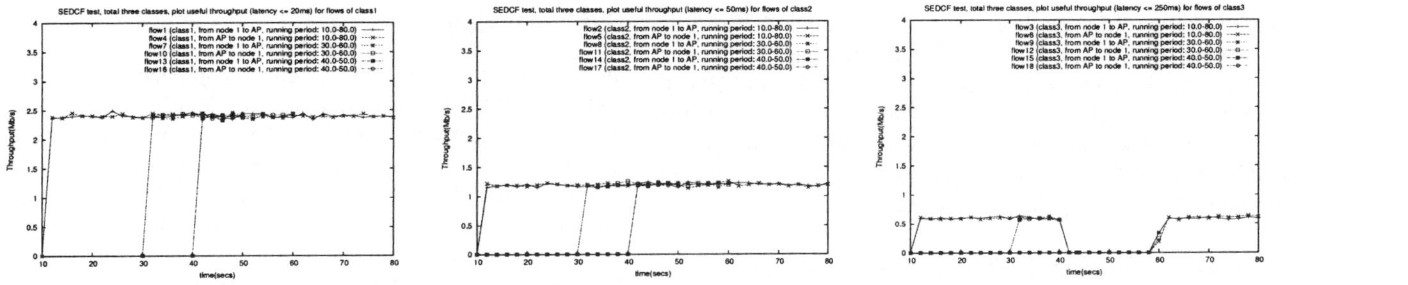

(a) **(b)** **(c)**

Figure 2. The useful throughput of the flows in three classes for SEDCF.

gorithm dynamically adjusts CAPs for each traffic class to match their QoS requirements. To achieve this goal, the algorithm follows two basic principles: (1) if the QoS of class i ($i < plevels$) is currently not satisfied ($macqis\_curr\_avg[i] > macqis\_threshold\_high[i]$), then the resources allocated to lower priority classes will be squeezed (by increasing the CAPs of these low priority classes); (2) if the QoS of class i ($i < plevels$) is better than required ($macqis\_curr\_avg[i] < macqis\_threshold\_low[i]$), then the resources of this class will be partly released (by increasing the CAPs of this class or decreasing the CAPs of lower priority classes). Enforcement of above principles leads to two major benefits: (1) the QoS of non best-effort traffic will be guaranteed if the total load of *these* traffic does not overload the network cell; (2) network resources are optimally utilized, which leads a higher aggregate throughput than with EDCF. Due to sapce limitations, we do not discuss the details of the algorithm further. One point worth to mention is, the algorithm does not have special requirements for the initial values of CAPs, an easy choice is to use the default values defined in [5].

4. Performance Evaluations

We have extended ns2 [8] to support SEDCF. Various simulations with different scenarios have been done to demonstrate the limitations of EDCF and effectiveness of SEDCF in supporting QoS. Due to limited space, we only report the result of one simple scenario here, the result of a more complex scenario is reported in [10]. For the scenario to be reported here, we assume that each node (including AP) operates at IEEE 802.11a PHY mode-6 [4], we list part of the IEEE 802.11a MAC/PHY parameters in table 2. With the scenario, there are 18 Constant Bit Rate (CBR) traffic flows between one wireless node and an AP. These flows belong to three traffic classes: $class1 - calss3$, among which $class1$ has the highest priority. Their characteristics are shown in table 2. We assume the tolerable maximum latency of the three traffic classes are 20ms, 50ms, and 250ms respectively. To show how SEDCF dynamically adapts the CAPs according to the load changes, we divide the 18 flows into three groups, each group has 6 flows. The three groups start at 10s, 30s, and 40s and ends at 80s, 60s, and 50s respectively, so that the total load of the network increases from 10s and 40s, and falls down from 50s to 80s.

Considering that the QoS requirements of the flows include both throughput (or call it bandwidth) and latency, we introduce a new concept *useful throughput* to demonstrate the QoS experienced at each flow. The difference between useful throughput and throughput is that the former one only counts the packets whose latency is less than some limit. We denote useful throughput as BW_{limit} in the context. For the scenario we considered, we investigate BW_{20ms}, BW_{50ms}, and BW_{250ms} for class 1, 2, 3 respectively. Their useful throughput are shown in Figure 1 and Figure 2 for EDCF and SEDCF respectively. We can see that, with EDCF, the QoS of class 1 and class 2 can be well satisfied during 10s-40s and 50s-80s, however, during 40s-50s, their useful throughputs decrease sharply; While with SEDCF, their QoS

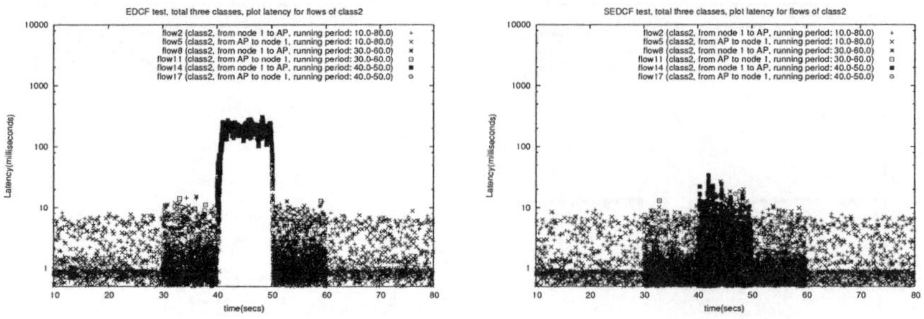

(a) (b)

Figure 3. The latency of the flows in class 2 for EDCF and SEDCF.

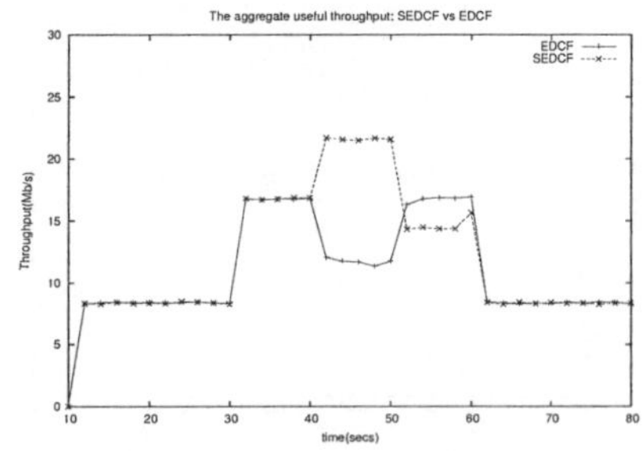

Figure 4. The aggregate useful throughput for EDCF and SEDCF.

can be guaranteed even during 40s-50s, although the price is sacrificing the QoS of the best-effort traffic (class 3). Figure 3 compares the latency experienced with EDCF and SEDCF (to save space, only the flows in class 2 are shown in the Figure). We also compare the aggregate useful throughput of EDCF and SEDCF in Figure 4. The SEDCF's average useful throughput is 10% higher than that of EDCF.

The CAPs used in the EDCF test are shown in Table 2, we use the same value to initialize the CAPs for SEDCF. We also tried some other CAPs for EDCF, for example, to make stricter priorities so that EDCF can guarantee the QoS of non best-effort traffics. However, first, such CAPs are quite difficult to select in advance, because of the difficulties to predict load variations; second, even if such CAPs can be achieved, the aggregate throughput of EDCF is much worse than SEDCF. In the SEDCF test, the values of F_{macqi}, α, and F_{sl} are 0.3, 0.5, and 0.2 respectively. The four thresholds ($macqis\_threshold\_low$, $macqis\_threshold\_high$, $macqil\_threshold\_low$, $macqil\_threshold\_high$) we used for class 1 are (0.001, 0.02, 0.02, 0.01); and those for class 2 are (0.005, 0.04, 0.01, 0.02).

IEEE 802.11a MAC/PHY parameters used in simulation			
Parameters			values
Date Rate			36Mb/s
Modulation			16-QAM
MAC Header			28 Bytes
Preamble Length			16μs
PLCP Header Length			4μs
SlotTime			9μs
Short Inter Frame Space (SIFS)			16μs
RTS Threshold			3000 Bytes
Fragmentation Threshold			3000 Bytes
Beacon Interval			100 ms
CAPs and flow characteristics of the three traffic classes			
Parameters	Class 1	class 2	Class 3
Contention Window Min (CWMin)	15	31	63
Contention Window Max (CWMax)	63	127	1023
Arbitration Inter Frame Space (AIFS)	43μs	43μs	52μs
Packet Size (Bytes)	1500	1500	1500
Packet Interval (ms)	5	10	20
Sending Rate (Kb/s)	2400	1200	600

Table 2. MAC/PHY/Channel Parameters and Flow Characteristics

5. Conclusions

In this paper, we present a novel approach called Smart EDCF (SEDCF) which utilizes the MACQIs and Z-scan algorithm to perform dynamic adaptation of channel access parameters to yield assured QoS over IEEE 802.11 wireless LANs. Simulation results obtained with different application scenarios demonstrate that SEDCF provides assured QoS for QoS-sensitive applications, and results a better overall performance than EDCF. Our future work will mainly focus on two issues: (1) currently, the selection of some parameters, for example, the thresholds for MACQIs, still partially relies on empirical experience, accurate mathematical model and analysis are needed so that parameters can be quickly and accurately selected; (2) efficient admission control mechanism is needed to make SEDCF more robust.

Acknowledgements

The authors would like to express their gratitude to Farshad Fotouhi and Monica Brockmeyer for their support in making the equipment avaible to us for conducting the experiments described in this paper.

References

[1] I. Aad and C. Castelluccia, Differentiation mechanisms for IEEE 802.11, IEEE Infocom 2001, April 2001.
[2] J. Deng and R. S. Chang, A priority Scheme for IEEE 802.11 DCF Access Method, IEICE, Vol. 82-B, No. 1, January 1999.
[3] IEEE 802.11 WG. "Part 11: Wireless LAN Medium Access Control (MAC) and Physical Layer (PHY) Specifications", IEEE 802.11 Standard, 1999.
[4] IEEE 802.11 WG. "802.11a, Part 11: Wireless LAN Medium Access Control (MAC) and Physical Layer (PHY) Specifications", Sep. 1999.
[5] IEEE 802.11 WG, 802.11e Draft 4.2, February 2003.
[6] S. Mangold et al., IEEE 802.11e Wireless LAN for Quality of Service, In Proc. of European Wireless, February 2002.
[7] Q. Ni et al., QoS issues and enhancements for IEEE 802.11 Wireless LAN, INRIA Research Report No. 4612, November 2002.
[8] The Network Simulator - ns-2, http://www.isi.edu/nsnam/ns.
[9] L. Romdhani et al., Adaptive EDCF: Enhanced Service Differentiation for IEEE 802.11 Wireless Ad Hoc Networks In Proc. of IEEE WCNC'03, New Orleans, and March 2003.
[10] L. Zhang, S. Zeadally, "A Novel Approach to Enhance QoS Support over Wireless LANs", to appear in the Proc. of WPMC 2003, Oct. 2003

A Reliable Region based Hierarchical Mobile Multicast Protocol for IPv6 Network*

Sun LiMin, Yong Liao, Wu Zhimei
Institute of Software, Chinese Academy of Sciences
P.O. Box 8718, Beijing 100080, China
Phone: (8610)62567838, Fax: (8610)62645410, e-mail: slm@iscas.ac.cn

Abstract: We propose a basic mobile multicast scheme in Mobile IPv6 called *previous network subscription* (*PS*). In *PS*, when a mobile host moves to a new foreign network, it will build a bi-directional tunnel with its previous network, so mobile host can continue to receive multicast packets soon after handover. We also propose a new reliable mobile multicast protocol called RRHMoM (Reliable Region based Hierarchical Mobile Multicast). It combines *PS* and remote subscription (RS). Hierarchical architecture is adopted in RRHMoM. The acknowledge mechanism combines NAK and ACK. Simulation results show that RRHMoM has better performance than existing reliable mobile multicast protocols.

Key Words: reliable multicast, mobile multicast, mobile IPv6, reliable mobile multicast protocol, RRHMoM.

1. INTRODUCTION

Many new applications in Internet, such as news dissemination, weather forecasting, and software distribution, require reliable data transmission from one transmitter to multiple receivers. Reliable IP multicast support is desired in Internet and many reliable multicast protocols have been designed.

In wireless networks using mobile IP, multicast packet is more inclined to lose because wireless medium is easier to be interfered. Another important fact is that mobile host continues to be handover between different subnets, which causes the multicast transmission path to change frequently and lots of packets may be lost during handover. Providing reliable multicast service to mobile host is a more challenging work.

The rest of this paper is organized as follows. In section 2, related works on reliable and mobile multicast is given out. Details of our protocol are presented in section 3. Simulation results and related analyses are given out in section 3. Section 4 is the conclusion.

2. RELATED WORKS

Mobile IPv6 [1] provides two basic multicast schemes that are called remote subscription and home subscription. In remote subscription, there should be a local multicast router in each foreign network and mobile host sends request messages to rejoin group by the multicast router in its current network after each handover. The local multicast router must be attached to the multicast delivery tree. Mobile host can receive and send packets directly from the foreign network by the shortest path. But the multicast delivery tree may be rebuilt frequently because of handover. In home subscription, mobile host sends and receives multicast packets by the unicast bi-directional tunnels between the mobile host and the home agents. The multicast delivery tree will not be reconstructed because of member location change. But the routing path for multicast delivery may be far from optimal.

Several mobile multicast protocols have already been proposed such as MoM (Mobile Multicast protocol)[2], RBMoM (Range-Based Mobile Multicast) [3], MA (Multicast Agent) [4], but there are less works done on reliable mobile multicast protocol.

RMMP (Reliable Mobile Multicast Protocol)[5] is based on remote subscription. The mobile agent in RMMP is not only the mobile manager but also the reliable multicast agent. After mobile host is handover to a new subnet, it reports the receiving state to the mobile agent on the previous subnet. If that mobile agent finds out that the mobile host has lost some packets, the mobile agent checks its cache to find out those packets and forwards them to mobile host by tunnel. RRBMoM (Reliable Range Based Mobile Multicast)[6] is extension of RBMoM (Range Based Mobile Multicast). The acknowledgement used in RRBMoM is periodical positive ACK. The sender and local servers act as recovery nodes. Mobile host sends ACK messages to its MAH (Multicast Home Agent). Then MHA fuses ACK messages from all receivers in its service region and sends the fused ACK message to upstream node. Error recovery packets are unicasted to corresponding receiver by MHA.

In this paper, we introduce a new basic mobile multicast scheme called *previous network subscription*. Previous network is the one that mobile host visited before it is handover to the current subnet. In *previous network subscription*, when a mobile host moves to a new foreign network, it builds a bi-directional tunnel with the mobile agent on previous network rather than that on home network as home subscription. Mobile host can continue to

* This work is supported by National Nature Science foundation of China under grants No.60272078 and 863 Hi-Tech Plan of China under grants No.2001AA112051

receive multicast packets soon after handover without the multicast delivery tree being reconstructed.

3. THE RRHMoM PROTOCOL

In this section, the details of RRHMoM are presented. We describe the system architecture, the error recovery, and the acknowledgement mechanism respectively.

3.1 Region Based Hierarchical Architecture

The RRHMoM protocol combines previous network subscription and remote subscription. The system architecture is a hierarchical one, as shown in Fig.1. There is only one Multicast Subnet Agent (MSA) in each subnet that provides reliable multicast service to all mobile hosts in that subnet. Several subnets form a service region in which there is only one Multicast Region Agent (MRA). MRA is a multicast router being attached to the multicast delivery tree. It acts as the access point for mobile host in its service region for connecting to the multicast backbone.

The MRA and MSA also act as recovery nodes. MRA is responsible for retransmitting lost packets to all MSAs in its service region. MSA is the recovery node that retransmits lost packets to mobile hosts in the subnet. MRA and MSA form a tree like hierarchical recovery system.

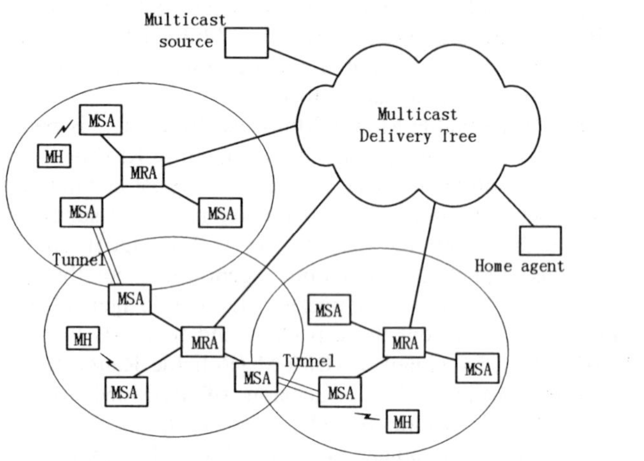

Fig.1 The hierarchical architecture adopted in RRHMoM

3.2 Error Recovery by Retransmission and Handover

There are two main methods to retransmit error recovery packet, unicast and multicast. In unicast retransmission, recovery node sends recovery packets to receivers that have lost packets by unicast. This kind of retransmission is suitable for that only a few receivers lose packets. In multicast retransmission, recovery node sends lost packets to receivers by multicast. It is efficient when there are lots of receivers having lost packets. These two methods can be adopted simultaneously. If the number of receivers requesting for a lost packet exceeds a threshold in a certain period of time, the recovery node retransmits that packet by multicast, or else, unicast retransmission is used.

Multicast retransmission can be classified into two groups according to the scope of multicast—global multicast and local multicast. Retransmission packets are sent to all group members in global multicast. That will waste network bandwidth and nodes' processing power, but the system is more robust. Local multicast retransmission just sends recovery packets to the region that packet losing has happened. Local multicast has the advantage of reducing network resource demand.

Because of the characteristic of wireless medium, mobile hosts in the same subnet are inclined to be interfered at the same time and lose packets. To deal with this situation in RRHMoM, MSA retransmits recovery packets to mobile receivers in its subnet by multicast. The usual connection between MRA and MSA is high quality wired lines that packet lose rarely occurs, so MRA retransmits recovery packets to MSA by unicast.

Handover usually causes multicast packet lose in RRHMoM protocol. After moving to a new foreign subnet, while the mobile host is registering its current care-of address to its home agent and correspondent nodes, it sends a MLD report message to the MSA in the new subnet to rejoin the desired multicast group. The mobile host also reports the max packet sequence number it has received to the MSA on the current subnet. If there are group members in the current subnet and current MSA caches the lost packets, mobile host can get lost packets and new packets from the MSA in the current subnet without any additional operations. If current MSA is receiving multicast packets of that group but has not the lost packets, it builds a tunnel to the MSA on mobile host previous visiting subnet (denoted as *p*MSA). Lost packets can be transmitted from *p*MSA to the current MSA and then to mobile host. If the mobile host is the first member of desired multicast group in the new subnet, the current MSA builds a tunnel to mobile host's *p*MSA and also sends a MLD report message to its MRA. MRA need only to forward multicast packets to the MSA if it has already joined that group, or else, the MRA should start to join the multicast group. After the current MSA receives multicast packets from its MRA and gets all lost packets from the *p*MSA, the tunnel between current MSA and the *p*MSA should be removed.

As the current subnet and the previous subnet are neighbored, the tunnel between them is usually much shorter than the tunnel between mobile host and its home agent. Because tunnels may have different delay (length) in home subscription, MLD pseudo-synchronization would be almost impossible to achieve [7]. This problem does not exist in previous network subscription because of short tunnel. Mobile host can not receive packets before the multicast router on current network rejoins multicast group in remote subscription, the service interruption time may be considerable. That interruption time is shortened in previous network subscription as mobile host can get packets through the short tunnel.

3.3 Acknowledgement in RRHMoM

Positive acknowledge (ACK) and negative acknowledge (NAK) are two main acknowledgement mechanisms in reliable multicast. In positive acknowledge, the sender need to maintain receiving states for all receivers. If sender sends out a packet and doesn't receive one receiver's ACK message after a certain time's waiting, sender should retransmits that packet. In negative acknowledgement, a receiver sends out a message to some recovery node to request the lost packet when it finds out that one packet has been lost. Positive acknowledge (ACK) mechanism can ensure that all receivers will miss no multicast packet, but be very poor in scalability. For negative acknowledge (NAK), the scalability to large number of receivers is not a problem, but the sender can not guarantee that all receivers can receive all packets if the sender has only finite cache memories.

In RRHMoM protocol, ACK and NAK combined acknowledgement mechanism is adopted. Because the typical connection between MSA and MRA is high quality wired lines, MSAs only send periodical ACK messages to their MRA in RRHMoM. As a MRA service for a limited number of MSAs, the ACK explosion is not a big problem. NAK mechanism is used in subnet so that mobile host can send out retransmission request messages to the MSA on the current subnet immediately after the mobile host discovers a packet losing, which can help to recover locally. To avoid NAK messages explosion, NAK suppression is used to prevent multiple mobile hosts in the same subnet to send out the same retransmission request messages.

If only periodical ACK acknowledgement is used, mobile host may move to another subnet before MSA finds out that some multicast packets is missed by the mobile host. That increases the retransmission delay and the retransmission cost. If NAK acknowledgement is used, the mobile host can send out retransmission request message to MSA immediately after mobile host detects a lost packet and the MSA can retransmit the lost packet promptly, so the retransmission delay and retransmission cost is reduced.

Compared with existing reliable mobile multicast protocols, RRHMoM protocol has the following advantages. First, because mobile host sends NAK messages immediately after it finds out a packet losing, the MSA on its current subnet can retransmit requested multicast packets in most case. The NAK mechanism and previous network subscription help to reduce both the retransmission delay and retransmission cost. Second, because of the region based hierarchical architecture and local recovery, the RRHMoM protocol is scalable and can be used in large network. Third, because of the open and shared property of wireless medium, multiple mobile hosts can be interfered at the same time. Multicast retransmission and NAK suppression in subnet are very useful for wireless bandwidth efficiency.

4. SIMULATIONS

We evaluate RRHMoM protocol by simulation, in which we focus on the packet losing during handoff and analysis the performance of RRHMoM in terms of retransmission cost, retransmission delay, and signaling cost.

4.1 Simulation model

The network used in simulation is a 12*12 mesh network. There are one MSA on each subnet. A MRA's service region is 4*4 square (the service of the MHA in RRBMoM is also set to 4*4). MRA is in the center of its service region. Multicast source is located in the center of the mesh network system. In the initial state, all group members are distributed randomly. During simulation, mobile hosts' group membership is unchanged. The interval between two ACK messages is 10 time units. A mobile host stays time in the same subnet for an average of 4 time units. The connection between MRA and MSA is high quality wired line and no packets will be lost on wired connections. Mobile host links to MSA by wireless medium. All wireless connections' packet losing rate is the same. When a mobile host is handover to a different subnet, one packet will be lost.

4.2 Simulation Results and Analyses

We compare the performance of RRHMoM with RRBMoM and RMMP. The packet losing rate of wireless medium is 10%, and the number of mobile member ranges from 2^0 to 2^{10}.

The change of average retransmission cost with the change of mobile member number is shown in Fig.2. Because NAK is not used in RMMP and RRBMoM protocols, the transmission path of recovery packets may be quite long. In RRBMoM, the mobile agents are not recovery nodes and all recovery packets are sent out by MHA. From Fig.2 we can see that the retransmission cost of RRHMoM protocol is the lightest and RRBMoM's retransmission cost is the heaviest.

The simulation result of average retransmission delay is shown in Fig.3. Because NAK messages are used in RRHMoM, the retransmission delay is the shortest and goes to 2 as the number of mobile host increases. RMMP and RRBMoM don't use NAK messages, so the retransmission delays are longer.

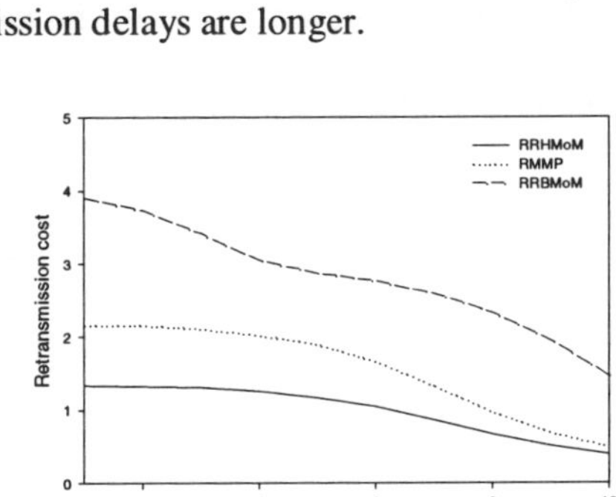

Fig. 2 Average retransmission cost of lost packet

We use the number of ACK and NAK messages to denote the signaling overload. The average signaling overload is shown in Fig.4. Because the mobile agent in RRBMoM does not cache multicast packets, and also because the retransmission request messages must go to MHA or even multicast source if MHA has not required packet, RRBMoM's signaling overload is the heaviest. RMMP does not use NAK messages too, but the retransmission request messages goes to the mobile agent on the previous subnet in the worst case, so its signaling overload is the lightest. Besides the periodically sent out ACK messages, NAK messages are also used in RRHMoM, but the average signaling overload is just slightly heavier than RMMP because of the NAK acknowledgement messages are suppressed.

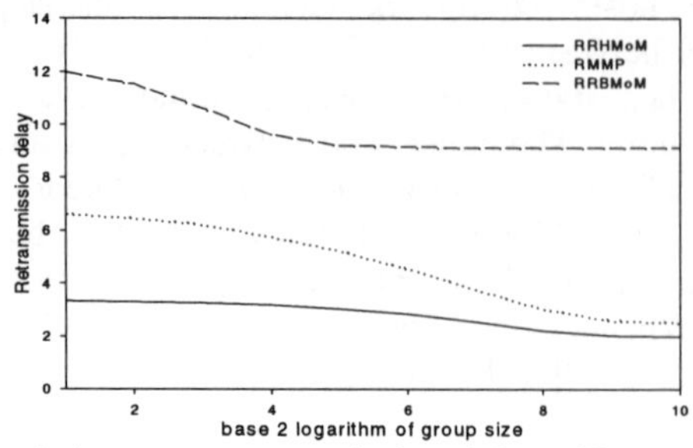

Fig. 3 Average retransmission delay of lost packet

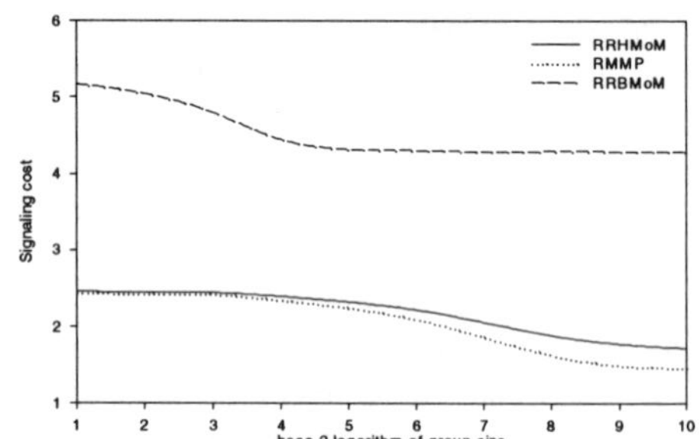

Fig. 4 Average signaling cost of lost packet

The multicast service disruption time caused by handoff is the duration from the time when mobile host stops receiving packet from the current subnet to the time when mobile host begins receiving packet from the next subnet. The multicast service disruption time can be denoted by join delay, which is the interval from mobile host's arrival to the new subnet to the time when mobile host can receive multicast packet again. The simulation result of average join delay is shown in Fig.5.

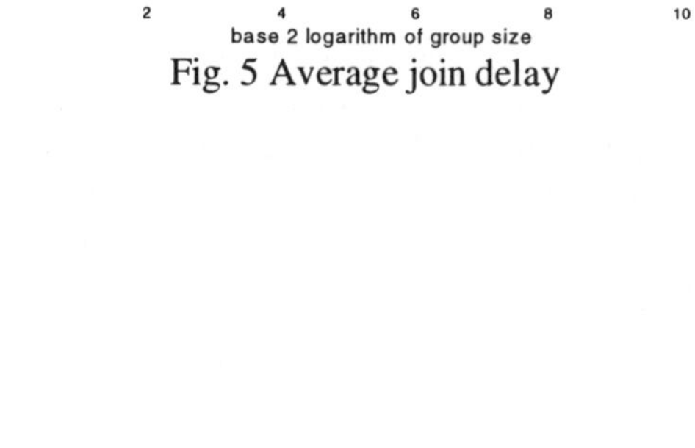

Fig. 5 Average join delay

When handover occurs in RRHMoM, if the current MSA is receiving packets of corresponding group, the join delay is 2 time units, or else the join delay is 4 time units because the packets comes from *p*MSA. The join delay declines and runs to 2 as the number of mobile host in the mesh network increases. Because mobile host needs to request to the MHA for multicast packets in RRBMoM, join delay in RRBMoM is larger than that in RMMP and RRHMoM.

5. CONCLUSION

In this paper, a reliable mobile multicast protocol called RRBMoM is proposed. The main features of RRBMoM can be summarized as region based hierarchical architecture, combining previous network subscription and remote subscription, retransmitting by multicast in a same subnet, combining ACK and NAK acknowledgement, and NAK suppression. Simulation result shows that RRHMoM exceeds existing reliable mobile multicast protocols in retransmission delay, retransmission cost, signaling overload, and multicast service disruption time.

REFERENCES

[1] D. Johnson, C. Perkins, J. Arkko. Mobility Support in IPv6 draft-ietf-mobileip-ipv6-21.txt. Mar. 2003.

[2] T. Harrison, C. Williamson, "Mobile Multicast (MoM) protocol: Multicast support for mobile hosts.," In: Proceeding of ACM Mobile'97. Budapest: IEEE Computer Society Press, pp. 151–160, 1997.

[3] Chunhung Richard Lin, Kai-Min Wang, "Mobile multicast support in IP networks," In: Proceedings of InfoCom' 2000. Tel Aviv: IEEE Computer Society Press, pp. 1664-1672, 2000.

[4] Yu Wang, W. Chen, "Supporting Multicast for Mobile Hosts," Mobile Networks and Applications, 2001, 5(6): 57-66.

[5] Y Jen-An Ke, Wanjiun Liao, "Reliable Mobile Multicast Protocol (RMMP): A Reliable Multicast Protocol for Mobile Networks Using Mobile IP," In Proc. Wireless Communications and Networking Conference (WCNC) 2000, Chicago, pp. 1488-1491, Sept. 2000.

[6] Chunhung Richard Lin, Cang-Jai Chung, "Mobile Reliable Multicast Support in IP Networks," ICC, pp. 1421-1425. 2000.

[7] C. Jelger, T. Noel. Multicast for Mobile Hosts in IP Networks: Progress and Challenges. IEEE Wireless Communications, vol. 9, page 58-64, 2002.

Impact of Security Architectures on Wireless Network Performance

Ray Hunt, John Vargo, Jenne Wong, University of Canterbury, New Zealand, Fax +64 3 364 2569
ray.hunt@canterbury.ac.nz, john.vargo@canterbury.ac.nz

Abstract

Wireless LANs have gained increasing market popularity in locations such as airports, cafés, universities, and businesses, but WLAN security remains an ongoing concern. The rapid deployment of WLANs has further emphasised the security vulnerabilities in the 802.11 standard which, in its initial deployment suffered from a number of security limitations. However mitigating these vulnerabilities implies an impact on network performance. In this paper, we investigate the network performance of WLANs when security technologies designed to improve the 802.11 standard are implemented. In particular we evaluated the performance of the 802.1X and VPN models including the various security configuration levels offered by each of these models. Experimental results indicated that the performance of various applications operating over the 802.1X security model performed better than over the VPN model. However each of these models offer a wide range of configuration options and operating conditions. These are analysed and the results discussed in this paper.

Keywords: Security, Mobile/Wireless Networks, IEEE802.1X and VPN Security Architectures, Performance, WEP, 3DES, MD5-Wireless Chap, IPSec, TLS

1. Introduction

Security of Wireless LANs has become a vital issue in recent times, particularly as the limitations of WEP have become apparent [1], [2]. There exists two key security architectures designed specifically for 802.11 [3] WLAN networks – 802.1X [4] and VPN [5]. These respective models are not a single security protocol but rather a framework from which around 10 different security options can be configured in accordance with an organisation's security policy requirements. Although these architectures have been implemented by a number of WLAN vendors as well as being supported by various OS vendors, little real performance data exists on these models (as well as data on the security levels within each model).

This paper reports on the establishment of a testing framework developed for the determination of the performance impact of each option. It is the intention that such real data can be used to assist in the selection of an appropriate security architecture in accordance with an organisation's security policy. In this study we used a single client although further research underway is currently evaluating performance with multiple clients – thus introducing contention. Further we kept the distance and noise level both minimal and constant. Handoff of authenticated and encrypted data streams between access points is a further variable that can be introduced.

Performance is necessarily dependent upon the application traffic overlaid on the network. This study examined two common types of application traffic – file transfer (FTP) and interactive transaction (HTTP) traffic utilising standard traffic generator tools. Further work is also underway to determine the performance impact on the various security levels by utilising different traffic profile generators. This study shows some very interesting results which are expressed in graphical format along with supporting explanation.

2. Security Configuration Architectures

2.1 802.1X Security Model

The 802.1X model consists of the 802.11 access mechanism using open and shared key authentication, WEP encryption, and the 802.1X port-based authentication. There are nine security levels defined in this model (802.1X Model column of Table 1). By combining 802.1X with 802.11 protocols (as security levels 6 to 9), the model provides a controlled wireless network with user identification, centralised authentication, and dynamic key management.

Level	802.1X Model	VPN Model (using firewall)
1	No Security (Default Installation)	
2	MAC authentication	PPTP Tunnelling & CHAP
3	WEP authentication	IPSec Tunnelling & CHAP
4	WEP (40 bit) authentication & encryption	PPTP & CHAP
5	WEP (128 bit) authentication. & encryption	IPSec & CHAP
6	IEEE 802.1X EAP-MD5	IPSec & EAP-TLS
7	IEEE 802.1X EAP-TLS	IPSec & CHAP & DES
8	IEEE 802.1X EAP-MD5 & WEP encryption	IPSec & EAP-TLS & DES
9	IEEE 802.1X EAP-TLS & WEP encryption	IPSec & CHAP & 3DES
10	-	IPSec & EAP-TLS & 3DES

Table 1 Security Levels in the 802.1X and VPN Models

The 802.11 access mechanism was tested for security levels 2 to 5 using a single subnet. Static key management and basic network access was facilitated by the access point. For security levels 6 to 9, the integration of 802.1X and 802.11 provided a dynamic key management and centralised authentication by the RADIUS server. Authentication methods chosen for the experiment were the EAP-MD5 [6] and EAP-TLS [7]; other proprietary authentication methods such as EAP-TTLS were not considered. This model does not support end-to-end security, as privacy and confidentiality are only ensured over the wireless link by the WEP, and not enforced on the wired counterparts. Such a solution requires the use of the VPN security architecture to be described in the following section.

Figure 1 shows the 802.1X model's logical flow and associated security levels and Figure 2 shows the architecture testbed.

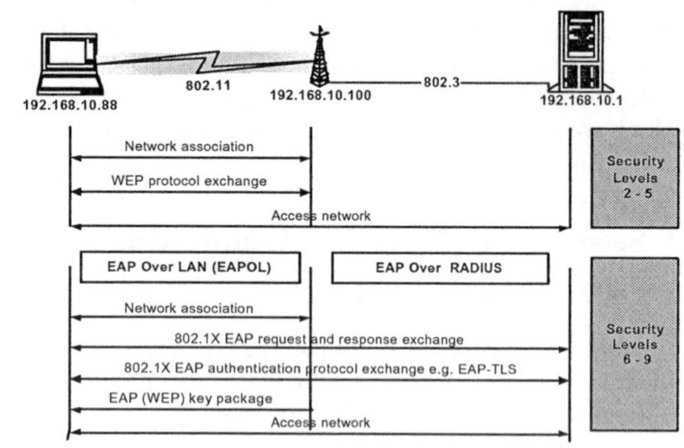

Figure 1 802.1X Model's Logical Flow

RADIUS server and X.509 digital certificate authorities were added to the basic network structure to provide 802.1X authentication support. The RADIUS server supports wireless user sign-on, and a X.509 certificate authority was used to issue certificates to users for EAP-TLS authentication.

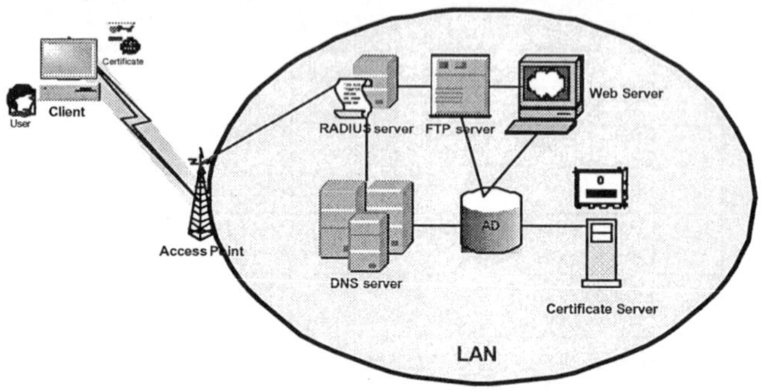

Figure 2 802.1X Model Implementation Testbed

2.2 VPN Security Model

The VPN model uses IPSec [8] to support an end-to-end secured communication from wireless to wired links. Although most of the testing was based upon the IPSec mechanism, PPTP – being a popular VPN tunnelling architecture – was also tested against IPSec [9]. PKI was selected over preshared keys to identify a user, and X.509 certificates were used for distributing and authenticating the keys. User authentication alternatives selected were EAP-MD5 (Wireless CHAP) and EAP-TLS. Ten security levels are specified as shown under VPN Model column of Table 1.

The VPN model can group the security levels into two parts; Levels 2 to 4 require authentication and tunnelling using either PPTP or IPSec before and after the firewall. Levels 5 to 10 require the IPSec protocol suite with a firewall to carry out authentication and encryption.

User authentication and tunnelling options were tested before and after the firewall installation in security levels 2 to 5. From there onwards, IPSec was the sole security protocol used with different user authentication alternatives. Security level 3 and levels 5 to 10 provided both user and device authentication. Encryption mechanisms such as DES and 3DES were used to ensure end-to-end data protection. The network was divided into subnetworks and treated the wireless subnetwork as an extranet. Figure 3 shows the VPN model's logical flow and associated security levels and Figure 4 shows the architecture testbed.

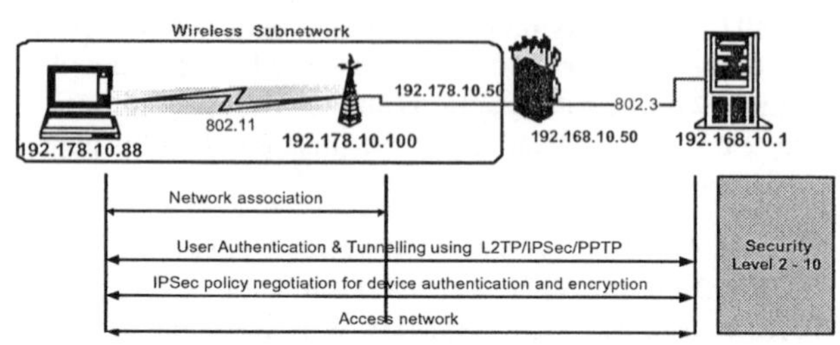

Figure 3 VPN Model Logical Structure

In the VPN model, additional components were introduced into the network: RADIUS server, VPN server, X.509 certificate authority, and a firewall. The RADIUS server and certificate authority were installed during the 802.1X model experiment; thus, only a VPN server and a firewall were newly implemented. The VPN server was configured to be the RADIUS client and a firewall was set up between the access point and the VPN server.

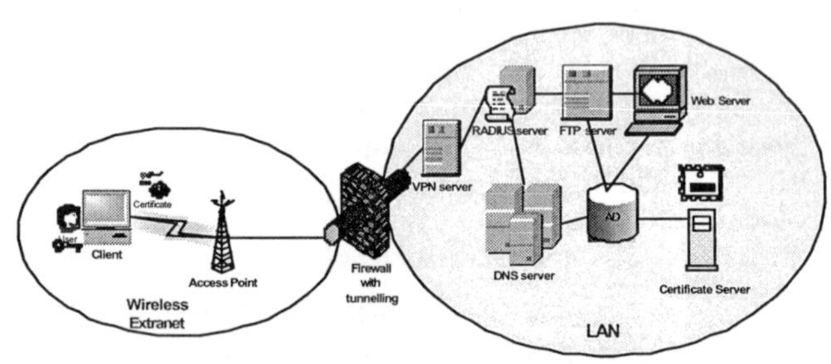

Figure 4 VPN Model Implementation

There are two types of authentication methods deployed in the experiments which provide direct comparison with the authentication methods deployed in the 802.1X model:

- *User* authentication based on open-standards – Wireless CHAP and EAP-TLS
- *Device* authentication using PKI with X.509 digital certificates

3. Performance Test Results
3.1 HTTP and FTP Testing Scenarios

As the security level (Table 1) gets higher, the general trend is increased response times and decreased throughputs. The graphs shown in Figures 5 to 8 present overviews of the security mechanisms' impacts on performance. The 802.1X model provides better response times and throughputs than the VPN model; the IPSec-based VPN model provides end-to-end security that results in higher performance overheads. We found that FTP performed better than HTTP because the later requires more interaction between the server and the client. However, some security mechanisms impacted FTP transactions more than the HTTP[1].

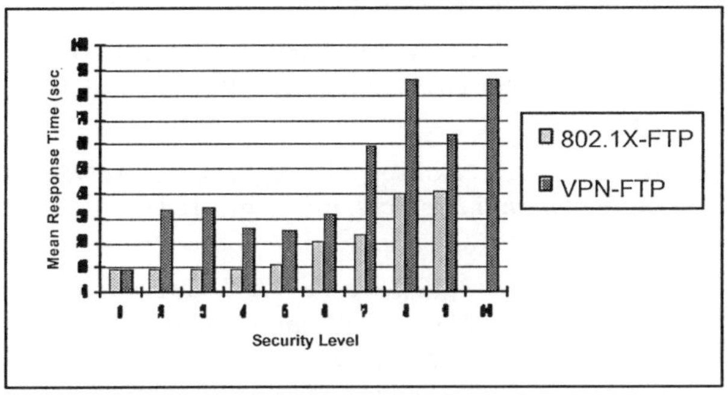

Figure 5 FTP Mean Response Times for 802.1X and VPN Models

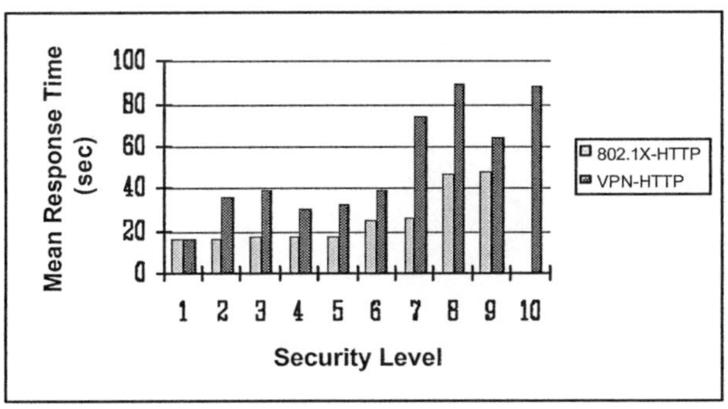

Figure 6 HTTP Mean Response Times for 802.1X and VPN Models

An inverse relationship was found in both the 802.1X and VPN models between response time and throughput: as response time increased throughput decreased. See Figures 5 and 7 for FTP traffic and Figures 6 and 8 for HTTP traffic.

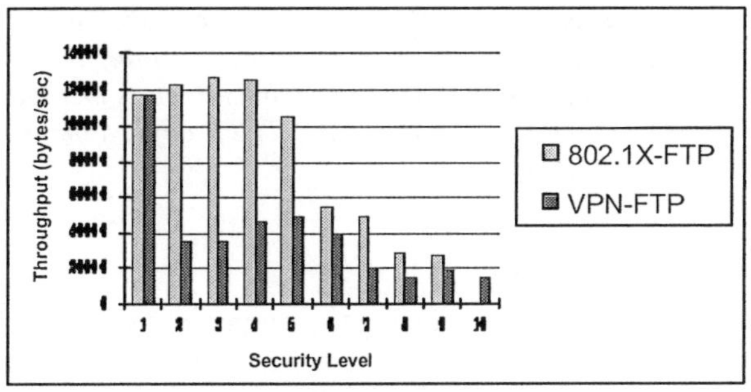

Figure 7 FTP Mean Throughputs for 802.1X and VPN Models

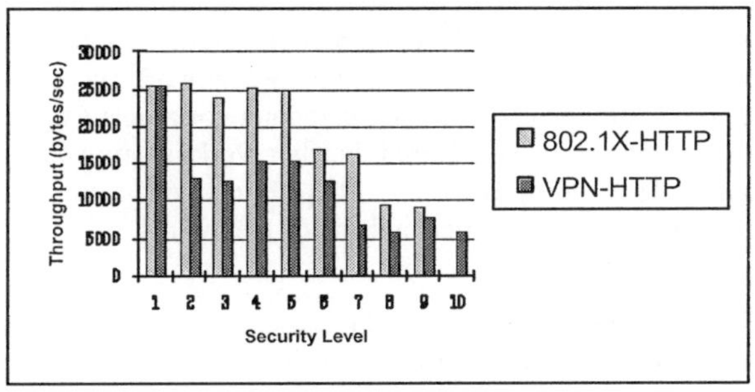

Figure 8 HTTP Mean Throughputs for 802.1X and VPN Models

These results are limited to a single cell WLAN. Integrating these models with a backbone network would provide further information on the effect of different traffic types and loads transferred over the network.

The two measured indicators of performance: throughput and response time, were affected by the security mechanisms deployed. We can generalise our findings as follows:

- *Deploying MAC and WEP authentication created no overheads.*
- *Different authentication methods created different levels of performance overhead; EAP-TLS generated the longest delay and decreased throughput. A performance comparison of the authentication mechanisms can be summarised as follows:*

  EAP-TLS> EAP-MD5/CHAP>WEP>MAC
- *Tunnelling produced large overheads; IPSec overheads > PPTP overheads.*
- *WEP encryption impact on performance varied; key length affected only response times. However, when WEP encryption was used in conjunction with 802.1X-based authentication, network performance was dramatically degraded.*
- *Deploying DES cryptographic methods introduced large overheads, however, there was little difference*

[1] For example, EAP-MD5 authentication had less impact on HTTP performance than WEP.

between 3DES and DES, especially when used with certificate-based authentication.

- *The interaction of authentication and encryption generated different results from those resulting from adding encryption to the same authentication methods for FTP and HTTP traffic. EAP-TLS produced the most adverse impact.*

4. Conclusions

WLAN systems are beginning to incorporate a variety of new security architectures. It is important that an appropriate security mechanism be selected in order to comply with an organisation's operation and performance requirements. Polices and resources developed for remote dial-up users may be helpful because of the similarity between a wireless and a dial-up client. Both are unknown users that must be authenticated before network access is granted, and the use of an untrusted network means that strong encryption is required. Further work is being carried out by Task Group IEEE 802.11i [10] in an effort to enhance the security mechanisms of the 802.11 architecture.

Our experimental evaluation investigated the performance impact incurred with various security mechanisms, and found that the more secured a network became, the higher the performance impact. The VPN model incurred greater performance degradation than the 802.1X model as we expected; the VPN model provided end-to-end security with double authentication (device and user), a stronger encryption method as well as better key management and tunnelling technology. The general pattern for various application protocols carried over a network also followed our expectation that request-reply (HTTP) applications created more performance overheads than FTP-based applications, as the interaction between two endpoints increased due to security negotiation and management. Our evaluation of different security levels showed that the higher the security level the greater the performance impact. There were exceptions however – for example adding MAC address authentication, WEP authentication and encryption (40-bit), and 3DES encryption in the VPN model actually created minimal impact and yet at least provided a basic level of security.

Further research is currently underway to investigate the performance of WLAN networks with multiple clients operating under contention as well as to determine the effects of different traffic generation profiles. Other characteristics under investigation include different cryptographic methods and performance resulting from secure handoff between WLAN (802.11) and WWAN (cdma) networks.

References

1. Fluhrer, S., I. Mantin, & A. Shamir. (2001). *Weaknesses in the Key Scheduling Algorithm of RC4.* Eighth Annual Workshop on Selected Areas in Cryptography, August.
2. Stubblefield, A., J. Ioannidis, & A. D. Rubin. (2001). Using the Fluhrer, Mantin, and Shamir *Attack to Break WEP*: AT&T Lab Technical Report TD-4ZCPZZ.
3. ANSI/IEEE. Std 802.11 (1999). *Part 11: Wireless LAN Medium Access Control (MAC) and Physical Layer (PHY) Specifications.* New York. First Edition. IEEE, Inc. ISBN 0-7381-1658-0. 20 August.
4. IEEE Std. 802.1X (2001). *Port-Based Network Access Control.* New York. IEEE, Inc. ISBN 0-7381-2627-5. 25 October.
5. Halpern, J., S. Convery, & R. Saville. (2001). *SAFE VPN: IPSec Virtual Private Networks in Depth.* Cisco System Inc. www.mnemonic.no/linker/pdf/IPSec_VPN_in_Depth.pdf
6. Dierks, T., & C. Allen. (1999). *The TLS Protocol Version 1.0*, RFC 2246. IETF.
7. Simpson, W. (1996). PPP Challenge Handshake Authentication Protocol (CHAP). RFC1334. IETF. August.
8. Kent, S., & R. Atkinson. (1998). *Security Architecture for the Internet Protocol.* RFC 2401. IETF, November.
9. Rincon, R. B. (2002). *Secure WLAN Operation and Deployment in Home and Small to Medium Size Office Environment.* Master's Thesis, Department of Engineering and Computer Science. Technische Universitat Berlin. 6 March.
10. TGi. (2003). *Task Group 802.11i*: IEEE, Inc. www.ieee802.org/11.

Asymmetric UMTS for Spectrum Efficient Asymmetric Services Delivery

L.Vignali[1], F. Malavasi[1], D.Grandblaise[2], D.Lacroix-Penther[3], J-P.Javaudin[3]
[1] Ericsson Telecomunicazioni Italy,
email: mailto:[luca.vignali, filippo.malavasi]@ericsson.com
[2] Motorola European Communications Research Lab, France,
email: david.grandblaise@motorola.com
[3] France Telecom R&D, DMR/DDH Laboratory, France
email: [dominique.lacroix,jeanphilippe.javaudin]@francetelecom.com

Ericsson Telecomunicazioni
Via Pio La Torre, Vimodrone, Milano, Italy
Phone: +39 02 26594 691
Fax: +39 02 26594 666

Abstract: The connection to the Internet by full-mobile users, is expected to be of crucial importance for the success of radio mobile systems as UMTS. Such systems must support services like Web Browsing, the most popular Internet application today, which is characterized by high bit rate bursty asymmetric traffic and like Video Call, evolution of traditional speech service, which are characterized by high sensitivity to delays and more symmetric traffic.
While the resulting traffic is highly asymmetric, mobile air interfaces like UTRAN FDD have been initially designed as systems supporting asymmetric user traffic in an overall system symmetric traffic scenario.
This paper, outcome of the ongoing EU founded project OverDRiVE, focuses on downlink enhancements of the UTRAN FDD air interface and proposes specific resource management of additional radio spectrum in order to handle efficiently asymmetric traffic.

Key Words: Asymmetric Traffic, OFDM, Resource Management, Spectrum Management.

1. INTRODUCTION

The traffic to be handled by UMTS, in the near future, is characterized by asymmetry in the downlink, as explained in section 2. As the UMTS has been designed mainly for symmetric traffic, while supporting asymmetric services, some effort is needed to enhance the system towards support for more extreme ratios of asymmetry in the traffic. In particular we address the enhancements of the air interface of the UTRAN FDD and the management of the additional downlink spectrum already identified for IMT2000. High Speed Downlink Packet Access and Orthogonal Frequencies Division Multiple Access evolution are covered in section 3. New algorithms for Radio Resource Management, devoted to arrange asymmetric traffic considering the additional downlink spectrum, are described and compared in section 4 before concluding the paper.

2. DEALING WITH TRAFFIC ASYMMETRY

The type of the service has an important impact on the asymmetry aspect of the associated traffic, which also depends on the level L of the radio communication system the asymmetry is observed (e.g. L can be user, cell, system) and on the observation period.
A consequent Traffic Asymmetry Factor (*TAF*) definition is provided by the following equation:

$$TAF(S,T,L) = \frac{DL\_BITS(S,T,L)}{UL\_BITS(S,T,L)} \quad \textbf{(1)}$$

The equation expresses the concept that the considered service S, observation period T and level L affects the *TAF* measure. The following table provides some values for services relevant to this study for the single user case [1].

Services	Speech or Video Call	Medium Multimedia	High Multimedia
TAF	1/1	1/0.026	1/0.005

Table 1 – Traffic Asymmetry Factor (TAF) for different services, over some minutes time scale for a single user.

When the observation level is the cell or the system, the equivalent traffic depends on the mix of different service profiles (symmetric and asymmetric) thus the global traffic is expected to be asymmetric. It is important to consider that enlarging the observation level for the *TAF* calculation smoothes the impact of the observation period.
Different techniques can be envisaged to handle these different asymmetry factors and variations occurring at these different levels. Hence, it is expected that some techniques are more applicable to the link level (to deal with rapid variations) whereas some other techniques are expected to be more appropriate for the system level (to deal with slow variations). They mainly consist in:

- Case 1: **Increasing spectrum efficiency** by the means of a new physical layer like the HSDPA (High Speed Downlink Packet Access) or OFDM (Orthogonal Frequency Division Modulation) technique. This approach considers the current allocated and fixed spectrum bands dedicated to UMTS FDD system operations. No modification on the spectrum allocation is required.

- Case 2: **Increasing spectrum bands availability** dedicated to UMTS FDD operations. This consists in adding some additional downstream spectrum bands for the UMTS FDD forward link operations.

The UMTS Forum Spectrum Aspects Group (SAG) is currently studying traffic features of future 3G systems. Regarding the feasibility studies on viable deployment of UTRA in additional and diverse spectrum arrangements have been initiated and are not frozen [2] [3].

It must be noted that the complete Asymmetric UTRAN FDD picture can be obtained by the union of the two described Cases.

In the following sections the enhancements of UTRAN FDD are considered as relevant example of Case 1, while additional spectrum management strategies are proposed and evaluated as innovative solutions of Case 2.

3. ENHANCEMENTS OF UTRAN FDD

In order to improve spectrum efficiency, while providing bursty and downlink asymmetric data service (like web browsing), 3GPP has introduced in Release 5 the **High Speed Downlink Packet Access (HSDPA)** [4], which includes advanced techniques typical of next generation mobile systems [5] like **Adaptive Modulation and Coding (AMC)** and **Hybrid ARQ (HARQ).**

The *temporal statistics multiplexing* gain is used to allocate to the maximum number of users with the required amount of radio resources at the right time with the right QoS for a fixed amount of spectrum. This is very well appropriate when the users request some services based on packet data, leading to a 50% ~ 150% [6] spectrum efficiency gain compared to Release 99 UMTS.

As the mobile radio systems evolve and a more integrated multi-radio system is envisaged [7], **Orthogonal Frequency Division Multiplexing (OFDM),** which is the technology already used in DVB-T [8] and 802.11a [9], is discussed as another option for enhancing UTRAN access method in this paper.

Actually OFDM for enhancing UTRAN was first proposed at 3GPP in February 2001 where a study group has been opened to study whether OFDM is technically feasible. The working assumption is a stand-alone bearer that would use an extra bandwidth of 5MHz, in downlink only. The most important characteristics of OFDM can be found in [10] [11] [12]. While in the remaining part of the section, a new OFDM waveform called IOTA [13] is proposed and compared to traditional OFDM modulation.

In OFDM-IOTA the interferences among the sub-carriers (due to Doppler effects and phase noise) are limited because the IOTA waveform modulating each sub-carrier is very well localized in time and frequency. So OFDM-IOTA provides the same robustness to multi-path effects than classical OFDM without requiring a cyclic prefix.

Figure 1 shows physical layer simulation results of two OFDM waveforms, the usual OFDM waveform, and the new OFDM-IOTA, without cyclic prefix. The advantages of OFDM-IOTA waveform are:

- Higher spectrum efficiency than the traditional one
- Better fitting in the UMTS 5Mhz bandwidth.

Simulations are performed considering the same spectrum efficiency (6,75Mbit/s in 5MHz) with different radio channels, according to parameters in table 2. Constellation is QPSK on each carrier, and turbo coding is used.

	FFT	mod. carriers	samp. rate	CP
OFDM	600	350	7,68 MHz	1/16
OFDM-IOTA	512	300	7,68 MHz	0

Table 2: OFDM physical layer parameters

The following figure shows that a gain on the signal to noise ratio C/N can be obtained while providing the same quality (Bit Error Rate), in all the relevant simulated radio channels. It must be noted that, as the cyclic prefix is not present in OFDM-IOTA waveform, a more robust coding (3/4 instead of 4/5) can be used, leading to a lower C/N needed to provide the same quality. For example: with a BER equal to 10^{-4} the C/N variation ranges from about -1.5dB in the Vehicular A channel to about –2dB in the pedestrian channel.

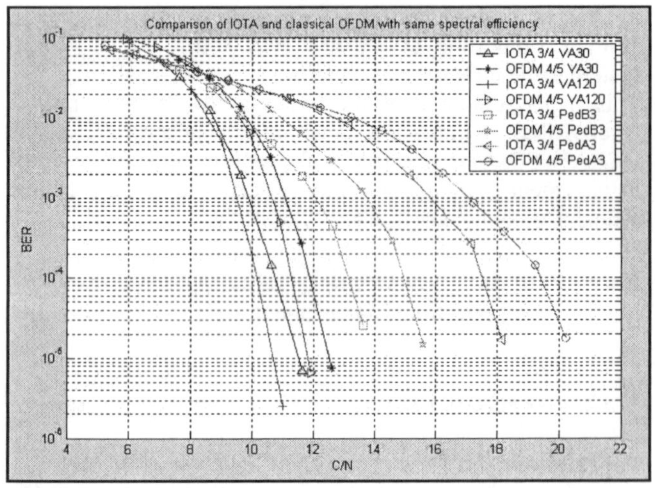

Figure 1: Physical Layer performances for OFDM, and OFDM-IOTA waveforms.

It can be noted that in the case of a same coding rate, OFDM-IOTA still outperforms classical OFDM from 0.2 to 0.4 dB. Thus the gain in throughput corresponds to the prefix cyclic overhead of classical OFDM.

4. ADDITIONAL SPECTRUM MANAGEMENT

In 2000, the WARC 2000 has identified three worldwide additional spectrum bands to support the additional downlink traffic provided by asymmetric services [14]. These additional (or extended) bands are described in Table 3.

In this paper we performed investigations in order to evaluate

Radio Resource Management (RRM) algorithms, considering two downlink 5MHz frequency bands per each uplink 5MHz frequency band. The 2.5 GHz band is hereafter considered as relevant extension for UTRA FDD: as discussed in [3] no additional propagation loss is considered compared to the 2GHz band, so the same cell deployment is assumed for both frequencies.

Extended bands for IMT-2000 services	Frequencies and Bandwidth
900 MHz band	806-960 MHz (154 MHz)
1.8 GHz band	1710-1885 MHz (175 MHz)
2.5 GHz band	2500-2690 MHz (190 MHz)

Table 3: Extended bands identified by ITU (WARC 2000) for IMT 2000 Services.

We considers a mix of Video Call (VC) Service (associated with a 64kbps Radio Bearer) and of Web Browsing Service (384kbps Radio Bearer). The *TAF (Session_duration, Link, Web)* is equal to 20 while it is equal to 1 for a Video Call service.

We model Mobile Terminals (MT) mobility, slow fading and Rayleigh fading, inner and outer loop power control, soft handover. No Admission Control is performed but MTs are dropped when they experience low quality for a certain amount of consecutive frames in the downlink.

As the considered services have different QoS requirements, the dropping acts when 10 or 30 frames are received with low quality for the Video and Web case respectively. Low quality is defined as Transport Block Error Rate (BLER) greater than five (Video) or ten (Web) times the BLER target (fixed at 1% for Video and 3% for Web services).

Uplink dropping has not been activated to better evaluate the algorithms, but logging of uplink dropped MTs has been done to evaluate if frequency band asymmetry can support the simulated traffic asymmetry.

We compare three different RRM algorithms:

- RANDOM assigns MTs to the selected bands randomly: each session begins and terminates in the same frequency band.
- SERVICE assigns different services to different frequency bands, to exploit the inherent QoS differences of Web and Video Call services.
- IFHO[1]. A transmitted power triggered Interfrequency Handover is proposed to balance the load on the two frequencies.

The comparison of results (Table 4) of **RANDOM and SERVICE** algorithms shows that, separating services is a good strategy, because the service QoS requirements and traffic characteristics are very different to each other. It can be observed that **the dropping probability is reduced in many cases by a factor 10**, while maintaining comparable mean BLER. Note the higher *TAF* in RANDOM, because dropping concerns mainly Video Call.

[1] The IFHO algorithm is proposed as an enhancement of RANDOM algorithm.

By comparing **RANDOM and IFHO** cases, it can be concluded that the IFHO algorithm can effectively provide a trunking efficiency gain. **The maximum gain, with dropping reduced by a factor 5**, can be obtained with 4 Video Call and 4 Web sessions average per cell, while, due to the inherent difficulty of the load prediction when bursty traffic is present, it performs worse in the case 12 Video, 4 Web.

The average BLER reported in Table 4 confirms that the quality perceived by not dropped users is close to the BLER target.

Table 5 shows that the maximum Uplink dropping probability is lower than the maximum Total downlink dropping percentage (bold in Table 4) for all the considered Traffic Asymmetry Factor (TAF). For example, in the RANDOM case (see Table 4 and Table 5), the Uplink dropping is 1.44 % maximum, while in Downlink ranges from 5.1 to 19.4, even with high traffic asymmetry (*TAF* up to 4.2). This result heavily depends on the QoS definition and consequent dropping strategy,

Table 5 shows the effect of the SERVICE algorithm. The Video Call sessions are not dropped (0%) because the power in downlink is sufficient to serve all the incoming sessions, while the Web services suffer a bit from power shortage. However, Web sessions are less sensitive to a bursty low quality link, because they can trades errors with delays, while delay sensitive application like Video Call can not. The final result is an improved global quality compared to RANDOM and IFHO.

5. CONCLUSIONS

In this paper we face the issue of asymmetric traffic handling by native symmetric wireless systems like UMTS. The topic is of utmost importance given the Wireless Internet scenario envisaged for the near future.

Both, the solution to improve downlink spectrum efficiency, especially for bursty traffic (Case 1), and strategies to use efficiently additional downlink frequency bands (Case 2) are first separately and then jointly addressed.

New solutions have been proposed for both cases. The performance of OFDM access technology with innovative cyclic prefix suppression are assessed and compared to standard OFDM modulation. Different Radio Resource Management algorithms have been proposed and compared in a multi-service scenario (Web Browsing and Video Call service) characterized by asymmetric traffic and by the availability of additional downlink spectrum.

ACKNOWLEDGEMENTS

This work has been performed in the framework of the IST project IST-2001-35125 OverDRiVE [15], which is partly funded by the European Union. The OverDRiVE consortium consists of Ericsson (co-ordinator), DaimlerChrysler, France Telecom, Motorola and Radiotelevisione Italiana, as well as RWTH Aachen, University of Bonn and the University of

Surrey. The authors acknowledge the contributions of their colleagues in the OverDRiVE consortium.

REFERENCES

[1] UMTS Forum, Report N°5 "Minimum spectrum demand per public terrestrial UMTS operator in the initial phase, 1998

[2] ITU-R M.1036-1, Frequency Arrangements for Implementation of International Mobile Telecommunications 2000 (IMT-2000) in the bands 806-960 MHz, 1710-2025 MHz, 2110-2200 MHz and 2500-2690 MHz

[3] 3GPP TR 25.889, Technical Specification Group Radio Access Network; "Viable deployment of UTRA in additional and diverse spectrum arrangements; Feasibility Study" (Release 6)

[4] 3GPP TS25308 "High Speed Downlink Packet Access (HSDPA); Overall description; Stage 2", Ver. 5.2.0.

[5] W.W.Lu, "4G Mobile Research in Asia", IEEE Communications Magazine, March 2003, Vol. 41, No. 3.

[6] T.E. Kolding, et alii, "High Speed Downlink Packet Access: WCDMA Evolution", IEEE VTS news, Feb. 2003.

[7] R. Toenjes et alii, "OverDRiVE: Spectrum Efficient Multicast Services to Vehicles", IST Mobile Summit, Thessaloniki, 16-19 June, 2002..

[8] ETS 300 744, "Digital Video Broadcasting (DVB); Framing structure, channel coding and modulation for digital terrestrial television (DVB-T)", *ETSI*, March 1997.

[9] IEEE 802.11a, "IEEE Standard for Telecommunications and Information Exchange Between Systems - LAN/MAN Specific Requirements - Part 11: Wireless Medium Access Control (MAC) and physical layer (PHY) specifications: High Speed Physical Layer in the 5 GHz band", *IEEE*, February 2000.

[10] B.Hirosaki, "*An orthogonally multiplexed QAM using the Discrete Fourier Transform*", IEEE Trans. on Communications, Vol.29, No.7, July 1981

[11] Wavecom, France Telecom, "Advantages of using OFDM for enhancing UTRAN", Tdoc R1-02-1256, 3GPP RAN1 Meeting #28 bis, Espoo, Finland, Octobre 7th-9th 2002.

[12] Eric Lawrey "The suitability of OFDM as a modulation technique for wireless telecommunication, with a CDMA comparison", master thesis 1997, James Cook University of North Queensland.

[13] B. Le Floch, M. Alard and C. Berrou, "Coded Orthogonal Frequency Division Multiplex", Proceedings of the IEEE, Vol. 83, N° 6, June 1995

[14] ITU-R M.1036-1, Frequency Arrangements for Implementation of International Mobile Telecommunications 2000 (IMT-2000) in the bands 806-960 MHz, 1710-2025 MHz, 2110-2200 MHz and 2500-2690 MHz

[15] OverDRiVE Home Page, www.ist-overdrive.org

Sessions/cell		RANDOM			SERVICE			IFHO		
VC	Web	Drop	BLER	TAF	Drop	BLER	TAF	Drop	BLER	TAF
4	4	6.2	1.28	2.0	0.46	1.03	1.98	1.67	1.35	1.95
8	8	9.5	1.53	2.3	0.46	1.55	1.9	4.82	1.5	1.96
12	12	**19.4**	1.74	2.5	2.03	2.12	1.74	**15.8**	**2.1**	2.26
4	12	5.1	**2.42**	4.2	**2.18**	**2.7**	2.9	4.1	2.07	3.4
12	4	5.9	2.37	1.4	0.59	1.03	1.33	8.8	0.99	1.36

Table 4 – RRM algorithms comparison in terms of session dropping percentage and mean BLER of the not-dropped sessions. The Traffic Asymmetry Factor (TAF) is considered over all simulated time and all the system for both services.

	RANDOM	SERVICE	IFHO
UL Drop	1.44 %	2.1 %	1.26 %
Total DL Drop	19.4 %	2.18 %	15.8 %
Web DL Drop	4.6 %	4.3 %	1.5 %
VC DL Drop	33 %	0 %	28 %
Total DL BLER	2.42 %	2.7 %	2.1 %
Web DL BLER	3.48 %	4.2 %	3.2 %
VC DL BLER	0.8 %	0.63 %	0.93 %

Table 5 – Maximum dropping probability in UL, maximum dropping percentage and maximum BLER in DL with per service differentiation.

ARCHITECTURE DESIGN OF EXTENSIBLE MULTI-MEDIA MICRO-BROWSER SOFTWARE

TOSHIHIKO YAMAKAMI

2-8-16 Sarugaku-cho, Chiyoda-ku, Tokyo 101-0064 JAPAN,
E-mail: yam@access.co.jp

The mobile Internet is one of the emerging computer communication domains. The browser-enabled mobile handsets open a new opportunity for the multimedia services. The author presents the issues and challenges in the mobile multimedia communication software. The author discusses the architecture design criteria for the extensible multi-media browser software. With constraints in the mobile handsets, the multimedia communication software on the mobile handsets is expected to be more network-centric. The author presents the three different mobile multi-media evolution scenarios and discusses the resource-constraint open extensible architecture.

1 Introduction

The mobile Internet is one of the emerging computer communication domains. It is expected that the billion devices will have the Internet capability in 2006 [3]. Since the launch of the micro-browser embedded mobile phones in the late 1990s, the Internet users with mobile handsets prevail. For example, the mobile Internet users using mobile handsets reached 59 million at the end of 2002 in Japan. This accelerates the demand for the reliable and extensible multimedia software. In this paper, the author discusses the design challenges for the extensible multimedia micro browser software from the architecture viewpoint.

2 Challenges and Issues

2.1 Related Works

With the growing needs in the mobile Web comes an increasing need to build scalable software architecture to cope with the need for multi-media web access in a constraint environment. The mobile and wireless embedded system needs energy efficient design needs optimization in both hardware and software. The emerging demands on such systems increase research and industrial attention in this domain. The variety of the underlying environment prevents the integrated architecture design studies. Energy-aware OS research is studied in a range of fields recently. OS-directed power management is done for energy-aware scheduling by [2]. Yuan [7] studied dynamic voltage scaling systems for the mobile environment. The mobile portable device emergence stimulates the various energy efficient system designs like [5]. The past research focuses on the low-level system design because the traditional embedded software is closed and components are under control of a single task manager. The software in the mobile handsets almost approaches to the level to deal with any existing Internet multimedia in an incremental addition of software components. The browser position to leverage mobile multimedia services was discussed for the network-centric applications [6]. However, the integrated architecture design to cope with the open multimedia application demands is rarely studied in the past literature.

2.2 Issues in the Embedded Software Design

In 1999, it was uncommon to expect the micro browser-based Internet access because people considered the mobile handsets too restrictive. The mobile handsets had the four major constraints for the Internet access: (a) small display size, (b) limited computer resource and memory resource, (c) limited user interface for input, and (d) limited network reliability. For the micro browser software design, the memory limitation was hard for the code footprint size and the working area. The compactness was the top priority for the browser design. Some of the mobile handsets are equipped with 100 times larger memory and 100 times faster CPU in the today's mobile environment. Even with this magnitude of scale change, the micro browser architecture has to cope with the constraints. The unchanged issues include: (a) portability to the wide range of the execution environments, (b) customizability to the wide range of the handset hardware capabilities, (c) battery power constraints, and (d) display and input method constraints. The network software architecture design for mobile handsets is getting more complicated. The complexity of mobile Internet services is increasing. Some of the mobile handsets use 200 MHz CPU. However, there is still a significant gap between the mobile handset execution environment and that of PCs. In addition, the many features invented in the PC Internet are still designed without any awareness of the constraints in the mobile environment. Some of the multimedia functions and security functions require the high speed processing to make the turn around time visible for end users within the user-tolerate range. There is no way except the increasing the clock rate. This leads to the extensive exploration of the dynamic clock or voltage management in the mobile systems. It adds even more complexity and diversity to the underlying environment for mobile multimedia network software. The case-by-case design for the embedded software prevents the integrated architecture design. The mobile multi-media web is approaching the take-off stage, however, the constraints on the mobile embedded systems still stay, e.g. battery constraint. In addition, the gap between low-end mobile handsets and high-end ones continue to grow. This comes to a need to the software design scalability, extensibility and constraint-awareness. The portability, extensibility, resource constraints continue. At the same time, the software design needs consideration to the open-ended mobile multimedia applications with the maximum utilization of the increased resources.

3 Design Criteria for the Multimedia Micro browser Software

The growing demand of multimedia software in the mobile handsets creates the new requirements:

- extensible to multimedia markup languages
- easy integration to the external multimedia application software
- reliability and quality assurance of the software, and
- resource efficient in the combination of the external multimedia application software

For example, the camera-equipped mobile handset reached 15 millions at the end of 2002 and the number is still growing. Such enhancement requires a new application and connectivity to the external environment. To support these new requirements, the modules in the micro browser play important roles because the core capabilities for networking, rendering, and multimedia encoding/decoding can be reused in those external multimedia applications. The XHTML browser and multimedia mail reader both need the additional XML-based functionality like SMIL or [1] SVG [4] to cope with the emerged multimedia service demands.

Design criteria for the multimedia micro browser software includes:

- modularity,
- extensibility in the micro browser, and
- extensibility to the external multimedia applications

The modularity means: (a) testability, (b) robustness in execution, and (c) adaptability to the variety of the handset capabilities. The extensibility in the micro browser means the easy integration of the additional multimedia XML languages, additional presentation capabilities and script execution. These design criteria reflect the conflicting requirements from the mobile multimedia in order to converge into the rich multimedia Internet environment with the inherit constraints in the embedded system engineering.

4 An Extensible Multimedia Micro browser Architecture

The network software architecture for the embedded environment is depicted in Fig. 1. It enables modular and extensible architecture. It provides a set of multimedia enabling components that are accessible for external applications. XML-based multimedia languages, such as SMIL and SVG can be easily implemented using the existing multimedia enabling components with the minimum additional memory. The basic components like network access, XML parsing, and multimedia rendering are implemented in the enabling components that makes the multimedia service extension easier. In addition, it can make use of the performance-tuned components. As the more capabilities are embedded in the mobile device chipsets, the new architecture enables the selection of the window system components. In the previous stage, the window function was embedded

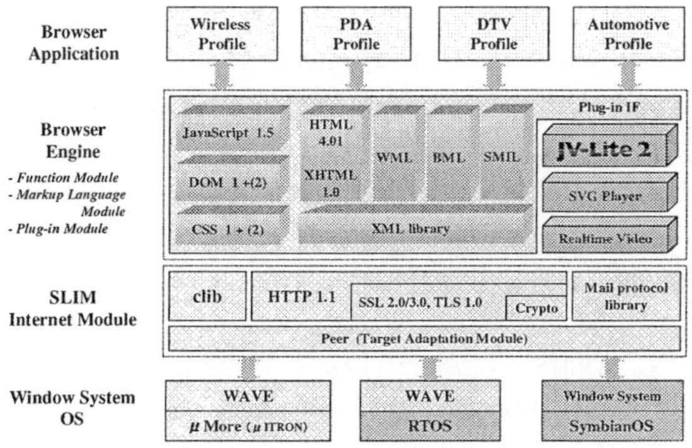

Figure 1. NetFront™ Architecture (V 3.0)

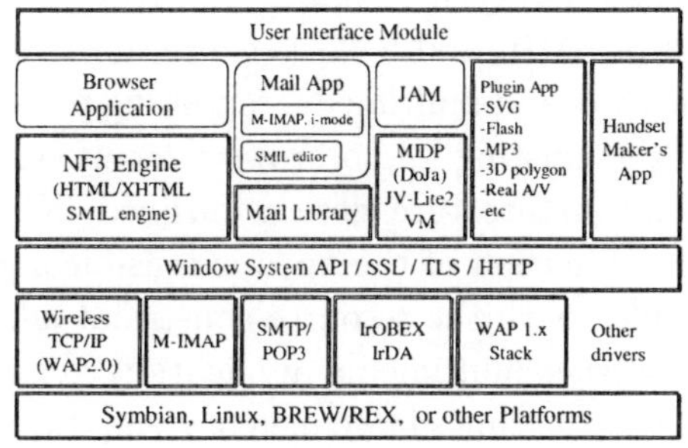

Figure 2. Internet-related Application Platform

in the architecture. Now, it is outside of the architecture to cope with the variety of the window system capabilities in the underlying environment. Also, the porting layer was encapsulated to SLIM part to encapsulate the platform dependent interfaces. Clib is a C library used in the components. It provides the independency for the underlying execution environment. No OS calls are necessary to port this embedded network software architecture. In addition, this part enables the performance tuning when necessary. It should be noted that there are three multimedia evolution directions in the today's mobile multimedia environment. The first one is the enhanced rich multimedia treatment in the existing framework. The object element in the mark up language is one of the possible extension targets. The object element is an XHTML inherit extension mechanism. The second one is the integration of the XML applications. XML was originally designed to increase extensibility in the markup languages. Vector graphics, animation, and voice are the possible extensions in this direction. The third one is the integrated multimedia experience for the users. It is natural that the all applications on the mobile handset will evolve into the rich multi-media environment. The processing of the multimedia content will be inevitable in mail messaging, instant messaging, presence service, services provided by scripting languages or game services. In the past environment, each application was tightly coded for the compactness in space and execution. The integrated multi-media experience needs will drive the reuse of the multi-media component in the constrained environment. The Internet-related application platform is depicted in Fig. 2. Each component should be designed to cope with three different evolutions.

5 Discussions

The early stage micro browser software put more emphasis on the code compactness and runtime efficiency than the code modularity. The error handling, language parsing, rendering and communication processing were integrated in the one process to utilize the code space and work area size. The whole browser engine could run on a single process for monolithic design to squeeze the possible compactness. The new architecture facilitates the followings:

- extensible to various multimedia markup languages and multimedia scripting,
- high reliability for software quality management, and
- modules reusable for external multimedia applications for performance and resource efficiency.

The extensible architecture has the following drawbacks:

- the software needs increased workspace and code footprint as a standalone base,
- the software needs overhead for the external use, and
- execution overhead for the modular implementation and context protection.

The architecture enables 100 Kbytes implementation of SMIL and SVG because most of the communication and multimedia rendering processing can reuse the existing micro browser facilities.

The architecture design outlined in this paper is a combination of the open extensible components embedded with the resource-constraint adaptability. The Internet multimedia application can easily consume the 100 times large memory-empowered mobile handset. To aim at the Internet convergence, the module design needs unchanged resource-awareness. The resource constraints in the mobile environment will not make the mobile-specific software architecture disappear. One of such fundamental constraints is battery constraint. The PC Internet is evolving into the broadband era; therefore, the Internet features are still designed without awareness of the mobile multimedia constraints. The hybrid requirement brings the two-sided design in the micro browser architecture, in which the open component architecture with the encapsulated resource awareness inside each component.

6 Conclusion

The emerged mobile Internet is rapidly moving toward the mobile multimedia services. To cope with the growing demands, the micro browser software architecture design needs extensibility and scalability as well as the portability and customizability. With the convergence to the Internet multi-media, the software design encounters a new challenge. The basic constraints still continue even with the high power CPU and increased memory. This hybrid requirement brings the mobile multi-media browser environment into the center between the legacy embedded systems and open Internet systems. The design criteria for the extensible multimedia browser architecture are presented. The increase trend in the mobile multi-media web poses the three different challenges: (a) improvement in the existing environment, (b) integration of the rich multi-media, and (c) consistent multi-media user experience in a wide range of applications. The rich set of network components in the micro browser is useful for the component kits for these evolutions. The author provides the modular extensible multimedia software architecture starting from the micro-browser architecture to cope with these three evolution paths. It is an initial step towards the multi-media processing component-based software architecture in the extensible and monolithic mobile embedded systems.

References

1. Ayars, J. et al(Eds): Synchronized Multimedia Integration Language (SMIL 2.0), W3C Recommendation 07 August 2001, available at http://www.w3.org/TR/2001/REC-smil20-20010807/ (Aug. 2001)
2. Bellosa, F.: The benefits of event: driven energy accounting in power-sensitive systems, Proceedings of the 9th workshop on ACM SIGOPS European workshop: beyond the PC: new challenges for the operating system, pp. 27–42 (Sep. 2000)
3. Cerf, V.: Beyond the Post-PC Internet, CACM, Vol. 44, No. 9, pp. 34–37 (Sep. 2001)
4. Ferraiolo, J.(Ed): Scalable Vector Graphics (SVG) 1.0 Specification, W3C Recommendation 04 September 2001, available at http://www.w3.org/TR/2001/REC-SVG-20010904/ (Sep. 2001)
5. Simunic, T. and Vikalo, H. and Glynn, P. and De Micheli, G.: Energy efficient design of portable wireless systems, Proceedings of the 2000 international symposium on Low power electronics and design, ACM Press, pp. 49–54 (Aug. 2000)
6. Yamakami, T.: Leveraging Information Appliances: A Browser Architecture Perspective in the Mobile Multimedia Age, in *Advances in Multimedia Information Processing - PCM 2002, Third IEEE Pacific Rim Conference on Multimedia* (Eds. Chen, Y.-C., Chang, L.-W., and Hsu, C.-T.), LNCS 2532, Springer, pp. 1–8 (Dec. 2002)
7. Yuan, W. and Nahrstedt, K.: Mobile and Wireless System: Integration of dynamic voltage scaling and soft real-time scheduling for open mobile systems, Proceeding of the 12th International Workshop on Network and Operating Systems Support for Digital Audio and Video, pp. 105–114 (May 2002)